Redefining Nature

EXPLORATIONS IN ANTHROPOLOGY
A University College London Series

Series Editors: Barbara Bender, John Gledhill and Bruce Kapferer

Redefining Nature

Ecology, Culture and Domestication

Edited by
Roy Ellen and Katsuyoshi Fukui

Routledge
Taylor & Francis Group
LONDON AND NEW YORK

First published 1996 by Berg Publishers

Published 2020 by Routledge
2 Park Square, Milton Park, Abingdon, Oxon OX14 4RN
605 Third Avenue, New York, NY 10017

First issued in hardback 2021

Routledge is an imprint of the Taylor & Francis Group, an informa business

Library of Congress Cataloging-in-Publication Data

A catalogue record for this book is available from the Library of Congress.

British Library Cataloguing-in-Publication Data

A catalogue record for this book is available from the British Library.

Typeset by JS Typesetting, Wellingborough, Northants.

Publisher's Note
The publisher has gone to great lengths to ensure the quality of this reprint but points out that some imperfections in the original copies may be apparent.

ISBN 13: 978-0-367-71918-0 (hbk)
ISBN 13: 978-1-85973-135-2 (pbk)

To Hal Conklin

Contents

List of Figures

List of Tables

List of Colour Plates

Preface

All of the essays especially revised for this volume were first presented at a MOA symposium held in Kyoto and Atami during March 1992, entitled 'Beyond nature and culture: cognition, ecology and domestication'. What is remarkable about the published collection, and what distinguishes it from anything comparable, is not only the academic distinction of the various contributors, but the way in which the volume brings together the combined insights of cognitive and ecological anthropology on the one hand, and the approaches and concerns of ethnography and scientific biology on the other. Although we have maintained a tripartite grouping of chapters for this book, the degree of interconnection and integration of the subject matter is so great that virtually any grouping, or none at all, would have seemed plausible. Although the focus is a broadly anthropological one, the issues are of self-evident interest to a wide readership cutting across some conventional boundaries between the sciences and humanities. Indeed, the contributions are relevant to anyone who in their scholarly or professional work makes use of, or who deprecates, the analytical and conceptual status of nature:culture or cognate distinctions.

The original symposium was funded by a grant from the MOA (Mokichi Okada Association) International. We would like to thank MOA for their generosity, and for their warm hospitality in Atami. The symposium was organized by Katsuyoshi Fukui, with the support of an executive committee comprising Tomoya Akimichi, Ryutaro Ohtsuka and Mitsuo Ichikawa; and by an advisory group composed of Junzo Kawada, Sadao Sakamoto and Yutaka Tani. All of these individuals have also contributed to this volume. Vigdis Broch-Due, Harold C. Conklin, Shun Sato and Kenji Yoshida gave papers which are not included in the published collection, though their participation at the symposium was immensely valuable. Tatsuo Kobayashi, Emiko

Namihira and Manabu Tsukamoto were present as guest observers and we are grateful to them for their presence and active participation in the concluding session, and to Hiroshi Matsuda.

Most of the editing involved in producing this volume has taken place at the University of Kent at Canterbury (UKC), and we would like to thank Jan Horn of the Department of Sociology and Social Anthropology for her expertise and efficiency in preparing the papers to a high standard and in dealing with a variety of administrative matters. Some of the original figures have been redrawn with the help of the UKC Printing Unit (Lesley Farr and Chris Lancaster), and we are pleased to acknowledge their assistance in this. Finally, we thank Barbara Delaney, Marion Berghahn, Kathryn Earle, Sara Everett and Robert Parkin for their patience and cooperation in ensuring that a complex international collaborative project, long in gestation, eventually reached fruition.

Katsuyoshi Fukui, Kyoto
Roy Ellen, Canterbury

Chapter 1

Introduction

Roy Ellen

I

Background[1]

Humankind has evolved over several million years by living in and utilizing nature, by transforming and assimilating it into culture. Indeed, the biological success of our species has been widely accepted to rest upon its abilities to influence, manipulate and completely change this thing called nature.[2] Yet, by the sixties of the present century, the idea that culture is divorced from, and necessarily in confrontation with, nature was being challenged by the experience of non-Western 'holistic' philosophies, by advances in environmental biology, by the recognition of the damage done by environmentally unfriendly practices, and in a new generation of more profound anthropological studies of ecology. The emphasis was now firmly on people as parts of larger systems, on culture *in* nature, on the cultural *construction* of nature, and on species co-existence and sustainable development.[3]

Anthropology is increasingly concerned with such matters, as they become more urgent for *Homo sapiens* as a whole, and as they become routinized as parts of political programmes and management styles. The present volume seeks to review the current condition of the concept of nature as an analytical device, and the way it features in anthropological explanation. Whereas it has now become conventional for ethnographers and historians to assert, and in some cases demonstrate in convincing detail, the cultural, ideological and moral construction of nature, the real challenge is to examine the implications of such epistemological relativity for the objective practices of scientists of all kinds, and for those who attempt to build on these to implement change in the lives of people outside the Academy. We need to ask, for example, what deconstructionism means for the way scientists and administrators deal practically with issues which are seemingly no longer clear cut, and to what extent we are able to separate our scientific selves, for whom concepts have precise and understood contingent characters, from our broader cultural selves, where other meanings interfere. Can person and environment ever be anything but implicate in each other? Is the 'environment' the same as the environment? To what degree can we anyway cope at a practical level with the multiple ordering of realities, or do we need the re-assuringly straightforward certainties of Cartesian dualism? A relativist discourse of nature and culture is much easier to handle for those who are in a position to treat all their data as texts, who deny or have no interest in explicit comparison or pan-human generalization. It is much more difficult if we wish to translate the import of such ideas into terms that are understood and productive in the work of 'natural' scientists and those in the applied professions who use their insights and models of the world, or if we seek to explain how it is that humans seem to share a particular experience of the world sufficiently to be able to find the things they talk about recognizable.

The present book examines these issues through twenty contributions based on research conducted among a wide range of peoples throughout the world, from the prehistoric and historic past as well as the ethnographic present. The book is organized around three themes: nature as a cultural concept, the implications for the nature–culture distinction of studies of the relations between specific domesticates and human populations, and the continuing plausibility of the idea of cultural adaptation to the environment. The first is concerned with the negotiable character of the nature–culture dichotomy, and seeks to examine approaches to understanding it which go beyond conventional unreflexive assumptions. The second theme builds on the observation that particular species of domesticated plants and animals have evolved through mutual interaction with human populations, and focuses on the various cultural strategies involved. The third theme focuses on the extent to which it is possible to speak of culture and society as 'reflecting', 'influencing', 'managing', 'regulating', 'controlling' and adapting to environmental relations.

Nature as a Cultural Concept

That nature is culturally construed and defined - even 'constructed' – has become a commonplace in anthropology and the history of ideas. By 'culturally defined' we mean here made meaningful through practical engagement and cognition (categorization, labelling, intellection and sensation). Few would now dissent from the view that nature, and the extent to which it exists as a discrete idea at all, varies between different populations, according to different levels of discourse, and over time. There are now many analyses of the cross-cultural variation of the concept of nature within the ethnographic present. The present edition contains some further examples, and it is no wonder that, given its collaborative character, we are especially intrigued by the problems of translating between Japanese and European traditions. There has also been a steady growth of studies of how conceptions of nature change over time, in both the realm of scholarly ideas [e.g. Collingwood 1945; Glacken 1967; Horigan 1988; Williams 1976: 184–9], and in the lives of ordinary folk [e.g. Dove 1992; Thomas 1983]. However, there is no particular consensus as to what firm conclusions might be drawn from all this.

The hard relativist position of many contemporary anthropological writers owes much to the insistence of Edmund Leach [1964: 34–5] that the environment is perceived as a continuum which does not contain any intrinsically separate things, reminding us of John Locke's blank sheet upon which children impose a 'discriminating grid which we discover through language'. This is a view which has not stood the test of time and has been challenged by a later generation of psychologists and anthropologists, yet we find it echoed clearly enough by Carol MacCormack [1980: 6]:

> The 'myth' of nature is a system of arbitrary signs which relies on a social consensus for meaning. Neither the concept of nature nor that of culture is 'given', and they cannot be free from the biases of the culture in which the concepts were constructed.

Similarly, Marilyn Strathern [1980: 177], in the same volume, reiterates the position:

> The point to extract is simple: there is no such thing as nature or culture. Each is a highly relativized concept whose ultimate signification must be derived from its place within a specific metaphysics. No single meaning can in fact be given to nature or culture in Western thought; there is no consistent dichotomy, only a matrix of contrasts.

But in the same passage Strathern is careful to voice a certain detached agnosticism. 'The question then becomes', she continues, 'how large a part of *the total assemblage* of meanings must we be able to identify in other cultures to speak with confidence of their having such notions?' She declines to answer her own question, preferring to 'bypass . . . the questions of underlying structures, and the usefulness of these terms for apprehending the workings of the human mind'. Instead, she addresses those styles of interpretation *'which impute to other people* the idea of nature–culture as a more-or-less explicit entity in their mental representations. Whatever status these concepts have in "rationalist" discourse there has been a demonstrable "empirical" appropriation of them' [Strathern 1980: 176–7 emphases added].

One way of attempting to answer Strathern's question is to examine the evidence for any underlying cognitive propensities which might generate the variety of images we intuitively interpret as conceptions of nature, even though these may vary between places and times in their degree of prominence and combinatory properties. We might, for instance, argue [e.g. Ellen in press] that three such propensities are sufficiently widespread and primal: the identification of *things* and perceived patterns between them, the contrast between self and other, and the recognition of some inner essence or force. Even if we conclude that no such pan-human cognitive template exists, the rubric is helpful in reviewing the variable characteristics of conceptions of nature.

The Thinginess of Nature

This refers to the human propensity to identify things (natural kinds), which, when inventorized and aggregated through resemblance, are seen as part of a whole we call nature. Thus, one of the main sources of data on how humans perceive and interact with the environment is the language we use to describe it, and the categories we infer from this. Such an approach has been integral to a particular branch of anthropology since the 1950s, and one of its pioneers, most influential advocates and practitioners has been Harold Conklin. In the paper which he presented at the Kyoto phase of the conference, Conklin [Conklin 1992; see also Conklin 1986] discusses the properties distinguishing everyday vernacular forms of environmental categorization, how linguistically reflected local systems capture the combined natural and artefactual status of domesticated plants and animals, and what we can learn from attending to the way people talk about these complex entities. Using the results of recent ethnobiological field research in Ifugao, northern Luzon, he explores these issues with respect to the multifaceted and interlocking local classifications of rice. Hundreds of expressions designate and characterize the myriad forms, functions, uses and meanings of this cultigen. For example, more than 50 terms contrastively differentiate parts of the plant, more than 70 specify uses other than as a food staple, more than 125 refer to distinct techniques of cultivating, storing, processing and handling; and more than 250 express characteristics of diagnostic significance in recognizing local (landrace) cultivars. Women seed selectors, traditional specialists in varietal identification, naming, and classification, provide excellent information for this collaborative ethnobotanical account of all locally distinguished rice-type categories.

That our images of environmental perception are dominated by the visual sense was evident from Conklin's paper. But recognition of this has often led to an underestimation of the role of the other senses. We perceive the environment as much through smell, taste, touch and hearing as through vision, even though science is dominated by visible or vision-based images of the invisible. Moreover, there is no particular reason to assert that 'thingification' is a cognitive propensity associated only with sight. For example, spoken language depends on the ability of the speech organs to produce, and the ear to recognize, particular kinds of *discrete* sounds (phonemes), just as sight requires the drawing of effective visual boundaries. The papers by Junzo Kawada and Feld pay particular attention to this. Kawada compares the interpenetration of culture and nature in the acoustical universe of the Japanese, French and Mosi (Burkino Faso). He examines differing degrees of arbitrariness in the meanings attached to sounds. Thus, in Japanese and Mosi verbal expressions the use of ideophones is preeminent, including those for non-acoustical sensations, whereas in French these are virtually absent. Instrumental sound communications range from pan-human signs such as alarms, through culturally modified melodic and rhythmic patterns, to highly conventionalized forms of communication, such as Mosi drum language. The interpenetration between nature and culture in the sound universe is exemplified by instrumental interpretation of natural sounds, such as bird vocalizations, which are common in Japanese and French, but virtually absent in Mosi. However, Japanese interpretations are usually associated with a limited number of species, mostly based on the idea of the rebirth of human beings who have died in unfortunate circumstances. Thus, sound relationships of this kind might best be understood as part of an idiom of human-animal metamorphosis.

Taking his cue from an observation by Roy Rappaport, that effective constructions of the environment may be less those that are objectively correct than those which invest parts of nature with value 'beyond themselves', Feld looks at two dimensions of the interrelatedness of adaptation and aesthetics amongst the Kaluli of Papua New Guinea: the first is the ecology of sound, the second the cartography of song. The natural soundscape of the Kaluli rain forest features the calls of some 130 species of bird, as well as the sounds of many frogs and the rhythms of many kinds of insects. Additionally, there is the sonic presence of creeks, streams, waterfalls, pools and other forms of water. These sound patterns index the time of day, seasons of the year, vegetation cycles, migratory patterns, heights and depths of forest, and many other aspects of the environment. In relation to this, Feld explores what Kaluli perceive and know about natural diversity in their world through these sounds, and how their own vocal and instrumental music is inspired by, modelled on, and performed through them. Similarly, Kaluli song texts are organized as maps of place names. The sequences of these are textually co-articulated with names of trees and vegetation, waterways and birds. Feld explores how nature (place) is quite literally 'placed' in the memory, and how its codification and evocation in the formal genre of song poetics intensifies the relationship between the feelingful, experiential dimension of cultural identity, and the adaptive dimensions of ecological knowledge and awareness.

The Otherness of Nature

In Western notions, nature is most obviously recognizable as what is 'out there', what is not ourselves and 'that which can take care of itself'. We find the same idea in many other, otherwise strikingly different cultural traditions. But the indicative content of that natural other may vary. In contemporary Europe images of nature are quintessentially of forest and mountains, of wilderness, and except for those who live off the sea, essentially terrestrial. But ideas of what is the best exemplar of that natural other vary. For the Satawalese, described here by Tomoya Akimichi, it is predominantly the sea. Though the sea is more intractable than the land when it comes to physical modification, it is not immune, and is subject to elaborate forms of cognitive re-structuring. Moreover, cultural images of natural otherness do not necessarily comply with biological history, as the chapter by Charles Frake in this volume well demonstrates. Frake shows us how what might be considered the domain of nature and things natural is the consequence of pre-meditated cultural actions and dramatic re-interpretations. He is concerned with the narratives which the people of a part of Norfolk in England employ to make sense of and orientate themselves within a particular landscape. The dominant idiom is one which refers to 'the subtle traces of past human endeavour enthusiastically discerned and interpreted by those in the present with present purposes in mind'. The Norfolk landscape is a product of past activity that requires constant cognitive attention and behavioural intervention to preserve and reconstruct what is valued in contemporary images of the past. This reading of the landscape is a cultural practice motivated not only by social entailments and political or economic agendas, but also by the experiential enrichment afforded by a meaningful engagement with the past of one's place.

And what works for the English countryside would appear to work also for Japanese industrial forest, for which John Knight

[1994] has demonstrated an evocation of the same wild qualities as those associated with natural forest. An even more dramatic example of the attribution of wildness to what is historically a tamed landscape is found in that epitome of natural otherness, 'jungle'. This is the English borrowing of the Urdu word *jongal* and has come to mean impenetrable tropical forest. Its history, however, as Michael Dove [1992: 241] suggests, is more complex. In what is now Pakistan, forest was removed by humans in order to create *jangala*, a savanna vegetation suited to the keeping of livestock. Only later, as *jongal*, did it come to mean forest in any sense. Thus, what are categorized as nature and culture not only shift and merge, but may even change places.

Sometimes the distinction between nature and culture may seem wholly irrelevant. Thus, it has been observed that the concept of nature as an oppositional other is inappropriate for describing hunter-gatherer world views. This is well exemplified in the review provided here by Tim Ingold. According to Ingold, hunters and gatherers deny with marked consistency that there is a rigid separation between the world of human persons and that of non-human agencies and entities, between society and nature. Anthropological accounts, however, typically present this view as entailing a particular social and cultural construction of nature, thereby reproducing the very dichotomy that, in other contexts, is recognized as peculiar to the Western tradition. Two consequences follow. First, the 'perceived' environment is seen to differ from the 'real' environment as the product of the ordering of sense data according to an imposed cultural schema (a theme to which I shall be returning later in this introduction). Secondly, the practical operations of hunting and gathering are reduced to interactions within a given, physical world, as implied in the notion of 'foraging'. In short, the perception of the environment is culturalized and the activities of production are naturalized. Ingold argues for an alternative theory, more consistent with the view of hunters and gatherers themselves, where the environment is only revealed to the perceiver through an active process of engagement, rather than constructed through an ordering of sensations passively received.

The argument that nature is an emergent property of culture is taken still further by Peter Dwyer. In New Guinea, ecological intensification (e.g. population increase and a growth in the importance of agriculture) is associated with gradients in land use. Increasingly, the domain of the invisible (e.g. the world of spirits) is connected with these gradients. Inasmuch as nature is 'other than culture', the comparative data suggest that in low intensity systems, like that of Kubo hunter-horticulturalists of the lowland interior relying heavily on wild sago and living in dispersed settlements, the analogue of nature *is* the invisible world and that, with increasing intensification, emerging geographic connections between the invisible and the visible predispose to an identification of the 'other' with some components of a perceived external nature. The artefactual Western notion of wilderness, now largely divorced from the invisible, represents an extreme position on this continuum.

Nature as Essence

The third cognitive propensity which routinely emerges when we examine concepts which are nature-like is the attempt to essentialize a vague and unknown phenomenon. This may involve recognition of a force exogenous to human will but which can to varying degrees be controlled, or it may define an irreducible core or an essence within things, such as the fluids and pulses associated with life; or it may be simultaneously, or varyingly, both, in equal or unequal measure. Thus, we have nature as opposed to nurture, instinct as opposed to reason, wildness as opposed to control, rawness as opposed to refinement [Strathern 1980: 217, following Mathieu]. We can see other examples of the same thing in culturally equivocal descriptions of nature as variously robust and fragile, benign and malign, capricious and ordered, perverse and tolerant, eternal and ephemeral [Schwartz and Thompson 1990: 2–13]. These qualities do not simply allow us to locate things in the world but to morally evaluate them as well. To say something is unnatural is to condemn it as reprehensible. This is not to say, however, that what is reprehensible is therefore to be understood as the logical complement of nature, meaning culture. This would not work in English semantics. On the other hand, to say that something is uncultured is not to mean that it is natural in this sense. It is all rather confusing. We biologize things to give them greater credence in moral arguments, as with ethnicity, and create the same effect by taking what is natural and incorporating it into culture. Thus, as Löfgren [1985: 211] has shown for recent Swedish bourgeois traditions, an abhorrence of 'natural ways' does not mean that there is no longing for the 'natural way of life'. One and the same person can think with animals in different and conflicting ways, partly because they are, paradoxically, both within and beyond us [Willis 1975: 9]. Science begins by distancing humanity from the natural world, and logically ends up by incorporating the first into the second. The new sensitivity and empathy towards animals associated with the animal rights movement bring together science and sentiment, as, for example, in current primate research [Reynolds 1994]. Löfgren [ibid.: 216] sums all this up in terms of cultural contradictions arising from the polarity between spheres of production and consumption in bourgeois culture: in the first the laws of science and rationality prevail, in the second sensuality and sentimentality.

Among the chapters presented here, that which best exemplifies this ambiguous propensity to essentialize is Hiroyasu Tomoeda's, which seeks to dissect a particular cultural construction of nature, this time through an examination of the fertility rituals of camelid herders in the high central Andes of Peru. Tomoeda argues that the very structure of language transforms the conventional tripartite relationship between gods, humans and animals into the bifurcate relationship 'I/you' in ritual songs, thus requiring us to rethink radically the appropriateness of conventional distinctions between nature and culture. This is especially as it is through this dyadic relationship that the concept of vital energy which permeates both environmental and social relationships is recharged.

II

The cognitive geometry of thing, other and essence provides a framework which may serve as the basis, in particular cases, of elaborate systems of meaning, most obviously those built around the opposition culture:nature. But the degree to which nature and culture *are* opposed, and the degree to which such an opposition becomes a device for cosmological coherence varies.

Nature need not be a clearly bounded concept, indeed it may most often not be. In Japan, the complex and multifaceted idea bound up in the characters *shizen* (自然) and *tennen* (天然) captures, depending on situation, the opposite of artificiality, the opposite of humanity, some inner fundamental property, common sense (*tao*), objects in the environment (such as trees), and creative processes not under human control (such as earthquakes). At the same time, it is accepted that there are strong spiritual connections between, say, mountain (*yama*) and village (*sato*) which cut right across any inherent dichotomizing tendency which these ideas might seem to be suggesting [cf. Asquith and Kalland, in preparation; Berque 1986; Shaner 1989; Tellenbach and Kimura 1989].

Where it is possible to extrapolate a 'difference' between nature and culture, this may not best be represented as an opposition [Strathern 1980: 179, following Wilden]. The nature in Frake's Norfolk landscape is not only objectively anthropogenic, but the intricacy of its interconnections with cultural history give it a comforting homely feel. Where the opposition nature:culture is transparent, the terms may not even be labelled. Amongst the Nuaulu, described in this volume by Roy Ellen, they exist as convenient covert categories strongly inferred from a series of contrasts apparent in ritual practice and mythic narrative, the interconnections between which cannot easily be spoken of in any other way. Whether labelled or not, they need not be part of a consistent set of corresponding analogical pairs, such as left:right, male:female, nor logically comparable to such classifications [Goody 1977: 64]. True, gender is strongly linked to nature:culture distinctions, but it need not be; and when the linkage is demonstrable, it is often contradictory. Thus, the idea of nature as passive and mute and therefore female, while culture is active, abstract and male is common enough, but many peoples do not conform, either all or part of the time (e.g. the Gimi [Gillison 1980]). Nature and culture often cannot be resolved into a single dichotomy [Strathern 1980: 178], and are not always the most appropriate extrapolations. The Hagen terms *mbo* and *romi* can be glossed as a matter of convenience as planted:uncultivated; belonging to people:belonging to wild spirits, though neither obviously fits the Western nature:culture distinction nor correlates necessarily with the male:female contrast [ibid.: 205]. There is, then, no necessary link between 'the natural world' and 'human nature'. For Hageners nature and culture are not concepts of order; 'the tension between the terms is different' [ibid.: 218].

European conceptions of nature are no easier to grasp. We have already seen this with respect to the essential qualities with which it is associated in contemporary thought, and as it has been opposed to different doctrines at different points in history so its meanings have altered [MacCormack 1980: 20]. We shall explore this further in the next section, explicitly in relation to scientific thought. But however nature is culturally constructed, and whatever (if any) its underlying cognitive template, in the way we use it is always a synergy of the utilitarian and the aesthetic, the pragmatic and the symbolic, and knowledge of it can never be independent of relations with it [Norgaard 1987: 118]. Or, to put it slightly differently, how 'society views nature is in part a function of how society has affected nature. Nature and the cultural conceptions of nature develop together; they co-evolve' [Dove 1992: 246].

The Cultural Construction of Nature in Science

To delineate variations in the cultural construction of nature is arguably easy and – some would say – increasingly self-indulgent. The real problem arrives when we observe that the analytic concepts of science (including anthropology) are themselves rooted in folk categories and therefore subject to the same critique. Not only are nature and culture folk concepts, they also form the basis of a complex series of dichotomies in science which serve both as descriptive and interpretative devices, and as ideological props. Many have contributed to this view in recent years, but few as persuasively as Dona Haraway [1989, 1991; see also Jordanova 1986]. In order to deal with the way in which nature is used in professional scientific discourse (and I shall focus here primarily on biology and anthropology) we must distinguish between several levels of assumption. The first is that nature really exists out there in the world in a positivist sense, and that science offers us a realistic model of how it is different from culture. The second assumption is that nature itself is out there but that science (including folk science) can only apprehend it through shifting cultural lenses. The third assumption is that even if nature is not *out there*, the contrast between nature and culture is a distinction which the human mind is predisposed to make. Of course, some will say that if the latter is true then so is the former, or at least we can never know. Let us take the first assumption first.

Historical accounts of changing modalities of nature in Western thought and science are hardly new. Collingwood [1945], for example, provides us with a surprisingly contemporary interpretation of the shifts and continuities between Greek, Renaissance and modern views. In the conventions of a presentist history of ideas it has become usual to locate the origin of the idea that nature is independent of our perception of it, and relations with it, in Greek *physis* and *nomos*: that which is unconnected with human agency and that which results from the human capacity to conceive and execute. But the popular pre-Christian Greek view of nature was essentially animist, in many respects resembling that of animist peoples recorded for the ethnographic present. Outside the treatises of individual philosophers, the Greek world of nature was saturated by mind. By contrast, the Renaissance view denies the Greek idea of nature as an organism, and asserts that it is both devoid of innate knowingness and more akin to a machine, subject to laws produced by an exogenous intelligence [ibid.: 5]. This is more in keeping with one interpretative stream in Greek thought, linking Plato with Aristotle [Lloyd 1992: 10]. But the concept of nature as a *scientific* category, as well as one crucial to political discourse, is basically one of the enlightenment [Williams 1976: 187], though it begins to find its modern expression towards the end of the eighteenth century. For Collingwood [1945: 9] the modern view is based on an analogy between processes of the natural world as studied by natural scientists, and human affairs as studied by historians, with an emphasis on process, change and development. Hence, we have *natural* history.

In effect, then, nature and culture became reified as scientific concepts. A tradition of objective knowledge emerged in biology (and later in anthropology) in a way which made us think that knowledge of nature was independent of our relations with it. By the seventeenth century it had become something which could be quantified, mechanized and dehumanized [Westfall 1992], 'distanced from the world of sense and common sense' [Torrance 1992: vi]. What some may think even more remarkable is the way in which this Western tradition has spread and taken hold in other parts of the world which have been influenced by Western science and thought, including Japan. However, such a view of the emergence, growth, spread and global acceptance of a European culture of science and its associated conception of nature is problematic. The way in which scientists have used the word nature has never been wholly consistent, and it has become increasingly apparent that what was once thought to be a clear distinction between nature and culture is at the very least ambiguous. So, while it is often remarked that oriental and occidental conceptions of nature are in certain over-arching symbolic senses opposed, such a view is not necessarily inconsistent with uniformity at the level of practical experience [Bruun and Kalland 1995: 16]. There may, in other words, be a disjunction between the timeless qualities of cosmologies organized through symbolic logic and the pragmatic, experiential character of everyday subsistence [Croll and Parkin 1992b: 16; Ellen 1994: 456]. The extent to which we can really claim that Japanese working within the tradition of Western science have simply internalized for practical purposes a Western opposition between nature and culture, man and environment, is an assertion which really requires much further careful investigation. Certainly, we assume, for all intents and purposes, that there is now one academic culture in which there is a sufficiently shared understanding of what nature is for science to work as a global discourse. But even though we must make it, there are problems with this assumption. Let us examine some of these.

To begin with, humans modify the world around them on an enormous scale, and have done so through co-evolutionary interactions for many thousands of years. Effectively, all landscapes with which humans routinely interact are therefore cultural: and our environment is every bit as much what is made socially as what is not. Frake's chapter supplies a nice illustration of this, but in a more esoteric realm high energy physicists, in their attempts to understand the underlying character of matter, quite literally produce 'new' nature in a particle accelerator [Nothnagel 1994], and in this context nature is manifestly *not* something which can 'take care of itself'. And perhaps most significantly from the point of view of this book, the literature on domestication provides us with many examples which make nonsense of any hard and fast distinction, none more so than the proposal that the conservation and regulation of 'wild' animals is essentially the same as domestication [Macbeth 1989]. How strange, then, that in another version of the biological imagination (that of classical evolutionary taxonomy) domesticated animals and non-endemics are, somehow, *not* the real thing. The complexities of biological reality, enhanced by the insights of modern ecology and genetics, make drawing the frontiers between organism and environment, between what is cultural and what is natural, almost impossible.

Secondly, whether some *thing* is natural or cultural may depend on the level of abstraction in our arguments, our methodology, or on time phase or context, not on any intrinsic qualities. At different levels nature and culture are identified in different ways: while at the most abstract level Amazonia as a whole may be seen as nature in contrast to the urban sprawl of São Paulo, at a local level nature is remnants of rain forest as opposed to secondary forest and cultivated areas. Similarly, technology is neither intrinsically nature nor culture. A tool may begin as a stone unmodified by humans, but through perception of its functions it becomes cultural. Again, a plant or animal may become food (culture) without any material change ever having taken place; food is thus simultaneously cultural and natural. Much the same can be said of bodily techniques: walking, eating, defecating, and so on.

Thirdly, conceptions of nature of a broadly Western kind, with which individual scientists of diverse cultural backgrounds may operate, are often, notwithstanding, affected by local folk or philosophical traditions. The significance of this has only emerged in recent years through the ethnographic and historical study of comparative culture. For example, in Christianity, movement across the animal-human boundary was, quite literally, a 'transgression', an abomination against God's creation, whereas in Buddhism and Shintoism, by contrast, there is no sharp demarcation. Indeed, in Shinto natural phenomena and species may be deified. Japanese downplay oppositional contrast and emphasise the fuzziness between opposites of nature and culture [Ohnuki-Tierney 1987], and there is a pervasive folk ideology that Japanese are 'closer to nature' than those in the West. Animals can assume human form, which has led to much greater freedom to anthropomorphize in primate studies [Asquith 1986a, 1986b]. In a sense, this parallels the breaking-down of the nature–culture opposition among Western primatologists as a result of the recognition of the close molecular similarities between humans and great apes. Animal rights are no longer restricted to ideas of the proper human stewardship of dumb creatures, but involve liberating them, intellectually and physically [Reynolds 1994]. Primatologists now extend consciousness and free will and culture to non-human animals, even if that culture is not quite what most anthropologists have in mind.

The problem is that in the real world finding out how things work must utilize modes of expression drawn from our subjective lives. This is so even if what is held to define science is its counter-intuitiveness rather than common sense [Wolpert 1992]. In this respect science is no different from the pragmatic knowledge by which all peoples live. As Tomoya Akimichi well illustrates in his chapter, Satawalese navigational knowledge – the basis for making routine life-and-death decisions – is not only practical and technical but involves domains often thought to be impractical and unrealistic. Orientation involving use of an indigenous sidereal compass and landfall techniques relying on the homing behaviour of sea birds are the foremost cases. However, such techniques are used in conjunction with others which presume the existence of mythical islands, fabulous species such as yellow whales and the ghosts of floating canoes. It would be misleading to polarize such knowledge as either functional or metaphorical. Rather, we must understand that in this domain natural and supernatural are parts of a unified system of knowledge. In one form or another, such views are present in all systems of ritual and belief, but they are noticeably and strangely absent from contemporary development planning.

The Satawalese case emphasises just how important it is for anthropologists and other human scientists seeking to elucidate one form of knowledge to relate it to the other. We work with not one, but a series of interacting models. These difficulties have led, in some branches of science, to a more sophisticated appreciation of nature as an object of enquiry, and in this volume Dwyer shows how the relativist conception of nature presents a challenge to conventionally trained biologists whose methodology treats environment as an independent variable. Similarly, Tim Ingold argues that our understanding of the environment, and of responsibilities towards it, has been distorted by equating 'environment' with an image of nature that pervades modern

Western thought and science. More generally, many [e.g. Horigan 1988] have abhorred the rigid oppositional uses of nature and culture, noting that these obscure more than they reveal.

Nature and Culture as Analytic Categories in Anthropological Theory

The history of anthropology is – in one way or another – the history of the categories nature and culture. By posing the relationship between the two and using it to legitimate democracy, Lévi-Strauss claimed that Rousseau had invented the subject [Freeman 1983: 30]; and to speak the language of nature and culture is for many inheritors of the British structural-functionalist consensus to relive the classic controversies of the nineteenth century. But in contemporary anthropology we know and use nature and culture in two ways which bear little resemblance to Rousseau's concerns: as objective analytic categories, and as a perceived universal cognitive opposition. As translated into the rhetoric of naturalism versus culturalism, these are part of a long and complex debate. This set of chapters does not seek to resolve, or even specifically to address, this debate, except in so far as the debate itself has perpetuated and reified concepts of nature and culture. The nature and culture of the title are those addressed in the problematic of ecological anthropology; not the nature and culture within us, but the interaction between human individuals, populations and social collectivities on the one hand, and the natural resources (the environment) which humans manipulate and consume in the conduct of their lives, on the other.

In early social evolutionism the nature–culture distinction was largely ignored, and one looks in vain in the writings of, say, Herbert Spencer, for any allusion to it. And so it was to continue in certain branches of psychology and in the eugenics of Francis Galton. For Tylor, like Marx, the history of mankind was part and parcel of the history of nature [Horigan 1988: 14], not apart from it, and certainly not against it. With Franz Boas, however, things began to change: nature became simply that to which the laws of biology applied, while culture became that which biology could not explain. Thus the stage was set for that 'unrelenting struggle between two doctrines', between 'two fervently held half-truths, each insufficient in scientific terms, which had originated amid the theoretical confusions of the late nineteenth century, the one overestimating biology and the other overvaluing culture' [Freeman 1983: 33, 35]. Moreover, as Horigan [1988: 4] has pointed out, the nature–culture opposition took on a special role in the social and human sciences by ensuring their legitimacy, providing them, in culture, with an object of study and a principle of disciplinary demarcation. Hence the importance of defending the autonomy of culture becomes part of the justification for 'the unrelenting struggle'. Every social anthropologist who asserts that there is no need to take heed of biological explanation is re-asserting the nature–culture opposition, even if the terms are not used. Yet what is obvious is that much anthropology since the second world war has assumed that the issue was an essentially dead one. True enough, social anthropologists are occasionally shocked and embarrassed by ethologists and sociobiologists and others claiming that nature intrudes on human behaviour in more than just a passive, facilitating sort of way, but these are minor irritants. Instead, in much socio-cultural anthropology nature:culture came to mean something else: a cultural distinction itself – in a sense *the* cultural distinction – which serves to organize a high proportion of what we experience of the world. This approach found its inspiration in the work of Lévi-Strauss, for whom nature–culture was a timeless, value-free model, a mechanism through which the mind organizes the world around itself, an 'artificial creation of culture' [Lévi-Strauss 1969: xxix]. In more recent times, suggests Horigan [1988: 33], it is also the idea which underlies the attachment of Marshall Sahlins – among others – to a purely symbolic and dualist definition of culture.

The history of anthropological studies of environmental relations reflects these general theoretical developments, but also materially influenced them. By insisting on the environment being outside the object of anthropological scrutiny, Boas forced the pace of possibilism; but it was empirical studies of hunter-gatherers and pastoralism which contrariwise showed the artificiality of eliminating nature from understanding central issues of social organization and cultural classification. In a different way, Conklin [e.g. Conklin 1957], in persuading us to take seriously the environmental classifications of Hanunóo swidden farmers, gave credence to Lévi-Strauss's prototypical nature–culture distinction. Again, Roy Rappaport was trying to dissolve the hard culture–nature distinction he saw in Julian Steward's programme of cultural ecology. In the 'ethno-ecological' phase which Conklin's work heralded – indeed he appears to have invented the word [Conklin 1954a, 1954b] – nature becomes bifurcate, a folk nature reflected in the ideas and discriminations of local people and 'real' nature on to which these sometimes significantly different conceptual worlds were mapped. Such a view becomes explicit in the dualistic approach of Vayda and Rappaport in which cognized (emic) models are kept rigorously distinct from analytic (etic) models and become part of the means by which humans adapt to nature and regulate it [Vayda and Rappaport 1968]. Thus the systems approach which enshrined the organism-environment distinction necessarily upheld a particular version of the nature–culture contrast. For this reason, claims Dwyer (this volume), Rappaport was paradoxically on the wrong side of the divide. Indeed, the objectivist, adaptionist version of ecological anthropology has been vigorously criticized since the mid-seventies. On the one hand, there have been those who have placed increasing causal priority on individual cultural constructions, an approach which eventually disappears into post-modernism. On the other hand, contemporary evolutionary ecology, with its inspiration drawn from methodological individualism, has no particular place for the concepts of nature and culture, for the distinction between intentionality and unintentionality, or that between cultural and natural selection. In fact, it has no need of the conventional adaptionist programme at all. Thus, the nature–culture distinction is constituted in different ways in different theoretical approaches in ecological anthropology, bridging three levels of analysis: one in which the environment is a biological given, one in which it is bifurcate (analytic and folk), and one in which all senses of nature or environment are 'constructed' and negotiable. What is distinctive about recent work in cultural anthropology, ecological anthropology and sociobiology is that what we might mean by nature has again become a central issue, for modernists and postmodernists alike.

So what is paradoxical about the ecological tradition in anthropology is that while starting off rooted in the naturalism/ culturalism debate, and while still fostering an objectivist agenda, it has been able to show just how insecure and contingent the nature–culture distinction is. It has done this by demonstrating empirically the extent to which the natural has been influenced by the cultural, by a systemic approach which abhors simple dualisms, and by illustrating how the most esoteric elements of culture might intricately affect our use of the environment and even regulate our relations with it. Thus, in his contribution to this volume, Mitsuo Ichikawa shows how such issues force us to adopt a co-evolutionary and historical stance in understanding relations between the Mbuti and other elements in the Ituri ecological system. The equatorial rain forest of the Zaire basin has been the habitat of hunting and gathering peoples for many centuries. Ichikawa shows how the Mbuti have made use of their profound knowledge of the forest: as plants used for food, medicine, poison, tools and construction, and for ritual and other non-material purposes. Plants are also shown to be indirectly useful as sources of food for other species which the Mbuti hunt or make use of. Ichikawa argues that despite their heavy dependence on forest resources, the Mbuti have not obviously degraded plant and animal populations. Indeed, they have actually improved the forest as a human habitat through marginal modifications. Understanding the character of the co-existence maintained between Mbuti and their environment provides important data for reconsidering what we mean by nature, and what the implications are for planned state conservation, for the sustainable use of forest, and for maintaining the integrity of Mbuti culture itself.

Relations between Specific Domesticates and Human Populations

The inadequacy of the distinction between what we conventionally call nature and culture is no better exemplified than through the examination of particular domesticates, species which owe their current genetic composition to close encounters with human populations which harvest them for food and other products. When domesticated animals appear in human subsistence systems we see this as the intrusion of culture; when animals domesticate humans it is natural. From an ecological viewpoint the process is the same. Domestication as a process raises intriguing issues of intention, selection and co-evolution, and poses in an acute form the difficulty of the terms nature and culture in scientific discourse.

Sadao Sakamoto elucidates the process by which cultural preferences in the early stages of the domestication of six cereal cultivars found only in east and southeast Asia led to the convergent co-evolution of waxy endosperm variants. He argues that what favoured the selection of waxy endosperm mutants in various kinds of rice and millets was a well-established preference for sticky foods such as yam, taro and banana. He further adduces that a pre-existing waxy endosperm food culture favoured the selection of waxy varieties of maize when this was introduced into the area in the fifteenth or sixteenth centuries. Sakamoto's is a classic ethnobotanical study which illustrates the difficulties of separating an understanding of the genetic process of domestication from a knowledge of detailed culinary ethnography. His paper illustrates the importance of ritual in the development of new domesticates and the essential conservativeness of innovation: waxy varieties of rice mimicking waxy varieties of tubers.

The origin of diversity in cultivated plants has long been understood to be either the result of natural selection modified by human interaction or the consequence of deliberate human actions. But such views cannot explain how and why such diversification occurred in the first place and was maintained under earlier agriculture. Taking examples from the ensete (*Ensete ventricosum*) zone of southwestern Ethiopia (Ari), Masayoshi Shigeta examines how new landraces are created and how and why diversity is maintained. Shigeta argues that 'cognitive selection', the categorization of the external appearances of plants, must be distinguished from utilitarian selection, that based on the usefulness of particular landraces. He demonstrates the existence of the former as a qualitatively different kind of selection, devoid of value judgement, but with an indispensable role in establishing new types in a repository of landraces. The maintenance of genetic diversity is exemplified by 'folk in-situ' conservation, though why diversity should be maintained is more complex. Shigeta rejects the utilitarian view of landrace diversity because it does not accurately reflect actual patterns of use, and instead seeks to explain why it should be that cultural values should underwrite diversity. Such an approach demonstrates a social mechanism for maintaining genetic diversity, while avoiding teleology and tautology.

James Boster shares Shigeta's view of the importance we must attach to cognitive selection in explaining the origin of domesticates. He begins by reviewing research which strongly implies that diverse groups of informants agree in the categorization of avifauna in a way which suggests a pan-human perceptual strategy for the selection of those features of a collection of organisms that yield the most informative classification. These results, taken together with those of other ethnobiological and cognitive researchers, suggest that many features of human cognition are best understood as having evolved to interpret and discriminate other products of evolution. Focusing on the relationship between the classification, cultivation and selection of Aguaruna cultivars of manioc, Boster reviews research that argues that the procedures cultivators use to distinguish crop varieties have the effect of selecting for increased variability in the features they use to distinguish the crops: cultivators change the world in the process of understanding it in such a way as to make their procedures for understanding more appropriate. Thus, human cognition has co-evolved with the natural world: the human mind has evolved to understand the natural world and leaves its mark as it does so.

The chapter by Paul Richards, like those of Sakamoto, Shigeta and Boster, emphasises the significance of aesthetic and moral factors as well as practical ones in the evolution of particular domesticates. Richards argues that a consideration of a group of rice types grown by the Mende and Temne peoples of Sierra Leone provides an opportunity to rehearse an important range of local ideas about the beginning and underpinnings of social life, and to discover the place of both elephants and rice agriculture within the moral economy of the peoples of the Upper Guinea coast in West Africa. The paper surveys what is known of the origins and domestication of African rice (*Oryza glaberrima*), speculates on the beginnings of agriculture (based on rice) and overseas trade (based on ivory) in the headwaters of the Niger, and considers the social significance of the survival of *O. glaberrima* cultivars for the present.

Like Boster and Shigeta, Katsuyoshi Fukui demonstrates the crucial role of folk classification in the co-evolution of human populations and domesticates, but in this case using data from the Bodi of southwest Ethiopia. Bodi colour cognition is largely based on their classification of cattle, and colour-patterns are used in their naming practices and cattle coats identified with developmental stages. Bodi folk genetics of coat-colour is complex and used for breeding purposes. Every part of their world view is affected by the language of colour-patterns and the metaphors associated with them, and it is argued that this has co-evolved with the diversification of cattle coats.

Whereas Fukui focuses on the way humans select domesticated stock through a particular folk classification, Yutaka Tani is concerned with how humans make use of and modify the social relations between animals in the process of domestication. He argues that in contrast to the hunter, pastoralists distinguish individual members in the flock not only by sex, age, mother-offspring lineage and so on, but also by individual features, often treating each individual differently. It is these differences at the interventional and cognitive level which constitute the core of techniques which have been developed during the domestication process. Tani reconstructs two stages in the process with reference to the interventions of shepherds in the Mediterranean and Middle East. He identifies several features which may be specific to the sheep-keeping nomads of this region, and concludes that in contrast to the traditions of animal rearing in central and south Asia, the underlying ideological construct in their techniques is the idea of the domestic animal as serf.

The chapter by François Sigaut reintroduces a theme raised by Tim Ingold, that *environments* have no existence in themselves, the concept only making sense as a collectivity of interrelated variables relative to the needs of particular organisms, in this case humans, and that these can only be approached via an examination of culture, especially *techniques*. Thus, 'climate' does not mean the same for wheat-growers, wine-growers, car-drivers, sea fishermen, travel agencies, and so on, in the same way that iron does not exist for people without metallurgy. Sigaut explores some of the relationships between technological traditions and environmental factors which have developed around staple foods such as cool-season cereals in the West, rice in the East, millets and other crops.

The chapter by David Harris provides a fitting conclusion to this group of papers, examining in broad terms the characterization of relationships of domestication. He begins by reiterating what is now well-demonstrated empirically, that it is difficult to sustain simple subsistence categorizations of the kind 'foraging' versus 'farming', but observes that understanding is still handicapped by the ambiguity of terms such as 'domestication' and 'cultivation'. He argues for an evolutionary approach which permits more precise definitions of such concepts and the replacement of simple wild/domesticated distinctions with a continuum of human-other species interaction. Harris defines a range of human subsistence strategies and their capacity for intensification. These include those which focus on the seeds of grasses and herbaceous legumes, the nutty seeds of trees and shrubs, plant roots and tubers, and the meat and other edible products of such animals as the social ungulates, aquatic mammals, reptiles and fish. He questions why certain groups entered into domesticatory relationships with humans and others did not. He concludes by examining how particular subsistence economies evolved which combine plant and animal foods in associations of varying nutritional effectiveness and ecological resilience, paying particular attention to the origins of food production in southwest Asia compared with elsewhere.

Nature and the Problem of Cultural Adaptation

The ease with which it is possible to demonstrate the cultural relativity of nature in science and folk science, and the complex problems raised by the blurring of nature and culture in modern studies of domestication should not lead us to underestimate the plausibility of scientific claims of individual and population adaptation, and the necessity in such models to accept some working definition of the environment or nature as being 'out there'. The more extreme forms of post-modernism negate the possibility of studying how people relate to their physical and social surroundings using scientific methods capable of leading to generalizations – given that each cultural construction is unique and non-comparable. In their less extreme forms, they still tend to privilege individual constructions above the social, cultural and population level of analysis. Such approaches are useful in focusing on the decision-making processes of individuals, but reduce our understanding of how individuals, as parts of social groups, produce and reproduce themselves, how they make decisions about the use of what they call nature or natural resources, and how these things are transformed by such use into social entities themselves and become critical elements in social relationships.

I shall return to this fundamental problem in the final section of the Introduction. All that need be said here is that if we do not distinguish between representation and represented, we are sucked into solipsism – the view that the self is the only knowable thing. The chapters by Ryutaro Ohtsuka, Emilio Moran, Michael Dove, Tomoya Akimichi and Roy Ellen illustrate, in their different ways, the productivity of such a qualified 'scientific' position and the explanatory advantages of not yielding to unbridled otherness. Traditional Gidra subsisted by sago-gathering, hunting, fishing and swidden cultivation. In the 1980s they numbered less than 2,000, but genealogical and demographic data indicate that they had a mean annual rate of population increase of 2 per cent during the last few decades of the pre-contact period. Ohtsuka argues that the reproduction rate (the rate of inter-generational population replacement for females) varied considerably between inland and riverine groups, in the former exceeding 1.0 and in the latter being below 1.0. This profile is judged to derive from differential patterns of subsistence, nutrition and incidence of malaria. Gidra oral traditions state that they originated inland and gradually dispersed into riverine areas, a view supported by Ohtsuka's demographic estimates and descriptions of present-day adaptive strategies. But although the new environment is disadvantageous when measured in terms of rate of population increase, the horticultural practices which it can sustain and the new possibilities for cash-cropping have raised its carrying capacity.

The reason why populations such as the Mbuti and Gidra are able to co-exist with tropical rain forest is that their use of it is low in impact. But whereas in the Mbuti case this is because extraction is largely of plant-parts and selective animal predation, in the case of the native Amazonians described by Emilio Moran it is due to the small areas cleared and the practice of abandoning the cultivation of annual crops within three years. Moran examines the management strategies of native Amazonians which accelerate the process of forest succession and evaluates them in terms of their site specificity, cultural appropriateness and potential for extrapolation to unlike areas. These strategies are useful and adoptable beyond the boundaries from which they originate. Most managed fallows select for plants of economic value and are intimately connected to a mythology which supports their valuation. While it is not required that such strategies be culturally explicit, fallow management is often a conscious strategy, and Moran investigates whether this is more common where regrowth is faster, as one might predict on pragmatic grounds. To get a 95 per cent biomass regrowth some areas require as little as eight years, while others are believed to require no less than 100 years due to soil differences.

Many anthropological studies have argued that pragmatic conservation may be achieved through ritual practices. In his contribution, Michael Dove suggests that the techniques of bird augury employed by the Kantu' and other indigenous peoples of Borneo help to randomize behaviour in selecting swidden sites and minimize a tendency towards systematization. The intra- and inter-household diversity in swidden strategies which this promotes ensures a successful adaptation to a complex and uncertain environment. Dove examines the implications for this system of augury through the questions raised by Gregory Bateson on the abrogation of human choice, and argues that in this case it represents an appeal not to a lesser intelligence but to a greater one, the goal of which is to overcome the lack of humility and systemic thinking in human affairs. This requires a distancing of society from its own circumstances, which is made possible, in part, by the cultural construction of a parallel spirit reality, the contemplation of which gives the Kantu' a perspective on their own affairs they would otherwise lack. This same perspective can be found in Tomoya Akimichi's chapter on the significance of navigational knowledge among the Satawalese of the central Caroline islands.

But a major problem in any analysis of cultural adaptation arises from the observation that practices which are objectively efficient in maintaining the biological viability of a human population are not always the same as those culturally prescribed for the reproduction of society. Roy Ellen addresses this issue by providing a critical review of different approaches to the analysis of hunting efficiency and a detailed analysis of the ethnography of Nuaulu hunting on the eastern Indonesian island of Seram.

The first part of the paper presents Nuaulu hunting as a set of social relations which not only make some contribution to daily protein requirements, but which – through patterns of meat distribution, exchange and ritual – constitute a necessary part of that process through which 'houses', clans and a traditional authority structure are reproduced. These demands place stress on the ecological system without regulating it, and those ritual prohibitions which do limit extraction do so in ways which have a minimal impact on rates of human predation. More important in conserving the meat supply are storage, time constraints and the need to attend to the provisioning of other key resources, especially palm sago. The second part of the paper examines the technical and ecological efficiency of Nuaulu hunting inputs in the context of what is known of the distribution and population dynamics of game species on Seram.

A Response to the Deconstruction of Nature

How can we bring together the various strands of the argument explored here? The meanings of nature are evidently myriad, multivalent and shifting, both between different populations and within them. In the foregoing sections it has sometimes been appropriate to generalize about dominant cultural ideas of nature, but as Lloyd [1992: 1] has shown for the ancient Greeks, Murray [1992: 29] for the European Middle Ages, and Strathern [1980] for the modern Hageners, different ideas of nature are associated with different contexts, or linked to different philosophical traditions. Indeed, as individuals, we constantly attach to nature different moral nuances, draw its boundaries in different ways depending on how it suits us. For some, an algal bloom of non-human origin on the surface of water is pollution, as it resembles detergent effluent; for others, flowers which are very much a consequence of human interference and which are ecologically noxious may be seen not only as non-polluting but as intrinsically part of the beauty of nature. Nature provides us, in the words of Schwartz and Thompson [1990: 6–13], with 'plural rationalities'. Given the extensive critical debate of the last ten years it would be reasonable to assume that we will never again be able to use the concept of nature without thinking twice . . . or more. We now understand the importance of the language we use, and about the tyrannical necessity of dichotomies. We should not privilege our own concepts above those of our informants without first asking why we should do so and what the consequences are of so doing.

This is not only crucial for the private practice of anthropologists, but has significant implications and applications for other disciplines and professions. The problematic and contingent – the fundamentally anti-monosemic – character of our constructions of nature is still not understood by many practitioners of science where such an understanding has an obvious impact on their mode of analysis. And the matter is an urgent and crucial one where science impinges on policy. Definitions of nature in the discourse on the new reproductive technologies [e.g. Strathern 1992], on transplant and prosthetic surgery, are increasingly acknowledged as being of practical moral import, though it is palpably contested conceptual territory. And it is also at the heart of environmental politics [Dahl 1993], development and conservation practice [Croll and Parkin 1992a; Ellen 1994]. It is not only science which works with its own and often conflicting definitions of nature; it is also true of states, bureaucracies and their agencies, which often have to navigate the interface between scientific concepts and political pressures. We need to examine the extent to which official definitions of nature simply legitimate those of the morally and politically powerful and the degree to which they combine the definitions of different constituencies. We need to ask how particular definitions of nature serve the interests of particular groups, whether these be the conservation lobby, the Roman Catholic Church, or indigenous peoples who see advantages in reinventing a particular tradition of nature – the ecological Eden model [Marcus and Fischer 1986: 129]. In the necessity of international organizations, such as the UN Environmental Programme and WWF, to develop a shared working language we can see a new route to the globalization of concepts of nature which will have practical consequences for the kinds of knowledge we generate and for the lives of millions of ordinary people.

Yet, we must be careful not to get caught up in a web of reflexiveness which will ultimately prevent us from explaining anything. Recognition of infinite relativity (even if we can surmount the problem of solipsism) is no basis for scientific comparison, intellectual communication or practical action. It may or may not be the case that 'we cannot in principle distinguish between the constructed nature of our intelligible world and the "independent" structure of the brute world' [Margolis 1991: 3]; but by talking about the cultural construction of nature there must be some assumption that concepts of nature exist. The more you talk about nature the more you create a meta discourse which relies upon its existence, and the more you give it life as a ruling concept; in trying to get *beyond* nature and culture we reify an opposition. The possible alternative, the idea of 'the total interconnectedness of things', is not something which best organizes day-to-day information we need in order to act on the world. The philosophical language of ecology in which nature is dissolved is remote and not easy to grasp for many people, and we can see our inherent cognitive tendencies at work in the way in which 'ecology' has become the same as environment in British political discourse; the deliberately inclusive exclusive, the persuasively monist dualist. The human mind requires shape and discreteness in order to perceive the universe. There is a similar problem with the seductive claim of Gibsonian psychology that environment is constituted in relation to the organism that dwells in it, for relational 'affordances' grounded in practice are almost always first translated into abstract cultural knowledge suitable for storing and recall which is not necessarily relational in character, at least not in the initial perceptual sense. It is this reorganized knowledge which feeds back into practices of perception. We do not learn them as affordances. Most of what we do is habitual, not novel, and we act or react to culturally situated cues with culturally routine responses. It is as if we cannot avoid a concept very much like nature to make sense of the world, and that if we try to dispense with it we will have to invent something remarkably similar to replace it. And this is not simply a folk compulsion, it is fundamental to the pursuit of science.

For some, nature is very real, itself the product of human cognitive evolution – a kind of *Ur*-nature which differs from its various cultural manifestations. In this view – argued most persuasively in this volume by Boster – natural forms have a different perceptual character from cultural forms, and nature is essentially coterminous with, and shaped by the discontinuities of, the biological world [see also Berlin 1992 on the 'inherent structure of biological reality' (p.8) and 'the unconscious recognition of "nature's plan"' (p.261)]. But even if nature is a semantic fiction, it is a very convenient one. A set of conventions concerning nature as an object of scrutiny had gained such widespread currency in international scientific circles by the beginning of the twentieth century that it became very difficult to dislodge. It just so happened that science as a global cross-cultural practice consolidated at a particular point in history. Having established the rules, rules which worked well enough, as evidenced by the productivity of scientific work, there was no particular point in changing them, and once invented, changing them would have been extremely difficult. The reason why the Western paradigm has apparently become the ruling one may, therefore, not necessarily be because it is demonstratively a more truthful way of perceiving the world, but because it had historical infrastructural priority. Similarly, nature–culture, like other dichotomies, is useful or misleading, not true or false; a simplifying model for organizing thought, not a way of the world [Gould 1991 (1987): 9]. As Stephen J. Gould reminds us, Goethe realized that some dichotomies must interpenetrate, and do not struggle to the death of one side, because each of their opposite poles – say, inhaling and exhaling – captures an essential property of any intelligible world [ibid.: 19].

At the heart of our discussions about nature lies a logical tautology. Individual *Homo sapiens* use symbolic culture as part of their extended phenotype to maintain and increase their adaptiveness, but the very existence of symbolic culture separates human self-ontology from an existential environment (nature, the natural world, the other . . . whatever we wish to call it). In other words, the emergence of culture necessitated a bifurcation between experience and representation, and all cognition ultimately derives from experience. In the most fundamental sense, culture is the symbolic representation of experience, and evolution has effected an a priori dualism in our dealings with the world. We construct symbolic models of the world, the most primary being the self-ontology model which presupposes self-other dualism. In this view, resembling that recommended to us by John Locke, the human environment comprises the sum total of the objects of our perception, which is partly existential and partly cognitively realized. This existential-cultural dualism is the core of the self-world or nature–culture dualism, and is the result of evolutionary adaptation. We can therefore have – and at this point I feel justified in reintroducing the inverted commas – 'nature' and nature. The opposition between nature and culture is not spurious [*contra* Horigan 1988: 6], though the distinction may well be tautological, in that it is invoked only to explain itself [ibid.: 110]. But to say this is an observation of its formal status, and is not fundamental to its existential character. Nature–culture dualism is created through the symbolic construction of experience and is *in nature* only in so far as *Homo sapiens* is also in nature. As we have seen, food (both raw and cooked), techniques and bodies are all simultaneously in nature and in culture. Culture emerges from nature as the symbolic representation of the latter. As culture is a subclass of nature (the most inclusive class) nature cannot be fully specified using ordinary language, which is a kind of symbolic culture. Thus, the tautology of cultural adaptation and cultural meaning [see Ingold 1992: 39, 53] is really an instance of the conundrum of Epimenides, that because all Cretans are liars then Epimenides (himself a Cretan) must also be a liar. As Bertrand Russell observed, such statements are internally contradictory because they include themselves within their scope. If culture gives meaning to nature, then nature gives meaning to culture, then culture gives meaning to nature (humans adapt), and so on ad infinitum. The 'paradox results from considering non-nameability and indefinability as elements in names and definitions' [Russell 1956: 61]. And 'whatever involves all of a collection (meaning, culture, nature) must not itself be part of the collection' [ibid.: 63]. Culture is subsumed with nature both empirically and within the logical constructs of ordinary language, and the opposition of nature and culture is therefore a pseudo-problem arising out of reflexive symbolic constructs (ordinary language) within culture itself.

Notes

1. This introduction is based in part on notes taken and circulated at the Kyoto and Atami conference at which the papers collected here were originally presented. Some of the

text also follows closely abstracts of papers submitted by the various participants. I can, therefore, fairly claim that it is not all my own work, and that I am indebted to all those who contributed to the discussion. I am particularly grateful to John Peacock, whose thinking and words are reflected in my final remarks on nature-culture as a pseudo-problem.

2. The problematic status of the words nature and culture, and the concepts which accompany them in different contexts, is now so thoroughly acknowledged in the scholarly literature that placing them in ironic quotation marks or in italics, or in some other font, to highlight this, is wholly redundant. Indeed, in a book of this kind, constantly to resort to such textual devices quickly becomes irritating, confusing and absurd. Therefore, outside quoted passages, they are not used.
3. It is not surprising, therefore, that the conference at which these papers were first presented started from a recognition that many of the old certainties bound up with conventional Western scientific notions of nature no longer hold. This was deemed to be so both in the sense of nature as a cultural concept which can be shown to be endlessly malleable, and – even – in its objectivist sense, where the discoveries of modern ecological and evolutionary biology, and of environmental anthropology, show that the boundary between what is natural and cultural is inherently fuzzy and dynamic. Such a view is well exemplified in the various volumes of the National Museum of Ethnology series entitled *Living on the earth*, and edited by Fukui [Fukui 1995; Hori in press; Kakeya 1994; Komatsu in press; Ohtsuka 1994], and it was this project which provided the original conference framework.

Bibliography

Asquith, P.J., 'Anthropomorphism and the Japanese and Western traditions in primatology', in J.G. Else and P.C. Lee (eds), *Primate ontogeny, cognition and social behaviour*, Cambridge: Cambridge University Press, 1986a

——, *The monkey memorial service of Japanese primatologists*, Honolulu: University of Hawaii Press, 1986b

Asquith, P.J. and A. Kalland (eds) in preparation, *The culture of Japan's nature*, London: Curzon Press.

Berlin, B., *Ethnobiology classification: principles of categorization of plants and animals in traditional societies*, Princeton, New Jersey: Princeton University Press, 1992

Berque, A., *Le sauvage et l'artifice: les Japonais devant la nature*, Paris: Gallimard, 1986

Bruun, O. and A. Kalland (eds), *Asian perspectives of nature: a critical approach*, Copenhagen: Curzon Press, for Nordic Institute of Asian Studies, 1995

Collingwood, R.G., *The idea of nature*,Oxford: Clarendon Press, 1945

Conklin, H.C., 'An ethnoecological approach to shifting agriculture', *Transactions of the New York Academy of Sciences* 17, 1954a, pp.133–42

——, 'The relation of Hanunóo culture to the plant world', New Haven: PhD dissertation, Yale University, 1954b

——, *Hanunóo agriculture: a report on an integral system of shifting cultivation in the Philippines*, Rome: Food and Agricultural Organization of the United Nations, 1957

——, 'Des orientements, des vents, des riz . . . pour une etude lexicologique des savoirs traditionnels', *Journal d'Agriculture Tropicale et de Botanique Appliquée* 32, 1986, pp.3–10

——, 'Language and environment: the nature of folk categorization and the classification of domesticates', University of Yale: Unpublished conference paper, 1992

Croll, E. and D. Parkin (eds), *Bush base, forest farm: culture, environment and development*, London: Routledge, 1992a

——, 'Cultural understandings of the environment', in E. Croll and D. Parkin (eds), *Bush base, forest farm: culture, environment and development*, London: Routledge, 1992b

Dahl, G. (ed.), *Green arguments and local subsistence* (Stockholm Studies in Social Anthropology 31), Stockholm: University of Stockholm, Department of Anthropology, 1993

Dove, M.R., 'The dialectal history of "jungle" in Pakistan: an examination of the relationship between nature and culture', *Journal of Anthropological Research* 48(3), 1992, pp.231–53

Ellen, R.F., Review of *Bush base, forest farm: culture, environment and development*, E. Croll and D. Parkin (eds), *Bulletin of the School of Oriental and African Studies* 57(2), 1994, p.456

——, 'The cognitive geometry of nature: a contextual approach', in G. Palsson and P. Descola (eds), *Nature and society: anthropological perspectives* [in press], London: Routledge.

Freeman, D., *Margaret Mead and Samoa: the making and unmaking of an anthropological myth*, Cambridge, Mass.: Harvard University Press, 1983

Fukui, K. (ed.), *Co-existence between nature and humans: co-evolution between genes and culture* (*Living on the earth*, 4), Tokyo: Yuzankaku Publishing Corporation, 1995 (in Japanese)

Gillison, G., 'Images of nature in Gimi thought', in C. MacCormack and M. Strathern (eds), *Nature, culture and gender*, Cambridge: Cambridge University Press, 1980

Glacken, C.T., *Traces on the Rhodian shore: nature and culture in Western thought from ancient times to the end of the eighteenth century*, California: University of California Press, 1967

Goody, J., *The domestication of the savage mind*, Cambridge: Cambridge University Press, 1977

Gould, S.J., *Time's arrow, time's cycle: myth and metaphor in the discovery of geological time*, London: Penguin, 1991 [1987]

Haraway, D.J., *Primate visions: gender, race and nature in the world of modern science*, New York: Routledge, 1989

——, *Simians, cyborgs and women: the re-invention of nature*, London: Free Association, 1991

Hori, N. (ed.), *Exploring the space for survival: the philosophy of space utilization* (*Living on the earth*, 5), Tokyo: Yuzankaku Publishing Corporation, [in press], (in Japanese)

Horigan, S., *Nature and culture in Western discourses*, London: Routledge, 1988

Ingold, T., 'Culture and the perception of the environment', in E. Croll and D. Parkin (eds), *Bush base, forest farm: culture, environment and development*, London: Routledge, 1992

Jordanova, L. (ed.), *Languages of nature: critical essays on science and literature*, London: Free Association Books, 1986

Kakeya, M. (ed.), *The socialization of the environment: the cognition of nature as a means of survival* (*Living on the earth*, 2), Tokyo: Yuzankaku Publishing Corporation, 1994 (in Japanese)

Knight, J., 'Wild timber? Japanese mountain forests in local perspective', Oslo: Paper presented at the European Association of Social Anthropologists meeting, 1994

Komatsu, K. (ed.), *The Urgestalt of culture: nature as cosmology* (*Living on the earth*, 1), Tokyo: Yuzankaku Publishing Corporation, [in press], (in Japanese)

Leach, E.R., 'Anthropological aspects of language: animal categories and verbal abuse', in E. Lenneberg (ed.), *New directions in the study of language*, Cambridge Mass.: MIT Press, 1964

Lévi-Strauss, C., *The elementary structures of kinship*, London: Eyre and Spottiswoode, 1969

Lloyd, G.E.R., 'Greek antiquity: the invention of nature', in J. Torrance (ed.), *The concept of nature: the Herbert Spencer lectures*, Oxford: Clarendon Press, 1992, pp.1–24

Löfgren, O., 'Our friends in nature: class and animal symbolism', *Ethnos* 8, 1985, pp.184–213

MacCormack, C., 'Nature, culture and gender: a critique', in C. MacCormack and M. Strathern (eds), *Nature, culture and gender*, Cambridge: Cambridge University Press, 1980

Macbeth, H., 'Nature/nurture: the false dichotamies', *Anthropology Today* 5(4), 1989, pp.12–14

Marcus, G.E. and M.M.J. Fischer (eds), *Anthropology as cultural critique: an experimental moment in the human sciences*, London: University of Chicago Press, 1986

Margolis, J., *The truth about relativism*, Oxford: Blackwell, 1991

Murray, A., 'Nature and man in the middle ages', in J. Torrance (ed.), *The concept of nature: the Herbert Spencer lectures*, Oxford: Clarendon Press, 1992, pp.25–62

Norgaard, R.B., 'Economics as mechanics and the demise of biological diversity', *Economic Modelling* 38, 1987, pp.107–21

Nothnagel, D., 'The (re)production of nature: the nature–culture interface in physics', Oslo: Paper presented at the European Association of Social Anthropologists meeting, 1994

Ohnuki-Tierney, E., *The monkey as mirror: symbolic transformations in Japanese history and ritual*, Princeton, New Jersey: Princeton University Press, 1987

Ohtsuka, R. (ed.), *The cultural adaptation to resources: the ecology of co-existence with nature* (*Living on the earth*, 3), Tokyo: Yuzankaku Publishing Corporation, 1994 (in Japanese)

Reynolds, V., 'Primates in the field, primates in the lab: morality along the ape-human continuum', *Anthropology Today* 10(2), 1994, pp.3–5

Russell, B., 'Mathematical logic as based on the theory of types', in R.C. Marsh (ed.), *Logic and knowledge*, London: Macmillan, 1956

Schwartz, M. and M. Thompson, *Divided we stand: redefining politics, technology and social choice*, London: Harvester Wheatsheaf, 1990

Shaner, D.E., 'The Japanese experience of nature', in J. Baird Callicott and R.T. Ames (eds), *Nature in Asian traditions of thought: essays in environmental philosophy*, Albany: State University of New York Press, 1989, pp.163–182

Strathern, M., 'No nature, no culture: the Hagen case', in C. MacCormack and M. Strathern (eds), *Nature, culture and gender*, Cambridge: Cambridge University Press, 1980

——, *After nature: English kinship in the late twentieth century*, Cambridge: Cambridge University Press, 1992

Tellenbach, H. and B. Kimura, 'The Japanese concept of "Nature"', in J. Baird Callicott and R.T. Ames (eds), *Nature in Asian traditions of thought: essays in environmental philosophy*, Albany: State University of New York Press, 1989, pp.153–162

Thomas, K., *Man and the natural world: changing attitudes in England, 1500–1800*, London: Allen Lane, 1983

Torrance, J. (ed.), *The concept of nature: the Herbert Spencer lectures*, Oxford: Clarendon Press, 1992

Vayda, A.P. and R.A. Rappaport, 'Ecology: cultural and non-cultural', in J. Clifton (ed.), *Introduction to cultural anthropology*, Boston: Houghton Mifflin, 1968

Westfall, R.S., 'The scientific revolution of the seventeenth century: the construction of a new world view', in J. Torrance (ed.), *The concept of nature: the Herbert Spencer lectures*, Oxford: Clarendon Press, 1992, pp.63–93

Williams, R., *Keywords: a vocabulary of culture and society*, London: Fontana/Croom Helm, 1976

Willis, R., *Man and beast*, St Albans: Paladin, 1975

Wolpert, L., *The unnatural nature of science*, London: Faber, 1992

Part I

Nature as a Cultural Concept

Chapter 2

Human Dimensions in the Sound Universe

Junzo Kawada

I

Human dimensions in the sound universe can be recognized in the interaction of two complex spheres of relations. One set of interactions involves human representations of sonic and non-sonic phenomena, and these further interact with non-human or 'natural' spheres. The mode of these interactions differs from culture to culture. In this paper I compare them in three cultures, one from sub-Saharan Africa (the Mosi of Burkina-Faso), one from Western Europe (French culture) and another from East Asia (Japanese culture). The comparisons take into account a series of distinctions, which I have set out in Table 2.1.

Table 2.1. Typology of sonic and non-sonic phenomena (1–9) and modes of interaction among them (a–l)

1 'biological' voices of humans (cries of pain, anger, joy, sorrow, etc.)
2 verbal sounds I:
based on direct sonic effects
3 verbal sounds II:
based on sound symbolism – onomatopoeia
4 verbal sounds II:
based on sound symbolism – 'figurative' ideophones
5 verbal sounds III:
based on conceptual meaning

6 human non-vocally produced sounds (sounds produced by musical instruments, by working tools, by hand claps, etc.)
7 rhythmic movements of the body
8 non-acoustic sensations (visual, olfactory, gustatory, tactile, and other non-acoustic sensations)
9 sounds in nature (animal, bird and insect sounds; sounds produced by natural phenomena, like thunder, wind, rain, etc.)

Modes of interaction among phenomena 1–9

a rhythmic bodily movements during labour or dance by means of the inciting of sounds produced by instruments or by a measured hand clap
b representation of non-acoustic sensations by means of instrumental sounds
c representation of natural sounds by means of instrumental sounds
d transmission of verbal sounds with conceptual meanings by means of instrumental sounds, like drum or whistle sounds (emission)
e reception of (d) sounds
f inciting rhythmic bodily movements and the artificial sounds of labour, or dance, by means of verbal sounds
g inciting the production of verbal sounds by means of human sounds and rhythmic bodily movements
h representation of natural sounds by means of verbal sounds with conceptual meaning
i **kuchishooga**, or the Japanese system of musical notation using ideophones
j representation by means of 'figurative' ideophones
k representation by means of onomatopoeia
l representation of non-acoustic sensations by means of verbal sounds with conceptual meanings

In Figures 2.1, 2.2 and 2.3, the verbal sounds are numbered in order of the degree of non-arbitrariness of sound–meaning association, following the notation provided in Table 2.1. If the arbitrariness of sound-meaning association in a verbal sound is greater, I have qualified it as more 'arbitrary' and less 'naturally motivated'.

In comparing these three cultures, I deliberately take account

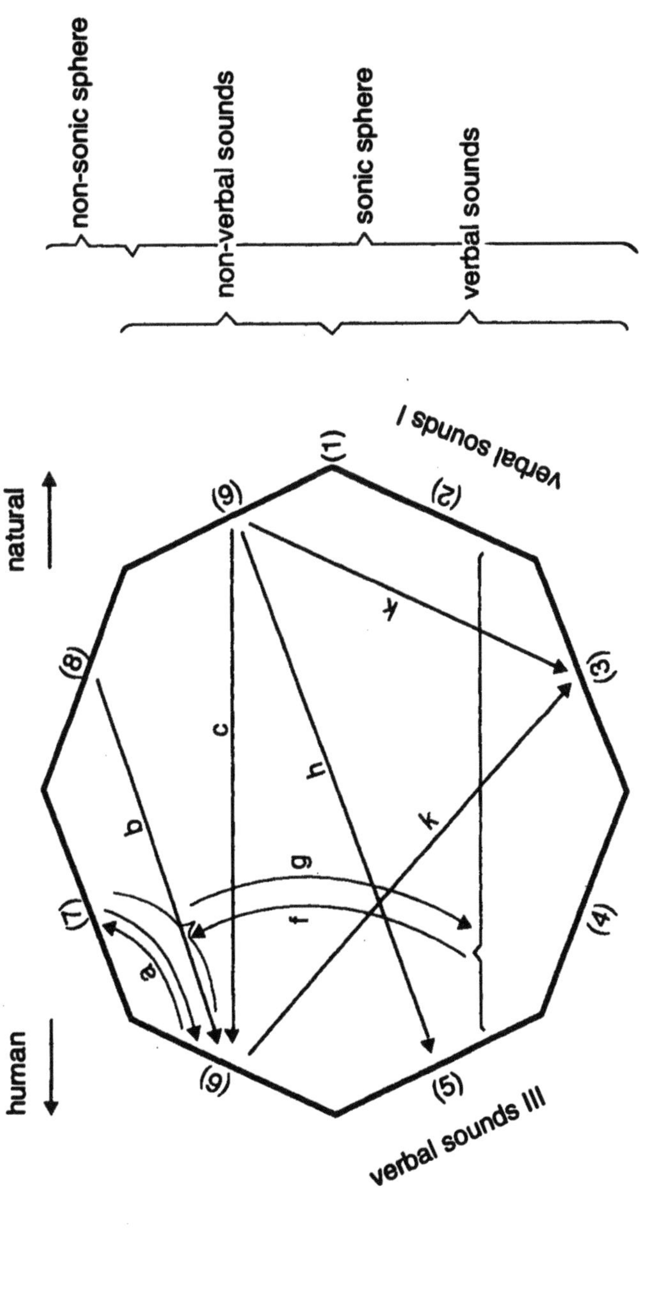

Figure 2.1. Verbal and non-verbal sounds in the French sound universe

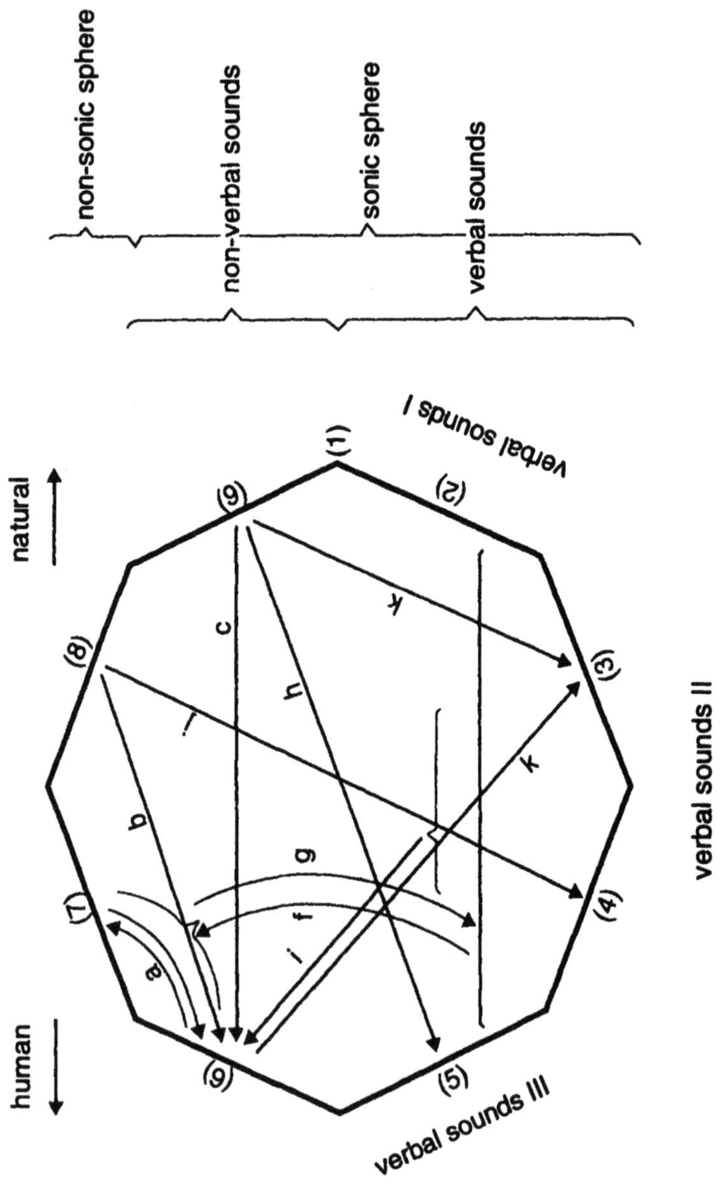

Figure 2.2. Verbal and non-verbal sounds in the Japanese sound universe

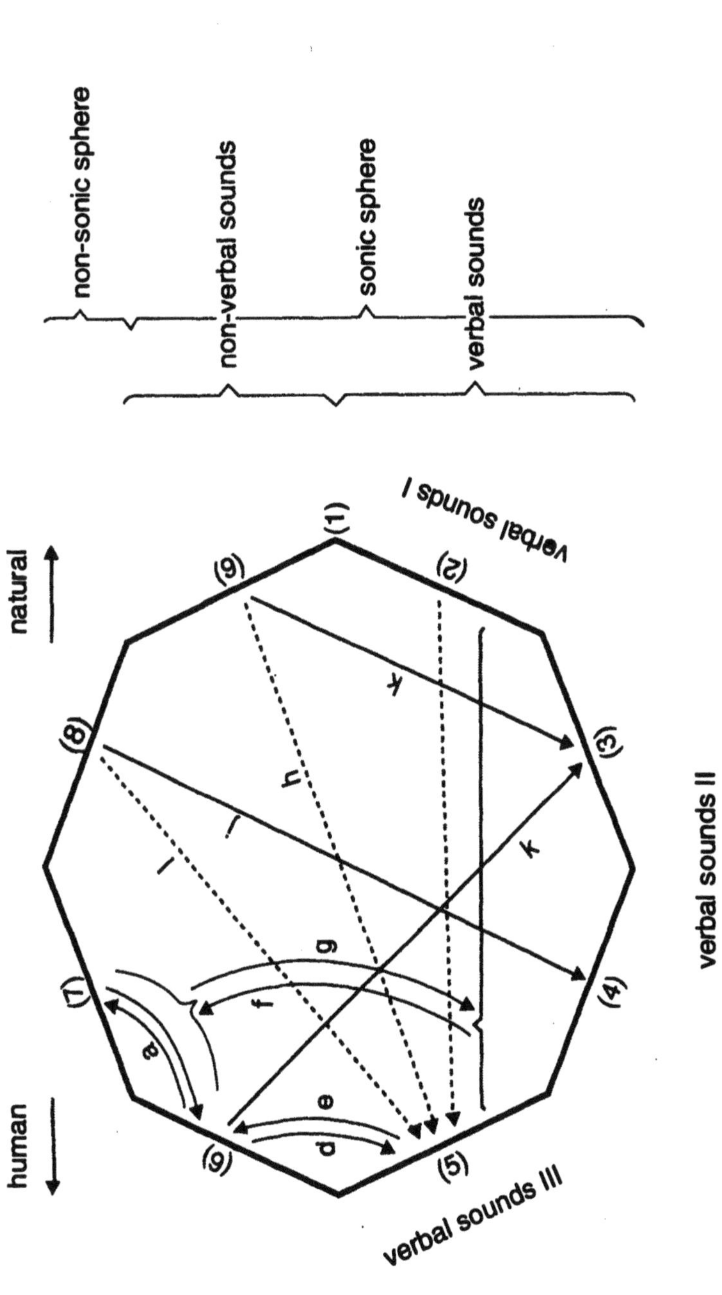

Figure 2.3. Verbal and non-verbal sounds in the Mosi sound universe

of historical changes and regional differences within each of them, but as they differ fundamentally from each other the heuristic value of this kind of comparison, even though it is tentative, may serve to lessen the minor variations within each culture.

II

Of the many points I might have mentioned, I will discuss here two which seem to me essential to an understanding of nature–culture interactions in the sound universe: first, the description or representation of nature by means of instrumental and vocal sounds, especially the problem of how sound symbolism and ideophones are to be treated; and secondly, the verbal interpretation of animal sounds, in particular those of birds.

The description or representation of nature by instrumental sounds is found in French and Japanese culture, while representation by means of verbal sound symbolism is more common in Mosi and Japanese culture. Japanese instrumental music is full of imitative or ideophonic expressions.

Taking examples from different kinds of instruments, among the well-known cases I might cite are the **kakebachi** of **hosozao-shamisen** (a three-stringed lute, with a thin neck, played with a plectrum), describing a group of birds (*Charadriiformes*) calling in flight, in the piece entitled **Shikino Yamamba**; or the **mushino aikata** of **hosozao-shamisen**, imitating the sounds of insects, especially of **matsumushi**, *Xenogryllus marmoratus*, a kind of cricket, in the piece **Akino irokusa**. The piece **Shikano toone**, which employs the sound of two **shakuhachi**, an end-blown bamboo flute with five holes (four front holes and one back hole), expresses the calls exchanged between a stag and a hind.

As for the music of **kotsuzumi**, a two-headed hour-glass-shaped drum with tuning cords beaten with the bare right hand, we can cite the following examples: **Namigashira**, in which the sounds of **kotsuzumi** imitate the murmuring of sea waves; **Uguisuno taniwatari**, where the sounds of this instrument reproduce the voices of **uguisu**, *Cettia diphone*; and **Kodama**, the sounds of woodcutting in a deep mountain.

In the music of **koto**, a thirteen-stringed long zither played with three picks fixed on the fingers of the right hand, there are also passages in which the sound of **hototoguisu**, *Cuculus poliocephalus*, is imitated.

In this respect, an interesting example is **yukioto**, or 'the sound of snow falling', a way of drumming **oodaiko**, a big, two-headed barrel-type drum beaten with two wooden sticks. The head is covered with a thick piece of cloth, which describes the snow falling at a **kabuki** theatre and thus stresses the visual impression with an acoustic one. We can find here a parallel between the perception/expression pattern of the Japanese sound universe and the ideophones used to describe different types of silence, like **shinshinto**, for example: 'The snow falls (or the night advances) **shinshinto**', or **hissorito** and **shiinto**: 'In the deep mountain where the silence dominates **hissorito (shiinto)**'. In such expressions a soundless phenomenon or a non-acoustic sensation is described by instrumental or verbal sounds.

This kind of association of two sensations from different spheres, which I like to call 'synaesthesia in the sound universe', can be found in everyday Japanese life. A good example is the case of **fuurin**, a sort of wind-bell, which the Japanese hang in the window in summer and which engenders a feeling of coolness at the sound of it tingling in the wind.

In French culture, examples of the description of natural sounds and other natural phenomena by means of instrumental sounds are not so common as in Japanese culture, but nonetheless they can be found.

The Song of Birds (Le chant des oyseaulx) by Clément Janequin in the sixteenth century is a fine example of the instrumental as well as vocal reproduction of bird sounds. It contains, among many onomatopoeia, some words with conceptual meaning (verbal sounds (5) in Table 2.1), like *joly* (pretty) or *coqu* (cuckold) which imitate the voices of certain birds. The word *cocu* (cuckold) itself may be derived from 'cuckoo', a bird of parasitic and polyandrous habits. The sounds of birds in this family have attracted the attention of many peoples in the world; in many languages the verbal onomatopoeia for the cuckoo's call has become the etymological origin of its name.

The use of musical instruments to reproduce bird sounds occurs in a more systematic and zealously 'ornithological' manner with the twentieth-century French composer Olivier Messiaen. *Birds Waking (Réveil des oiseaux)*, *Exotic Birds (Oiseaux exotiques)* and *Catalogue of Birds (Catalogue d'oiseaux)* are some of his best-known works of this sort. As a pious Catholic he regarded birds as singing the praises of God.

But a more representative and stylized way of using instrumental sounds to create impressions of the outer world has been an implicit, if not always explicit, element of French music since François Couperin and Jean-Philippe Rameau. We can trace this through Jules Massenet or Gabriel Fauré, to those modern composers often described as 'impressionists': for example, Claude Debussy or Maurice Ravel. Among the most famous pieces we can cite 'describing' nature through instrumental sound, especially birds, is *The Nightingale in Love (Le rossignol en amour)* by Couperin and *The Swan (Le cygne)* by Camille Saint-Saëns.

This kind of description of natural phenomena by means of instrumental sound is totally absent in Mosi culture. The absence of a descriptive tendency is also seen in dance, another important artistic domain of Mosi culture. The same is more or less true in many other sub-Saharan African cultures, where the passion for symbolic '*ex*pression' seems to surpass the taste for creating a descriptive '*im*pression' of the outer world.

This feature of Mosi dance is in striking contrast to the descriptive or mimetic character of Japanese dance, where it is even conceptualized as a dancing technique called **ate-buri** (**ate** means literally 'affected' and **buri**, 'gesture'). To describe, for instance, a mountain, the rising moon or sea waves by means of hands with a Japanese fan or a Japanese handkerchief is commonly seen. This feature may be understood in relation to the importance of the expressive use of hands in Japanese dance – and to some extent in other Asiatic dances, of which Indian Bharata Natyam presents a good example – whereas in Mosi dance, and in sub-Saharan African dance in general, hands do not play a significant role.

III

Description of natural phenomena through verbal sound and symbolism is elaborate and widespread in Japanese and Mosi culture. In French culture it is far less common, and as far as it concerns the use of ideophones for non-sonic phenomena, is practically absent. Some rare examples of the latter include **gnan-**

gnan, which means 'flabby, spineless', **chi-chi**, that is 'in an affected manner', and **dare-dare**, or 'double quick'.

This aspect of the sound universe is particularly interesting, as it reveals the nature–culture interrelation in a double sense. First, it includes the interrelationship between natural phenomena (sonic and non-sonic) on the one hand, and verbal, and therefore cultural, sounds on the other. Secondly, it consists also in the interrelation of the arbitrary, or culturally determined sound–meaning association, and the non-arbitrary, or naturally motivated sound–meaning association.

The first sense does not need explanation. I will give some brief comments on the second, referring to Figures 2.1–2.3.

The human voice identified in (1) refers to non-linguistic 'biological' vocalizations. It includes cries of pain, anger, joy, sorrow, and so forth. It lies in an intermediary domain between verbal and non-verbal sounds, and as it is not very culturally conventionalized, it can be understood by a receiver from a different culture, and in some cases, like the cry of agony, even by the receiver from a different species.

If expressed by verbal sounds, or in a culturally conventionalized way, such vocalizations may still be understood transculturally, as in the case of interjections of pain, like 'ouch' in English, *aïe* in French, *wai* in Mosi, and *aittsu* in Japanese. But if expressed by verbal sounds with conceptual meaning (5), as for example in 'I have a stabbing pain in my toe', it is impossible for a non-anglophone receiver to understand the meaning of the verbal sounds.

The human voices identified in (2), or 'verbal sounds I', are verbal in the sense that they consist of the phonemes of a given language. They have neither the conceptual meaning of (5), nor a definite symbolic meaning like onomatopoeia and other ideophones (3 and 4). However, they still transmit a certain message through direct acoustic effects, as instrumental sounds do in 'musical' communication.

For example, in Mosi word-play, one player proposes a series of amusing verbal sounds without meaning, and the opponent must give a phrase with conceptual meanings, reproducing the tones of some of the last syllables of the proposed phrase. The first player might say '**m lɩ́k n lɩ́ge**', a phrase which has no meaning, and his opponent replies reproducing the tones of the last four syllables of the proposed phrase, such as '**sàgbõaas yík**

n kúli', meaning 'the man who came to ask for a dish stood up and went out'. Then the first player says '**lìg lɩg lɩ́g lìnglam**', and his opponent replies reproducing the tones of the last three syllables, say '**sàgbõaas rób yànglam**', meaning 'the man who came to ask for a dish is crouching there'. Then the first player says '**m lɩ́g m lɩɩga**', and his opponent replies '**sàgbõaas wáag m laaga**', meaning 'the man who came to ask for a dish broke my dish', reproducing the tone of the last four syllables of the proposed phrase. Again the first player says '**lìg lɩg lɩ́k lìg lîgr lɩ́g lɩ́ɩrè**', to which the reply is '**sàgbõaas yîk làf pág m zágnóorè**', meaning 'the man who came to ask for a dish stood up and shut the door of my hut', and so on. These proposed phrases with verbal sounds devoid of meaning should be pronounced in an amusing way.[1]

In this kind of word-play in Mosi, which like many other African languages is a tone language, the key of the play is to reproduce the same tonality of the proposed phrase. Rhyming is not necessary, though it occurs sometimes, as can be seen in the cases just cited.

In a Mosi folktale, in which a singing fish is favoured by the king, the song of the fish is expressed by verbal sounds, such as **firika lieng lieng**. These have neither lexical nor symbolic meaning, but must be pronounced in an amusing and attractive way so that the audience understands the king's special favouritism toward this fish.[2]

Examples of verbal sounds of this kind (Table 2.1: 2) can be often found in Japanese folk songs, for example, **zuizui zukkorobashi, ojomma jomma** or **shimbai kombai**. They are also found in some modern popular songs, such as **oppekepe oppekepe oppekepeppo peppopo**, or **sui sui suudararatta**, or in a song recently in vogue, **urara urara ura urade, urara urara uraurayo**.

In the French sound universe it is difficult to find examples of this kind, but the phrase **londérira**, for example, sung in a working song for the buffing of silk-worm cocoons, is in the same vein. As already noted, these verbal sounds have no definite symbolic value which is valid in other contexts, unlike ideophones. However, it is possible that their acoustic effects, formed in specific cultural and linguistic contexts, might be based on a sound symbolism which is partly universal. I shall return to this later.

In the verbal sounds indicated in Table 2.1 as (3), which correspond to so-called onomatopoeia, the sound–meaning association is more arbitrary than those of (2). However, in general they are less arbitrary than (4), in the sense that they imitate certain non-verbal sounds by means of verbal sounds. By contrast, 'figurative' ideophones (4) have verbal sounds which represent, in a conventionalized fashion, the impression of a certain non-acoustic sensation. Nevertheless, it is obvious that even onomatopoeia, based on sound/sound association, are more or less culturally conventionalized ways of reproducing the outer sounds. It is well known that in many languages there is an association between the verbal sound [k] or [g] and the onomatopoeia of the voice and name of birds of the family *Corvidae*, commonly called 'crow' in English, and, in our three cultures, 'corbeau' in French, **gããbgo** in Mosi, and **karasu** in Japanese. But the onomatopoeia which imitate the vocalization of a cock differ widely even among Indo-European languages of Europe, as in English 'cock-a-doodle-do', French 'cocoriko' and German 'Kikeriki'.

In class (4) verbal sounds, which I call 'figurative ideophones', the sound–meaning association is more arbitrary than with onomatopoeia. This is because they represent a non-acoustic sensation. They are less arbitrary or more 'motivated' than class (5) verbal sounds because they transmit the meaning through sound symbols, though not through the medium of conceptualized *signifié*, as in class (5) verbal sounds.

Nevertheless, in many 'figurative' ideophones acoustic impressions are not entirely excluded, as exemplified by the case of ideophones which describe the ways in which a person walks, an object falls, or a person gets angry. Thus, Mosi **soara soara** and Japanese **sootto** express the manner of walking with stealthy steps, **zãi zãi** in Mosi and **hashi hashi** in Japanese express 'walk with vigorous steps', while **bãna bãna** in Mosi, and **yobo yobo** in Japanese, mean 'walk unsteadily' without force, where there might be some elements derived from acoustic impressions.

IV

We cannot recapitulate here the long and varied arguments concerning sound or phonic symbolism [Kawada 1988b: ch.3], but my view can be summarized in two observations. First, verbal sounds can be classified into four spheres according to different degrees of sound/meaning arbitrariness. These are indicated in Figures 2.1–2.3. As we must recognize interactions, the spheres may be considered continuous. Secondly, and derived from the first observation, apart from the universal symbolic value of some verbal sounds, problems of sound symbolism must be examined first in the context of particular languages and cultures.

To return to the first observation. Although interactions between verbal sounds (2) and (3), and between (3) and (4) are important, it is equally important to consider interactions between verbal sounds (3), (4) and (5). Of the three sound cultures treated here, verbal sounds (4) are found practically only in Mosi and Japanese cultures. Given the absence of written records in Mosi, there is much uncertainty as to etymological questions. In Japanese, however, many etymological studies have been done verifying that class (3) and (4) verbal sounds are extremely common, also occurring frequently in the written form, even in modern literature. The novels of Yasunari Kawabata, for instance, are full of ideophonic expressions.

In Japanese, we can easily pair related words of verbal sounds 3, 4 and 5, of which we do not know etymologically whether the ideophone 3 and 4 is derived from a word with conceptual meaning 5, or whether the second is derived from the first. The following are just some of many examples. In the pairs of words, the first is the ideophone, the second a verb, adjective or noun followed by an English translation:

ira ira–iradatsu to get irritated	**une une–uneru** to wind
koro koro–korogaru to roll	**sube sube–suberu** to glide
tsuya tsuya–tsuya luster	**nuku nuku–nukumori** warmth
fusa fusa–fusa tassel	**moya moya–moya** haze
yusa yusa–yusuru to sway	**yuttari–yutori** comfort

Even if the sound–meaning association here seems to be arbitrary, we cannot deny the possibility that these verbal sounds were chosen, even latently, because of the symbolic suitability of their sounds to that meaning.

Without excluding such a possibility, I cite the following examples where the ideophone might be derived from a word of which the sound–meaning association is considered arbitrary. The second term of each pair is an ideophone which might have derived from the first:

utatane to doze > **uto uto**
uku, or **ukitatsu** to float, or to become light-hearted > **uki uki**
ochitsuku to be calm > **ochi ochi**
kutabireru to get tired > **kuta kuta**
yawarakai soft > **yanwari**

In contrast, we can cite words where the etymology of an ideophone, including onomatopoeia, is certain, but the symbolic effect of their sound is now almost forgotten, especially when these words are written in **kanji**, or characters of Chinese origin, such as **tataku** (to beat) derived from the ideophone **tata** describing the manner and sound of beating; **tsutsuku** (to pick) derived from **tsutsu**; **fuku** (to blow) from **fu**; **sasayaku** (to whisper) from **sasa**, and so on.

In European languages too we can find numerous examples of words of ideophonic origin, if we include so-called onomatopoeia in the category of 'ideophones' in the wider sense. This is the case with 'whisper' in English, just mentioned in connection with its homologue in Japanese. In French, the equivalent of 'whisper', the verb 'chuchoter', is of onomatopoeic origin.

Apart from words of which the ideophonic origin is apparent, there are many European words where the ideophonic origin may no longer be consciously recognized in current use, as with English 'ban', 'banish' and 'banner', corresponding to the French words *ban, bannir* and *bannière*. These are all derived from the onomatopoeia 'bang', imitating the violent beating sound of a drum or gong.

In Mosi, etymological questions cannot be clearly resolved due to lack of written documents. Moreover, as spoken Mosi is not standardized in relation to an orthography or writing system, vocalization of verbal sounds is at once individualistic and expressive. Even words for which the sound/meaning association is clearly arbitrary can be pronounced in ways which rely heavily upon the sonic effects of an individual vocal style. Consequently, interpreting interrelations between the verbal sounds (3), (4) and (5) is potentially much more complicated than in Japanese. Some examples of the many Mosi words whose ideophonic origin cannot clearly be discerned follow. It is commonly recognized in many languages that the name of an animal often derives from an ideophone. Examples in Mosi include the word **katre**, 'hyena', possibly derived from the ideophonic verbal sounds **katre katre** (corresponding to the Japanese ideophone **gikushaku**), describing the clumsy walking manner of this animal, which is endowed with long, uneven front legs and short hind legs. With the pig, the verbal sound **kulkuli** describes the turning-round movement of this animal, which on the African savannah is slim and agile in its movements. But it is also likely that some acoustic impression of its voice may be involved.

The ideophone **põsi** describes a wooden stick breaking, and the verb 'to break' is **põse.** The weary manner in which an old person walks is expressed as **bãna bãna**, while the adverb 'slowly' is **bãnem bãnem**, the adjective 'weak' or 'pitiful' **bãnga**, and the verb 'to be weakened' **bããna**.

Thus, in the interrelationship of type (3), (4) and (5) verbal sounds, the problem of the secondary formation of sound symbols is worth reconsidering. As the Japanese examples show, the formation of an ideophone may be possible on the basis of the repeated use of an originally arbitrary sound–meaning association. A part of these verbal sounds may be detached and used as an ideophone, often in reduplicated form, to stress the sonic effect.

V

From the discussion so far my second observation should now be evident. That is, just as sound effect, sound symbol and arbitrary sound–meaning association – the three basic elements of vocal communication – are interrelated, so sound symbolism as an intersection of nature and culture must be examined in the context of specific languages and cultures. This is in contrast to an interpretation favouring universals, as found in the writings of several linguists, from Maurice Grammont to Roman Jakobson.

I recognize, of course, the value of the distinctive feature theory of phonemes, presented by Jakobson in collaboration with Gunnar Fant and Morris Halle. As 'the ultimate distinctive entities of language' [Jakobson, Fant and Halle 1963: 3], these features are always valid, and I refer to them. Nevertheless, as I have argued in detail on another occasion [Kawada 1988b: ch.3], these features are determined at the level of phonemes and appear in sound symbolism mostly in a combined state, where their symbolic value is always contextual.

The analysis of sound symbolism must proceed by investigating similar verbal sounds which have similar meanings, and by tracing correspondences between these sounds and meanings in different languages and cultures. To take one example, in Japanese, to describe the sun shining intensely, we use the ideophone **kan kan**, and we say 'the sun shines **kan kan**'. In Mosi the corresponding ideophone is **bas bas**. But to emphasise rather the heat of the sun's rays, we use the ideophone **jiri jiri**, the corresponding ideophone in Mosi being **wim wim**.

In Japanese, the verbal sound [ka] is a combination of a compact and tense consonant /k/, qualified also as 'fortis', and /a/, also a compact vowel, and is used to form other ideophones, in other contexts, to express something dry, hard or empty. On the other hand, the Mosi verbal sound [bas] is a combination of a diffuse, grave and lax consonant /b/ with a vowel /a/ and /s/, a diffuse, acute, tense and strident consonant. The symbolic value of **bas bas** in Mosi is concerned with the quantitative aspect of what is described, and this ideophone is used, in other contexts too, always to stress great quantity.

The ideophone **jiri jiri** describes something burned at a high temperature, and the acoustic impression of burning is felt as well. But the Japanese verbal sound [jiri], a combination of a compact, grave and lax consonant with a diffuse and acute vowel, has two other symbolic fields. One is psychological, and the ideophone **jiri jiri** expresses a state of irritation in a person, whose heart is 'burnt' in a figurative sense, and the Japanese verb **jirasu** 'to irritate' is derived from this sound symbolism. Another important aspect of the symbolic value of this ideophone is its use to express a slow but steady advancement. Among Japanese words with a conceptual meaning, we can find this sound symbolism in cases like the verb **nijiriyoru**, meaning 'to crawl on one's knees to' or 'to edge up', and in a substantive like **nijiriguchi**, which designates the narrow entrance of the room for the tea ceremony, through which the guests enter crawling on their knees.

The corresponding Mosi ideophone, **wim wim**, is used to describe something being stimulated greatly like the burning heat of the sun's rays, mentioned above, as well as the hot taste of red pepper, which is expressed in Japanese by the ideophone **piri piri**, also comprehensible in Mosi. More broadly, the symbolic field of [wi] in Mosi overlaps largely that of [pi] in Japanese. For example, in Mosi one says 'the lightening flashes **wila wila**'. This phrase can be expressed in Japanese with the ideophone **pika pika**. In Mosi one says 'He speaks **wilem wilem**', or 'fluently', and in Japanese, **pera pera**. The piercing cry of birds is described by the Mosi ideophone **wila wila**, and the whistle is called in Mosi **wiiga**, while in Japanese these sounds are described by the ideophone **pii pii**.

At the same time, the Mosi verbal sound [pi] has, both in other contexts and in words that are not ideophones, a similar symbolic value to Japanese [pi] or [pa]. For example, **pira pira**, a Mosi ideophone describing the noise produced when something hard and dry is broken by force, corresponds to the Japanese ideophone **pari pari**. The Mosi verb **pio**, 'to break something abruptly', can best be described by the Japanese ideophone **pachin**, the Mosi verb **piisi**, 'to tap lightly and repeatedly', by the Japanese ideophone **pita pita**, the Mosi verb **pinda**, 'to shine very brightly', by the Japanese ideophone **pika pika**, and so on.

On the other hand, the Mosi verbal sound [w] can in other contexts express a different symbolic value, which corresponds to that of various verbal sounds in Japanese (given in parentheses), such as **wigri**, 'dry and hard' (**kari kari**), **wiku** 'skinny to an extreme' (**gari gari**), **wisi** 'to smoke without flame' (**busu busu**) and **wita** 'to burn with flame' (**mera mera**).

We have seen that the symbolic field covered by the sound symbolism of one language may overlap partly that of another language, but we cannot find one-to-one correspondences between respective symbolic fields of verbal sound in different languages.

Though not universal, broadly similar principles of sound symbolism can be ascertained in different cultures. The sound symbolism of [pi] is a combination of a highly diffuse consonant /p/, with very weak phonetic power (which has been said to produce the sense of 'smallness'), and a diffuse, but acute and plain vowel /i/, which has been said to produce the sense of 'thinness, height, lightness and sharpness'. It is true that in various languages, and in many words which apparently did not derive from an ideophone, the sound [pi] is associated with the above-mentioned values, although as always we can find exceptions like *pierre* (stone) in French, and 'pig' in English. As we have seen, in Mosi the sound [wi], and in Japanese [ka] and [me], largely cover the symbolic fields covered by [pi] in other languages.

Another principle which seems to be logically 'universal' is the occasional association of the physiological mechanism used in pronouncing certain verbal sounds with the meaning expressed by them. For instance, 'to yawn' in French is *bâiller*, in Mosi **yaamde**, in Japanese **akubi**, always with an open vowel at the first syllable, as it is concerned with an action of a largely open oral cavity. The same is true for verbs such as *souffler* (to blow) in French, **fuuse** in Mosi, and **fuku** in Japanese, associated with breathing. **Moge** (to chew), which takes the ideophonic form **mog mog** in Mosi, **mogu mogu** in Japanese, is a combination of a bilabial consonant /m/, a soft palatal consonant /g/ and two back and close and half-close vowels /u/ and /o/. These are suitable for expressing the movements of the mouth with lips shut. In Mosi the same verb **moge** is used to mean 'to chew', which has an analogous mimetic feature to the French verb *sucer* and the corresponding Japanese verb **shaburu**. In the same way, *chuchoter* (to whisper) in French and **sasayaku** (the verb deriving from a mimetic word **sasa**) in Japanese are similar, in the sense that both reproduce the movements of mouth and breath. In Mosi, the action of whispering is expressed by the verbal sound **walem**, possibly because of the previously noted symbolic value of [wa], but also perhaps because it represents the low pitch of voice rather than the manner of speaking.

VI

The preceding arguments show that in the domain of verbal sounds, the two types of nature–culture interrelationship vary, even within my limited comparative investigation of three cultures. Generally speaking (and also taking account of aspects indicated within the octagonal figures but not fully discussed here for reasons of space), I can say that in Japanese sound culture, the nature–culture interrelationship is pre-eminent both in the field of arbitrariness of the sound–meaning association and in cultural representations of nature. Both Mosi and Japanese sound cultures indicate a similar continuous character in the arbitrariness of sound–meaning associations. But in the realm of nature–culture relations, in instrumental sounds, Japanese culture shares some common traits with French culture, though not with Mosi culture. In both the Japanese and Mosi languages, differentiations in arbitrariness or in the natural and cultural aspects of sound–meaning associations is continuous; in other words, the differentiation itself is 'natural'. In this respect, it is worth noting that in both Japanese and Mosi, rules of rhyme do not exist in poetry, while in French, the rhyme, or the technique of combining and of repeating definite segmental features of certain words has been artificially elaborated.

Japanese poetic style has been based, from the most ancient written poems of the eighth century to modern times, and regardless of whether they concern war or love, solely on quantitative criteria, that is the regular combination and/or repetition of the verse of a definite number of mora, in most cases five and seven morae. Taking the pause into account, this regular repetition constitutes an eight-pulse rhythmic pattern which is commonly found in many spheres of Japanese traditional music. Consequently, it is easy to compose a poem in traditional styles, **haiku** or **waka**. The popularity of poetical composition is a remarkable feature of Japanese culture. This feature may have contributed to the development of the verbally conceptualized perception of nature in Japan, to which I refer below.

As for Mosi, as in many other sub-Saharan African languages, poetic rules at the linguistic level do not exist, except for occasional use of antithesis; but as the poem is always sung, the musical rhythm gives style to the sung verse.

Another point to be noted concerning the verbal sound universe of these three cultures is that the Mosi and Japanese cultures display opposite traits as to the relationship between verbal and instrumental sounds. In the traditional Japanese verbal notation for music, called **kuchishooga**, ideophones designate the detailed playing methods as well as the sounds – especially the timbre – of respective musical instruments. By contrast, in Mosi drum language, which reproduces the supra-segmental features of the verbal message in a conventional way, it is the instrumental sound that transmits the verbal sound, while any verbal notation system employing ideophones (like **kuchishooga**) is absent.[3] The *solfège*, practised in French culture, differs greatly from these two in that it is a verbal notation based on a completely arbitrary association between two features, the segmental and the supra-segmental – especially the pitch – of the vocal sound.

VII

Bearing a close relationship to what has been discussed above, my second observation (see pp.44–46 above), reveals another aspect of the nature–culture relationship in the sound universe, namely the character of verbal interpretation of animal voices, especially those of birds. By 'verbal interpretation', I mean one that is made in a more or less mimetic manner, without an arbitrary application of the imagined verbal message to an animal vocalization. Unlike what we may suppose from the preceding arguments, this kind of interpretation is almost absent in Mosi culture, but it exists in French culture and is especially prominent in Japanese culture.

In this respect, French culture differs greatly from that of Japan. The materials found in collections of folklore [Hertz 1970; Perbosc 1988; Sébillot 1984] contain a wide variety of animal vocalizations, ranging from many species of wild birds, domesticated mammals and fowl, to some insects like cicadas and crickets, which are mimetically interpreted in verbal messages. Wild mammals are not found except for the fox. All of these species are familiar to people living in the country. The contents of these interpreted messages are often announcements or warnings to farmers concerning their seasonal work. In many other cases they are dialogues between characters in a comedy, involving either members of the same species or from two or more species. In general, the interpretation is witty, and some of them treat sexual matters.

In the case of Japanese culture, by contrast, the interpretations of vocalizations are especially numerous for a relatively limited number of bird species: *Cuculidae* like **hototogisu** (*Cuculus poliocephalus*), *Columbidae* like **yamabato** (*Streptopelia orientalis*) and *Charadriidae* like **chidori**, which covers several species of plover. These vocalizations are interpreted as messages from a person metamorphosed into a bird. What is characteristic in these metamorphoses is that they result from some tragic conflict among kin, who as birds continue to cry out their sorrowful messages.[4]

There are also many examples of announcements or warnings, and comic dialogues between birds, similar to the French cases. But a further difference between Japanese and French culture in this respect is that in Japan this kind of interpretation of bird vocalization has been treated in literature since the beginning of written literature, from the eighth century [Yamaguchi 1989].

The problem must be examined continuously for each of the three cultures in the wider context of human–animal relations, in terms of their occurrence in hunting and/or the raising of animals, as well as in their representative forms in folklore and literature; whether they be described in terms of temporary or enduring human–animal metamorphoses, or as human or animal rebirth, or as human–animal marriages [Kawada 1987].

In Japanese culture we find certain conventionalized ways of representing nature, typically expressed in words like **kikinashi**, a stylized way of interpreting natural sounds, especially bird and insect voices; and **mitate**, a conventional way of considering something fictitious, mimetic or artificial as real, authentic or natural. These conventions might have been formed on the basis of the everchanging and transmigratory concept of the universe, but might also result from the derivative situation of Japanese culture. These two concepts reveal some of the essential features of Japanese aesthetics as well as of Japanese ideas of nature.

VIII

Of the sound universes examined here, French and European cultures in general seem to have tried artificially to purify musical as well as verbal sounds. As for musical instruments, the effort has been made to exclude noises and to determine precisely the pitch of sounds. As for verbal sounds, Europeans have tried to banish the 'childish' and 'natural' mimetic character of verbal sounds, but have elaborated the rules for the combination and repetition of verbal sounds in poetry. A highly artificial use of sound symbolism in musical composition is found in the work of some modern Western composers, such as Arnold Schönberg, Luciano Berio and Charles Amirkhanian. French composers do not seem to have been very active in this respect, although in the eighteenth century, Jean-Jacques Rousseau's pioneering ideas combined verbal and musical expressions in both theory and practice.

In contrast, and as already noted, there has been consistency in Japanese rules of poetry since the eighth century, based solely on the temporality of the verbal sound – an eight-pulse elementary unit including pause. Furthermore, the popularity of poetry and the ease of poetic composition in Japan, along with the conception of the universe mentioned in the previous section, seem to have contributed to more general verbal conceptions, in which **mitate** and **kikinashi** appear as the key terms.

Notes

1. The Mosi language (called **moore** in their own language, but for the sake of simplicity referred to here as 'Mosi') is transcribed according to the system adopted by the national language committee of Burkina-Faso. The italic marks the lax vowels as: *e*/e, *i*/i, *u*/u; the nasal is marked by a tilde; the tone is marked only where necessary, as ´ rising, ` falling, ^ rising-falling, and it is not marked if the tone remains at the same level; except that in this language with the 'terraced level tonal system', a tone lower than the preceding one is marked as ״. The tone is marked on the consonants m and n when they are syllabic.
2. For recordings of these Mosi word-plays, refer to Kawada 1982, 1988a: Theme 1: 3–4.
3. For a cross-cultural study of **kuchishooga** and drum language, see Kawada 1986.
4. Many tales of this type are presented under the heading **kotori zensho**, 'previous lives of birds', in Seki 1979.

Bibliography

Hertz, R., 'Contes et dictons recueillis sur le front parmi les poilus et d'ailleurs (Campagne de 1915)', in *Sociologie religieuse et folklore*, Paris: Presses Universitaries de France, 1970

Jakobson, R., C.G.M. Fant and M. Halle, *Preliminaries to speech analysis: the distinctive features and their correlates*, Cambridge: MIT Press, 1963

Kawada, J., *Sound universe of the savannah*, as disk-album (Tokyo: Toshiba-EMI, 1982) or cassette-book (Tokyo: Hakusuisha, 1988a), (in Japanese, with a detailed table of contents in French)

——, 'Verbal and non-verbal sounds: some considerations on the basis of oral transmission of music', in Y. Tokumaru and O. Yamaguti (eds), *The oral and the literate in music*, Tokyo: Academia Music, 1986, pp.158–72

——, 'Men and animals in the folk-tale', in *Studies in Folk-Narrative* No. 10, Tokyo: The Society for Folk-Narrative Research of Japan, 1987, pp.20–39 (in Japanese)

——, *The voice*, Chikuma Shobo: Tokyo, 1988b (in Japanese)

Perbosc, A., *Le langage des bêtes: mimologismes populaires d'Occitanie et de Catalogne*, Garae/Hesiode: Carcassonne, 1988

Sébillot, P., *Le folklore de France: la faune*, Paris: Edition Imago, 1984 [1904]

Seki, K., *The complete compilation of Japanese folk-tales*, Vol. 9, Tokyo: Kadokawa-Shoten, 1979 (in Japanese)

Yamaguchi, N., *Hear the cry 'chinchin' of plovers*, Tokyo: Taishukan, 1989 (in Japanese)

Chapter 3

A Poetics of Place: Ecological and Aesthetic Co-evolution in a Papua New Guinea Rainforest Community

Steven Feld

Because knowledge can never replace respect as a guiding principle in our ecosystemic relations, it is adaptive for cognized models to engender respect for that which is unknown, unpredictable, and uncontrollable, as well as for them to codify empirical knowledge. It may be that the most appropriate cognized models, that is, those from which adaptive behaviour follows, are not those that simply represent ecosystemic relations in objectively 'correct' material terms, but those that invest them with significance and value beyond themselves.

Roy Rappaport , *Ecology, Meaning, and Religion* [1979: 100–1]

One way to imagine the potency of 'nature' as a cultural construction is to imagine the appropriateness of the word 'aesthetic' in each place where Rappaport uses the word 'adaptive' in his essay on ecology and cognition. To do that this paper reviews two intertwined dimensions of a mutualism of adaptation and aesthetics among the Kaluli people of Bosavi in Papua New Guinea: the ecology of natural sounds as a human musical ecology; and, the conceptualization of place as a cartography of human song and lament. In the first instance the mutualism begins with the natural soundscape of the Papuan rain forest in the Bosavi region. Here the calls of some one hundred and thirty species of birds, as well as the sounds of many frogs, the rhythms of cicadas and insects, the sonic presence of creeks, streams, waterfalls, pools and other waterway formations are obvious quotidian presences. For Kaluli people, these sound patterns are indexically heard as the time of day, seasons of the year, vegetation cycles, migratory patterns, heights and depths of forest, and many other markers of place as a fused human locus of time and space. What Kaluli perceive and know about natural diversity in their world articulates often through attention to these sounds and through elaborate conceptual and cognitive indicators of the centrality of sound to experiential truth.

At the same time, Kaluli vocal and instrumental musical sounds are inspired by, modelled upon and performed with these environmental sounds. And the evocative powers of these musical performances, as well as their interpretation and affective response, are modelled from the same pervasive perceptual and epistemological primacy of sound. Here, ecological and aesthetic co-evolution means that the music of nature is heard as the nature of music. Moreover, this iconized or constructed 'naturalness' is brought into alignment by an overarching cosmological framework for the interpenetration of nature and culture. This framework is located in the dualism of birds, who both coordinate the local natural-historical sense of being in space-time, and whose presence continuously announces the presence of spirits. To each other the birds appear as people, interacting as such and sounding out through talk. To Kaluli birds 'show through' as a metaphoric human society whose colours, behaviours and sound categories (whistling, singing, talking, saying their names, crying, making noise) thoroughly fuse the 'natural-historical' and 'symbolic' dimensions of imagination and engagement [Feld 1990c: 44–85].

Turning to the ways physical place, sensed and sensible, is imaginatively coded in a cartography of songs and lamentations, the emphasis will shift to how Kaluli song and lament texts are organized as maps of place names. The sequences of these names are textually co-articulated with names of trees and vegetation, creeks and waterways, and birds. Here we can concretely understand how places are quite literally 'placed' in memory, and how their codification and evocation in the formal genres of song and funerary-lament poetics intensifies the expressive relationship between the biographical, feelingful, experiential dimension of cultural identity, and the adaptive dimensions of ecological knowledge and awareness.

People and Place

The ethnographic dimensions of this paper draw from field research[1] among the Kaluli, one of four subgroups of two thousand Bosavi people who live in the tropical rain forest of the Great Papuan Plateau in the Southern Highlands Province of Papua New Guinea. On several hundred square miles of lowland and mid-montane forest land, at an altitude of about two thousand feet, they hunt, fish, gather, and tend land-intensive swidden gardens that yield sweet potatoes, taro, pandanus, pumpkin, bananas, and many other fruits and vegetables. Their staple food is sago, processed from wild palms that grow in shallow swamps and creeks branching off larger river arteries that flow downward from Mount Bosavi, the collapsed cone of an extinct volcano reaching to eight thousand feet [Feld 1990c; B.B. Schieffelin 1990; E.L. Schieffelin 1976; E.L. Schieffelin and R. Crittenden 1991 provide general ethnographic, linguistic, musical and historical details on the people and region]. Bosavi people live in distinct longhouse communities; in each, many people still reside in a single communal house, comprising some fifteen families or sixty to eighty people. Social life for the village is centred around the house, where primary face-to-face interaction occupies most of the time people are not in their gardens, on the trails, visiting relatives in other communities, or staying at small garden homes or sago camps for major food-processing activities.

This is a classless and small-scale society. Traditionally, no occupational specializations, stratifications, ranks, professions or ascribed or achieved statuses formed the basis for social differentiation. This was also a generally egalitarian society in matters economic and political. With no appointed or elected leaders, speakers, chiefs, bosses or controllers, Kaluli people hunted, gathered, gardened and worked to produce what they needed, taking care of themselves and their associates through extensive cooperation in food sharing and labour assistance. There was little accumulation of goods, rewards or prestige, and no highly valued jobs or roles automatically rewarded by material objects. The egalitarian dynamics here involved both a lack of centralized social institutions and a lack of deference to persons, roles, categories or groups based on power, position, or material ownership. Rather, extended family size, verbal agility, particularly the ability to talk and sing others into one's sphere of personal influence, and energetic actions to mobilize labour assistance accounted for many instances of emergent male stratification over the shorter and longer courses of social action. Lest it seem that 'egalitarian' here carries any vestigial romanticism for 'universal social equality', it can also be pointed out that fundamental gender-based inequalities have always characterized Bosavi social life. Bridewealth negotiations and marital arrangements constitute the most obvious arena where men cultivate authority and make decisions that expressly control women. Such gender inequalities are widely reported among societies whose economic, political, and general social organization is marked by general egalitarianism [Flanagan 1989]. Forms of gender opposition and complementarity clearly appear now, as they did traditionally, in a tense balance with emergent hierarchy and dominance in the overall picture of Kaluli egalitarianism.

Emergent hierarchy developed dramatically around the social changes that have recently refigured Kaluli life, beginning with the advent of colonial government contact, particularly by the late 1950s. But it was evangelical missionization which brought sweeping changes to the Bosavi area, beginning in the mid-1960s with the building of an airstrip, then consolidating from the early 1970s with the first resident fundamentalist missionaries and local pastors in each village. A new wave of national-government impact followed Papua New Guinea's independence in 1975. Into the 1980s and 1990s, the presence of a second airstrip, a hospital, schools, aid post stations, a mission station, government development personnel, and particularly local pastors has introduced increasingly complex forms of deference based on differentiated wealth, particularly with a cash base. These changes have had a significant and escalating impact on the nature of stratified social categories in Bosavi life. Currently the Bosavi area is in the throes of a more complex set of changes that implicate cultural and ecological futures. Oil and gas projects are already transforming the region, and gold and timber development plans are also being explored. With these have come the chaotic responses that large infusions of cash, development of roads and ensuing patterns of out-migration always bring to such small communities in remote zones.

Acoustic Ecology as Aesthetic Adaptation

Does it require deep intuition to comprehend that man's ideas, views and conceptions, in one word, man's consciousness, changes with every change in the conditions of his material existence, in his social relations and in his social life?

Karl Marx and Frederick Engels,
Manifesto of the Communist Party [1975: 503–4]

A commonplace orientation of cultural ecology is to the processes by which societies adapt to environments. Generally this entails understanding how cultural configurations emerge, transmute, reproduce, change and sustain. Emphases on technology, economics and the control and regulation of resources are central to these concerns, as is the role of ecological interactions within a broader understanding of the biological and cultural interface of adaptation. One rarely sees or hears talk of aesthetics within these discourses on cultural ecology, co-evolution and adaptation, matching the equally scant attention to cultural ecology in the discourses on cultural expressive systems. Indeed, among all domains of ideational, symbolic and hermeneutic approaches to culture, materialists and ecologists have tended to view the aesthetic-expressive lens as superstructural and epiphenomenal. And for their own part, cultural aestheticians have equally tended to view the ecological and adaptational lens as reductionist and positivist. Such caricatured discursive polarization has had intellectually unproductive consequences. And even if that point is widely recognized and acknowledged by many more anthropologists than it was ten or fifteen years ago, there is still little work that takes the ecological-aesthetic interface beyond a low-level 'reflectionist' paradigm toward an intellectual challenge worthy of extended theoretical or empirical scrutiny. Parallel polarizations can be found as easily in discourses on geography as in cultural anthropology. Even within the diverse literatures of humanistic geography [Buttimer and Seamon (eds) 1980; Jeans 1979; Silk 1984; Tuan 1989, 1990] rapprochement of systematic and experiential perspectives is admired more than achieved.

One way to pose the challenge of a new dialogue is through clarification of how an approach to aesthetics in a cultural and phenomenological framework can articulate with ethnographic and materialist agendas. Aesthetics is here taken, to use Robert Plant Armstrong's terse phrase, as an ethnographic location of 'form incarnating feeling' [1975: 11], and specifically where works of what Armstrong called 'affecting presence' [1971] are witnessed as 'a direct presentation of the feelingful dimension of experience' [1975: 19]. Like Armstrong and other phenomenologists, my approach here sidesteps aesthetics as an asocial philosophical discourse aimed at abstractly illuminating conditions of virtuosity, excellence in execution and beauty. Rather, the ethnographic intrigue is to locate what Kaluli find affecting and moving about mundane and intensified experiences, to explore the vicissitudes of what they interpret as powerful and non-ordinary, whether deriving from casual or heightened realms of sociability. Aesthetic here means to recall the Greek notion of 'sensuous perceptions', albeit ethnographically formulated as fully articulated social facts.

Reviewing similar issues from a more materialist stance, Raymond Williams traced the limiting dimensions of 'aesthetic' as a historically specific discourse in the West:

> . . . aesthetic, with its specialized references to art, to visual appearance, and to a category of what is 'fine' or 'beautiful', is a key formation in a group of meanings which at once emphasized and isolated subjective sense-activity as the basis of art and beauty as distinct, for example, from social or cultural interpretations. It is an element in the divided modern consciousness of art and society: a reference beyond social use and social valuation which, like one special meaning of culture, is intended to express a human dimension which the dominant version of society appears to exclude. The emphasis is understandable but the isolation can be damaging, for there is something irresistibly displaced and marginal about the now common and limiting phrase 'aesthetic considerations', especially when contrasted with practical or utilitarian considerations, which are elements of the same basic division. [Williams 1983: 33]

Williams' own attempt to create a more suitable analytic category, under the rubric 'structures of feeling' [1977: 128–35], formulates a more socially situated aesthetics as the concepts and practices that animate senses of evocation. 'Evocation' illuminates a universal dimension of human existence: that all people sense and mark certain experiences as special, acknowledging the ways they evoke behaviours, moods and feelings heightened in awareness, yet centred in a local sense of identity. Transposed to any local situation, one may empirically investigate what is found compelling in expressive forms and their performed presentations, and how such practices are rationalized, valued and internalized as core experiences. Such embodiments, local premises of style, are central to understanding an aesthetic imagination in action.

The central circumstance in such an exploration of how Kaluli aesthetics challenges an ecological perspective concerns the local imagination and performance of linkages between poetic, musical, choreographic and visual-material expressive systems on the one hand, and perceptions of the structures and contents of the rainforest environment on the other. Three continuously intersecting patterns organize such linkages; in simple terms, inspiration, imitation and appropriation. Inspiration involves imagining and creating expressive forms symbolically interpenetrated with environmental knowledge and appreciation. Imitation involves the mimesis or the iconic or indexical suggestion of actual elements – usually more referential or overtly identifiable features – of the environmental sensorium in expressive forms and their associated perceptual reception. Appropriation is the more thorough incorporation or conventionalized stylization in expressive forms of larger patterns of either formal structures or evocative features of environmental order. The structures of such transcorporation usually invite application of the notions of metaphor or trope.

In an attempt to explore these patterns systematically among Kaluli people, my research [reported in 1988, 1990 a,b,c, 1991b; sound recordings are 1982, 1985, 1991a] has explored these three ecological nature-expressive culture linkage patterns across four general areas of musical and verbal arts. These have been (1) myth, (2) song texts and vocal style, (3) instrumental sounds and performance style, and (4) musical theory and verbalization. In the case of myth this work has concerned the representation of the voice of weeping as the voice of a fruit dove; with text it turns to song. In the area of song texts and vocal style the work has illuminated how and why Kaluli sing with, to and about birds, water and insects. Emergent sound play and a poesis of onomatopoeia directly relates to singing with nature as well as about it and like it, interlinking ethnoecology and ethnopoetry. In the area of instrumental sounds and performance style the work has analysed how drums literally and metaphorically incorporate bird voices and through them become the voices of spirit children. Refraction of the bird voice through the spirit/drum voice uncovers an elaborate set of ideas about the spatialization and density of natural and instrumental sound. In the area of theory and verbalization, centrally the linguistic mediation of musical concepts vis-à-vis actual musical practices, the work has analysed the technical metalanguage and polysemy linking the semantic fields of water and sound, as well as the broader range of conceptual metaphors, like the notion of 'flow', related to style and performance. It is in this last area that the conceptual and practical dimensions of aesthetics are more poignant in their cross-modal significance, linking musical thought to realms of the visual, verbal and choreographic, as well as back to ecological modelling of the rain forest.

One of the key notions in this area, central to grasping a Kaluli aesthetics, is **dulugu ganalan** 'lift-up-over sounding' [Feld 1988]. This is the term that prescribes and describes natural sonic form for Kaluli, and given my primary interest in the realm of sound, I first thought it was a notion that referred exclusively to that domain. Instead, this term is a spatial-acoustic metaphor, a visual image set in sonic form and a sonic form set in visual imagery. By calling attention to both the spatial ('lift-up-over') and temporal ('sounding') axes of experience, the term and process explicitly presuppose each sound to exist in fields of prior and contiguous sounds. In practice this is quite the case because the antithesis of 'lift-up-over sounding' is unison. All 'lift-up-over sounding' sounds are dense, heavily blended, layered; even when voices or sound types momentarily coincide, the sense is that the unison is either accidental or fleeting, and indeed, it is entirely by chance.

The essence of this 'lift-up-over sounding' idea involves two components. One is part-relations that are simultaneously in-synchrony while out-of-phase. 'In-synchrony' means that the overall feeling is of togetherness, of consistently cohesive part coordination in sonic motion and participatory experience. Yet the parts are also 'out-of-phase', that is, at distinctly different and shifting points of the same cycle or phrase structure at any moment, with each of the parts continually changing (even competing) in degree of displacement from a hypothetical unison.

A second component concerns timbre, the building blocks of

sound quality, and texture, the composite, realized experiential feel of the sound mass in motion. Timbre and texture are not mere ornaments; a stylistic core of 'lift-up-over sounding' is found in nuances of textural densification – of attacks and final sounds; decays and fades; changes in intensity, depth and presence; voice colouration and grain; interaction of patterned and random sounds; playful accelerations, lengthenings, shortenings; and the fission and fusion of sound shapes and phrases.

In the forest, sounds constantly shift figure and ground; examples of continually staggered alternations and overlaps, at times sounding completely interlocked and seamless, are abundant. One hears no unison in nature; presence and absence of sound and changes in direction and dimension co-ordinate space as intersecting upward and outward. For Kaluli this is the naturally coherent model for sound-making, whether human, animal, or environmental, a constant textural densification constructed from a 'lift-up-over sounding' that is simultaneously in-synchrony yet out-of-phase. Interlock, alternation and overlap, as a fused locus of natural and human sound production, embodies simultaneous icons of competition and cooperation. In other words, sounds constantly interact in a tense quest for primacy, one only fleetingly standing out from the others, conveying the sense that any primacy is fluid, momentary, and as quickly lost as it is gained.

Kaluli apply this acoustic figure/ground idea not just to the 'naturalness' of the rainforest soundscape and to the vocal and instrumental music they create that is of a piece with it, but to the realm of visual form as well. The face-painting styles associated with Kaluli ceremonial costume masks involve a singular figure and ground principle realized in both a shiny/dull texture contrast and black/red colour contrast. These contrasts visually mirror sonic 'lift-up-over' and are discussed in that way. Comments on colour pictures taken at earlier ceremonies or from other parts of Papua New Guinea also spontaneously elicited this conceptual designation and idea. Hence we have in face-painting a visual homology-resemblance to the sonic principles of 'in-synchrony yet out-of-phase' and 'textural densification'.

A similar homology is evidenced in ceremonial costumes. Costumes mix many types of materials: layering possum fur, frame headpiece with white cockatoo feathers; painted body (face, arms, stomach, legs) in red, black, and white; shell necklace surrounded and centred by woven cross bands reaching under the arm; flapping feathers strung from bamboo in arm, belt and knee pieces; waist belt with attached crayfish-claw rattle in rear, emerging through palm streamers densified with cordyline top pieces. Costumes project layered density. The sound emanates from shells and streamers in motion as the dancer bobs up and down, 'lifted-up-over' by the drum and rattle or by voices and rattles. The performer's voice 'lifts-up-over' the costume-as-waterfall so as to sound like the voice of a bird that dances in place by a waterfall. Thus there is a visual-bodily-sonic interaction of textural densification in costume, dance and sound that merges with a visual-bodily-sonic in-synchrony and out-of-phase sensation.

A final 'lift-up-over' touch that completes costumes is called the **tamin**, a single white cockatoo feather, or a pair of same, placed in the top centre of the dancer's head-dress. This **tamin** feather (a name derived from the verb for 'lead' or 'go first') is attached to the end of a long pliable piece of bamboo that springs up from the head-dress. As the dancer bobs up and down, the **tamin** swings backwards and forwards in a long arc continuously outlining a backward then up-and-over pendulum. This swaying arc flows in direct synchrony with the dancer's vertical movement, but is also out-of-phase by virtue of mapping the opposite (horizontal) motional plane. The **tamin** then 'lifts-up-over' the rest of the costume both materially and motionally. It completes the sense that all costume materials are in layered multiple visual-sonic-motion figure and ground relationships. Hence we have another homology linking the 'lift-up-over sounding' of the dance costume to that of body painting.

When applied to the sound world of the rain forest, 'lift-up-over sounding' refers to the fact that there are no single discrete sounds to be heard. Everything is mixed into an interlocking soundscape. The rain forest is like a world of co-ordinated sound clocks, an intersection of millions of simultaneous cycles all refusing ever to start or stop at the same point. 'Lift-up-over sounding' means that Kaluli people hear their rain-forest world as overlapping, dense, layered. Whether it is the more specific overlapping of avian antiphony (like the duetting of the New Guinea Friarbird and Brown Oriole) or the way forest heights and depths are signalled by the swell of cicadas and constant refiguring of creek and river hisses wherever one walks, Kaluli often note, both on the trail or in the longhouse area, that the forest is 'lift-up-over sounding'. And Kaluli apply the same principle to their own music when they say that their voices 'lift-up-over' like the trees of the forest canopy. Or that sounds of drums or axes 'lift-up-over' like tumbling waterfalls into swirling waterpools.

In sum, the 'lift-up-over sounding' idea takes in nature (birds, insects, water), human music (vocal and instrumental form and performance style), body painting and costume (figure/ground shifts and textural density), choreography (sound and costume in motion), work and interactional sociability (overlapping speech and song coordination; the term for 'conversation' in Kaluli is 'lift-up-over talking'). And it unites them such that visual, motional and sonic (musical, verbal, natural) dimensions of style are conceptually and practically united through active engagement and participation, linking feelingful experience and everyday knowledge and action.

To turn to the significance of this pattern, two notions, of trope and synaesthesia, are helpful for understanding the cohesive sensory, perceptual, emotional, feelingfully practical dimensions of aesthetics packed into the Bosavi concepts and practices of 'lift-up-over sounding'. Moreover, these notions are helpful to understanding the naturalization of music as an ecologically modelled system and the culturalization of nature as an aesthetic system. Roy Wagner speaks of trope as 'a single phenomenon or principle [that] constitutes human culture and cultural capacity... the phenomenon is coherent and pervasive, organizing conditions for the perception of meaning over the whole scale-range of cultural forms' [Wagner 1986: 126]. Likewise, Lawrence Marks glosses synaesthesia as 'intersensory correspondences ... the transposition of sensory images or sensory attributes from one modality to another ...' [1978: 8]. 'The synaesthetic, like the metaphoric in general, expands the horizon of knowledge by making actual what were before only potential meanings' [ibid.: 254]. Synaesthesia and trope concern concomitant sensation, simultaneous joint perceptions. An analogic icon for Kaluli ceremonial and artistic synaesthesia thus might be the experience of the rain forest itself, where in a sensually involuntary and culturally conventional manner, features of sound, texture, space and motion are interrelated. In the tropical rain forest height and depth of sound are easily confused.

Lack of visual-depth cues couple with the ambiguities of different vegetation densities and ever-present sounds, like water hiss, to make depth often sensed as height moving outward, dissipating as it moves. 'Lift-up-over sounding' precisely yet suggestively codes that ambiguous sensation: upward *feels* like outward.

What of course is central in all this connection is the mysterious world of Bosavi birds, which is the world of Kaluli spirit reflections, of 'voices' that are also 'gone reflections' (**ane mama**) between visible and invisible worlds. The primacy of sound in the rain forest makes music the domain of nature for which Kaluli have so much natural-historical, mythological and ornithological interest. Thus it seems fitting that in Bosavi there was a tendency for prolific composers also to be expert ornithologists. Notions like this, inextricably linking the transformative play of nature and culture, are in no way singular to Bosavi, and various commentators on Papua New Guinea have reported them in a variety of ways for a wide range of societies. Perhaps one of the most repeated renditions comes from an anecdote, reported in the voice of quaint colonial journalism, of Colin Simpson, an Australian writer who produced a well-known trilogy about Australian exploration in Papua New Guinea. In the opening to a chapter about birds of paradise, he writes [1954: 183]:

> Between the natives and the birds there is a marked and intimate relationship – though it is perhaps too predatory on the native side to be called an affinity. The native ornaments himself with bird plumage. In his dancing he imitates some of the postures of the birds that perform courtship dances. In his songs some of the birds' calls are recognizably introduced. And the natives make dancing lawns which – particularly at Mt. Hagen where trees are planted and buttressed in much the same way as the bird surrounds a sapling with a circle of built-up moss – resemble the playgrounds of the Gardener Bower Bird. This was pointed out to me by Fred Shaw Mayer who has more first-hand knowledge of the birds and animals of New Guinea than anyone I have met or ever hope to meet.
>
> He once said to a man at Mt. Hagen, 'You people must have copied your dancing grounds from the Gardener Bower Bird'.
>
> 'Oh, no', the native said. 'The bird copied us.'

The Cartography of Song and Lament

The events of one's life take place, *take place*. How often have I used this expression, and how often have I stopped to think what it means? Events do indeed take place; they have meaning in relation to the things around them. And a part of my life happened to take place at Jemez. I existed in that landscape, and then my existence was indivisible with it. I placed my shadow there in the hills, my voice in the wind that ran there, in those old mornings and afternoons and evenings. It may be that the old people there watch for me in the streets; it may be so.

N. Scott Momaday, *The Names* [1976: 142]

A humanistic and artistic appreciation of the universal prominence of place in human lives and activities may still be most easily found in literary expression, and indeed, forcefully indicated in the literary works of indigenous peoples, like American Indian author N. Scott Momaday. And while such accounts present alternative, complementary or oppositional discourses to ethnographic writing, the general tendency to recognize the special significance of place has come to be more taken for granted in academic discourses of anthropology too, even though ethnographers lag far behind humanistic geographers in actually theorizing place. Heidegger's [1962] theory of dwelling, a 'topology of being', is well figured in geographical versions like Yi-Fu Tuan's [1974] *Topophila*, but has only more recently been taken up in ethnographic studies, like James Weiner's [1991] *The Empty Place*. From such an ethnographic standpoint the key issues around the prominence of place are how spaces are transformed and 'placed' through human action, and, more crucially, how places embody cultural memories and hence are substantial sites for understanding the construction of social identities.

The issue here is as much in the theorization of memory as of place. Edward Casey's phenomenological study of remembering [1987] argues that vicissitudes of memory go considerably beyond temporal recall and retrospection to a significant second dimension, that of place. He analyses how place memory has been greatly overlooked both in philosophical discourses and in cognitive psychology.

> It is the stabilizing persistence of place as a container of experiences that contributes so powerfully to its intrinsic memorability. An alert and alive memory connects spontaneously with place finding in it features that favour and parallel its own activities. We might even say that memory is naturally place-oriented or at least place-supported. Moreover it is itself a place wherein the past can revive and survive; it is a place for places meeting them midway in its own preservative powers, its 'reservative' role. Unlike site and time, memory does not thrive on the indifferently dispersed. It thrives, rather, on the persistent particularities of what is properly *in place*; held fast there and made one's own. [1987: 186–7, emphasis in original]

Casey views place as reflexively locative and evocative:

> place is selective for memories; that is to say, a given place will invite certain memories while discouraging others . . . memories are also selective for place; they seek out particular places as their natural habitats . . . places are congealed scenes for remembered contents; and as such they serve to situate what we remember. [ibid: 189] [Moreover] this 'down home' sentiment is not only a matter of feeling at ease in a given place but of feeling at ease in a place that has become one's own in some especially significant way. 'One's own' does not imply possession in any literal sense; it is more deeply a question of appropriating, with all that this connotes of making something one's own by making it one with one's ongoing life'. [ibid.: 191–2]

But there is more than holding and appropriating, for place too has its position in and as 'affecting presence'. Like Armstrong's 'form(s) incarnating feeling(s)', Casey sees potentials for place as evocative embodiment.

> The relationship between emotion and expression is close indeed, and it is therefore not surprising to discover that the expressiveness of landscapes is linked to their inherent emotionality. This link is especially evident in the case of 'special places', which bring with them, as well as engender, an unusual emotional claim and resonance. The power of such places to act on us, to inspire (or repel) us, and thus to be remembered vividly is a function of such emotionality – but only as it finds adequate expression in the features of landscapes. [ibid.: 199]

Moving to the more specifically poetic dimensions of place, Keith Basso's recent writings provide several keys to locating these broader philosophical notions of place, memory and biography within a linguistic and ethnographic approach to place through place-*names*. He writes, 'place-name terminologies provide access to cultural principles with which members of human communities organize and interpret their physical surroundings' [1984a: 79]. Basso illustrates how Apache place-names are not just descriptions of locations but are always replete with allusions to activities, danger, historical events. As such, Apache place-names do more than just narratively anchor events and people: 'instead of describing . . . settings discursively, an Apache story-teller can simply employ their names and Apache listeners, whether they have visited the locations or not, are able to imagine in some detail how they might appear' [1984b: 32]. Recalling Edward Sapir's notion of Algonquian verb stems as 'tiny imagist poems' and remarking on how Western Apache place-names are simultaneously so dense and compact, descriptive and evocative, Basso cites a Western Apache acquaintance, Benson Lewis, who made this comment about a place visited: 'It's name is like a picture' [1984b: 27].

Locating the resourceful and poetic potential of place-names, Basso notes,

> placenames are arguably the most highly charged and richly evocative of all linguistic symbols. Because of their inseparable connection to specific localities placenames may be used to summon forth an enormous range of mental and emotional associations – associations of time and space, of history and events, of person and social activities, of oneself and stages in one's life. And in their capacity to evoke, in their compact power to muster and consolidate so much of what a landscape may be taken to represent in both personal and cultural terms, placenames acquire a functional value that easily match their utility as instruments of reference. [1988: 103]

Moreover, 'Poets and songwriters have long understood that economy of expression may enhance the quality and force of aesthetic discourse, and that placenames stand ready to be exploited for this purpose' [ibid.].

The affective quality and aesthetic force Basso speaks of can also be coded poetically in more vague or sparse place imagery.

For example, Melvin Dixon's recent book [1987] on geography and identity in African-American literature discusses oppositional creations, uses of place to alter displaced marginality. Images of journeys, conquered spaces, imagined havens, wandering, wilderness, underground, mountain tops, and places of refuge are part of an expressive transformation from rootless to rooted. These images break down the physical realities of 'settlement' and reinvent, in their stead, a placed identity. 'The wilderness, the underground, and the mountain top are broad geographical metaphors for the search, discovery, and achievement of self. They shape Afro-American literary history from texts that locate places for physical and spiritual freedom . . .' [1987: 3].

Dixon's perspective on slave songs illuminates how place is ultimately about position, in this case, options for where to be located when insider and outsider positions are set in the key of domination and power. Aloneness, alienation, integration and linkage are thematically part of this way that placeness and personness are critical literary constructs.

> Music creates a landscape, defines a space and territory the singer and protagonist can claim. The slave songs initiate pilgrimages and other self-creating acts, including resistance and escape, that ultimately defeat the inertia of place and identity upon which the institution of slavery had thrived. By seizing alternatives through poetry and music, slaves charted journeys to many kinds of freedom. The symbolic geography in the slave's religion told where and how to reach the territory of freedom. [ibid.: 14]

The range of issues raised by Casey, Basso and Dixon figure prominently in genres of song and lament in Bosavi. Place-names, what Kaluli generically call **hen wi** or 'land name', and imagined symbolic geography, what Kaluli call a **tok** or 'path', operate throughout songs sung in both ceremonial and everyday leisure or work settings as well as in laments, sung-wept at funerals and other moments responding to intense loss. Each genre of Kaluli song involves different contexts for performance, occasions for use, ceremonial and instrumental regalia, and elaborateness of staging. All of these genres involve, as a key textual common base, paths of place-names that feature prominently in the text. Song paths may start far away and end where you are singing, or start where you are singing and lead far or close by. They may circle the land vaguely or closely. They may be dense with names of places and sparse with corresponding trees and creeks. Or they may only name particular progressions of waterways, or small shrubs. Some just insert the sounds of those waterways in place of the names. Song paths may follow the course of bird flight-paths, meander alongside creeks, stick to more major walking paths, or just follow the course of hills, ridges and high places. There are a great variety of strategies for poeticizing the movement through place in Kaluli songs. But all songs have 'paths' and in the approximately one thousand songs that I have recorded over the last twenty years there are about six thousand places named, prominently lands,waters, mountains, and valleys. Simple songs like those of the **heyalo** genre can have at least three or four named places in a given text. Complex songs like those of the **gisalo** genre will tend to have between ten and thirty-five place-names to a song.

A rudimentary example of the location of biographical placing and memory work in the poetics of a single song is indicated by the **heyalo** genre. All **heyalo** songs have two parts; the first is named **mo**: 'trunk', the second **dun**: 'branches'. The 'trunk' is what would be called a 'refrain' in Western terminology, i.e. same melody and text repeating intact in alternation with the second part of the song. The 'branches' are what would be called 'verses' in Western terminology, i.e. same melody formula with text changes, alternating with stable refrain. So the form of a 'trunk' and 'branches' or refrain and verses song is as follows: A + B1 + A + B2 + A + B3 + A + B4, etc. Note the resemblance pattern here to moving up a tree; the song continually 'branches' off from the 'trunk', always to return. This is another example of how the structure of song is iconically related to fundamental images in nature.

'trunk'

iya:u	iya:u (*Ptilinopus ornatus*, Ornate fruitdove)
'ni imolo sole'	'I'm staying empty'
fo:fo:ndolobo-o	alone, with no family
'ne ada:lomade'	'I have no ada:' (older sister/younger brother)

'branch' 1

Sululeb ilikiyo	staying at Sululeb (longhouse of clan Bono:)
iya:u ganalabe	**iya:u** already (just) called out
bol huyayo:	by the **bol** tree standing alone
'bas ne tafo-ane'	'brother-in-law you went and left me'

'branch' 2

So:lo: mugun ilikiyo	staying at So:lo: mugun (So:lo: stream ridge)
iya:u ganalabe	**iya:u** already (just) called out
odage fo:wo:	by the fruited **odag** tree there
'bas ne tafo-ane'	'brother-in-law you went and left me'

'branch' 3

Yolo do:mo ilikiyo	staying at Yolo domo (hill over Yolo creek)
iya:u ganalabe	**iya:u** already (just) called out
sila:m fuseyo:	by the **sila:m** tree with yellowed leaves
'bas ne tafo-ane'	'brother-in-law you went and left me'

The key structural dimension of Kaluli poetics is parallelism, the use of repetition with slight variation at one or more linguistic (e.g. phonological, syntactic, semantic) levels, where one occurrence or aspect of a feature sets up the next and the second recalls the first, thus creating figures and grounds. For example, a poetic feature of the 'trunk' is the parallelism of quoted voice marked by textual position, rhyme and, melody contour. The song begins with the mimetic character of a bird name. This signifies that the song is being sung from the point of view of that bird; that the voice of the song is an **iya:u** voice. This is a plaintive image because the bird gives its name to the sound of a crying voice. The onomatopoeic syllable **ya:-** indicates crying. In performance the vocalization of the syllable **ya:-** and the following **-u** are elongated and enunciated with a downward melodic curve. This mixture of textual onomatopoeia and elongated falsetto descent in melody signal the voice of the bird.

The second line then shifts from the onomatopoeia of bird sound to the 'bird sound words' of poetry; it is a present voice, opening with a first person vocative. The verb employed is highly marked to indicate both physical hunger and a sense of isolation. An English term of similar poetic weight might be 'yearning'. The third line is more distant and descriptive, as if to step outside the frame of the personal voice and say 'what that means is ______.' The fourth line shifts back to both the vocative and quoting frame, intensifying the image by linking the loss to the specific lack of an **ada:**. This is a very important social relationship, marked by a reciprocal relationship term (not a kinship term) that involves both deep affection and obligation [B.B. Schieffelin 1990: 112–35]. As is usually the case, the 'trunk' has no place-names or place imagery. Its imagery is all related to establishing thematic dimensions of loss by equating loss of food, family, relationship and **ada:** with the loneliness and abandonment of a crying bird voice.

Looking now at the form of the 'branches', the paradigmatically changing terms are in boldface. In line one the item is a land name + optional descriptive; in line three the item is a tree name + obligatory descriptive. (This slot could also be filled by various other items, e.g. a water name + shrub name, a water name + onomatopoeic term for its sound, a hill name + cloud or sunlight descriptive.) The progression of line one to three in every 'branch' fills out the sense of a place, linking names with features of landscape and soundscape. This progression is bracketed in lines two and four by verbal shifts, from the immediate recent past (**-abe**) in the second line switching to definite past (**-ane**) in the fourth. There is a play here of macro and microparallelism, with a progression in detail and concreteness of description from lines one to three, and a progression of experienced time from lines two to four. There is also parallelism across the 'trunk' and 'branches' sections. Both are four-line configurations with weak one/three and strong two/four end rhyme. Relationship terms appear in the final line of both sections creating a special kind of poetic closure. (The relationship term in the last line of the 'branches' could also be paradigmatically slotted to change with each repetition, using any kin or relationship term.)

As to the 'branches' movement of place-names and prominent features attached to them, Sululeb is a longhouse site name; the **bol** tree (a very large hardwood) there is **huyayo:**, standing alone by a cleared pathway. This tree prominently marks one of the two major paths in and out of the Sululeb longhouse area. It is the prominent site of roosting afternoon grackles, butcher-birds, or friar-birds in groups. The image is familiar to anyone who has walked into Sululeb, and there are multiple resonances to the notion of a lone tree as well. It indicates a path has been cleared, that people are nearby, that one is out of the density of the forest. It is also the most generic image of this longhouse site, home to some seventy people. So:lo: mugun is a ridge over the major creek that bounds the other main entry/exit path, at the opposite side of the village. The fruited **odag** tree there is an enormous buttressed fig that is visually prominent from numerous points on paths in either direction from the village. As one approaches voices of many spirit pigeons, doves, parrots and other fruit-eaters are apparent. Yolo do:mo is the hill above the Yolo stream, triangulating a sight-sound line to both previous place-names. It is the site of the house where my colleagues and I lived during our trips to Bosavi in the mid-1970s and 1980s. Just behind the house is a large breadfruit tree; the image here is of leaves **fuseyo:**, yellowing and dropping.

These three 'branches' are the kernel of a song that has many variants and elaborations (up to six other 'branches' could be sung). I know the song well because it is about me and was intended to make me sad. It was composed at a time when my two colleagues returned to the USA in 1977 and left me in Bosavi to continue my work alone. Because I was introduced to Bosavi people as the younger brother of Bambi B. Schieffelin, our relationship was that of younger brother/older sister; to Edward L. Schieffelin I was brother-in-law. Hence the forcefulness of those relationship images (**ada:** and **bas**) in the final line of the 'trunk' and 'branches'. The kernel of the song draws on the images of place that were most familiar to us and those who knew us; the prominent markers of passage in and out of the village, and the place where we lived. While it has specific resonances to me, it is also a song that has various significances to a number of people at Sululeb, people who worked closely with us.

The song is highly specific to time as well as to place and biography. The period after the fruiting of one kind of tree and during the decaying of another means a certain moment in vegetation cycles, presence of certain foods and absence of others. Those factors also call certain activities to mind, certain weather, the presence of certain birds, and the like. Place and time fuse. In short, the song immerses the listener into an interlinked set of sensations, of place, time, experience, relationship and biography, recalled forcefully through the anchor of place-names and descriptives. While numerous additional features of the song and its evocation could be traced here, the above sketch is enough to make it clear that each Bosavi song compresses a great deal of knowledge and feeling, and James Weiner's recent [1991] and complementary account of song poetics for a closely adjacent society, the Foi, opens up a theoretical line of inquiry about the generality of such a place poetics to Papuan ontologies.

The same kind of placed interbiographical condensation is not limited to composed songs; it is equally characteristic of **sa-ya:lab**, the sung-texted laments [Feld 1990b, 1990c: 86–129; performances can be heard on Feld 1985: B3,4] improvised as women weep during funerals. The important point here is that this sort of poetics is not dependent on conscious compositional practices; it can be improvised, and Kaluli women become known for their ability to improvise textual lamentation maps of places they shared with a deceased person. The process of performing **sa-ya:lab** wailing is a 'lift-up-over sounding' involving two or more women simultaneously voicing personal memories in a layered collaborative text-voicing, thereby situating their immediate emotions, their relationships and social biographies to each other, the deceased, and those listening. While each voice is personal and specific, the process of several women wailing at the same time produces a multiple intertextual discourse relationship. Here the spatio-temporal character of multiple voice utterances is indexical to a process of emergence as a cohesive text. This intertextuality as a discourse relationship is also the key to the social relationships shared by its producers, and to the emergent understandings and feelings evoked for their audience.

Two kinds of intertextuality are clear in Kaluli **sa-ya:lab** performances; the first variety is performance-internal, and the second variety is cross-performance. In performance-internal intertexts, multiple voices wail at the same time, and their texts interact by 'lifting-up-over' one another, continually staggering, alternating and interlocking such that in addition to the purely sonic layering of the four-tone descending melodic contour, the linguistic messages interact to build jointly produced themes. In the cross-performance variety the intertextuality extends over a period of a day or two, with different periods of wailing and different participants linking their texts.

Prominent in both kinds of intertextuality is the position of place-names. The formulaic phrase 'you and I were together at our ______' is highly characteristic of these laments, and women weep it often, sequentially linking places. As one mentions a particular connection to the deceased, perhaps a longhouse site, sago place, garden place or fishing place, the others may do the same. Even though the names are distinct, they are additively layered as marking the same types of connections, memories and shared social ties that have since been ruptured. Figure 3.1 is a map of place-names cited during just fifteen minutes of lamentation for a man named Bibiali of the Aso:ndo: longhouse community. Gania, Famu, Fofo and Hane wept for him and all spoke of things they shared with him and things they were already missing. In just this short time more than seventy place names are named, linking the deceased to the communities of Tabili and Aso:ndo: concretely through the voices of weepers, and the thoughts of others who were gathered in the longhouse listening.

Sa-ya:lab voice landscapes of memory. These are projective blue-printings of socially affective space on to the physical space of the environment. **Sa-ya:lab** performances not only 'map' biographies of feeling and relating; these performances laminate such affective fields to durative pasts and futures. The aesthetics of relating, of sharing a map, a **tok** of **hen wi**, is inherently intertextual, and produces an emergent recognition of both maximized individuality and maximized collective resonance. **Sa-ya:lab**, then, are compactings of thought and feeling that reproduce focal sociability while expressing the inevitable anger and grief of rupture and loss.

Like poetic song, **sa-ya:lab** weeping is more than a vehicle of memory. Voicing and witnessing are active modes of expressing and constituting memory, literally 'placing' it, and this involves a merger of practical and aesthetic knowledge and judgement, a way of imagining and enacting belonging, identity. The pain of dissolving social relations, of forgetting them and giving them up, takes place through the activity of constituting and reconstituting them, of remembering. The imagistic qualities of place-names and their sequence extends their uttering and witnessing from the sense of maps to the sense of journeys. A song or weeping text that travels a **tok**, a path of **hen wi**, place-names, is

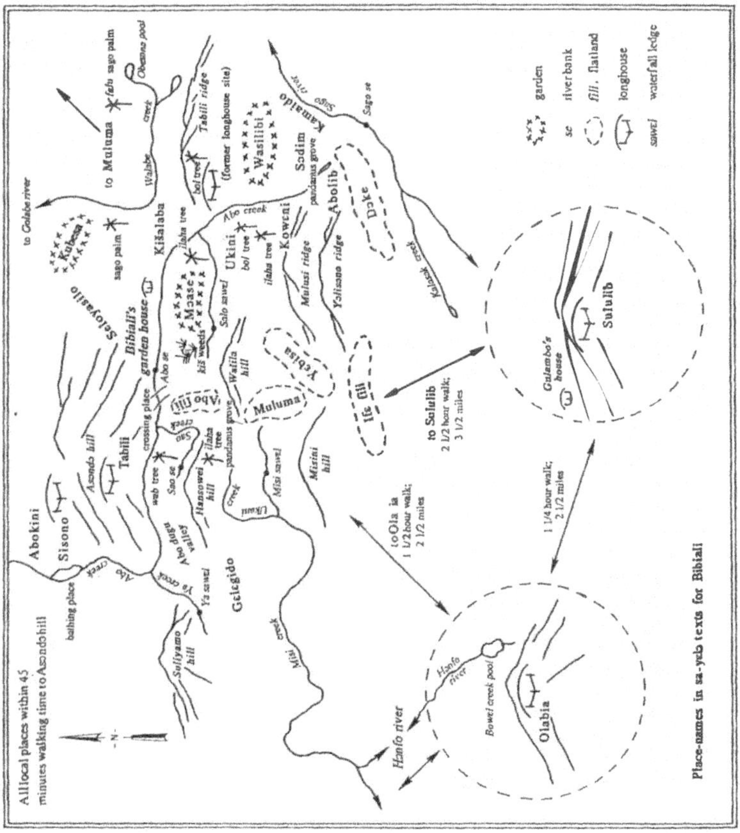

Figure 3.1. Map of place-names cited in lamentation

itself a particular kind of performance, one that is an indexical icon of the journey to, from, in and around about those places. It is an intensified microscopic grand tour of the feelings and experiences engrained in those places.

The time of the song or the weeping, literally its precise duration as a performance, creates a time-space for immersion, a momentary witnessing linked to the time of travel and experience, the durations of travel, of movement, of journeying. Here the journey is completely in the listener's head rather than out along those lands, but it is a journey nonetheless, and a powerfully integrative one because of the compact, corporeal, durative moment. What makes the moments from the start to the close of the song or weeping particularly powerful is this framing of introspective space, a space that is placed in lands and their relationships, a space that senses the experience of journey as the experience of biographical review. Knowing where you are is knowing who you are, and this is one of the ways in which place is an integrative site for memories, and hence a truly transformative locus of naturalized culture and culturalized nature in Bosavi expression.

Note

1. Field research in 1976–77, 1982, 1984, 1990 and 1992 was undertaken in the communities of Sululeb and Bolekini in Bosavi, Southern Highlands Province, Papua New Guinea. I wish to thank all of my hosts, and especially Jubi, Kulu, Ayasilo, Ho:nowo:, Hobole, Hasele, Gigio, Ha:gulu, Deina, Seyaka and Ulahi for their years of friendship and considerable efforts to make their world more understandable to me. For field research and recording support I am grateful to the Cultural Studies Division of the National Research Institute of Papua New Guinea (formerly the Institute of Papua New Guinea Studies, especially its Music Department), and in the United States, the National Science Foundation, the National Endowment for the Arts, the National Endowment for the Humanities, the Wenner-Gren Foundation for Anthropological Research, Rykodisc, 360° Productions, and the John D. and Catherine T. MacArthur Foundation.

Bibliography

Armstrong, R. P., *The affecting presence: an essay in humanistic anthropology*, Illinois: University of Illinois Press, 1971

——, *Wellspring: On the myth and source of culture*, Los Angeles: University of California Press, 1975

Basso, K., Western Apache Place-name Hierarchies, in E. Tooker (ed.), *Naming systems*, Washington: American Ethnological Society, 1984a, pp.78–94

——, 'Stalking with stories': names, places and moral narratives among the Western Apache, in E. Bruner (ed.), *Text, play and story: the construction and reconstruction of self and society*, Washington: American Ethnological Society, 1984b

——, 'Speaking with names': language and landscape among the Western Apache, *Cultural Anthropology* 3(2), 1988

Buttimer, A. and D. Seamon (eds), *The human experience of space and place*, New York: St Martin's Press, 1980

Casey, E. S., *Remembering: a phenomenological study*, Bloomington: Indiana University Press, 1987

Dixon, M., *Ride out the wilderness: geography and identity in Afro-American literature*, Urbana and Chicago: University of Illinois Press, 1987

Feld, S., *Music of the Kaluli*, 12-inch stereo disc with notes, photos, map, Boroko: Institute of Papua New Guinea Studies, 1982, IPNGS 001

——, *Kaluli weeping and song*, 12-inch stereo disc with notes in English and German, photos, map, transcriptions, Musicaphon/Music of Oceania series, Kassel: Bärenreiter, 1985, BM 30SL 2702

——, 'Aesthetics as iconicity of style, or, "lift-up-over-sounding": getting into the Kaluli groove', *Yearbook for Traditional Music* 20, 1988, pp.74–113 and cassette

——, 'Aesthetics and Synesthesia in Kaluli Ceremonial Dance', *Journal of Dance Ethnology* 14, 1990a, pp.1–16

——, 'Wept thoughts: the voicing of Kaluli memories', *Oral Tradition* 5(2–3), 1990b, pp.241–66

——, *Sound and sentiment: birds, weeping, poetics and song in Kaluli expression*, Philadelphia: University of Pennsylvania Press, 2nd edn, 1990c

——, *Voices of the rainforest: Bosavi, Papua New Guinea*, CD/ cassette, The World, Series producer: Mickey Hart, Salem, MA: Rykodisc RCD/RAC 10173, 1991a

——, 'Sound as a symbolic system: the Kaluli drum', in D. Howes (ed.), *The Varieties of sensory experience: a sourcebook in the anthropology of the senses*, Toronto: University of Toronto Press, 1991b

Flanagan, J., 'Heirarchy in Simple "Egalitarian" Societies', *Annual Review of Anthropology* 18, 1989, pp.245–66

Heidegger, M., *Being and time*, London: Basil Blackwell, 1962

Jeans, D.N., 'Some literary examples of humanistic descriptions of place', *Australian Geographer* 14(4), 1979, pp.207–13

Marks, L., *The unity of the senses: interrelations among the Modalities*, New York: Academic Press, 1978

Marx, K. and F. Engels, 'Manifesto of the Communist Party', in *Collected Works* vol.6, Moscow: Progress Publishers, 1975

Momaday, N.S., *The names: a memoir*, Tucson: University of Arizona Press, 1976

Rappaport, R., *Ecology, meaning and religion*, Berkeley: North Atlantic Books, 1979

Schieffelin, B.B., *The give and take of everyday life: language socialization of Kaluli children*, New York: Cambridge University Press, 1990

Schieffelin, E.L., *The sorrow of the lonely and the burning of the dancers*, New York: St Martins Press, 1976

——, and R. Crittenden, *Like people you see in a dream: first contact in six Papuan societies*, Stanford: Stanford University Press, 1991

Silk, J., 'Beyond Geography and Literature', *Environment and Planning* 2, 1984, pp.151–78

Simpson, C., *Adam in plumes*, Sydney and London: Angus and Robertson, 1954

Tuan, Y.-F., *Topophilia*, Englewood: Prentice-Hall, 1974

——, 'Surface phenomena and aesthetic experience', *Annals of the Association of American Geographers* 79(2), 1989, pp.233–41

——, 'Realism and fantasy in art, history, and geography', *Annals of the Association of American Geographers* 80(3), 1990, pp.435–46

Wagner, R., *Symbols that stand for themselves*, Chicago: University of Chicago Press, 1986

Weiner, J., *The empty place: poetry, space, and being among the Foi of*

Papua New Guinea, Bloomington: Indiana University Press, 1991
Williams, R., *Marxism and Literature*, New York: Oxford University Press, 1977
——, *Keywords*, London: Fontana, 1983

Chapter 4

A Church Too Far Near a Bridge Oddly Placed: The Cultural Construction of the Norfolk Countryside

Charles O. Frake

The first thing one notices about Cley is the position of the splendid parish church, more than a half a mile from the village, overlooking a large green . . . Yet even this is not the whole story, for the traveller cannot fail to notice that there are bridges across the old estuary at both Cley and Wiveton. That at Wiveton is, indeed, a well-preserved medieval bridge, probably of the fifteenth century.

W.G. Hoskins

Thus does Hoskins, a local historian, instruct the readers of his *Fieldwork in Local History* [1982: 151–2] how to read a bit of landscape found along the north Norfolk coast of England. The terms of these instructions – churches and bridges, relative locations, worthiness of notice, antiquity, preservation – are ingredients in a discourse of 'the countryside' that thoroughly pervades English life. Hoskins' traveller may or may not fail to notice this particular church and these two bridges, but anyone who has listened for long to the talk of the English cannot fail to notice the mention of such features of the landscape in the course of constructing entertaining stories, devising trenchant arguments and asserting their unique position in the world. We ask here how this discourse is constructed from the materials of the landscape and why it is so prominent in the lives of its tellers. The locale of our investigation encompasses the very church and bridges of Hoskins' lesson in fieldwork.

The English county of Norfolk juts out northeastward into the North Sea as an exposed flatland sheltered from wind and waves

Figure 4.1. The Norfolk landscape

only by a low ridge of sand dunes. If one stands on top of these dunes, turns one's back to the sea and looks inland over what would naturally be a rather flat and featureless landscape, one finds, with local guidance, much to see. True, there are no natural wonders – no Grand Canyons, no Niagara Falls, no Fujiyamas. Nor are there any great sites – no coliseums, no castles, not even any megaliths. The most prominent features are the flat-topped towers of medieval churches that arise in every direction from clumps of trees amidst the fields. One can see farmsteads and small villages here and there as well, but their sparsity and small size only serve to make the abundance of church towers all the more noteworthy. One looks in vain for a grid or patterned network of roads linking churches, villages and farms. What one sees from a distance are sinuous paths of hedges that line both field and road forming no discernible patterns and few straight lines. The most naturally appearing features of the landscape are the stretches of marshlands, wetland woods, waterways and lakes. But these too, locals will quickly point out, are the products of human activity: they are the remains of medieval peat diggings. This is a landscape that has been made remarkable, worthy of talking about, almost exclusively by rather subtle traces of past human endeavour enthusiastically discerned and interpreted by those in the present with present purposes in mind. The Norfolk landscape is a product of past activity that requires constant cognitive attention and behavioural intervention to preserve and reconstruct what is valued in contemporary images of the past. This reading of the landscape is a cultural practice motivated not only by social entailments and political-economic agendas but also by the experiential enrichment afforded by a meaningful engagement with the past of one's place.

The questions I pose about local interpretations of the landscape are sincerely asked out of a conviction that cultural constructions have, in their constructed form, much to teach us about ourselves as well as providing us much to admire. While fully prepared to point out the sometimes deceptive political-economic façades of culturally constructed edifices, this essay is not intended to contribute to the depressing climate of condescending deconstruction that pervades much of current human science, deconstruction which reduces real people to gullible dupes of a hidden malevolence that makes itself visible only to the discerning investigator. This particular investigator, more bumbling than discerning, owes his limited understanding of the topic at hand to the perspicuity and patience of those whose lives he presumes to investigate. The topic at hand, it must be emphasised, is the story of the countryside as told to us by the people living in it. Accounts of the past form a dominant component of this story. These accounts may well often be good history, but it is the 'story', not the 'real history', that is our concern. We leave the uncovering of 'real history' in the competent hands of local historians and archaeologists, keeping in mind that their accounts rather quickly become incorporated into 'the story'.

The story is something told, talked about and written about. It is a 'discourse'. The telling of the discourse of the countryside is a practical activity with practical ends in mind. It is a 'practice'. I will use the terms 'discourse' and 'practice' as appropriate labels for these phenomena. They should not be taken as flags of identity affirming affiliation with any of the several current perspectives in human sciences that style themselves the 'theory of discourse' or the 'theory of practice'.

With these declarations out of the way, it is time to proceed to the particulars of distant churches and misplaced bridges.

The Terms of the Discourse: Churches

> Norfolk and Suffolk can boast of a greater density of churches than any other county. In each county there is on the average one church to every 1,850 acres and it is amazing the number of churches you can see from the churchyard of Happisburg, which stands on a slight elevation and looks over the flat country inland.
>
> Cautley, *Norfolk Churches*

It is appropriate to begin with churches. Simply in their abundance, they are the most prominent feature of the Norfolk landscape. Few rural churches are, by themselves, particularly striking. They are generally small, squat-towered and unadorned. But they are everywhere. And they are old.

Accounts of why this is so, why this so quintessential English institution, the medieval church, so permeates the countryside, are easy to come by. During the late middle ages, the story goes,

Norfolk, largely because of the international wool trade, was a very prosperous corner of England. Norwich, the county capital, was England's second city. Coastal ports were booming. Parish churches, built by local gentry, were a display of this wealth. This standard history is well known. A common elaboration, one that seeks a monetary rather than a symbolic explanation, has it that churches were the original medieval tax shelter. Gentry could avoid paying tithes to the church by building their own church attended by themselves, their tenants and workers. The tiny village of Worstead encapsulates this local history. Once a major centre of the wool industry whose name survives worldwide as a kind of wool, it now consists only of a magnificent medieval church, where a local weaving club keeps its looms, a few ancient buildings with Flemish façades, and, of course, a pub. Post-medieval history passed Worstead and much of the rest of rural Norfolk by. The flat countryside could not supply the water power upon which the industrial revolution depended. It was left with many more churches than post-medieval populations, less numerous and increasingly dissident and secular, could support. The Norfolk parish church is a monument to a bygone era. It is a museum as well as a place of contemporary worship and service. These two roles can conflict, thereby providing a potential line of cleavage in local social alignments.

Adding to the mystery of the abundance of Norfolk churches is their location with regard to settlements. Hoskins sets up a puzzle in the beginning of his essay on field work in local history with his claim that 'The first thing one notices about Cley is the position of the splendid parish church, more than a half a mile from the village . . . .' The claim rings true, but not because such a church seemingly too far from its village is at all unusual in rural Norfolk. Indeed, isolated churches with nothing but fields about them are a common sight.[1] Hoskins' claim has merit because, in spite of the prevalence, if not, indeed, the typicality of the isolated situation, the position of the local church, whether isolated or not, is invariably a matter of notice. If it is isolated, it is remarkable *per se*: isolation requires explanation. Why put a church out in the middle of nowhere? If it is not isolated it is remarkable *de facto*: in a sea of isolated churches, how does one account for the non-isolated cases? Hoskins' explanation for the current isolation of the Cley church is that the village has moved. He provides convincing documentation for this move, which was motivated by the progressive silting up of the estuary that made Cley an important medieval port. True history it undoubtedly is. But it is also a well-known local story repeated in church pamphlets and tourist brochures. In local tellings, the story is embroidered with tales of Cley's past glory as a pirates' den, commercial entrepôt and base for the Icelandic fishery. Church and village pamphlets tell these stories in some detail, noting that during the flood of 1953 the sea broke through so that Cley church once again overlooked an expanse of water.[2] A disaster is read as a fleeting re-creation of an era of past glory. An entertaining alternative to the story of the moved village is the story of the lazy lord. When a local landowner built his church, he would place it near his manor for his convenience and ostentation rather than in the village of his dependants. A third explanation, that the village of the church has simply disappeared over time because of plague, enclosure and agricultural depressions, though probably true in many cases, is not commonly offered locally.[3] However, if the disappearance was dramatic, then one has a story. The most remarkable location a village and its church can have in Norfolk is under the sea. A number of coastal villages in Norfolk and Suffolk have succumbed to the continuous attack of the sea on the dunes. In these cases, the church tower, by local account, is invariably the last structure to disappear beneath the waves. Several of these places have become tourist attractions, even though there is generally nothing to see except the waves that allegedly roll over the ruins. When one stands on the dunes and gazes out to sea, one is encouraged to imagine, lying under water, the same flat-church-dotted landscape that lies behind one. Occasionally, even now, low tides and storm surges reveal the ruins of medieval structures. In one case they uncovered a graveyard with exposed skeletons.[4]

The most remarked-about external attribute of Norfolk churches is the tower. Towers come round or square. Round towers are said to be unusually common in Norfolk and its East Anglian neighbour, Suffolk.[5] Roundness thus requires explanation – a requirement made manifest by the 'Round Tower Church Society', based in Lowestoft, Suffolk, and devoted to the preservation and interpretation of roundness in church towers. The explanations exploit what is a major dimension of the discourse on the countryside: antiquity. Round towers, some say, mark a church as pre-Norman, Anglo-Saxon. Others say that the abundance of round towers in Norfolk and Suffolk is easily explained by the lack of building stone in East Anglia. Towers, round or square, as well as walls are built up of small stones with flint facings imbedded in mortar. With this kind of material it is easier, the argument goes, to make a tower round than square. Another story, often heard and well elaborated in a classic work on church architecture [Cautley 1949: 2–3], argues that round towers were originally watch towers built to warn of Scandinavian raiders. Subsequently, churches, as they were built, simply incorporated these convenient pre-existing structures. The church at Little Snoring presents a nice wrinkle on this story. It has a round tower separated from its church, but there are clear marks of another church that was once attached to the tower.

All of these explanations focus on the uniqueness and antiquity, and hence, in the terms of this discourse, the value of round towers. Square towers suffer only marginally as topics of admiration. Their antiquity, construction and height (or lack of it) are, for almost any church, matters of note and admiration. But there are revealing exceptions, as the Ordnance Survey guide to Norfolk waterways, echoing local sentiment, tells us: 'The actual village church of Burgh St. Peter lies twelve miles to the east, on a slight prominence between the River Waveney and Burgh Marshes. The charming thirteenth-century thatched nave is unfortunately dominated by a grotesque folly of a tower – four brick boxes piled up wedding-cake style. Built c.1800 on a sixteenth-century base, its pointed windows simply accentuate its awfulness' [Perott 1986: 72–3].

There are many other features of churches, external and internal, that are deemed worthy of comment and interpretation: entrances, windows, roofs, fonts, screens, benches, bronzes, etc., as well as typifications of overall architectural style as Saxon, Norman, Decorated or Perpendicular. This talk of churches reveals the principal dimensions that structure the discourse of the countryside: antiquity, authenticity, and distinctiveness unspoiled by 'modern' (i.e. post-medieval) alterations. We can see these dimensions at play when we turn to several other much talked about features of the Norfolk landscape. We turn next to a feature which, unlike the man-made church, a symbol of the 'cultural' heritage, is taken to be a product of 'nature'.

Terms of the Discourse: Broads

> In few regions does the human settlement depend more on the natural features of the area than in broadlands, for the story of Man in this part of East Anglia is largely the story of his reactions to a powerful and compelling physical environment. Despite his acquisition in recent centuries of new and potent means of subduing his surroundings, much of the broadland still remains untamed.
>
> R.R. Clarke [in Ellis 1965: 233]

The expanses of open water or 'Broads', connected by a network of channels, locally 'dikes', and rivers, bordered by marshes and wetland woods, locally 'carr', form a natural landscape, 'the Broadland', unique to Norfolk. The other 'natural' landscape of Norfolk, its North Sea coast of beaches, dunes, salt marsh and mud flats, is also much admired by locals and much visited by outsiders. I will limit discussion here to the Broads; they are distinctively Norfolk, and their story contains a number of interesting paradoxes for the discourse of the countryside. The coast was shaped by the sea, a force unarguably 'natural'. The provenance of the Broads, on the other hand, was, equally unarguably, not the forces of nature but the hands of humans. That the Broads were formed from medieval peat-digging was not firmly established until 1960 [Lambert, Jennings, Smith, Green and Hutchinson 1960], but this fact is now well known locally by both residents and visitors. The Norfolk Broads are 'natural' only in the sense that, whatever their origin, in their present state they are very definitely construed as part of 'nature' as opposed to the cultural landscape of cultivated fields, settlements and industrial sites. Less well known than their origin in human activity is the fact that it takes constant human intervention against the forces of nature to maintain the broads in their 'natural' state. Processes of natural succession operate to fill in the open expanses of water, first with marsh, then with wetland woods. In the past, there was apparently enough harvesting of thatching materials and clearing of channels to keep the lakes clear, although there have undoubtedly been many changes to their contours and expanse over the centuries. The more these facts become known and thereby incorporated into the discourse of the countryside, the more it becomes clear to users of the discourse that a 'natural state' is a human construction and, as such, the authenticity of its 'naturalness' is open to negotiation.

In recent times, the former uses of the Broads that held nature at bay have declined, while, at the same time, human activities have increasingly attacked nature on another and much more destructive front. Formerly, past human intervention served to keep natural succession, at least partially, in check; now human intervention threatens to alter the Broadland landscape irrevocably. Modern agriculture, encouraged to expand arable land by government subsidy, has drained large areas. It has also added pollutants to the waters. Houseboat renting tourists, attracted by the 'natural' beauty of the Broadland, at the same time threaten to overwhelm it. An agency of local government, the Broads Authority, was set up in 1978 to regulate use and restrict development. In addition, the National Trust, the Nature Conservancy and the Norfolk Naturalists' Trust, as well as many other smaller organizations, are involved in conservation and restoration activities. One can probably thank the tourist industry for all this institutional concern. Whatever their impact on the landscape, tourists have had the effect of spreading and popularizing the discourse of the countryside in the Broadland.[6]

Terms of the Discourse: Roads

The village lies folded away in one of the shallow valleys which dip into the East Anglian coastal plain. It is . . . approached by a spidery lane running off from the 'bit of straight', as they call it, meaning a handsome stretch of Roman road, apparently going nowhere.

Blythe, *Akenfield*

Roads are a remarkable feature of the English countryside, not as signs of modern progress, but as relics of the past. Roads that count in the discourse of the countryside are old roads. There are two kinds of old roads, Blythe's 'spidery lane' and the 'bit of straight'. England is famous for its narrow, winding, hedge-lined roads. Norfolk roads conform well with this image. One cannot travel for long without finding oneself perpetually in the bends of a one lane, shoulderless, depressed, hedge-lined track.[7] No need to worry about what side of the road to drive on! But once in a while the road will suddenly straighten. It won't widen as well, but at least one can now see what's coming. 'Do you know why this bit is straight?' I was once asked while being driven across central Norfolk by a local friend (this was the most direct route – we were not sightseeing). 'It's a Roman road', he explained. Later I looked up the road on the 1:25 000 scale Ordnance Survey map – the same map that prints 'Wiveton Bridge' in Gothic lettering to denote its medieval antiquity – and found printed in bold Roman capitals: 'Roman Road'. My friend, the driver, denied ever having seen this map: 'Been here all my life; don't need a map to tell me what kind of road I'm on'.

Of course, not all straight roads are Roman – there are modern highways, railway lines, and roads that follow former railway lines. The post-medieval field enclosures, discussed below, also produced some straight bits of road now several hundred years old. But a Roman road is an *old* straight road. Its pre-medieval antiquity makes it worth mentioning in the discourse of the countryside, and it makes its straightness forgivable. Otherwise, to be a proper feature in the image of the English countryside a road should be both sinuous and narrow. The value on narrowness seems quite firmly established despite the inconveniences of such roads in the modern world. The attitude is well expressed by Rackham [1986: 199] when, in describing what had been an embarrassingly straight and broad road in Cambridgeshire, he notes that, during early nineteenth-century enclosures, 'the opportunity was taken to narrow the road . . ..' In 1991, in the coastal village of Sea Palling, a local resident preparing a photographic display he was calling *The Rape of a Village* included – along with vivid scenes of the effects of development, neglect and storms – before-and-after photos of the disaster of the highway department's widening of the village street.

To construe a road as old, it is enough that its track, as it appears on a map, can be shown to be old. No one demands of a Roman or medieval road that it retain, in its pavement or its vegetation, actual physical remains of the past. Of course, it greatly helps to secure the place of a road in the discourse of the countryside if it displays some tangible markers of demonstrable antiquity. The bridge at Wiveton thus achieves its special importance, an importance acknowledged on the Ordnance Survey Map by spelling its name in bold Gothic letters to signal genuine medieval antiquity.

Terms of the Discourse: Hedges

. . . try to consider the English landscape without the hedge. It would not be the English landscape.

H.E. Bates[8]

The typical English countryside, seen on the covers of innumerable books and magazines, is a patchwork of tiny, irregularly shaped fields bounded by hedges. Not all English countryside in fact looks like this, and some of it probably never did. It is common knowledge that, in the Middle Ages, the typical field pattern in much of England was of large open fields and open commons associated with compact villages of cultivators, each of whom had rights to a number of strips of land scattered in the open field and for the use of the village common. Beginning after the Middle Ages and reaching a height in the eighteenth century, these open fields and commons were consolidated, with government support, by landowners and grouped into enclosed, hedge-lined, individual tracts. By the early nineteenth century the enclosure process was virtually complete throughout England. It was this newly enclosed landscape of Victorian times that became the 'typical English countryside'. This history, thus told, poses a threat to the integrity of the discourse of the countryside. If the typical countryside is not also an ancient part of the English heritage, its value is considerably diminished. Antiquity, as we have seen, is a fundamental value of the discourse.

There is, however, a retelling of the 'open field and enclosure' story that saves the countryside from the embarrassment of lost antiquity. According to this retelling, for which there is good historical and topographic evidence, there were areas of lowland England, the 'ancient countryside', that, in contrast to the 'planned countryside', were never in open fields.[9] The strong version of this story, which has very respectable scholarly credentials, claims that all of England was once enclosed, and then some areas, mostly in the Midlands, were later opened up in a foreshadowing of what is happening now with modern agriculture.[10] These nuances to the 'open field' story are well known in East Anglia which, according to maps reproduced in many guides to the English landscape, is divided by the boundary between the two types of field systems.[11]

What the revised story means for the reader of the landscape in Norfolk and Suffolk is that not all hedges are the same. A hedge can be an old remnant of the medieval closed countryside or it can be a new ('new', that is, in the context of deeply penetrating English views of the past) product of post-medieval enclosures. Given the fundamental maxim in the discourse of the countryside, 'the older the better', it goes without saying that a landscape with old medieval hedges is more valued than a 'newly' constructed patchwork of hedged fields. But it is also clearly the case that hedges are better than no hedges. Is, then, a post-medieval enclosure-produced hedged field to be more valued than the more ancient, but unhedged medieval field that preceded it? In the discourse of the countryside does being hedged rank above being ancient? Enclosure hedges presumably make the countryside look more like it did in early times before open fields, but at the same time they mark a 'recent' transformation of a more ancient landscape. Should one regret the loss of the authentic medieval open-field landscape and, with it, the pristine rural life of the pre-capitalist peasant farmer? Or should one view the enclosures as an improvement of the countryside, a reconstruction of a still more ancient past, albeit, like all reconstructions, not the same as the 'real' thing? These questions are not superficial ones. The issue is not merely one of where to locate one's nostalgic dreams. These questions go to the heart of the discourse, revealing ambiguities that enable it to be employed as a political weapon by both sides in contemporary disputes whose outcomes are seen as seriously impacting the lives of people and the fate of the countryside.

The Power of the Discourse: Origins

Although the hedge is now eulogized in England, and symbolizes the countryside we all have a right to enjoy, not so long ago it was hated by large sections of the people as a symbol of tyranny and despotism.

Williamson and Bellamy, *Property and Landscape*

Talking about hedges, as the English frequently do, leads one quite suddenly into tangled thickets of political controversy. Within those thickets we are left to ponder about the users and creators of the discourse which led us there. We then confront the grand question: from where does the discourse acquire its power?

The discourse of the countryside, for all its emphasis on antiquity, is of relatively recent origin. The term 'countryside' itself, now omnipresent, only began to be used in its current sense at the beginning of the twentieth century.[12] It is instructive to look at what is a major topic of the contemporary discourse, the past enclosures, and ask what was the nature of the discourse that, in the past, accompanied this major transformation of the English landscape. We have numerous accounts from both sides of the debate at that time. The enclosures, often government sponsored (the 'parliamentary enclosures'), caused considerable unrest among the closed-out populace. There were protests and abortive rebellions, the most notorious of which went under the name of a legendary 'Captain Swing'. Norfolk protests occurred as late as 1870.[13] The victims of this agricultural transformation, of course, did not have a prominent voice in what survives of the discourse of that time. They did, however, have their champions among the literati, who argued eloquently against the injustice and cruelty of the enclosure process. But their protests were not justified by expressions of regret for a lost landscape. In no case, it would appear, did anyone protest that the authentic medieval landscape was being destroyed. On the contrary, it was the other side, the supporters of this great transformation, who argued that a notable improvement in the appearance of the landscape was taking place. E.W. Bovill, writing for the *Country Book Club* in 1962, echoes early arguments supporting the enclosures: 'Until late in the eighteenth century, in all such [grain-growing] parts of England, close to the village lay the traditional open field. . . . Beyond and around this *bleak and hedgeless field*, without shelter for cattle or crops, lay seemingly limitless wastes of scrubby woodland, moor, or marsh, loosely termed the common. . . . So long as the open-field system survived, progress was impossible. . . . In many a village an intransigent minority of one [!] blocked the road to progress . . . recourse usually had to be made to a private Act of Parliament . . . '[Bovill 1962: 2–3 (emphasis mine)].

In the discourse of that time, enclosures not only improved the scenic value of the former open-field country, they also represented an improvement over ancient countryside that had never been in open fields. 'The fields on the left', wrote Cobbett on a ride from Royston to Huntingdon in 1822, 'seem to have been enclosed by an act of parliament; and they certainly are the most beautiful fields that I ever saw. . . . Divided by quickset hedges, exceedingly well planted and raised.' Like his contemporary Mary Mitford, who spoke of 'old irregular hedges', we can still often tell a pre-enclosure hedge from a later one by its crookedness and the varied material composing it [both cited in Bovill 1962: 5]. Now 'old irregular hedges', with their 'crookedness' and 'varied material', are to be cherished; then they were a blight on the landscape. Cobbett himself bitterly opposed the enclosures. He rode around England in the early nineteenth century documenting their consequences for the poor [Cobbett 1967; Williamson and Bellamy 1987: 115]. But he seems to have had no objection to their consequences for the landscape. Others at the time vehemently attacked the ancient, enclosed landscape as inferior to that produced by enclosures. According to the *Farmer's Magazine* of 1800:

> The first circumstance that strikes an observant stranger, accustomed to the regularity of modern inclosure, is the great inequality in the sizes of field, in the anciently inclosed districts of England, the crookedness of the hedges, and the strangely irregular shapes of the inclosures. But the object which I have particularly in view, in these strictures, and which merits the severest reprehension, is, that almost everywhere the hedges have been permitted to run wild, overspreading and irregular breadth of surface with a belt of useless and cumbrous brush-wood, while they scarcely half fence the fields, which they disfigure greatly, and the value of which they lessen very materially . . . these crooked belts of brush-wood, *improperly styled hedges*. . . . (quoted in Bovill 1962: 7–8; emphasis mine).

What are, in our time, taken to be the old, authentic, real hedges were, then, not properly hedges at all![14]

The battles fought several centuries ago over the impact of changing agricultural practices are being refought now. But now both the perceived effects of agricultural practice and the nature of the countryside being transformed have been reversed. Then, changes in agricultural practice closed up an open countryside; now, agricultural practice is opening up a closed countryside. And now, the arguments for and against change are framed in a new discourse of the countryside with shared values quite different from those of the past. Far from being scorned, irregularity and heterogeneity are now valued as signs of authentic antiquity. Far from being seen as an improvement to the landscape, agricultural practice is now viewed as a destructive threat. The new discourse that appeals to the value of the landscape, the discourse of the countryside, is now a tool for the other side, the side opposed to the modern great agricultural transformation, a transformation which, ironically, is seen as taking England back to a landscape of broad, open, hedgeless fields. But of course these are not, however much they may look like it, the authentic open fields of the Middle Ages, but the products of modern mechanized agriculture.

The new discourse of the countryside, although achieving prominence only since World War II, had its beginnings at the end of the nineteenth century in conjunction with the appearance of the Ancient Monuments Act, the Society for the Protection of Ancient Buildings, the National Trust for Places of Historic Interest and Natural Beauty, the British Naturalists' Association, the Society for the Promotion of Nature Reserves, the Society for Checking the Abuses of Public Advertising, the Commons Preservation Society, the National Footpaths Preservation Society, the Society for the Preservation of the Wild Fauna of the Empire, the Selbourne Society for the Protection of Birds, Plants and Pleasant Places, and, to promote the discourse, *Country Life* magazine.[15] In retrospect, the beginning of the discourse of the countryside is also seen as the time of the beginning of the end of the authentic countryside of current image: 'Since that time and especially since 1914, every single change in the English landscape has either uglified it or destroyed its meaning, or both' [Hoskins 1982: 298]. As the old living landscape was killed, it was necessary to recreate it in discourse, in an image of the 'countryside'. As a discourse that appeals to a constructed image of the past to shape arguments for reconstruction and preservation under conditions of present use, it generates serious paradoxes over issues of antiquity and authenticity. These paradoxes leave the discussion vulnerable to manipulation, reinterpretation and even deconstruction in the context of contemporary political conflicts.

The Power of the Discourse: Contestants

Farming created the countryside and gave it its fascinating diversity, and now farming is destroying it

Williamson and Bellamy, *Property and Landscape*

A simplistic view of political alignments in the contemporary battle over the countryside pits the farmer against those who cherish and seek to maintain the heritage of the English countryside. The 'farmer' in question is not the old, authentic, small-scale farmer, attuned to nature, ploughing his hedged field with his trusty horse. This farmer is the modern, mechanized, rapacious, capitalist farmer interested only in maximizing profit. It is necessary for a teller of the discourse of the countryside to thus distinguish kinds of farmers, for farmers are both the heroes and villains of the story. Whatever one's view of the 'true' English countryside, it has been shaped by farming out of some previous state that was also shaped by farming.

Who are these farmers, the heroes and villains of the story? In England as in America, the ideal image of the farmer and the farm differs greatly from the current reality. The English image and realities are, however, somewhat different from the American ones. In the English image, there is no presumption that the 'farmer' owns the land he farms. The farmer may well be a 'tenant' (but definitely not a 'tenant farmer' in the American image) who rents the land from a landowner. Even less is there a presumption that the farmer works his land with his own hands. A 'farmer' is not a 'farm worker'. All that is presumed is that the farmer occupies and runs the farm. This image of a farmer maps on to the image of the three-class structure of English rural society: the land-owning gentry, the farmers and the workers. The realities of who owns what and who does what are of course much more complex. They are also difficult to discover. In fact, these difficulties achieve almost mythic proportions as part of the discourse of the countryside. It would seem that 'who really owns the land' is one state secret the British have been able to guard with impenetrable security. There have been two major attempts to breach this security. The first, and most successful, was the Domesday survey of the eleventh century ordered by William the Conqueror. The other was the 'new Domesday' ordered by Lord Derby in the 1870s to disprove reformists' claims that only 30,000 people owned almost all the land in Britain. Much to Lord Derby's chagrin this survey showed that eighty per cent of the land was under the control of only 7,000 landowners. A further attempt was made in 1974–9 by the short-lived Royal Commission on the Distribution of Income and Wealth. Before being abolished by the Thatcher Government, the Commission's brief statement on wealth in land concluded: 'The paucity of comprehensive up-to-date information on land ownership is remarkable. In the absence of a survey yielding data on the lines of the 1873 survey, it is difficult to carry our analysis any further.' Whether they own land or not, farmers in Norfolk wield political power out of proportion to their numbers. 'Although farmers comprised only two per cent of the adult population of North Norfolk District, they made up twenty-six per cent of the membership of the District Council in November 1985.'[16]

The farmer, more than the landlord and the farmworker, is seen as belonging to the land. However villainous some people may think his role to be, he is unarguably an 'authentic' part of the countryside. In this respect the farmer is at opposite poles with the tourist. The Norfolk countryside, especially the coast and the Broads, has been a destination for outsiders on holiday since the nineteenth century. These visitors are by no means homogeneous in their interest in the countryside. There is the amusement-arcade set at the tackier coastal resorts, and there are those who rent houseboats on the Broads to use as mobile platforms for continuous beer-guzzling. But there is also a large contingent attracted by the appeal of the countryside, not only the 'natural' landscape of the Broads and the coast, but also the cultivated countryside of farmland, small villages and magnificent country homes. There is a large local industry catering to the needs and interests of these visitors.

As elsewhere, the tourist in Norfolk presents a terrible dilemma. He and she are not authentic parts of the landscape but intrusions. Yet it is the 'authenticity' of the Norfolk countryside that attracts them. The importance of tourism provides an argument for preserving the attractiveness of the countryside; at the same time, the more successful the attraction, the more the intrusion of outsiders threatens the authenticity upon which the attraction is based. Also, the more farmers turn their attention to increasing the tourist appeal of their farms, barns and homes, the more their activity loses its authenticity. They become not farmers but museum curators.

Between the authentic but destructive modern farmer and the intrusive but countryside-loving visitor lie the majority of the residents of rural Norfolk. Many are 'new-comers', retirees and white-collar commuters attracted to a 'country-life' lifestyle. Others may have old roots in Norfolk, but follow occupations far removed from farming: teachers, clerks, architects, artists, tourist facility operators, secretaries, etc. It is among this contingent, whose lifestyle is seen to be threatened by modern farmer and tourist alike, that the discourse of the countryside has its strongest appeal. Their voice in local politics is becoming increasingly heard in Norfolk.[17]

Each side in the political conflict over the countryside has an array of institutions, governmental, quasi-governmental and independent, dedicated to supporting its cause. Almost all of these institutions make use of the discourse of the countryside, fashioning this rather flexible instrument to support diverse arguments. When one asks, further, who belongs to what institutions, some surprises are in store. Take one man in the Broadland area.[18] He is both a farmer and a landowner with long-standing roots in Norfolk. His landholding amounts to about 300 acres of dry land and marsh, much of which he has drained. He is the chairman of the development committee of the district council, the local government body regulating the commercial and residential development of agricultural land. He is vice-chairman of the Drainage Board, which plans and oversees land drainage in the Broads, a process which transforms wetlands into arable farm land. At the same time, this farmer is also a member of the Broads Authority, the government agency responsible for the preservation and restoration of the Broads. When one looks at specific individuals, it is not always clear who is on which side, even though the sides themselves may be clear enough. It is also difficult to judge whether men like this Norfolk farmer acquire and wield their political power solely in pursuit of their own selfish economic interests or whether they see themselves as responsible and self-sacrificing stewards of sensible land use.

The Discourse Challenged: A Fraudulent Image?

The landscape is the prime anachronism of the national heritage – essentially a vast museumised ruin.

Lowenthal, 'British National Identity and the English Landscape'

Farmers and others with a vested interest in the local landscape rarely directly challenge the rational of the discourse of the countryside. Rather, they mould and manipulate it to support their own arguments. 'We're just making the landscape more like it really was in medieval times', a farmer might argue in appealing to the discourse's value on authentic antiquity. Increasingly, however, the very foundations of the discourse of the countryside have come under attack, not from locals with an economic axe to grind, but from outside intellectuals, academics and journalists, who write for a popular audience. These writers deconstruct the discourse by demonstrating that the image of the countryside it promotes is mostly mythical. Whatever basis it had in reality was a recent (again by English standards of 'recent') construction of eighteenth- and nineteenth-century enclosures together with the extensive building of landscape parks by the gentry. These parks were fashioned by such famous (or infamous) gardeners as the legendary 'Capability' Brown, who could construct a private countryside out of the community open fields and commons of past landscapes. Brown's 'capability' lay in the power to transform an image into reality, albeit an artificial one. David Lowenthal [1991a: 213] has recently claimed,

> The now hallowed visual cliché – the patchwork of meadow and pasture, the hedgerows and copses, the immaculate villages nestling among small tilled fields – is in fact quite recent; only after the pre-Raphaelites did the recognizably 'English' landscape become an idealized medieval vision, all fertile, secure, small-scale, seamed with associations.

Lowenthal continues, in the lines leading up to this sections' epigraph: 'The landscape . . . is pervasively antiquated. Virtually every familiar feature was created for purposes now outdated, for needs now redundant, by means no longer to hand, in circumstances that no longer hold.' A popular version of these arguments has also appeared in the newsstand magazine *History*

Today.[19]

The argument of the countryside deconstructionalists is essentially that the expression 'authentic past' is an oxymoron. By singling out, marking, preserving and restoring features of the landscape as relics of the past, we thereby alter the meaning of the landscape as a sign of the past. We cannot notice the past without changing it. These arguments are spelled out by Lowenthal [1979; 1985; 1991a, 1991b] in several papers and a recent book. They are echoed, rather more vehemently, by political essayists such as Patrick Wright [1985] and Robert Hewison [1987]. None of these attacks on the legitimacy of the discourse of the countryside are made in support of modern agriculture, industrialization or suburban development. On the contrary, the tenor of this genre is anti-development. It portrays the modern developed world as the destroyer of the richness of everyday life. In compensation people are forced to seek refuge in the pleasant prospects of a past countryside, a search fostered by the state in a hegemony of heritage. It is not clear, in fact, what these writers want their readers to do, other than tear their hair out while contemplating Lowenthal's [1991a: 222] disquieting vision that 'the heritage landscape is less and less England, more and more "Englandland", Europe's all-engulfing offshore theme park'.

'In no other Country . . . ': The Discourse Defended

But no other country can produce anything which, like stitchery, binds together the varying patterns of the landscape in such a way that the pattern is made infinitely more beautiful.

H.E. Bates[20]

In spite of the modern deconstruction of its myth, the discourse of the countryside retains a devoted band of avid defenders. To this band, the main enemies are not the deconstuctors of the discourse among the literati, but the destroyers of the real landscape, the farmers. Philip Lowe [1989], in a paper explicitly titled 'The Rural Idyll Defended', argues that this actual destruction has made the vision of the countryside more realistic. The real threat is now seen not as one presented to an abstract ideal of the countryside by the spread of urban industrialism and suburbia. One now sees the real countryside being destroyed from within the countryside itself. The deconstructionists, however, have – probably inadvertently – strengthened the arguments of the farmers by arguing that the historical image of the countryside is a myth. The landscape is always changing. What is wrong with changing it once again?

Against such arguments, the defence of the image of the countryside has been mounted on three fronts: history, identity, and what might be called experiential enrichment – or more simply 'beauty'.[21] Hoskins, Rackham, Williamson, Bellamy and many others have argued, sometimes impassionately, that there is real history, going back to the Middle Ages and to antiquity, embedded in the current landscape.

> We have to contend . . . with tenaciously held popular misconceptions about the history of the landscape . . . people still write articles on the basis that the whole of the rural landscape has always been changing and the hedges, woods, and so on that we have now are no more than the passing effect of fashions in agricultural practice. This doctrine is usually presented either in the form of the 'Enclosure-Act Myth', the belief that our present hedged and walled landscape is merely the result of agricultural reorganization in the eighteenth and nineteenth centuries, or of the 'Capability Brown Myth,' the notion that the landscape we have inherited is no more than an extension to the countryside at large of the romantic ideas expressed in landscape parks. [Rackham 1985: 68–9]

It is, this defence goes on, not only that the countryside is truly historic, but also that it is so in ways that are peculiarly English. 'In no other country . . .' is a phrase that prefaces both this section's epigraph and also many other assertions about the English countryside. 'We [English]', Hoskins [1982: 32] proclaims, 'live in a country that is richer than any other in visible remains of the past'. Lowenthal [1991a: 213] himself adopts this rhetoric: 'Nowhere else is landscape so freighted with legacy. Nowhere else does the very term suggest not simply scenery and *genres de vie*, but quintessential national virtues.' Lowenthal may not intend to value this state of affairs. But he asserts it, and many others take it as a positive value. The image of the countryside is a fundamental part of the English national heritage. It may be a myth, but it is, say its defenders, a myth that gives real meaning to real landscapes. It may be a myth, but it is a shared myth of English identity.

A final defence of the discourse of the countryside is that the image of the countryside it promotes, whatever its 'authenticity', is an image that brings satisfaction to those who see the countryside through its lens. The English countryside, partly in reality and surely in image, is, defenders declare with considerable conviction, a beautiful and pleasurable thing to behold. As real countrysides are destroyed, 'there is', Rackham [1986: 25] complains, 'the loss of beauty, especially that exquisite beauty of the small and complex and unexpected, of frog-orchids or sundews or dragonflies'. Investigators should not make judgements of this sort, only report them. But having made the report, I must confess some sympathy with its sentiment.

In 1940, the German air force failed in its attempt to destroy Britain. It did succeed, however, in photographing it. In doing so, the Germans inadvertently collected some critical evidence for the future defence by the English of the peculiarly English image of the English countryside. This German evidence has been incorporated by Rackham [1986: 25] in words that provide a fitting conclusion to this essay on the English discourse of the countryside:

> It is not just through the rosy spectacles of childhood that we remember the landscape of the 1940s to have been richer in beauty, wildlife, and meaning than that of the 1980s. It was, and the Luftwaffe aerial photographs prove it.

Acknowledgements

The research reported here is part of a larger on-going historical and ethnographic project on the Cultural Construction of the Past in Rural East Anglia, which is being conducted by myself and my colleague, Ezra Zubrow. The research is supported by the Samuel P. Copen Chair of Anthropology of the University at Buffalo, State University of New York. Joanne Coury, my wife and colleague, has played an important role in the inception and progress of this project. Figure 4.1 was prepared by Terence Loan. Our work in Norfolk has had invaluable support from local people in all walks of life. They are too numerous to mention by name here. We are deeply grateful for their patience, good humour and friendship.

We hope that they will receive this essay in the spirit in which it is intended: as a tribute to their lives and their 'countryside'.

Notes

1. Some data on isolated churches: the church at Worstead is located right in the centre of a tiny village square, but of the thirty churches located within a five-mile radius of Worstead church, seventeen are isolated at least as far as the Cley church is from its village. The small market town of Reepham has three churches (one now in ruins) next to each other off the town square. (People say that in medieval times it also had fourteen pubs, now reduced to three plus a brewery.) In a five-mile radius of Reepham, the Ordnance Survey map marks twenty-three churches, eight of them isolated. Of course, 'isolation' is a relative thing. Cley church is, after all, only half a mile from its village – certainly an easy walk. No church in Norfolk is so isolated that it is beyond easy walking distance from some settlement. I doubt that one could walk for an hour in any direction from any point in northeast Norfolk without – barring marshes, lakes and the North Sea – encountering at least one medieval church.
2. Brooks 1984: 7. See also Cozens-Hardy 1936 and Mellor 1989.
3. There is a fourth possibility: the church has always been isolated. However, no one has yet offered such an explanation to me.
4. Eccles on the northeast coast of Norfolk is a locally well-known case. The Suffolk case of Dunwich is described by Parker [1978].
5. Cautley [1949: 2–3] says of round towers: 'These are a peculiarity of East Anglia, there being 119 in Norfolk, 41 in Suffolk and 8 in Essex. They are all built of rubble with flint facings.'
6. See Dymond 1985 for a discussion of these efforts and an evaluation of their effectiveness.
7. The English word for 'shoulder' of a road is 'verge'. But the two words are not exactly equivalent. The English can talk and write at length about the vegetation of the verges along roads that, to an American, seem utterly without shoulders.

8. Quoted in Mossman's [1978: 26] popular *Shell Book of Rural Britain.*
9. There is a variety of labels for these contrasting kinds of landscape. Rackham, in several popular works [1985, 1986], uses 'ancient' as opposed to 'planned'. Others, for example Williamson and Bellamy [1987], use 'woodland' as opposed to 'champion'.
10. See Williamson and Bellamy 1987 for a scholarly, but popularly oriented, presentation of the consequences for the modern landscape of the different medieval agricultural practices. The subject of medieval field systems is a complex and controversial one on which there is a large scholarly literature. See Dodgshon 1980 for an introduction to the issues and the literature.
11. See the maps in Rackham [1985: 69] and Williamson and Bellamy [1987: 16]. Not unexpectedly, on closer investigation, the geographical and temporal dimensions of various field systems in Norfolk produce a picture much too complex for such simple cartographic display; see Dymond's [1985] history of the Norfolk landscape.
12. See the discussion by Betjeman in Fawcett 1976.
13. Bovill 1962: 28–45; Newby 1987: 38–47; Williamson and Bellamy 1987: 114.
14. There are several guide books to Norfolk from the early nineteenth century. Although they sometimes mention the 'pleasant prospect' of a coastal village and offer praise of manor parks, there is very little reference to the country-side as an object of discourse. See Blomefeld 1805–9, Cooke 1822, Crowell 1818, Hood 1938, Robberds 1826, Woodward 1842.
15. This list is selected from listings by Lowenthal [1985: 104] and Lowe [1989: 114].
16. Quotations from United Kingdsom 1979; see also Shoard 1987: 97, 149; Williamson and Bellamy 1987: 209.
17. In rural areas of England close to London, such as the villages of Surrey, the preservation minded, 'country-life' new comers now dominate local politics and the local economy. See Connel 1978; Rogers 1989.
18. This case is discussed by Newby, Bell, Rose and Saunders [1978] and by Shoard [1987: 148–9].
19. Lowenthal 1991b. It is perhaps a comment on British popular

culture that *History Today* occupies, in Britain, a place similar to that of *Psychology Today* in America, the contrast being a preoccupation with the past rather than with one's psyche.

20. See note 12.
21. See especially Putnam's 1990 discussion 'Myths of the Countryside: Obstacles to Progress or Bastions of Defence?'

Bibliography

Blomefeld, F., *An essay toward a topographical history of the county of Norfolk*, 10 vols, London: W. Bulmer, 1805–9

Blythe, R., *Akenfield*, New York: Dell, 1973

Bovill, E.W., *English country life 1780–1830*, London: The Country Book Club, 1962

Brooks, P., *Cley: living with memories of greatness*, Norfolk: Poppyland, 1984

Cautley, H. M., *Norfolk churches*, Ipswich: Norman Adlard, 1949

Cobbett, W., *Rural rides* [1830], Harmondsworth: Penguin, 1967

Connell, J., *The end of tradition: country life in central Surrey*, London: Routledge and Kegan Paul, 1978

Cooke, G.A., *A topographical and statistical description of the county of Norfolk*, London: Sherwood, Neely, and Jones, 1822

Cozens-Hardy, B., 'The maritime trade of the Port of Blakeney, which included Cley and Wiveton – 1587 to 1590', *Norfolk Record Society* 8, 1936, pp.19–37

Crowell, T. K., *Excursions through Norfolk illustrated with engravings*, London: T. Davison, 1818

Dodgshon, R. A., *The origin of British field systems: an interpretation*, London, New York: Academic Press, 1980

Dymond, D., *The Norfolk landscape*, London: Hodder and Stoughton, 1985

Ellis, E.A. (ed.), *The Broads*, London: Collins, 1965

Fawcett, J., (ed.), *The future of the past*, London: Thames and Hudson, 1976

Hewison, R., *The heritage industry: Britain in a climate of decline*, London: Methuen, 1987

Hood, C.M. (ed.), *The chorography of Norfolk: an historicall and chorographicall description of Norfolk by John Norden* [Early 17th Century], Norwich: Jarrold and Sons, 1938

Hoskins, W.G., *Fieldwork in local history*, London: Faber and Faber, 1982
Lambert, J.M., J.N. Jennings, C.T. Smith, Charles Green and J.N. Hutchinson, *The making of the Broads: A Reconsideration of their Origin in the Light of New Evidence*, London: The Royal Geographic Society (Research Series no. 3)
Lowe, P., 'The rural idyll defended', in G. Mingay (ed.), *The rural idyll*, London: Routledge, 1989, pp.113–31
Lowenthal, D., 'Age and artifact: dilemmas of appreciation', in D. Meinig (ed.), *The interpretation of ordinary landscapes: geographical essays*, Oxford: Oxford University Press, 1979
——, *The past is a foreign country*, Cambridge: Cambridge University Press, 1985
——, 'British national identity and the English landscape', *Rural History* 2 (2), 1991a, pp.205–30
——, 'Heritage and the English landscape', *History Today* 41, September, pp.7–10, 1991b
Mellor, D., *The Glaven valley: historical jottings*, Cromer, NR: Cheverton and Son, 1989
Mossman, K., *The Shell book of rural Britain*, Newton Abbot: David and Charles, 1978
Newby, H., *Country life: a social history of rural England*, London: Weidenfeld and Nicholson, 1987
Newby, H., C. Bell, O. Rose and P. Saunders, *Property, paternalism and power: class and control in rural England*, London: Hutchinson, 1978
Parker, R., *Men of Dunwich: the story of a vanished town*, London: Collins, 1978
Perrott, D. (ed.), *The Ordnance Survey guide to the Broads and Fens*, London: Nicholson/Ordnance Survey, 1986
Puttnam, D., 'Myths of the countryside: obstacles to progress or bastions of defence?', *Royal Society of Arts Journal* 138, 1990, pp.625–36
Rackham, O., 'Ancient woodland and hedges in England', in S.R.J. Woodell (ed.), *The English landscape: past, present, and future*, Oxford: Oxford University Press, 1985
——, *The history of the countryside*, London: J.M. Dent and Sons, 1986
Robberds, J.W, *Geological and historical observations on the eastern vallies of Norfolk. Plan of the eastern vallies of Norfolk with the principal Roman stations and military roads in that*

district, Norwich: Bacon and Kinnebook, 1826

Rogers, A.W., 'People in the countryside', in G. Mingay (ed.), *The rural idyll*, London: Routledge, 1989

Shoard, M., *This land is our land: the struggle for Britain's countryside*, London: Paladin, 1987

United Kingdom, Royal Commission on the Distribution of Income and Wealth, *Report Number 7*, Command 7595, London: HMSO, 1979

Williamson, T., and L. Bellamy, *Property and landscape: a social history of land ownership and the English countryside*, London: George Philip, 1987

Woodward, S., *The Norfolk topographer's manual*, London: Nichols and Son, 1842

Wright, P., *On living in an old country: the national past in contemporary Britain*, London: Verso, 1985

Chapter 5

Hunting and Gathering as Ways of Perceiving the Environment

Tim Ingold

That nature is a cultural construction is an easy claim to make, and it is one that figures prominently in recent anthropological literature. It is not so easy, however, to ascertain what might be meant by it. One of my principal objectives in this chapter is to demonstrate that this claim is incoherent. To illustrate my argument I shall consider the anthropological treatment of those peoples classically regarded as operating within a natural economy, namely societies of hunters and gatherers. Comparing this treatment with the understandings that people who actually live by hunting and gathering have of themselves and their environments, I shall show that the latter systematically reject the ontological dualism of that tradition of thought and science which – as a kind of shorthand – we call 'Western', and of which the dichotomy between nature and culture is the prototypical instance. I propose that we take these hunter-gatherer understandings seriously, and this means that far from regarding them as diverse cultural constructions of reality, alternative to the Western one, we need to think again about our *own* ways of comprehending human action, perception and cognition, and indeed about our very understanding of the environment and of our relations and responsibilities towards it. Above all, we cannot rest content with the facile identification of the environment – or at least its non-human component – with 'nature'. For the world can only be 'nature' for a being that does not belong there, yet only through belonging can the world be constituted, in *relation* to a being, as its environment.

Nature, Culture and the Logic of Construction

Let me begin by outlining what I take to be a commonly adopted position within social and cultural anthropology. I admit that this has something of the character of a 'straw man', and I am indeed setting it up in order to knock it down. Nevertheless, it is one that has proved remarkably resilient, for reasons that I hope will become clear as we proceed.

Of all species of animals, the argument goes, humans are unique in that they occupy what Shweder [1990: 2] calls 'intentional worlds'. For the inhabitants of such a world, things do not exist merely 'in themselves', as indifferent objects, but only as they are given form or meaning within systems of mental representations. Thus to individuals who belong to different intentional worlds, the same objects in the same physical surroundings may mean quite different things. And when people act towards these objects, or with them in mind, their actions respond to the way they are already appropriated, categorized or valorized in terms of a particular, pre-existing design. That design, transmitted across the generations in the form of received conceptual schemata and manifested physically in the artificial products of their implementation, is what is commonly known as 'culture'. The environments of human beings, therefore, are culturally constituted. And when we refer to an environment – or more specifically to that part of it consisting of animate and inanimate things – as 'nature', then this too has to be understood as an artefact of cultural construction. 'Nature is to culture', writes Sahlins, 'as the constituted is to the constituting' [1976: 209]. Culture provides the building plan, nature is the building; but whence come the raw materials?

There must indeed be a physical world 'out there', beyond the multiple, intentional worlds of cultural subjects, otherwise there would be nothing to build with nor anyone, for that matter, to do the building. Minds cannot subsist without bodies to house them, and bodies cannot subsist unless continually engaged in material and energetic exchanges with components of the environment. Biological and ecological scientists routinely describe these exchanges as going on within a world of nature. It is apparently necessary, therefore, to distinguish between two kinds or versions of nature: 'really natural' nature (the object of study for natural scientists) and 'culturally perceived' nature (the object of study for social and cultural anthropologists). Such distinctions are indeed commonplace in anthropological literature: examples are Rappaport's between the 'operational' models of ecological science, purportedly describing nature as it really is, and the 'cognized' models of native people; and, perhaps most notoriously, the much used and abused distinction between 'etic' and 'emic' accounts [Ellen 1982, ch.9; Rappaport 1968: 237–41; cf. Ingold 1992a: 47–8].

In the formula 'nature is culturally constructed', nature thus appears on two sides: on one as the product of a constructional process, on the other as its precondition. Herein, however, lies a paradox. Many anthropologists are well aware that the basic contrast between physical substance and conceptual form, of which the dichotomy between nature and culture is one expression, is deeply embedded within the tradition of Western thought. In other words it is recognized that the concept of nature, in so far as it denotes an external world of matter and substance 'waiting to be given meaningful shape and content by the mind of man' [Sahlins 1976: 210], is part of that very intentional world within which is situated the project of Western science as the 'objective' study of natural phenomena [Shweder 1990: 24]. And yet the notion that there are intentional worlds, and that human realities are culturally constructed, rests on precisely the same ontological foundation. The paradox may be represented as follows:

- Culture
  - Culture
  - Nature (culturally perceived)
- Nature (really natural)

If the concept of nature is given within the intentional world of the Western scientist, the concept of culture must – by the same token – be given within the intentional world of the Western humanist. Each, indeed, presupposes the other. Not only, then, must the concept of nature be regarded as a cultural construct, but so also must that of culture. As MacCormack puts it: 'Neither the concept of nature nor that of culture is "given", and they cannot be free from the biases of the [European] culture in which the concepts were constructed' [1980: 6]. The fact that 'culture' appears twice in this statement at once alerts us to a basic contradiction. For the references, in the second part of the statement, to culture and to the logic of construction take as 'given' the very concepts that, in the first part of the statement, are said to be historically relative. Nor can the problem be contrived to disappear by trying to have it both ways, as Hastrup does when she suggests that instead of regarding nature as 'either a relative cultural category or an objective physical framework around culture', it might better be seen as 'both-and' [1989: 7]. For then culture too must be both-and, both an objective categorical constructor and a relative category constructed. To attempt to apply this logic is at once to be caught in the vortex of an infinite regress: if the opposed categories of 'nature' and 'culture' are themselves cultural constructs, then so must be the culture that constructs them, and the culture that constructs *that*, and so on *ad infinitum*. And since, at every stage in this regress, the reality of nature reappears as its representation, 'real' reality recedes as fast as it is approached.[1]

In what follows I shall argue that hunter-gatherers do *not*, as a rule, approach their environment as an external world of nature that has to be 'grasped' conceptually and appropriated symbolically within the terms of an imposed cultural design, as a precondition for effective action. They do not see themselves as mindful subjects having to contend with an alien world of physical objects; indeed, the separation of mind and nature has no place in their thought and practice. I should add that they are not peculiar in this regard: my purpose is certainly not to argue for some distinctive hunter-gatherer worldview or to suggest that they are somehow 'at one' with their environments in a way that other peoples are not. Nor am I concerned to set up a comparison between the 'intentional worlds' of hunter-gatherers and Western scientists or humanists. It is of course an illusion to suppose that such a comparison could be made on level terms, since the primacy of Western ontology, the 'givenness' of nature and culture, is implicit in the very premisses on which the comparative project is itself established (see Figure 5.1).

What I wish to suggest is that we reverse this order of primacy and follow the lead of hunter-gatherers in taking the human condition to be that of a being immersed from the start, like other creatures, in an active, practical and perceptual engagement with

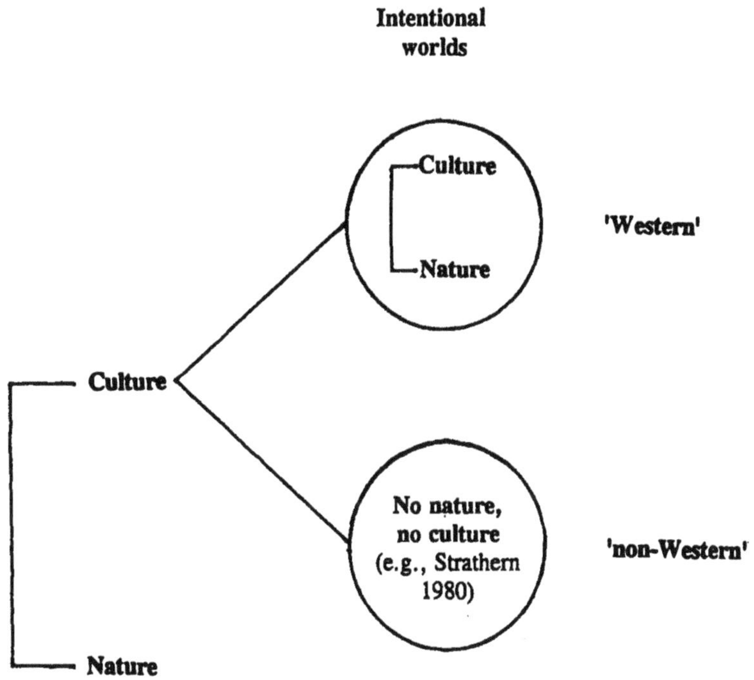

Figure 5.1. A comparison between 'non-Western' and 'Western' intentional worlds (assumes the primacy of the Western ontology, with its dichotomy between nature and culture, or between physical substance and conceptual form).

constituents of the dwelt-in world. This ontology of dwelling, I contend, provides us with a better way of coming to grips with the nature of human existence than the alternative, Western ontology whose point of departure is that of a mind detached from the world and which has literally to formulate it – to build an intentional world in consciousness – prior to any attempt at engagement. The contrast, I repeat, is not between alternative views of the world: it is rather between two ways of apprehending it, only one of which (the Western) may be characterized as the construction of a view, that is, as a process of mental representation. As for the other, apprehending the world is not a matter of construction but of engagement, not of building but of dwelling, not of making a view *of* the world but of taking up a view *in* it [Ingold 1991b].

In the following three sections, I shall move on to examine in more detail how this contrast has been played out in the context of Western anthropological studies of hunters and gatherers. First, I shall consider how certain tropical hunter-gatherer peoples perceive their relations to their forest environment. Secondly, I shall look at the way northern hunters, in particular the Cree of northeastern Canada, understand their relations to the animals they hunt. Thirdly, drawing on ethnographic material from Aboriginal Australia and subarctic Alaska, I shall consider the way hunters and gatherers perceive the landscape. I conclude by showing how anthropological attempts to depict the mode of practical engagement of hunter-gatherers with the world as a mode of cultural construction of it have had the effect, quite contrary to stated intentions, of perpetuating a naturalistic vision of the hunter-gatherer economy.

Children of the Forest

In his classic study of the Mbuti Pygmies of the Ituri Forest, Turnbull observes that the people recognize their dependence on the forest that surrounds them by referring to it as 'Father' or 'Mother'. They do so 'because, as they say, it gives them food, warmth, shelter and clothing, just like their parents', and moreover, 'like their parents, [it] gives them affection' [Turnbull 1965: 19]. This form of reference, and the analogy it establishes between the most intimate relations of human kinship and the equally intimate relations between human persons and the non-human environment, is by no means unique to the Mbuti.[2] Precisely similar observations have been made among other hunter-gatherers of the tropical forest, in widely separate regions of the world. For example, among the Batek Negritos of Malaysia, according to Endicott, the forest environment 'is not just the physical setting in which they live, but a world made for them in which they have a well-defined part to play. They see themselves as involved in an intimate relationship of interdependence with the plants, animals and **hala'** (including the deities) that inhabit their world' [1979: 82]. The **hala'** are the creator beings who brought the forest world into existence for the people, who protect and care for it, and provide its human dwellers with nourishment. And again, among the Nayaka, forest-dwelling hunter-gatherers of Tamil Nadu, South India, Bird-David found a similar attitude: 'Nayaka look on the forest as they do on a mother or father. For them, it is not something "out there" that responds mechanically or passively but like a parent; it provides food unconditionally to its children' [Bird-David 1990: 190]. Nayaka refer to both the spirits that inhabit the landscape and the spirits of their own predecessors by terms that translate as 'big father' and 'big mother', and to themselves in relation to these spirits as sons and daughters.

What are we to make of this? Drawing an explicit parallel between her own Nayaka material and the ethnography of the Batek and Mbuti, Bird-David has put forward the challenging argument that hunter-gatherer perceptions of the environment are typically oriented by the primary metaphor 'forest is as parent', or more generally by the notion that the environment *gives* the wherewithal of life to people – not in return for appropriate conduct, but unconditionally. Among neighbouring populations of cultivators, by contrast, the environment is likened to an ancestor rather than a parent, which yields its bounty only reciprocally, *in return* for favours rendered. It is this difference in orientation to the environment, she suggests, that most fundamentally distinguishes hunter-gatherers from cultivators, and it is upheld even when the former draw (as they often do) on cultivated resources and when the latter, conversely, draw on the 'wild' resources of the forest [Bird-David 1990]. In a subsequent extension of the argument, and drawing once again on Mbuti, Batek and Nayaka ethnography, Bird-David [1992a] proposes that hunter-gatherers liken the unconditional way in which the forest transacts with people to the similarly unconditional transactions that take place among the people of a community, which in anthropological accounts come under the rubric of 'sharing'. Thus, the environment shares its bounty with humans just as humans share with one another, thereby integrating both human and non-human components of the world into one, all-embracing 'cosmic economy of sharing'.

But when the hunter-gatherer addresses the forest as his or her parent, or speaks of accepting what it has to offer as one would from other people, on what grounds can we claim that the usage is metaphorical? This is evidently not an interpretation that the people would make themselves. Nevertheless – taking her cue from Lakoff and Johnson [1980] – Bird-David argues that these key metaphors enable them to make sense of their environment, and guide their actions within it, even though *'people may not be normally aware of them'* [1992a: 31, my emphasis; also 1990: 190]. One senses an inconsistency here. On the one hand, Bird-David is anxious to offer a culture-sensitive account of the hunter-gatherer economy as a counterpoint to the prevailing ecologism of most anthropological work in this field. On the other hand, she can do so only by imposing a division of her own, which forms no part of local conceptions, between actuality and metaphor. Underwriting this division is an assumed separation between two domains: the domain of human persons and social relations, wherein parenting and sharing are matters of everyday, commonsense reality; and the domain of the non-human environment, the forest with its plants and animals, relations with which are understood by drawing, for analogy, on those intrinsic to the first domain. In short, hunter-gatherers are supposed to call upon their experience of relations in the human world in order to model their relations with the non-human one.

The theoretical inspiration for this analytical tactic comes from Gudeman [1986], so let us now look at how he approaches the matter. Starting from the assumption that 'humans are modellers', Gudeman proposes that 'securing a livelihood, meaning the domain of material "production", "distribution" and "consumption", is culturally modelled in all societies' [1986: 37]. Entailed in the notion of modelling is a distinction between a 'schema' which provides a programme, plan or script, and an 'object' to which it is applied: thus 'the model is a projection from the domain of the schema to the domain of the object' [ibid.: 38]. Comparing Western and non-Western (or 'local') models of livelihood, Gudeman suggests that in the former, schemas taken from the 'domain of material objects' are typically applied to 'the domain of human life', whereas in the latter the direction of application is reversed, such that 'material processes are modelled as being intentional' [ibid.: 43–4]. But notice how the entire argument is predicated upon an initial ontological dualism between the intentional worlds of human subjects and the object world of material things, or in brief, between society and nature. It is only by virtue of holding these to be separate that the one can be said to furnish the model for the other. The implication, however, is that the claim of the people themselves to inhabit but one world, encompassing relations with both human and non-human components of the environment on a similar footing, is founded upon an illusion, one that stems from their inability to recognize where the reality ends and its schematic representation begins. It is left to the anthropological observer to draw the dividing line, on one side of which lies the social world of human modellers of nature, on the other the natural world modelled as human society.

In the specific case with which we are concerned, hunter-gatherers' material interactions with the forest environment are said to be modelled on the interpersonal relations of parenting and sharing: the former, assigned to the domain of nature, establish the object; the latter, assigned to the domain of society, provide the schema. But this means that actions and events that are *constitutive* of the social domain must be *representative* of the natural. When, for example, the child begs its mother for a morsel of food, that communicative gesture is itself a constitutive moment in the development of the mother-child relationship, and the same is true for the action of the mother in fulfilling the request. Parenting is not a construction that is projected *on to* acts of this kind, it rather subsists *in* them, in the nurture and affection bestowed by adults on their offspring. Likewise, the give and take of food beyond the narrow context of parent-child ties is constitutive of relations of sharing, relations that subsist in the mutuality and companionship of persons in intimate social groups [Ingold 1986: 116–17; cf. Price 1975]. Yet according to the logic of the argument outlined above, as soon as we turn to consider exchange with the *non-human* environment, the situation is quite otherwise. For far from subsisting in people's practical involvement with the forest and its fauna and flora in their activities of food-getting, parenting and sharing belong instead to a construction that is projected on to that involvement from a separate, social source. Hence, when the hunter-gatherer begs the forest to provide food, as one would a human parent, the gesture is not a moment in the unfolding of relations between humans and non-human agencies and entities in the environment, it is rather an act that says something *about* these relationships, a representative evaluation or commentary.[3]

In short, actions that in the sphere of human relations would be regarded as instances of practical *involvement* with the world come to be seen, in the sphere of relations with the non-human environment, as instances of its metaphorical *construction*. Yet those who would construct the world, who would be 'modellers' in Gudeman's sense, must already live in it, and life presupposes an engagement with components not only of the human but also of the non-human environment. People need the support and affection of one another, but they also need to eat. How, then, to stay with the same argument, do hunter-gatherers deal, actually rather than metaphorically, with non-human beings in the practical business of gaining a livelihood? They cannot do so in their capacity as persons, since non-human agencies and entities are supposed to have no business in the world of persons save as figures of the anthropomorphic imagination. Hence the domain of their actual interaction with the non-human environment in the procurement of subsistence must lie *outside* that of their existence as persons, in a separate domain wherein they figure as biological objects rather than cultural subjects, that is as organisms rather than persons. This is the *natural* domain of organism-environment interactions, as distinct from the *social* domain of interpersonal relations. In Figure 5.2 (upper diagram) this result is indicated schematically.

There is a profound irony here. Was not the capital objective to counteract that 'naturalization of the hunter-gatherer economy' which, as Sahlins comments [1976: 100], has formed the received anthropological wisdom, in favour of an account sensitive to the nuances of local culture? Yet what we find is that such naturalization is entailed in the very stance that treats the perception of the environment as a matter of reconstructing the data of experience within intentional worlds. The sphere of human engagement with the environment, in the practical activities of hunting and gathering, is disembedded from the sphere within which humans are constituted as social beings or persons, as a precondition for letting the latter stand to the former as schema to object. The consequences are all too apparent from the conclusion towards which Gudeman moves in bringing his argument to a close:

> In all living societies humans must maintain themselves by securing energy from the environment. Although this life-sustaining process amounts only to a rearranging of nature, a transforming of materials from one state or appearance to another, humans make something of this activity. [1986: 154]

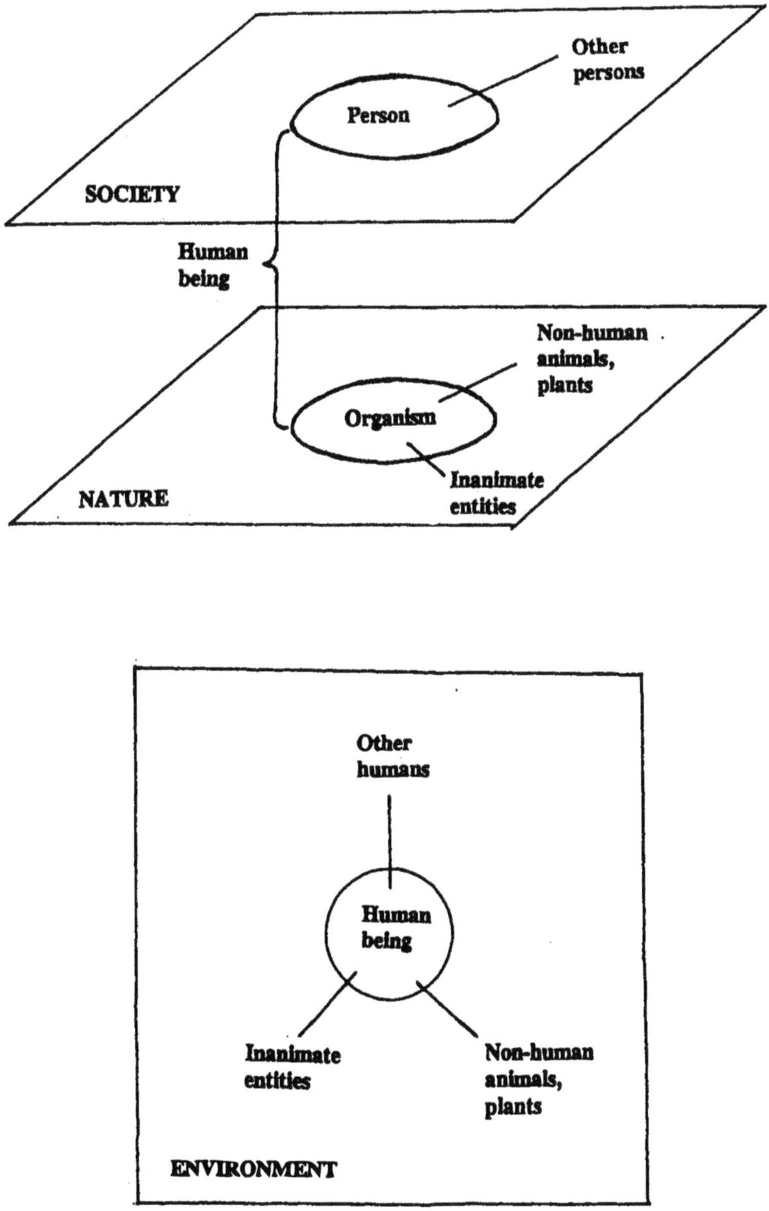

Figure 5.2. Western anthropological (above) and hunter-gatherer (below) economies of knowledge.

By his own account, then, the life-process of human beings, shorn of the diverse constructions that are placed upon it and that 'make something' of it, is nothing more than *a rearranging of nature.*

In this connection, we may recall Sahlins's attempt to treat 'economy' as a 'component of culture', which led him to contrast 'the material life process of society' to 'a need satisfying process of individual behaviour' [1972: 186 fn.1]. Hunting and gathering, in this account, are operations that take place in nature, consisting of interactions between human organisms with 'needs' and environmental resources. Only after having been extracted is the food transferred to the domain of society, wherein its distribution is governed by a schema for sharing, a schema inscribed in the social relations which the economic practices of sharing serve to reproduce [see Ingold 1988a: 275]. In the economy of knowledge, as conceived in general by Gudeman and specifically for hunter-gatherers by Bird-David, what applies to food applies also to sensory experience. That experience, gained through human organism-environment interactions, provides the raw material of sensation that – along with food – hunters and gatherers 'take home' with them. Carried over to the domain of interpersonal relations, it too is assimilated to a social schema, to yield a cultural construction of nature such as 'the forest is as parent'.

In Figure 5.2 this anthropological conception of the economy of knowledge is contrasted with that of the people themselves. In their account (lower diagram) there are not two worlds, of nature and society, but just one, saturated with personal powers and embracing humans, the animals and plants on which they depend, and the features of the landscape in which they live and move. Within this one world, humans figure not as composites of body and mind but as undivided beings, 'organism-persons', relating as such both to other humans and to non-human agencies and entities in their environment. Between these spheres of involvement there is no *absolute* separation, they are but contextually delimited segments of a single field. As Bird-David observes, hunter-gatherers 'do not inscribe into the nature of things a division between the natural agencies and themselves, as we [Westerners] do with our "nature:culture" dichotomy. They view their world as an integrated entity' [1992a: 29–30]. And so one gets to know the forest, and the plants and animals that dwell therein, in just the same way that one becomes familiar with other people, by spending time with them, investing in one's relations with them the same qualities of care, feeling and attention. This explains why hunters and gatherers consider time devoted to forays in the forest to be well spent, even if it yields little or nothing by way of useful return: there is, as Bird-David puts it, 'a concern with the activity itself' [1992a: 30], since it allows people to 'keep in touch' with the non-human environment. And because of this, people know the environment 'intimately, in the way one "knows" close relatives with whom one shares intimate day-to-day life' [Bird-David 1992b: 39].

That the perception of the social world is grounded in the direct, mutually attentive involvement of self and other in shared contexts of experience, *prior* to its representation in terms of received conceptual schemata, is now well established. But in Western anthropological and psychological discourse such involvement continues to be apprehended within the terms of the established dualisms of subject and object, persons and things. Rendered as 'intersubjectivity', it is taken to be the constitutive quality of the social domain *as against* the object world of nature, a domain open to human beings but not to non-human kinds [Willis 1990: 11–12]. Thus according to Trevarthen and Logotheti, 'human cultural intelligence is seen to be founded on a level of engagement of minds, or intersubjectivity, such as no other species has or can acquire' [1989: 167]. In the hunter-gatherer economy of knowledge, by contrast, it is as entire persons, not as disembodied minds, that human beings engage with one another and, moreover, with non-human beings as well. They do so as beings *in* a world, not as minds which, excluded from a given reality, find themselves in the common predicament of having to make sense of it. To coin a term, the constitutive quality of their world is not intersubjectivity but *interagentivity*. To speak of the forest as a parent is not, then, to model object relations in terms of primary intersubjectivity, but to recognize that at root, the constitutive quality of intimate relations with non-human and human components of the environment is one and the same.

Humans and Animals

According to Feit, the Waswanipi Cree of northeastern Canada 'say that they only catch an animal when the animal is given to them. They say that in winter the north wind, **chuetenshu**, and the animals themselves give them what they need to live' [1973: 116]. This idea, that the nourishing substance of animals is received by humans as a gift, is widely reported among northern hunting peoples, but in what follows I shall confine my remarks to studies of two other Cree groups. Among the Wemindji Cree, 'respectful activity towards the animals enhances the readiness with which they give themselves, or are given by God, to hunters' [Scott 1989: 204]. And for the Mistassini Cree, Tanner reports that the events and activities of the hunt, though they have an obvious 'common-sense' significance in so far as they entail the deployment of technical knowledge and skill in the service of providing for the material needs of the human population, are also 'reinterpreted' on another, magico-religious, level:

> The facts about particular animals are reinterpreted as if they had social relationships between themselves, and between them and anthropomorphized natural forces, and furthermore the animals are thought of as if they had personal relations with the hunters. The idealized form of these latter relations is often that the hunter pays respect to an animal; that is, he acknowledges the animal's superior position, and following this the animal 'gives itself' to the hunter, that is, allows itself to assume a position of equality, or even inferiority, with respect to the hunter. [Tanner 1979: 136]

In short, the animals figure for these northern hunters very much as the forest figures for such tropical hunter-gatherers as the Mbuti, Batek and Nayaka: they are partners with humans in an encompassing 'cosmic economy of sharing'.

Now Western thought, as is well known, drives an absolute division between the contrary conditions of humanity and animality, a division that is aligned with a series of others such as those between subjects and objects, persons and things, morality and physicality, reason and instinct, and above all, society and nature. Underwriting the Western view of the uniqueness of the human species is the fundamental axiom that *personhood as a state of being is not open to non-human animal kinds.* It is for this reason that we are able to conflate both the moral condition and the biological taxon (*Homo sapiens*) under the single rubric of 'humanity'. And for this reason too, we can countenance an enquiry into the animal nature of human beings while rejecting out of hand the possibility of an enquiry into the humanity of non-human animals [Ingold 1988b: 6]. Human existence is conceived to be conducted simultaneously on two levels, the social level of interpersonal, intersubjective relations and the natural ecological level of organism-environment interactions, whereas animal existence is wholly confined within the natural domain. Humans are both persons *and* organisms, animals are all organism.

This is a view, however, that Cree and other northern hunters categorically reject. Personhood, for them, *is* open equally to human and non-human animal (and even non-animal) kinds. Here, once again, is Feit on the Waswanipi:

> In the culturally constructed world of the Waswanipi the animals, the winds and many other phenomena are thought of as being 'like persons' in that they act intelligently and have wills and idiosyncrasies, and understand and are understood by men. Causality, therefore, is personal not mechanical or biological, and it is ... always appropriate to ask "who did it?" and "why?" rather than "how does that work?" [1973: 116]

This rendering of the Cree perspective is echoed by Tanner, who points to the significant implication of the idea that game animals live in social groups or communities akin to those of human beings, namely 'that social interaction between humans and animals is made possible' [1979: 137–8]. Hunting itself comes to be regarded not as a technical manipulation of the natural world but as a kind of interpersonal dialogue, integral to the total process of social life wherein both human and animal persons are constituted with their particular identities and purposes. Among the Wemindji Cree, qualities of personhood are likewise assigned to humans, animals, spirits and certain geophysical agents. As Scott writes: 'human persons are not set over and against a material context of inert nature, but rather are one species of person in a network of reciprocating persons' [1989: 195].

Though the ethnographic accounts offered by Tanner and Scott are in striking agreement, their interpretations are not, and it is revealing to explore the contrast between them. The problem hinges on the question of whether, when the Cree hunter refers to animals or to the wind as he would to human persons, he does so within the compass of what Feit, in the passage cited above, calls a 'culturally constructed world'. Tanner is in no doubt that they do. Thus he asserts that 'game animals participate simultaneously in two levels of reality, one "natural" and the other "cultural"' [1979: 137]. On the natural level they are encountered simply as material entities, organic constituents of the object world to be killed and consumed. On the cultural level, by contrast, they are 'reinterpreted' as anthropomorphic beings participating in a domain 'modelled on conventional Cree patterns of social and cultural organization' [ibid.]. In terms of this analysis, then, animals are constructed as persons through their assimilation to a schema drawn from the domain of human relations. This is entirely in accord with Gudeman's theory of the cultural modelling of livelihood, which I discussed in the previous section. Indeed, Gudeman draws for ethnographic support, *inter alia*, on Tanner's ethnography. 'The Mistassini Cree', he writes, 'construct their hunting and trapping activities as an exchange between themselves and animal spirits . . . and the exchange itself is patterned after ordinary human relationships, such as friendship, coercion and love' [Gudeman 1986: 148–9, citing Tanner 1979: 138, 148–50].

I have already shown, in the case of hunter-gatherer relations with the forest environment, how the constructionist argument is founded on an ontological dualism between society and nature, which in this instance reappears as one between humanity and animality. On one side, then, we have the world of human modellers of animals, and on the other the animal world modelled as human. If the people themselves profess to be aware of but one world, of persons and their relationships, it is because, seeing their own social ambience reflected in the mirror of nature, they cannot distinguish the reflection from reality. Now by all accounts, as we have seen, the dualism of humanity and animality, and the entailed restriction of personhood to human beings, is not endorsed by the Cree. This does not mean, of course, that they fail to differentiate between humans and animals. On the contrary, they are acutely concerned about such differences. For example, while humans may have sexual relations with certain other humans, and may kill and consume certain non-human animals, the consequences of categorical confusion – of sex with non-humans or killing fellow human beings – would be disastrous [Scott 1989: 197].

The point is that the difference between (say) a goose and a man is not between an organism and a person, but between one kind of organism-person and another. From the Cree perspective, personhood is not the manifest form of humanity; rather the human is one of many outward forms of personhood. And so when Cree hunters claim that a goose is in some sense like a man, or that the two are even consubstantial, far from drawing a figurative parallel across two fundamentally separate domains, they are rather pointing to the real unity that underwrites their differentiation. Whereas Western thought sets out from an assumed dichotomy between the human and the animal and then searches about for possible analogies or homologies, the Cree trajectory – as Scott explains – 'seems rather the opposite: to assume fundamental similarity while exploring the differences between humans and animals' [1989: 195]. To posit a 'metaphorical' equivalence between goose and man is not, then, to render 'one kind of thing in terms of another' [Lakoff and Johnson 1980: 5], as Western – including Western anthropological – convention would have it. A more promising perspective is offered by Jackson, who argues that metaphor should be apprehended as a way of drawing attention to real relational unities rather than of figuratively papering over dualities. Metaphor, Jackson writes, 'reveals, not the "thisness of a that" but rather that "this *is* that"' [1983: 132].[4]

It follows that the equivalence can work both ways. It is not 'anthropomorphic', as Tanner suggests [1979: 136], to compare the animal to the human, any more than it is 'naturalistic' to compare the human to the animal, since in both cases the comparison points to a level on which human and animal share a common existential status, namely as living beings or persons. The move, if you will, is not from the literal to the figurative, but from the actual to the potential – for personhood, at root, is the potential to become a human, a goose, or any other of the innumerable forms of animate being. From this perspective, it makes no significant difference whether one renders animal actions in human terms or human actions in animal terms. As Scott puts it:

> One might observe that a consequence of the sort of analogical thinking that I have been describing would be to anthropomorphize animals, but that would be to assume the primacy of the human term. The animal term reacts with perhaps equal force on the human term, so that animal behaviour can become a model for human relations. [1989: 198]

This same argument can be applied, *pari passu*, to the metaphor 'forest is as parent', considered in the last section. One could just as well say that 'parent is as forest', for the force of the metaphor is to reveal the underlying ontological equivalence of human and non-human components of the environment as agencies of nurturance.

What humans and non-humans have in common, for Cree as for other hunter-gatherers, is that they are *alive*. Ostensibly, and barring certain geophysical phenomena that Cree would regard as animate but that we might not, this is a conclusion with which Western thinkers would not disagree. Yet in Western biology, life tends to be understood as a passive process, as the reaction of organisms, bound by their separate natures, to the given conditions of their respective environments. This carries the implication that every organism is pre-specified, with regard to its essential nature, *prior* to its entry into the life process – an implication that in modern biology appears in the guise of the doctrine of genetic pre-formation. With this view, personal powers – of awareness, agency and intentionality – can form no part of the organism *as such*, but must necessarily be 'added on' as capacities not of body but of mind, capacities that Western thought has traditionally reserved for humans. Even today, now that the possibility of non-human animal awareness has arisen as a legitimate topic of scientific speculation, the basic dualism of mind and body is retained – for the question is phrased as one about the existence of animal *minds* [Griffin 1976, 1984; see Ingold 1988c]. Consciousness, then, is the life of the mind.

For the Cree, life has a different meaning. Scott tells us that 'the term *pimaatisiiwin*, "life", was translated by one Cree man as "continuous birth"' [1989: 195]. It is the creative unfolding of a total field of relations *within which* beings emerge and take on the particular forms that they do – humans as humans, geese as geese, and so on – each in relation to the others. Life is not the revelation of pre-specified forms but the process wherein forms are constituted. And every being, as it is caught up in this process and carries it forward, arises as an undivided centre of awareness and agency, an enfoldment, at some particular nexus within it, of the generative potential that is life itself. Thus personhood, far from being 'added on' to the living organism, is implicated in the very condition of being alive: the Cree word for 'persons', according to Scott, 'can itself be glossed as "he lives"' [1989: 195].

Organisms are not just *like* persons, they *are* persons. Likewise, consciousness is not supplementary to organic life but is, so to speak, its 'cutting edge' – 'on the verge of unfolding events, of continuous birth', as Scott [ibid.] renders the Cree conception.

Now, the ontological equivalence of humans and animals, as organism-persons and as fellow participants in a life process, carries a corollary of capital importance. It is that both can have points of view. In other words, for both the world exists as a meaningful place, constituted in relation to the purposes and capabilities of action of the being in question. Western ontology, as we have seen, denies this, asserting that meaning does not lie in the relational contexts of the perceiver's involvement in the world, but is rather laid over the world by the mind. Humans alone, it is said, are capable of representing an external reality in this way, organizing the data of experience according to their diverse cultural schemata. So when the Cree claim, as indeed they do, that the same events surrounding a hunt afford two possible interpretations, from the points of view respectively of the human hunter and of the animal hunted, the Western observer is inclined to regard the former as literal and the latter as figurative, 'as if' the animal were human and so could participate with 'real' humans in a common world of meanings. And this is precisely what Tanner does [1979: 136–7] when he re-presents to us – his readers – as a 'cultural' reality (as opposed to a 'natural' one) what the Cree originally presented to him as a 'bear reality' or 'caribou reality' (as opposed to a 'human' one). Note that the distinction between natural and cultural levels of participation is not one that the Cree themselves make. According to Scott, Cree has 'no word corresponding to our term "nature"', nor does it have any 'equivalent of "culture" that would make it a special province of humans' [1989: 195].

A creature can have a point of view because its action in the world is, at the same time, a process of *attending* to it. Different creatures have different points of view because, given their capabilities of action and perception, they attend to the world in different ways. Cree hunters, for example, notice things about the environment that geese do not, yet by the hunters' own admission [Scott 1989: 202], geese also notice things that humans do not. What is certain, however, is that humans figure in the perceptual world of geese just as geese figure in that of humans. It is clearly of vital importance to geese that they should be as attentive to the human presence as to the presence of any other potential predator. On the basis of past experience, they learn to pick up the relevant warning signs and continually adjust their behaviour accordingly. And human hunters, for their part, attend to the presence of geese *in the knowledge that geese are attending to them.* 'The perceptions and interpretations of Cree hunters', Scott observes, 'suggest that geese are quite apt at learning in what contexts to expect predation, at learning to distinguish predatory from non-predatory humans, and at communicating appropriate behavioural adaptations to other geese' [1989: 199].

In short, animals do not participate with humans *qua* persons only in a domain of virtual reality, as represented within culturally constructed, intentional worlds, superimposed upon the naturally given substratum of organism-environment interactions. They participate as real-world creatures endowed with powers of feeling and autonomous action, whose characteristic behaviours, temperaments and sensibilities one gets to know in the very course of one's everyday practical dealings with them. In this regard, dealing with non-human animals is not fundamentally different from dealing with fellow humans. Indeed the following definition of sociality, originally proposed by Schutz, could – with the insertions indicated in brackets – apply with equal force to the encounter between human hunters and their prey: 'Sociality is constituted by communicative acts in which the I [the hunter] turns to the others [animals], apprehending them as persons who turn to him, and both know of this fact' [Schutz 1970: 163]. Humans may of course be unique in their capacity to *narrate* such encounters, but no one can construct a narrative, any more than build a model, who is not already situated in the world and thus already caught up in a nexus of relations with both human and non-human constituents of the environment. The relations that Cree have with the latter are what we, as outside observers, call hunting.

Perceiving the Landscape

Life, of course, is an historical process, embodied in organic forms that are fragile and impermanent. Yet for terrestrial species, this process is carried on upon the surface of the earth, a surface whose contours, textures and features, sculpted by geological forces over immense periods of time, appear permanent and immutable relative to the life-cycles of even the most long-lived of organisms [Ingold 1989: 504]. This surface is what geology textbooks call the 'physical landscape'. How do hunters and gatherers perceive this aspect of their environment?

Among the Pintupi of the Gibson Desert of Western Australia, people say that the landscape was formed, once and for all time, through the activities of theriomorphic beings, ancestral to humans as well as to all other living things, who roamed the earth's surface in an era known conventionally as the Dreaming. The same idea is, in fact, current throughout Aboriginal Australia, but in what follows I shall confine my illustrative remarks to the Pintupi. According to Myers, Pintupi say that as ancestral beings travelled from place to place,

> [they] hunted, performed ceremonies, fought, and finally turned to stone or 'went into the ground', where they remain. The actions of these powerful beings – animal, human and monster – created the world as it now exists. They gave it outward form, identity (a name), and internal structure. The desert is crisscrossed with their lines of travel and, just as an animal's tracks leave a record of what has happened, the geography and special features of the land – hills, creeks, salt lakes, trees – are marks of the ancestors' activities. [1986: 49–50]

Such features are more than mere marks, however, for in their activities the ancestors did not leave a trail of impressions behind them, like footprints in the sand, while they themselves moved on. Rather, they metamorphosed *into* the forms of the landscape as they went along. Ever present in these forms, their movements are congealed in perpetuity.

On the land travelled by the ancestors in the Dreaming, people make their way in the temporal domain of ordinary life, pursuing their own everyday activities. Though the paths they take are not constrained to the lines of ancestral travel, in following tracks (as in hunting) and in making tracks themselves they replicate the original, creative movement of the ancestral beings, inscribing their own identities into the land as they go. As Wagner has put it, with reference to the neighbouring Walbiri people, 'the life of a person is the sum of his tracks, the total inscription of his movements, something that can be traced out along the ground' [Wagner 1986: 21]. And for the Pintupi, Myers writes that 'for each individual, the landscape becomes a history of significant social events ... previous events become attached to places and are recited as one moves across the country' [Myers 1986: 68]. There is thus a second level in the constitution of the landscape, one tied to the historical actions of ordinary human beings, as opposed to the 'transhistorical' actions of the ancestors [ibid.: 55]. On the first level, named places were created by the ancestral beings at the sites of their activities, or at points where they entered or emerged from the ground. Connected by the paths of ancestral travel, these places make up what Myers calls a 'country' – a term he offers as one possible rendering of the Pintupi word **ngurra**. But **ngurra** can also mean 'camp' – that is, the place temporarily constituted by virtue of the everyday activities of a group of people who happen to set up there. Such places, unlike the named places envisioned as the camps of the ancestors in the Dreaming, do not endure for ever. Each is identified with the particular people who live there, and will be avoided for many years after someone thus connected to the place has died. But 'despite these identifications ... camps are impermanent. Eventually they are overgrown and their associations forgotten, while significant new spaces are constantly being established' [ibid.: 56–7].

If persons inscribe their identities into the landscape as historically constituted, it is from the transhistorical level of the Dreaming that these identities are initially derived. Thus each person takes his or her primary identity from a particular named place, and is regarded as the incarnation of the ancestor whose activity made that place. That is why, as Myers notes [ibid.: 50], 'it is not unusual ... to hear people describe actions of the Dreaming in the first person'. For in speaking about my ancestor, I am speaking about myself. Throughout life, additional components of identity accrue through association with other named places, such as where one was initiated or where one has long resided, so that who one is becomes a kind of record of where one has come from and where one has been. It follows that the network of places, linked by paths of ancestral travel, is at the same time a network of relations between persons. When social relations are spoken of, as they often are, in terms of relations between places, the comparison does not draw a parallel across separate domains of society and the physical world but rather reveals that – at a more fundamental ontological level – these relations are equivalent. That level is the Dreaming. It is a level, however, that is not directly given to experience but rather revealed in the actions and events of the phenomenal world that are its visible signs [ibid.: 49].

We might sum up this Pintupi understanding of the landscape in the following four precepts. First, it is not a given substrate, awaiting the imprint of activities that may be conducted upon it, but is itself the congelation of past activity – on the phenomenal level, of human predecessors, but more fundamentally of ancestral beings. Secondly, it is not so much a continuous surface as a topologically ordered network of places, each marked by some physical feature, and the paths connecting them. Thirdly, the landscape furnishes its human inhabitants with all the lineaments of personal and social identity, providing each with a specific point of origin and a specific destiny. And therefore, fourthly, the movement of social life is itself a movement *in* (not *on*) a landscape, and its fixed reference points are physically marked localities or 'sites'. In short, the landscape is not an external background or platform for life, either as lived by the ancestors in the Dreaming or as relived by their ordinary human incarnations in the temporal domain. It is rather life's enduring monument.

What can we learn from the Pintupi? It could be argued, of course, that their ideas of the Dreaming – though not unique to themselves – are specifically Aboriginal ones affording no grounds for generalization beyond the Australian continent. Indeed, comparisons between Australia and other continents of hunter-gatherers are fraught with difficulty. Nevertheless, in order to indicate that there are genuine similarities in the ways that hunters and gatherers apprehend the landscape and their own position in it, I would like to refer briefly to another study from a quite different region of the world, Nelson's [1983] of the Koyukon of Alaska.

The Koyukon say that the earth and all the beings that flourish in it were created in an era known as the 'Distant Time'. Stories of the Distant Time include accounts of the formation of prominent features of the landscape such as hills and mountains [Nelson 1983: 16, 34]. An elaborate code of rules, brought down from the Distant Time, establishes forms of proper conduct that people are bound to follow; thus 'the Koyukon must move *with* the forces of their surroundings, not attempting to control, master or fundamentally alter them' [ibid.: 240]. As people move around in the landscape, in hunting and trapping, in setting up camp in one locality after another, their own life histories are woven into the country:

> The Koyukon homeland is filled with places ... invested with significance in personal or family history. Drawing back to view the landscape as a whole, we can see it completely interwoven with these meanings. Each living individual is bound into this pattern of land and people that extends throughout the terrain and far back across time. [ibid.: 243]

Places, however, can possess meaning at different levels. Some have a fundamental spiritual potency connected with the Distant Time story of their creation. Some, where people have died, are avoided for as long as the memory persists. Others, again, are known for particular hunting events or other personal experiences of encounters with animals. On all these levels – spiritual, historical, personal – the landscape is inscribed with the lives of all who have dwelt therein, from Distant Time human-animal ancestors to contemporary humans, and the landscape itself, rather than anything erected upon it, stands in memory of these persons and their activities [ibid.: 242–6].

Now let me turn to the anthropological interpretation of these ways of apprehending the landscape. Astonishingly, we find a complete inversion, such that meanings that the people claim to discover *in* the landscape are attributed to the minds of the people themselves and are said to be mapped *on to* the landscape. And the latter, drained of all significance as a prelude to its cultural construction, is reduced to *space*, a vacuum to the plenum of culture. Thus Myers can write, of the Pintupi, that they have 'truly culturalized space and made out of impersonal geography a home, a **ngurra**' [1986: 54]. A moment later, however, the Pintupi achievement reappears as an artefact of anthropological analysis: 'we will consider country *as if it were* simply culturalised space' [ibid.: 57, my emphasis]. The ontological foundation for this interpretative strategy is an initial separation between human persons as meaning-makers and the physical environment as raw material for construction; the 'culturalization of space' is then what happens when the two are brought into juxtaposition, such that social relations are mapped on to spatial relations. The Pintupi are said to superimpose the Dreaming, a 'distinctly Aboriginal cultural construction' [ibid.: 47], on to the 'real' reality of the physical landscape, causing the latter to recede from view, cloaked by the 'perceived' reality enshrined in the stories people tell of ancestral beings and their activities. This, of course, flatly contradicts Pintupi ontology, which is premissed on the fundamental *indissolubility* of the connection between persons and landscape, and on the assumption that phenomenal reality is open to direct perception whereas the order of the Dreaming is not, and can be apprehended only by way of its visible signs.

The same contradiction is apparent in Nelson's account of the Koyukon. His experience of the discrepancy between the Koyukon attitude to the environment and that derived from his own 'Euro-American' background led him, he tells us, to endorse the perspective of cultural relativism, whose basic premiss he sets out as follows:

> Reality is not the world as it is perceived directly by the senses; reality is the world as it is perceived by the *mind* through the medium of the senses. Thus reality in nature is not just what we see, but what we have *learned* to see. [ibid.: 239]

That we learn to see is not in doubt, but learning in this view entails the acquisition of cultural schemata for building *representations* of the world in the mind from data delivered by the senses. So the Koyukon, viewing the world in their mind's eye through the lens of received tradition, are supposed to see one reality; the Westerner, viewing it in terms of the concepts of scientific ecology, sees another. There is, Nelson concludes, no 'single reality in the natural world . . . absolute and universal'. Yet not only is the existence of such a 'real' reality implied in the very notion that perceived realities are representations, in the mind, of a naturally given world 'out there', but this mentalist ontology also flies in the face of what the Koyukon themselves, by Nelson's own account, are trying to tell us.

This is all about watching and being watched [ibid.: 14–32]. Knowledge of the world is gained by moving about in it, exploring it, attending to it, ever alert to the signs by which it is revealed. Learning to see, then, is a matter not of acquiring schemata for mentally *constructing* the environment but of acquiring the skills for direct perceptual *engagement* with its constituents, human and non-human, animate and inanimate. To adopt a felicitous phrase from the founder of ecological psychology, J.J. Gibson, learning is an 'education of attention' [1979: 254], a process not of enculturation but of enskillment. If the Koyukon hunter notices significant features of the landscape of which the Western observer remains unaware, it is not because their source lies in 'the Koyukon mind' [Nelson 1983: 242], which imposes its own unique construction on a common body of sensory data, but because the perceptual system of the hunter is attuned to picking up information, critical to the practical conduct of his hunting, to which the unskilled observer simply fails to attend. That information is not in the mind but in the world, and its significance lies in the relational context of the hunter's engagement with the constituents of that world. Moreover, the more skilled the hunter, the more knowledgeable he becomes, for with a finely honed perceptual system, the world will appear to him in greater richness and profundity. New knowledge comes from creative acts of discovery rather than imagining, from attending more closely to the environment rather than reassembling one's picture of it along new conceptual lines.

It will at once be objected that I have taken no account of that vital component of knowledge that comes to people through their instruction in traditional lore, for example in the stories of the Dreaming among the Pintupi and of the Distant Time among the Koyukon. Do not these stories, along with the accompanying songs, designs, sacred objects and the like, amount to a kind of modelling of reality, a representation of the world that native people might consult as Westerners would consult a map? I think not. People, once familiar with a country, have no need of maps, and get their bearings from attending to the landscape itself rather than from some inner representation of the same. Importantly, Myers notes that among the Pintupi the meanings of songs remain obscure to those who do not already know the country, and that individuals who are new to an area are first instructed by being 'taken around, shown some of the significant places, and taught to avoid certain sites' [Myers 1986: 150]. One might question what use songs, stories and designs could possibly have *as maps* if they are unintelligible to all but those who possess such familiarity with the landscape as to manage quite well without devices of this kind. I do not believe, however, that their purpose is a representational one. Telling a story is not like weaving a tapestry to *cover up* the world or, as in an overworn anthropological metaphor, to 'clothe it with meaning'. For the landscape, unclothed, is not the 'opaque surface of literalness'[5] that this analogy suggests. Rather, it has both transparency and depth: transparency, because one can see into it; depth, because the more one looks the further one sees. Far from dressing up a plain reality with layers of metaphor, or representing it, map-like, in the imagination, songs, stories and designs serve to conduct the attention of performers *into* the world, deeper and deeper, as one proceeds from outward appearances to an ever more intense poetic involvement. At its most intense, the boundaries between person and place, between the self and the landscape, dissolve altogether. It is at this point that, as the people say, they become their ancestors and discover the real meaning of things.

Conventional anthropological interpretation tends to range, on two sides of a dichotomy, peoples' practical-technical interaction with environmental resources in the context of subsistence activities, and their mytho-religious or cosmological construction of the environment in the context of ritual and ceremony. Hunters and gatherers are said to be distinctive, however, in so far as they do not seek physically to reconstruct the landscape to conform with their cosmological conceptions, but rather find these conceptions 'ready made' in the world as given. On these grounds they are supposed still to occupy a 'natural' rather than an 'artificial' or 'built' environment. Wilson sets out this view very clearly:

> The hunter/gatherer pins ideas and emotions on to the world as it exists: the landscape is turned into a mythical topographical map, a grid of ancestor tracks and sacred sites, as is typical among Australian aborigines . . .. A construction is put upon the landscape rather than the landscape undergoing a reconstruction, as is the case among sedentary peoples, who impose houses, villages, and gardens on the landscape, often in place of natural landmarks. Where nomads read or even find cosmological features in an already existing landscape, villagers tend to represent and model cosmic ideas in the structures they build. [1988: 50]

In rather similar vein, Rapoport writes that 'the organization of space cognitively precedes its material expression; settings and built environments are thought before they are built – as is shown by those cases where they are never built as for example among Australian Aborigines' [1994: 488].

Once again, we find that the view of the landscape as culturalized space entails the naturalization of hunting and gathering. Only as represented in thought is the environment drawn into the human world of persons, to become what Rapoport calls a 'setting'; thus the practical business of life is reduced to material interactions in an alien world of nature, in which humans figure as 'mere organisms'. Yet the people themselves insist that the real-world landscape in which they move about, set up camp and hunt and gather, is not alien at all but infused with human meaning – that this meaning has not been 'pinned on' but is there to be 'picked up' by those with eyes to see. They are, as their ethnographers have noted (with some surprise, else they would not have cared to remark on the fact), thoroughly 'at home' in the world. The Pintupi, Myers tells us, 'seem truly at home as they walk through the bush, full of confidence' [1986: 54]. And the lands of the Koyukon, according to Nelson, 'are no more a wilderness than are farmlands to a farmer or streets to a city dweller' [1983: 246]. As this statement implies, it is not because of his occupancy of a built environment that the urban dweller feels at home on the streets; it is because they are the streets of his neighbourhood along which he is accustomed to walk or drive in his everyday life, presenting to him familiar faces, sights and sounds. And it is no different, in principle, for the hunter-gatherer, as the inhabitant of an environment unscarred by human engineering. As I have remarked elsewhere, 'it is through *dwelling* in a landscape, through the incorporation of its features into a pattern of everyday activities, that it becomes home to hunters and gatherers' [Ingold 1991b: 16].

My argument is that the differences between the activities of hunting and gathering on the one hand, and singing, story-telling and the narration of myth on the other, cannot be accommodated within the terms of a dichotomy between the material and the mental, between ecological interactions *in* nature and cultural constructions *of* nature. On the contrary, both sets of activities are, in the first place, ways of dwelling. The latter, as I have shown, amount not to a metaphorical representation of the world, but a form of poetic involvement. But it is no different with the activities of hunting and gathering, which entail the same attentive engagement with the environment, and the same exploratory quest for knowledge. In hunting and gathering, as in singing and story-telling, the world 'opens out' to people. Hunter-gatherers, in their practices, do not seek to transform the world; they seek revelation. The intentions of non-human animals, for example, are revealed to Cree hunters in the outcomes of their endeavours. And Pintupi are forever alert to signs in the landscape that may offer new clues to ancestral activity in the Dreaming [Myers 1986: 67]. In short, through the practical activities of hunting and gathering, the environment – including the landscape with its fauna and flora – enters directly into the constitution of persons, not only as a source of nourishment but also as a source of knowledge.

But reciprocally, persons enter actively into the constitution of their environments. They do so, however, *from within*. For the Pintupi, the world was created in the Dreaming, but the Dreaming is *trans*historical, not *pre*historical. No generation is further removed from its ancestors than were its predecessors. The events of the Dreaming, though they occurred at particular places, are themselves timeless, each one stretched to encompass an eternity, or what Stanner [1965: 159] called 'everywhen'. And so the landscape, brought into being in these events, is movement out of time. People, as the temporal incarnation of ancestral beings, are not so much creators themselves as living on the *inside* of an eternal moment of creation. Their activities, which replicate on a much smaller scale the land-forming activities of the ancestors, are therefore part and parcel of the becoming of the world and are bound to follow the course set by the Dreaming: life, as the Pintupi say, is a 'one-possibility thing' [Myers 1986: 53]. Likewise, Koyukon are bound to the course of the Distant Time, and must move with it, never against it [Nelson 1983: 240].

In Western ontology, as we have seen, the landscape is not so much a course to be followed as a resistance to be overcome, a naturally given, material substrate that is 'humanized' by imposing upon it forms whose origins lie in a separate domain of the imagination, to yield an environment that is 'artificial' or 'built'. Hunter-gatherers are seen as people who have failed to achieve this, whose designs are not translated into action, settings thought but not built. Hunter-gatherer ontology, however, asserts the precise opposite. Action does not serve to translate pre-existent form from one domain (the mental) to another (the material); rather, form arises and is held in place *within* action: it is movement congealed. To express this point, we could do no better than borrow the words of the philosopher G.H. Mead. 'The object', he wrote, 'is a collapsed act' [1977: 97].

What Do Hunters and Gatherers Actually Do?

The received vocabulary of anthropological analysis offers but two alternative terms to denote in general the processes whereby human beings draw a subsistence from their environments: 'collection' and 'production'. The opposition between these terms goes back to Engels who, in a note penned in 1875, made it into the essential criterion of distinction between humans and other animals: 'The most that the animal can achieve is to *collect*; man *produces*, he prepares the means of life, in the widest sense of the words, which without him nature would not have produced' [1934: 308]. When, however, it came to giving concrete examples of production, Engels referred exclusively to the work of agriculture and pastoralism, instances in which the environment of plants, animals and the landscape itself had been demonstrably transformed through the imposition of human design [Ingold 1986: 71–2]. For Engels, production was synonymous with 'the transforming reaction of man on nature', by which humans have 'succeeded in impressing the stamp of their will upon the earth' [1934: 34, 179]. Yet human beings still living in the state of so-called 'savagery' had manifestly failed in this regard, having not even achieved that basic mastery over the forces of nature entailed in the domestication of plants and animals [Engels 1972: 39–41]. For Engels, as for many other leading scholars of his time – including Darwin and Morgan – savages figured as humans who had nevertheless not risen above a basically animal level of existence. Like animals, they were seen to be food-*collectors* rather than food-*producers*.

To this day, the human status of savages, now known more politely as hunters and gatherers, has remained equivocal, to say the least. Though no one would any longer deny them full membership of the human species, it is still commonly held that in deriving their subsistence from hunting and trapping 'wild' animals and gathering 'wild' plants, honey, shellfish and so on, they are somehow comparable in their mode of life to non-human animals (especially non-human primates) in a way that farmers, herdsmen and urban dwellers are not. Their life, it is often said, most closely resembles that of our hominid ancestors; and where human-animal comparisons are made, hunter-gatherers are almost invariably chosen as exemplars. Not so long ago, the American archaeologist Robert Braidwood saw fit to write that 'a man who spends his whole life following animals just to kill them to eat, or moving from one berry patch to another, is really living just like an animal himself' [1957: 22]. Though few today would put the matter so crudely, I believe the basic attitude betrayed in Braidwood's comment lives on in many anthropological circles, and that it is revealed above all in the abbreviation of hunting and gathering to 'foraging'.

I am not concerned here with the narrow sense of foraging in which it has sometimes been contrasted with collecting.[6] I mean rather to draw attention to the way in which 'foraging' has been adopted in a very general sense as a shorthand for 'hunting and gathering', ostensibly on the grounds of simple convenience. 'Forager', it is argued, is shorter than 'hunter-gatherer', and the term carries no unwarranted implications as regards the relative priority of animal and vegetable foods, or of male and female labour. But the concept of foraging also has an established usage in the field of ecology to denote the feeding behaviour of animals of all kinds, and it is by extension from this field that the anthropological use of the term is explicitly derived. Thus, introducing a volume of studies on 'hunter-gatherer foraging strategies', Winterhalder and Smith note that 'the subsistence patterns of human foragers are fairly analogous to those of other species and are thus more easily studied with ecological models' [1981: x]. And it is precisely the definition of human foragers as those who do *not* produce their food that legitimates the comparison: 'Foraging refers inclusively to tactics used to obtain nonproduced foodstuffs or other resources, those not directly cultivated or husbanded by the human population' [Winterhalder 1981: 16]. In short, it appears that humans can be only either foragers or producers; if the former, their subsistence practices are analogous to those of non-human animals; if the latter, they are not.

Barring the substitution of 'foraging' for 'collection' as an inclusive term for the activities of hunting and gathering, this formulation represents no advance on Engels's of over a century ago. The reason for this conceptual inertia is not hard to find, for it lies in that very separation of mind and nature, of the human world of ideas and the material world of objects, that is, as we have seen, the bedrock of Western thought as it has developed in both the natural sciences and the humanities. Quite simply, foraging describes an interaction *within* nature, production describes an imposition *upon* nature of ideal form. The producer is seen to intervene in natural processes, from a position at least partially outside them; the forager is supposed never to have extricated him- or herself from nature in the first place. To be sure, the argument runs, human foragers share with producers a capacity for culture that non-human foragers may lack – they do, therefore, have ideas and intentions. It is just that these ideas and intentions remain as models *of* the physical world rather than as blueprints *for* its actual transformation. Hunter-gatherers, it seems, construct the world in their heads but not on the ground; the world is 'made' by fitting it to a pre-existent design, not by physically changing it. As I myself put it in an argument that I would now disown, 'the environment of the hunter-gatherer *is not constructed but co-opted, it is not artificial but "natureficial"*' [Ingold 1986: 72–3, original emphases]. Or to recall Rapoport's comment, it is thought but never built.

I would now argue very differently, that the world as perceived by hunters and gatherers is constituted as such by virtue of their very mode of engagement with it, in the course of their everyday, subsistence-related practices. These practices cannot be reduced to their narrowly behavioural aspect, as strategically programmed responses to external environmental stimuli, as implied in the notion of foraging. Nor, however, can they be regarded as planned interventions in nature, launched from the separate platform of society, as implied in the notion of production. *Neither foraging nor production is an adequate description of what hunters and gatherers do.* As an alternative, Bird-David suggests 'procurement':

> Distinguished from 'to produce' and 'production', as also from 'to forage' and 'foraging', 'to procure' (according to the *Shorter Oxford*

> *Dictionary*) is 'to bring about, to obtain by care or effort, to prevail upon, to induce, to persuade a person to do something'. 'Procurement' is management, contrivance, acquisition, getting, gaining. Both terms are accurate enough for describing modern hunter-gatherers who apply care, sophistication and knowledge to their resource-getting activities. [1992b: 40]

This is a suggestion I would endorse. The notion of procurement nicely brings out what I have been most concerned to stress: that the activities we conventionally call hunting and gathering are forms of skilled, attentive 'coping' in the world, intentionally carried out by persons in an environment replete with other agentive powers of one kind and another. The point may be most readily summarized by referring back to Figure 5.2. In the upper diagram, representing the Western ontology, foraging would be positioned as an interaction in the plane of nature, between the human organism and its environment, whereas production would appear as an intervention in nature from the separate plane of society. In the lower diagram, representing the hunter-gatherer ontology, there is but one plane, in which humans engage, as whole organism-persons, with components of the environment, in the activities of procurement.

My argument has been that the 'naturalization' of the activities of hunting and gathering, as revealed in their apparently unproblematic designation as 'collecting' or 'foraging', is a product of the 'culturalization' of the perceived environment. In the case of hunter-gatherers of the tropical forest, we have seen how their perception of the forest environment, as being in some respects like a human parent, has been interpreted anthropologically as due to the application of a schema for metaphorically constructing it, and how, as a result, the forest itself and hunter-gatherers' interactions with it come to be excluded from the domain in which they relate to one another as persons. In the case of the northern hunters, we have likewise seen how the assumption that in their capacity as persons, humans can relate to animals only as the latter are represented within human intentional worlds, leads to the placement of real encounters of hunting beyond the bounds of these intentional worlds, in a separate domain designated as 'natural'. And finally, in examining Aboriginal perceptions of the landscape, we found that by treating the perceived world as culturalized space, the real-world landscape in which people live and move comes to be rendered as an indifferent and impersonal physical substrate, raw material for imaginative acts of world-making.

In short, a cultural constructionist approach to environmental perception, far from challenging the prevailing ecological models of hunting and gathering as foraging, actually reinforces them, creating by exclusion a separate logical space for organism-environment interactions wherein these models are appropriately applied. Those who oppose the designation of hunter-gatherers as foragers often do so on the grounds that it makes them seem just like non-human animals [e.g. Bird-David 1992b: 38], while not questioning the applicability of the foraging model to the animals themselves. However, by paying attention to what hunter-gatherers are telling us, this is just what we *should* be questioning, and in doing so laying down a challenge not only to cultural anthropology but to ecological science as well. We may admit that humans are, indeed, just like other animals; not, however, in so far as they exist as organisms rather than persons, as constituent entities in an objective world of nature presented as a spectacle to detached scientific observation, but by virtue of their mutual involvement, as undivided centres of action and awareness, within a continuous life process. In this process, the relations that human beings have with one another form just one part of the total field of relations embracing all living things [Ingold 1990: 220].

There can, then, be no radical break between social and ecological relations; rather, the former constitute a *subset* of the latter. What this suggests is the possibility of a new kind of ecological anthropology, one that would take as its starting point the active, perceptual engagement of human beings with the constituents of their world, for it is only from a position of such engagement that they can launch their imaginative speculations concerning what the world is like. The first step in the establishment of this ecological anthropology would be to recognize that the relations with which it deals, between human beings and their environments, are not confined to a domain of 'nature' separate from, and given independently of, the domain in which they lead their lives as persons. For hunter-gatherers as for the rest of us, life is given in engagement, not in disengagement, and in that very engagement, the real world at once ceases to be 'nature' and is revealed to us as an environment for people. Environments are constituted in life, not just in thought, and it is only because we live in an environment that we can think at all.

Acknowledgements

This paper has benefited greatly from discussions with Nurit Bird-David, Steven Gudeman, Mitsuo Ichikawa and James Weiner.

Notes

1. This argument is developed further in Ingold 1991a.
2. Subsequent ethnographic work among the Mbuti has, it should be noted, cast considerable doubt on the authenticity of Turnbull's somewhat 'romantic' account. Thus, Grinker [1992] fails to find indigenous conceptions that would correspond to the feeling for the forest that Turnbull imputes to the Mbuti. And Ichikawa [1992] observes that Mbuti attitudes towards the forest are, in reality, decidedly ambivalent: the forest is held to be the home of destructive as well as benevolent powers. But such ambivalence is equally characteristic of intimate relations in the human domain, which also have their undercurrent of negativity. However, by addressing the forest as 'Father', Ichikawa states, Mbuti 'are appealing to it for the benevolence normally expected from a parent' [1992: 41].
3. In responding to the criticisms of Abramson [1992] and myself [Ingold 1992b], Bird-David significantly softens this contrast. Following Gudeman [1992], she stresses the pragmatic – as against the cognitive – aspect of modelling, regarding it in the first place as a kind of activity or performance. Through performance, the model is actualized as lived experience. Considering the example 'a dog is a friend', she points out that by bestowing the affection due to a human familiar upon her dog – to which the dog evidently responds by showing every sign of affection for her – it actually *becomes* a friend, and is not merely 'thought of' as such [Bird-David 1992a:44]. To refer to the dog as her friend is thus to draw attention to an

underlying quality of relationship that can subsist just as well in gestures towards non-human as towards human familiars. This argument, though it comes close to agreement with that advanced in this chapter, by the same token departs significantly from the approach of Lakoff and Johnson [1980].

4. As Bird-David puts it, in connection with the friendliness of her dog (see note 3, above), the dog is not merely 'like' a friend, 'it *is* a friend' [1992a: 44].
5. I adopt this phrase from Ho [1991].
6. For example by Binford [1983: 339–46]; see Ingold [1986: 82–7] for a review.

Bibliography

Abramson, A., Comment on Nurit Bird-David, 'Beyond "The Original Affluent Society"', *Current Anthropology* 33(1), 1992, pp.34–5

Binford, L.R., *Working at archaeology*, London: Academic Press, 1983

Bird-David, N., 'The giving environment: another perspective on the economic system of gatherer-hunters', *Current Anthropology* 31, 1990, pp.189–96

——, 'Beyond "The Original Affluent Society": a culturalist reformulation', *Current Anthropology* 33, 1992a, pp.25–47

——, 'Beyond "the hunting and gathering mode of subsistence": culture-sensitive observations on the Nayaka and other modern hunter-gatherers', *Man* (N.S.) 27, 1992b, pp.19–44

Braidwood, R.J., *Prehistoric men* (3rd edition), Chicago: Chicago Natural History Museum Popular Series, Anthropology, 37, 1957

Ellen, R.F., *Environment, subsistence and system*, Cambridge: Cambridge University Press, 1982

Endicott, K., *Batek Negrito Religion*, Oxford: Clarendon Press, 1979

Engels, F., *Dialectics of nature*, Moscow: Progress, 1934

——, *The origin of the family, private property and the state*, New York: Pathfinder Press, 1972 [1884]

Feit, H., 'The ethno-ecology of the Waswanipi Cree: or how hunters can manage their resources', in B. Cox (ed.), *Cultural ecology: readings on the Canadian Indians and Eskimos*, Toronto:

McClelland and Stewart, 1973

Gibson, J.J., *The ecological approach to visual perception*, Boston: Houghton Mifflin, 1979

Griffin, D.R., *The question of animal awareness*, New York: Rockefeller University Press, 1976

——, *Animal thinking*, Cambridge, Mass.: Harvard University Press, 1984

Grinker, R.R., Comment on Nurit Bird-David, 'Beyond "The Original Affluent Society"', *Current Anthropology* 33(1), 1992, p. 39

Gudeman, S., *Economics as culture*, London: Routledge and Kegan Paul, 1986

——, Comment on Nurit Bird-David, 'Beyond "The Original Affluent Society"', *Current Anthropology* 33(1), 1992, pp.39–40

Hastrup, K., 'Nature as historical space', *Folk* 31, 1989, pp.5–20

Ho, M-W., 'The role of action in evolution: evolution by process and the ecological approach to perception', *Cultural Dynamics* 4(3), 1991, pp.336–54

Ichikawa, M., Comment on Nurit Bird-David, 'Beyond "The Original Affluent Society"', *Current Anthropology* 33(1), 1992, pp.40–1

Ingold, T., *The appropriation of nature*, Manchester: Manchester University Press, 1986

——, 'Notes on the foraging mode of production', in T. Ingold, D. Riches and J. Woodburn (eds), *Hunters and gatherers (1): History, evolution and social change*, Oxford: Berg, 1988a

——, 'Introduction', in T. Ingold (ed.), *What is an animal?*, London: Unwin Hyman, 1988b

——, 'The animal in the study of humanity', in T. Ingold (ed.), *What is an animal?*, London: Unwin Hyman, 1988c

——, 'The social and environmental relations of human beings and other animals', in V. Standen and R.A. Foley (eds), *Comparative socioecology*, Oxford: Blackwell Scientific, 1989

——, 'An anthropologist looks at biology', *Man* (N.S.) 25, 1990, pp.208–29

——, 'Becoming persons: consciousness and sociality in human evolution', *Cultural Dynamics* 4(3), 1991a, pp.355–76

——, 'Against the motion (1)', in *Human worlds are culturally constructed*, Manchester: Group for Debates in Anthropological Theory, 1991b

——, 'Culture and the perception of the environment', in E. Croll and D. Parkin (eds), *Bush base, forest farm: Culture, environment and development*, London: Routledge, 1992a

——, Comment on Nurit Bird-David, 'Beyond "The Original Affluent Society"', *Current Anthropology* 33(1), 1992b, pp.41–2

Jackson, M., 'Thinking through the body: an essay on understanding metaphor', *Social Analysis* 14, 1983, pp.127–48

Lakoff, G. and M. Johnson, *Metaphors we live by*, Chicago: University of Chicago Press, 1980

MacCormack, C., 'Nature, culture and gender: a critique', in C. MacCormack and M. Strathern (eds), *Nature, culture and gender*, Cambridge: Cambridge University Press, 1980

Mead, G.H., 'The process of mind in nature', in A. Strauss (ed.), *George Herbert Mead on Social Psychology*, Chicago: University of Chicago Press, 1977 [1938]

Myers, F.R., *Pintupi country, Pintupi self*, Washington, D.C.: Smithsonian Institution Press, 1986

Nelson, R.K., *Make prayers to the raven*, Chicago: University of Chicago Press, 1983

Price, J.A., Sharing: the integration of intimate economies, *Anthropologica* 17, 1975, pp.3–27

Rapoport, A., Spatial organization and the built environment, in T. Ingold (ed.), *Companion encyclopedia of anthropology: Humanity, culture and social life*, London: Routledge, 1994

Rappaport, R.A., *Pigs for the ancestors*, New Haven: Yale University Press, 1968

Sahlins, M.D., *Stone age economics*, London: Tavistock, 1972

——, *Culture and practical reason*, Chicago: University of Chicago Press, 1976

Schutz, A., *On phenomenology and social relations*, H. R. Wagner (ed.), Chicago: University of Chicago Press, 1970

Scott, C., 'Knowledge construction among Cree hunters: metaphors and literal understanding', *Journal de la Société des Americanistes* 75, 1989, pp.193–208

Shweder, R., 'Cultural psychology – what is it?', in J. W. Stigler, R. A. Shweder and G. Herdt (eds), *Cultural psychology: essays on comparative human development*, Cambridge: Cambridge University Press, 1990

Stanner, W.E.H., 'The dreaming', in W. A. Lessa and E. Z. Vogt (eds), *Reader in comparative religion*, New York: Harper and Row, 1965

Strathern, M., 'No nature, no culture: the Hagen case', in C.

MacCormack and M. Strathern (eds), *Nature, culture and gender*, Cambridge: Cambridge University Press, 1980

Tanner, A., *Bringing home animals*, London: Hurst, 1979

Trevarthen, C. and K. Logotheti, 'Child in society, society in children: the nature of basic trust', in S. Howell and R. Willis (eds), *Societies at peace*, London: Routledge, 1989

Turnbull, C.M., *Wayward servants*, London: Eyre and Spottiswoode, 1965

Wagner, R., *Symbols that stand for themselves*, Chicago: University of Chicago Press, 1986

Willis, R., 'Introduction', in R. Willis (ed.), *Signifying animals: human meaning in the natural world*, London: Unwin Hyman, 1990

Wilson, P.J., *The domestication of the human species*, New Haven: Yale University Press, 1988

Winterhalder, B., 'Optimal foraging strategies and hunter-gatherer research in anthropology: theory and models', in B. Winterhalder and E. A. Smith (eds), *Hunter-gatherer foraging strategies*, Chicago: University of Chicago Press, 1981

——, and E.A. Smith, 'Preface', in B. Winterhalder and E. A. Smith (eds), *Hunter-gatherer foraging strategies*, Chicago: University of Chicago Press, 1981

Chapter 6

The Invention of Nature

Peter D. Dwyer

you people try and dig little bit more deep
you bin digging only white soil
try and find the black soil inside

(Paddy Roe, an Australian[1])

Modern thought treats nature as separate from culture and has assigned ontological priority to the former. This is analogous to the separation of environment and organism that informs much of biology. The analogy is unsurprising. In a materialist world digital codes offer powerful analytical tools [Wilden 1972]. We are easily taught that nature is other than culture or that environment is other than organism; that cultures, like organisms, are emergent products. These understandings mesh comfortably with a tradition of thought that, for more than a century, has been underlain by an evolutionary perspective. I wish to revise, and to some extent up-end, this tradition. If we must adhere to a logic that is digital, then, I shall argue, in the domain of human affairs culture should be taken as prior, nature as emergent. The sad truth may be that the idea of 'wilderness' – that supposed last refuge of nature – is no more than an attempt to represent an imaginary place as a concrete symbol. 'Nature' as Westerners know it is an invention, an artefact.

These introductory words both contain and conceal a paradox. On the one hand 'nature' and 'culture' are opposed as discrete categories. On the other hand, they appear as moments on a continuum that may be either developmental or evolutionary. The paradox is well known: it concerns the tension between product and process, between entity and relationship [Wagner 1977: 386].

By and large, in anthropology, the paradox is ignored. The often elegant cross-cultural analyses of those who study symbols, seeking rules of transformation or generative grammars, remain, at base, static. By contrast, the rich empiricism of ecological studies, which often focus on specific populations or circumscribed questions, is driven by an evolutionary imperative; the terms of reference of evolutionary ecologists allow no alternative and, for the rest, the awkward, ill-defined or undefinable concept of 'adaptation as state of being' is ever-present [cf. Dwyer 1988].

Neither approach to the understanding of nature and culture is satisfactory. In each, ingrained categories of Western thought are assumed to underlie the mental processes, behaviour or environmental relations of other people. The assumption is not spelled out. In moving toward a different appreciation of nature and culture I will proceed as follows.

First, I shall briefly examine conventional understandings. My starting point here will be an analysis of the dance costume of Kubo people of the interior lowlands of Papua New Guinea. I shall then turn to the substance of my paper. Using information from three Papua New Guinean societies I shall depict for each not one, but two human geographies and ask if and how these interconnect. The two geographies are those of the visible and invisible worlds experienced by the people concerned. I shall then generalize to an alternate perspective on nature and culture. And finally, though briefly, I will turn to global issues. The Papua New Guinean material raises challenging questions concerning the relevance of ecological science to the study of people and the impact of conservation, as strategy and philosophy, on their lives.

The Kubo Dancer

Among Kubo, curing ceremonies, intercommunity prestations, initiations and the completion of new longhouses may be all marked by dancing. At these events one or two males, their bodies painted and elaborately costumed, dance to the slow, rhythmic beat of a drum that they themselves carry. They dance through the night until dawn and are accompanied by others, both male and female, who sing. The songs are nostalgic, of loss and longing, and the dancer, set apart from his audience, is withdrawn, his head hung, the tears painted beneath his eyes a concrete symbol of the collective mood of subdued sadness. Yet to those who watch, the dancer is beautiful, he is the Raggiana Bird of Paradise. The body painting, costume, drum beat and movement combine as a transcendent whole that reaches beyond the mundane world to communicate with spirits.

There are many possible interpretations of the dancer's costume (Figure 6.1). I shall refer to three.[2]

Figure 6.1. The Kubo Dancer

From the palm skirt, trailing on the floor of the longhouse, to the hornbill feathers that bounce at the top of the head-dress there is an orderly shift from aquatic to aerial domains. The skirt is a waterfall with cleansing properties, the drum is crocodilian and, at the rear, the dancer wears a rattle of crayfish claws. The bark belt and cape, the arm ornaments of palm fronds and fibre, and the soot and red and yellow ochres used to paint the torso are all elements and colours of the earth. The shell ornaments around the neck and on the chin are acquired by trade. Their ultimate places of origin are not known to Kubo. They denote relations that connect all people who move upon the earth. The band of cuscus fur across the forehead is from a cryptic, tree-dwelling species. People say that the animal looks like leaves against a background of sky. Next are plumes of the Raggiana Bird of Paradise that displays in the tree tops. Above these – and, I note, above the dancer's body – are feathers from three species that are usually seen above the forest. The apical hornbill feathers are enigmatic; they may have come from the spirit of a dead person. This reading of the costume is simple. Both natural elements and natural order are transformed as culture.

Another theme is evident, though, in these enlightened times, it is, perhaps, more risky. The dancer's skirt, front and rear, is made from the fronds of sago palms, palms that epitomize women's contribution to subsistence. The frontal skirt *is* that of a woman and the designs painted on the torso may represent female genitalia.[3] But, moving vertically to the ornaments of shell, bone, fur and feathers we encounter objects from the domains of trade and hunting, domains that are largely the prerogative of men. Even the decorative clusters worn on the upper arms are made from the fronds of black palm, the source of men's bows, and not from those of sago palms. We seem to have a secondary transformation, paralleling the first, even reinforcing it, from 'natural' female to 'cultural' male.

These interpretations are my own. By training I am a biologist and so it is not surprising that they are rooted in such basic concerns of biological science – sex, diversity of species and the proper ordering of niches. Nor is it surprising that I should perceive a flow from concrete source to abstract meaning, from literal to figurative. Anthropologists, however, are predisposed to grasp the immediacy of the figurative. Thus, Knauft [1985a] offers a different reading of the similar costume of Gebusi people.

From the outset each costume element is symbolic of the form in which a category of spirit people is manifest in the natural world. Knauft [ibid.: 257–60] writes:

> The items worn on the upper half of the body derive from the upper spirit world (tree tops and sky), while those worn on the lower half are from the lower spirit world (in the water and beneath the ground) . . .. The dance costume is a collage of all the true spirit beings in their natural-world forms. The Gebusi dance is in fact an enactment of a dance these spirits are simultaneously attending in their own world.

The transformation we see here is of elements retrieved from the natural world being employed as concrete symbols of spirit beings and unified 'into the living motion of spiritual harmony' [ibid.: 261].

These sorts of themes are a commonplace in anthropological writing. The domain of the familiar, here taken as 'nature', is the source of metaphors (or tropes) that inform understanding of the less familiar [Dwyer 1979]. The human body is a source of countless 'natural symbols'. Through metaphor, the contrastive pairs 'self and other', 'culture and nature', are linked, and meaning arises. The transformation always flows from the perceived raw materials of nature to the mysterious unity of culture. The process is one of cultural genesis.

We are drawn to the generalizations of Mary Douglas [1970], Claude Lévi-Strauss [1966] and Victor Turner [1974]. Yet I remain dissatisfied. Too often, I think, the artistic symmetries we perceive deflect us from important and challenging questions. We conceal the ambiguities and inconsistencies of our own interpretations. Thus, Knauft identified elements of the Gebusi dance costume as symbolic of the natural-world forms of spirits. But is it true that *all* spirit forms are represented? If they are not, then there is less unity than Knauft asserted. Or again, the representation of an underground spirit-being in the upper part of the costume disrupts the appealing logic of upper and lower worlds that Knauft proposed. My own interpretations fare no better. The transitions from aquatic to aerial, and from female to male, are less tidy than I suggested and the case for 'natural symbols' is seriously marred by the incorporation of trade items into the costume.[4]

I am not comforted by the thought that symbols allow multiple meanings or, as Fredrik Barth [1987] has asserted, that there is no striving for coherence. Such assertions may rationalize intractable data; they do not challenge underlying theoretical structures. The analyses are too simple. Nature, we argue, is transformed. But where precisely does the moment of transformation reside? Is there perhaps a zone of transformation, a 'culture-tone' (cf. ecotone) that is neither natural nor cultural? Or might we borrow from Ingold [1986, 1988] and ask whether the act of retrieval from nature – the intentionality that underlies that appropriative act – serves to mark the transition?

Ingold's concern was with the boundary between the social and the ecological. He proposed that the boundary marked 'the point . . . where purpose takes over from, and proceeds to direct, the mechanism of nature' [1988: 285]. I want to go further and dissolve the boundary. I have wondered whether, for Kubo, the component elements of the dance costume may never have been natural. The transformation we, as outsiders, witness may be one in which a collage of what were always cultural elements assumes a new and different status. The transformation may be of meanings, a transformation wherein cultural elements are related analogically to make communication possible with the world of spirits, a transformation wherein elements of the visible world, properly assembled, mediate the separation between Kubo people and *their* perceived 'other'.

A new approach is needed. Anthropological perceptions of nature have grown from a deep appreciation of and concern for the social. What might we learn of culture if our starting place is ecology or, as I prefer here, geography? It is this that I shall explore with reference to Papua New Guinean populations. But some definitions are needed. They must be sufficient but not inflexible. 'Commentaries' might be a better word.

Commentaries

The focal referent of 'geography' is the landscape itself. My concern is with the ways people live within that landscape, with the impress of them upon it and of it upon them. This, of course, is part of the intellectual endeavour called ecology but it is not the only, nor even the primary, concern of modern ecological studies. Indeed, by using the word 'geography' I seek to avoid the methodological requirement of ecologists that 'organism-environment' relations are uppermost. That dichotomy is not central to the analyses of this paper. Space and time will be integral to my geographic readings, but we should not forget the importance of human knowledge that emerges from and feeds into each landscape. Human geography cannot be held apart from human cognition.

I must distinguish also between the worlds that I call visible and invisible. In the first instance the visible world is material, the invisible world is immaterial. The former comprises plants, animals and people and the physical media within which or upon which these reside. The latter comprises mythological, spiritual and fabulous beings that are known to the people as, for example, creators, witches, ancestors, omnipotent forces or the disembodied essences of corporeal beings. In their 'purest' forms these beings and forces are usually unseen. Indeed it may be the case that to see them is to die; that the event may be inferred by the living but never directly reported.

But spiritual beings may be often manifest in the material world. They mark their passing by signs and they assume the forms, or inhabit the bodies, of particular physical entities such as individual rocks, plants, animals or persons. These manifestations must be classed within the visible world. They provide one channel for communication between the two worlds. There are other channels as well. In sleep or when dreaming the spiritual essence of a person may cross to the other world, in trance the bodies of mediums may speak with the voices of the spirits of the dead, and through ritual all people may participate in acts that, at least temporarily, make connections between visible and invisible worlds.

Papua New Guinea

I have lived with and worked among three Papua New Guinean populations (Figure 6.2). In 1972 I spent ten months at a village on the slopes of Mount Erimbari (2850 m) in Simbu Province. Leu village was at an altitude of 1975 m; the 100 people who lived

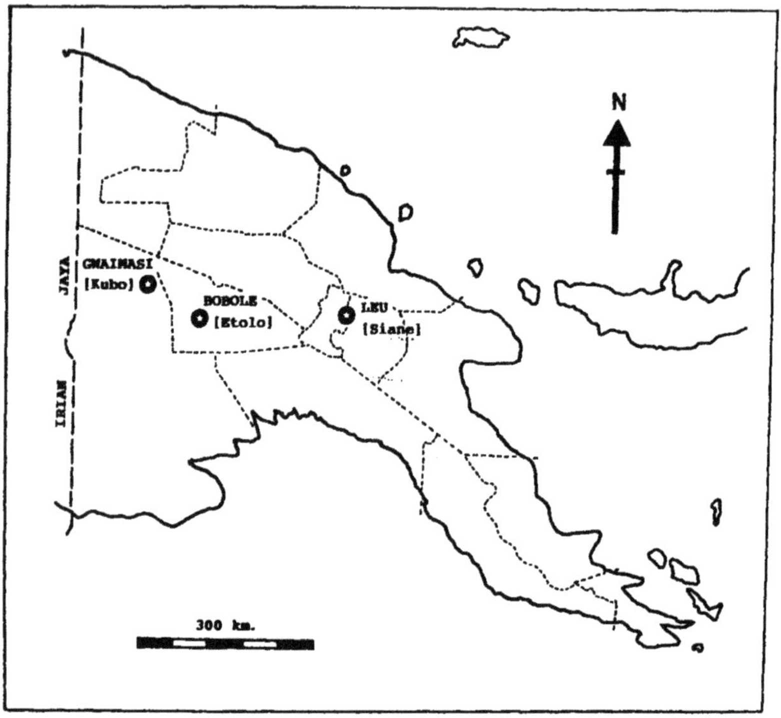

Figure 6.2. Papua New Guinea, showing locations of the three communities discussed

there were of Rofaifo clan and spoke the Siane language. Seven years later, in 1979–80, I lived for fifteen months with 109 Etolo speakers at the village of Bobole in the Southern Highlands Province. This village was at an altitude of 1100 m on the steep slopes of Mount Haliago (2689 m). More recently, in 1986–7, I spent fifteen months with twenty-five Kubo speakers at the small village of Gwaimasi, 80 m above sea level, in the interior of the Western Province. Kubo and Etolo are members of the same language family and cultural complex [Knauft 1985b; Shaw 1986].

In moving west along latitude 6°S, and from higher to lower altitudes, my research emphases shifted from the biological to the anthropological. And progressively, despite the lapse of fourteen years, my encounters were with people whose lives had been less affected by contact with missionaries or colonial and national governments.[5]

Subsistence modes in both traditional form and as observed by me were very different between these populations. The trend from east to west is of decreasing intensification: population densities decline, and reliance on domesticated resources is reduced as the importance of 'non-domesticated' resources increases. In parallel, the east-west axis or, better perhaps, the shift from higher to lower altitudes corresponds with reduced emphasis upon resources whose returns are delayed and with the expected concomitants of this: mobility increases, community composition is more fluid, disputes are more often resolved by departure than by negotiation, and both sharing and egalitarianism are more pronounced.

These differences between the three populations are etched into the local landscapes. They may be depicted most simply, though I hasten to add superficially, by viewing topographical maps published during the 1970s. The figures that follow have been extracted from those maps. My focus will be with the actual places at which I lived but I will reverse my own historical trajectory by taking Kubo first, then Etolo and finally Siane. The shift will be from less intensive to more intensive systems of production.

Kubo

The Kubo landscape is sparsely populated and particular assemblies of people often divide and relocate (Figure 6.3). At Gwaimasi, which did not exist when the topographical map was published, twenty-five people use 50 km². But communities are from three to six hours apart and crude population density is much less than 0.5 people per km². Only 1 per cent of the land comprises current gardens or second growth forest and, in most directions, one would travel several days before this value increased. The tiny regrowth patches seen on the map are near primary waterways and it is the latter that orient people within the landscape. Altitude is a minor consideration.

Aerial photography is not omniscient. When the land near Gwaimasi is viewed from the ground the apparent importance of gardens is diminished (Figure 6.4). An anastomosis of trails spreads across the land to connect village, gardens, sago-processing sites, hunting shelters and so forth. Other trails lead

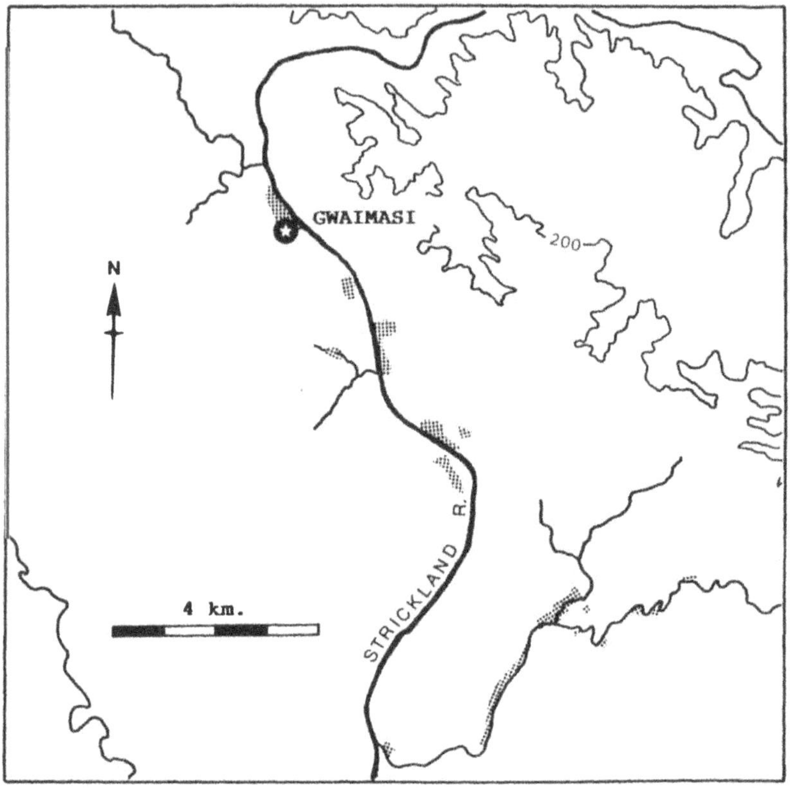

Figure 6.3. Map of Gwaimasi region (Kubo). Shading shows anthropogenic habitats recorded by PNG 1:100,000 Topographic Survey, Sheet 7385 (edition 1) Series T601, Printed 1979. The 200 m contour is marked.

to favoured fishing spots, tiny groves of fruit pandanus or to nut trees. Everywhere there are other signs of people: clumps of bamboo and okari and breadfruit trees indicate earlier gardens; the remains of bird hides, rotting logs or sago palms may be places where animals were captured; and some areas of inhibited regrowth show where canoes were made. With guidance other dimensions of understanding emerge. A fifteen-minute walk through the forest provides a list of a hundred species of plants and a companion list of their uses. At one place the collapsed frame of a shelter reminds people of a man's illicit elopement and, at another, a tuft of feathers prompts the story of a

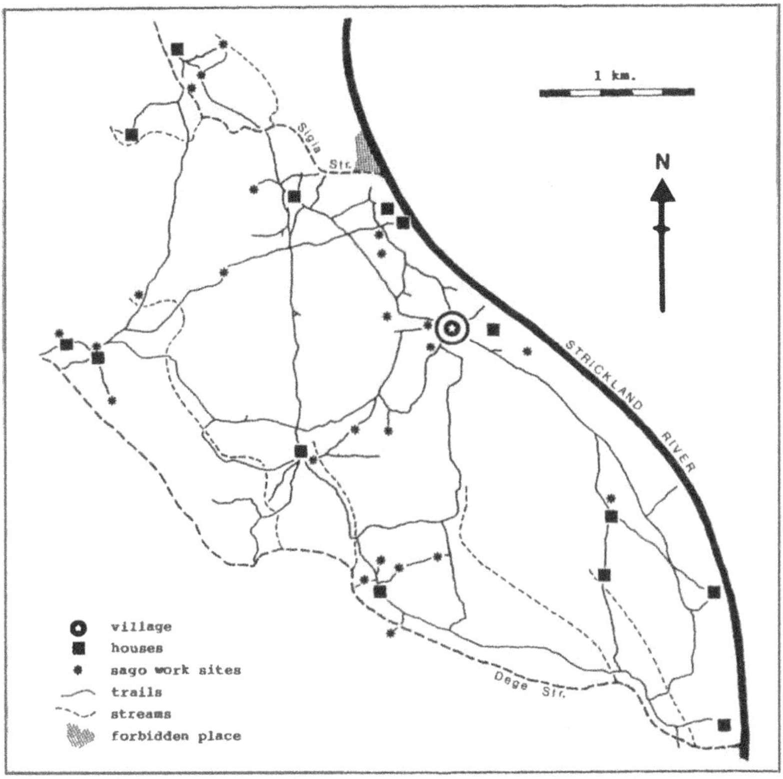

Figure 6.4. Walking trails associated with Gwaimasi village: 1986–7 (Only the area bounded by the Strickland River and the Sigia and Dege streams was comprehensively surveyed)

cassowary hunt. An ancient post hole recalls a longhouse where people once lived, the initiations held there or cannibal feasts, the sorcery that occurred there and the departures that followed. To Kubo the land is a web of past and present human action and interaction and it is of these events that men, women and children speak when they walk together in the forest.

Village life is not the preferred style of these people. They range widely through the day and, on 25 per cent of nights, sleep at bush or garden houses away from the village. The common theme of their activities is food production but the pace is relaxed, the demands on time and effort seldom great.

Different sorts of procurement activities are environment or season-specific.[6] Gardens are made on levee banks, sago processed in the back-swamps and pigs preferentially hunted in the foothills. But these activities are widely dispersed in space and evenly distributed in time. The location of favoured fruit and nut trees shifts from year to year and the dispersion of productive gardens changes continually. Women range as far as men; they visit the same environmental zones though they cannot visit all the same places. There are no sustained gradients of either aggregate land-use or gender-based association with land. Because different domestic units select different locations for the production of food and are often engaged in different activities, potential gradients are diffused.

If the depiction of land-use was founded in a classification of activity types, then the pattern would be a far-ranging and ever-changing mosaic. The dispersal in space and time would, I think, greatly exceed that reported by Ohtsuka [1983] for Oriomo Papuans. There is a single notable exception to this description. Contained within the landscape is a forbidden place, the **toi sa**. Except in special ritual circumstances, people do not venture there. Yet even this is a cultural artefact: it is the abode of the spirits of the dead. The **toi sa** serves to both introduce the invisible world and to characterize my description of it.[7] The resting place of the spirits of the dead is contained *within* Kubo territory. Indeed, the customary land of each clan contains its own **toi sa**.

The invisible world permeates the land. Fabulous beings are associated with specific environmental zones or even particular places; there is a giant hunter in the back-swamps, a huge eel in the Strickland River and a python without a tail that guards the **toi sa**. The spiritual essences of animals are unconfined by the habitats of their mundane forms and, in the form of animals, the spirits of the dead may be seen in both expected and anomalous places. These last, in spirit form, interact freely with the spiritual essences of people, sometimes assisting, sometimes harming, corporeal beings. Their journeys and encounters occur within the landscape of human action, for there is no other place. Only the threatening essences of some foreigners suggest a perception of geographic gradients. From the swamp lands to the south they come as spirit warriors, marking their presence and intentions with signs. From the mountains to the north they come on the wind, or follow large waterways, to infiltrate the air and insinuate their way through narrow gaps in the walls of houses. They too, however, travel wherever Kubo live.

The geography of the Kubo landscape is not characterized by strong gradients of use value or spiritual association. The visible and invisible worlds are co-extensive. In each, the significance of particular places is pre-eminent but always transient. Through time, places of current significance drift across the land. The two worlds converge in a mutual dynamic that facilitates their intercommunication. There is no sense in which the invisible has been positioned toward the periphery of the visible, and yet this is what we shall find elsewhere in Papua New Guinea.

Etolo

Etolo live to the east of Kubo, in mountainous terrain (Figure 6.5). For them 5 per cent of the land is overtly shaped by people, though, to the northeast and west-south-west, it is only a day's walk to landscapes that are more intensively managed. Etolo communities of thirty to a hundred people are one or two hours apart and always separated by forest. Crude population density is two people per km^2 though effective density in areas of primary usage may be 7.5 people per km^2. The locations of communities are constrained by altitude and peculiarities of drainage. Here, therefore, the orientation of people is toward both major streams – the torrents that fan out from Mount Haliago – and the precipitous ridges that gather at its summit.

Etolo food production is more intensive than that of Kubo.[8] This is evidenced by the facts that Etolo cultivate one and a half times as much garden land per head as Kubo, emphasise tubers rather than bananas, manage orchards containing hundreds of fruit pandanus, and operate long lines of mammal traps in positions that are fixed for many years. Their gardens are usually fenced, and abandoned plots are often maintained as foraging enclosures for pigs. Again, in contrast with Kubo, most sago palms have been planted by people. Forest products, both plant and animal, are less important than for Kubo though this statement needs qualification. The nutritional importance of wild animals to Etolo cannot be underrated; it is the quantitative representation of non-domesticated foods and the variety of species appropriated that are low relative to Kubo. In sum the

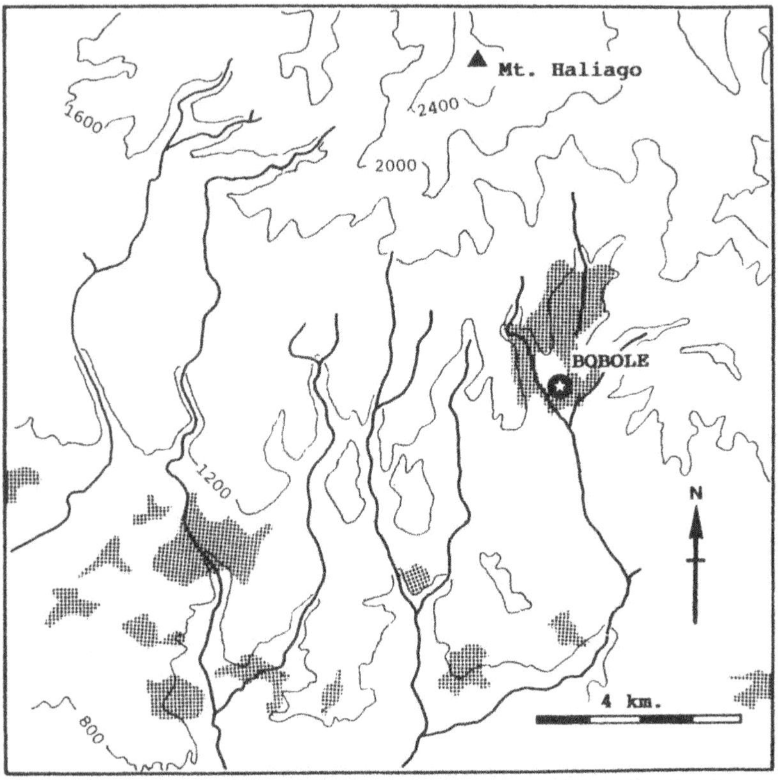

Figure 6.5. Map of Bobole region (Etolo). Shading shows anthropogenic habitats recorded by PNG 1:100,000 Topographic Survey, Sheet 7485 (Edition 1) Series T683, Printed 1974. Contour intervals are 400m apart.

landscape occupied by Etolo is more overtly managed, and the managed domain more permanent, than is the case among Kubo.

Etolo food-producing activities display strong gradients in both space and time (Figure 6.6). Hunting, trapping, gardening and processing of sago are the dominant activities. Each occurs within a restricted zone, with most hunting in primary forest above 1200 m altitude and most sago-processing below 900 m altitude. In addition, there are shifts through time in the sorts of activities that dominate the productive process. Hunting gives way to the repair of traps, new gardens are made and sago work follows. There is a second wave of gardening and the cycle then repeats. Elsewhere I have argued that this sequence is not preordained by seasonal constraints but, instead, is an outcome of choices made by people. I wrote that 'key subsistence tasks were organised such that people did similar things at similar times' [Dwyer 1990: 157]: by and large, they did them in similar places. The same could never be said of Kubo.

The timing, sequence and synchrony of Etolo subsistence activities, and the specificity of the zones where these take place, underline my depiction of food production along the axes of altitude and time. The significance of these axes is reinforced by

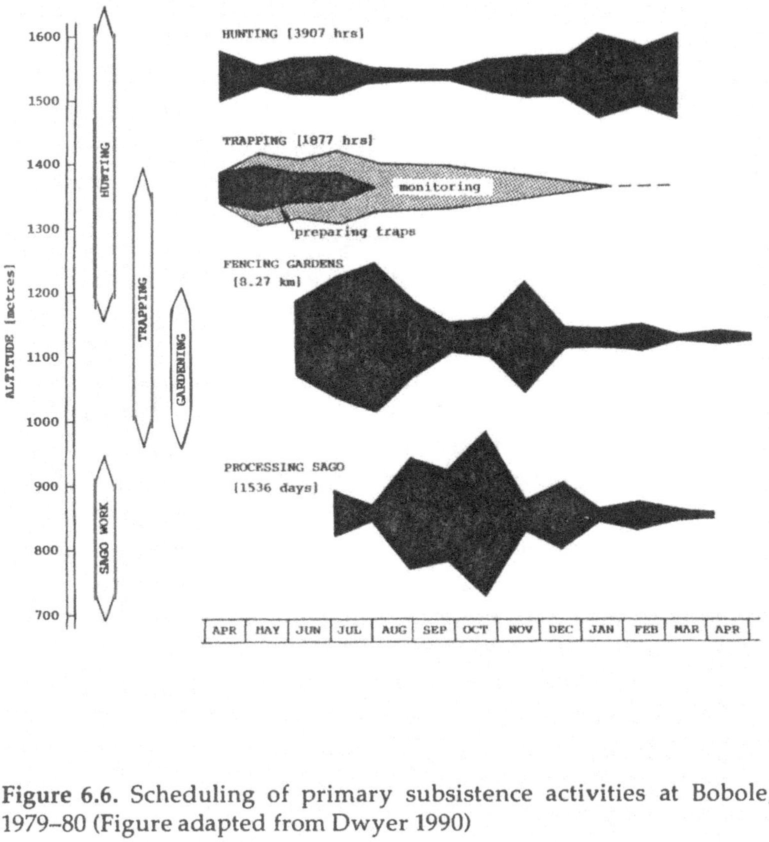

Figure 6.6. Scheduling of primary subsistence activities at Bobole, 1979–80 (Figure adapted from Dwyer 1990)

quantitative differences in the times allotted to specific productive modes. Gardening requires the greatest input and, with trapping, can be accomplished by people who reside at the village. To process sago, however, people move to lower altitudes for periods of a week or more. On average, these departures amount to about thirty days per person in twelve months. Hunting in the high forest is even less demanding with an annual average of perhaps eight days per person, though in this case nearly all input is from males older than fifteen years. Clearly, there are major differences in the intensity of both aggregate use and gender-specific use across the altitudinal span of the landscape.

The invisible world of Etolo is populated by several categories of spirit being [Dwyer 1990: 13–16; Kelly 1977]. The spiritual essence of people wanders alone when the body sleeps. There are witches that invade the bodies of some individuals and prey upon others. There are the spirits of the dead; forest spirits whose society invisibly mirrors that of people; and spirit bachelors that behave as vicious witches. These beings, assuming the forms or occupying the bodies of plants and animals, can appear anywhere within the landscape. But the forest spirits, in invisible form, dwell only in the high forest; people turn to these to identify witches and ratify subsistence decisions. Inasmuch as those decisions are ordered temporally, so too the role of forest spirits changes with the seasons.

When people die their spirits move into the bodies of various species of bird. And when the birds themselves die, perhaps by being killed, the spirits descend to rivers and occupy the bodies of large fish. The geography of the invisible, as experienced by Etolo, displays gradients of the sort depicted for the visible world. There is not, however, a neat one-to-one correspondence. It is the underlying theme, not the details, that is the same. The spirits of the dead transfer through time from an upper to a lower world, from the tree tops to the largest rivers, from where they may pass beyond the domain of people. The forest spirits occupy the place that people visit and use least often. Not all Etolo spirit beings behave in these ways but, in contrast to Kubo, the peripheries of the visible world assume importance as the abode of invisible beings.

Siane

Most of the land used by and known to Siane people of the highlands is anthropogenic (Figure 6.7). Less than 25 per cent of the area near Leu village can be classed as primary forest. Villages of 100 or more people are common, often only a kilometre or two apart. Population density is thirty to forty people per km^2. When people move across the land they skirt well-tended gardens, traverse *Imperata* grasslands and stands of *Casuarina* trees that were once gardens or, less often, encounter thickets of tall cane grass within which forest regenerates. Only

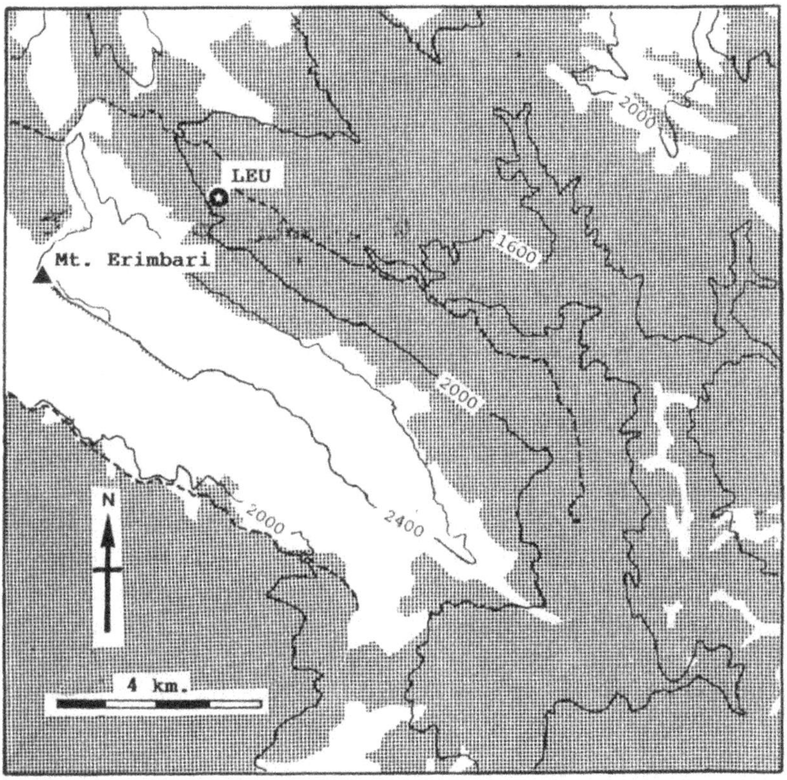

Figure 6.7. Map of Leu region (Siane). Shading shows anthropogenic habitats recorded on PNG 1:100,000 Topographic Survey, Sheet 7985 (Edition 1) Series T601, Printed 1978. Contour intervals are 400 m; dashes show vehicular roads

at the peripheries of clan territories, especially at higher altitudes, does the forest seem unmodified. It is this altitudinal gradient that dominates the land and conditions people's orientation to it. At Mount Erimbari the basal rock is cavernous limestone and large streams are not a surface feature.

The intensification of Siane food production is an order of magnitude greater than that of Etolo.[9] The people are gardeners and pig-husbanders whose impact on the land is long term and permanent. They are familiar with techniques of tillage, composting, drainage and irrigation, obtain ninety percent or more of their food from gardens, and direct surplus production to pigs. Different procurement systems are often closely associated in space, and the productive activities of males and females are distinct and tightly constrained.

These characteristics centralize the zone of primary food production to the vicinity of living places and establish strong gradients across the land in both use-intensity and gender association. Altitude delineates these gradients. Different garden types may be vertically zoned and, at Mount Erimbari, the high forest is a place where men hunt and, in particular seasons or suitable years, where fungi and pandanus nuts are harvested. The distance is not great but even the most enthusiastic hunters are unlikely to spend twenty days in the forest each year and will seldom remain there for two consecutive days. Average visitation rates per year may be only a few days.

In the high forests of Mount Erimbari lives a white cassowary that should not be seen and a society of benevolent humans. These latter spirit people are almost always invisible; when they are seen their form is human. Indeed, the visible and invisible worlds of Siane do not routinely intergrade. Animals may understand human speech or, at particular times and in particular places, represent messages sent by invisible beings. But few animals are likely to be either transformed spirits or occupied by spirits. An identification of flying foxes with ancestors is the notable exception, though, in addition, some capricious spirit beings appear as snakes and fireflies [Salisbury 1965].

Salisbury [1965] discusses Siane religion at some length. He reports a god who existed before people did, who rules over a land of the dead and 'takes the form of a circle of white light' [ibid.: 55]. The spiritual essence of people, pigs and some other animals animates them in life, can leave the body in sleep and appear as a white mucous secretion. After death these spirits persist for a time as ghosts, inhabiting the places they formerly occupied and appearing as humans in different form: 'they are white and insubstantial, emerge only at night and like cold and quiet while disliking strong smells and sexuality' [ibid.: 56]. Through this phase of their existence they may be vindictive. Eventually, however, they are 'incorporated into the undifferentiated body of ancestral spirits' who are well disposed toward people [ibid.].

Finally, there are what I will call 'spirit birds' [Salisbury 1965]. Their only manifestation in the visible world is as a variety of flute calls and associated songs that are performed by men. They may have no other form, material or immaterial. Their songs record that they live in uninhabited forest to the southeast or with their mother, who is linked to the sun.

The invisible world of Siane is unlike that of either Kubo or Etolo. The beings that inhabit it often assume forms that have no parallel in the visible world. They may reside far beyond the accessible domain of people or even in undifferentiated space. In the senses of both form and location they are peripheral beings who communicate with the living through intermediaries, through animals as messengers or bamboo flutes made by men. Similar themes are widespread across highland Papua New Guinea in regions where population density is high and agricultural production intensified [e.g. Lawrence and Meggitt 1965]. There are creators that appeared from distant or unknown places, made the material world, and went elsewhere. And there are societies of sky people, isomorphic with those of the earth, that, though they were the colonizing source of terrestrial populations or the originators of wealth, cannot be contacted by earthly beings.

Towards Theory

Comparison of the Kubo, Etolo and Siane permits several generalizations. Starting with the visible worlds experienced by these peoples, I will summarize primary themes and extrapolate from them towards theory.

Kubo, Etolo and Siane represent different points on a continuum of intensification. That continuum, however, should not be misrepresented as a unilinear, temporal trajectory. After all, each of the three constellations of intensification is the current end-point of an ever-changing clade. The increase in intensification is correlated with the appearance of strong gradients in both the use of land and gender-based associations with land. Few such gradients are seen among Kubo, and these are transient. By contrast, their expression among Siane is evidenced in nearly all aspects of daily life.

An outcome of the appearance of gradients of the sort described is the centralization of the productive domain. It adheres to the dwelling sites of people. Thus, the peripheries of local territories may be either little used or fulfil special purposes. Human impact on the environment is less overt in these places.

Often, the peripheral domain will be demarcated topographically. This might be expected, for example, where the relief of the land varies a great deal; it will be contingent upon the varying suitability of different altitudinal zones to particular systems of procurement. There is, however, no necessity here. Distance from a central place or use-intensity alone may be sufficient to demarcate the periphery. The peripheral domain may be contained within, or dispersed throughout, the central domain. Peripheries originate through differential use but, ultimately, they are cognitive constructs.

Inasmuch as the periphery of local territories is demarcated by use value, so it is likely that it will be demarcated lexically and, potentially, contrasted with the centralized productive domain. The literature seems full of examples but a major difficulty arises. The frequent equation of terms with, for example, 'domestic' and 'non-domestic', 'productive' and 'unproductive', 'village' and 'forest' or 'culture' and 'nature' may often entail imposition of the ethnographer's classificatory categories. Without extended exegesis such cases must be suspect.

Strathern [1980] provides a careful examination of Hagen terms that she translates as 'domestic' and 'wild'. Agricultural production and pig husbandry among these people of the Western Highlands of Papua New Guinea are more intensive than among Siane [Feil 1987]. But the 'domestic-wild' differentiation identified by Strathern 'is in terms of the essence or character of certain resources' [1980: 193]; it is not geographically bound though it is wedded to a distinction between 'social' and 'non-social'. Strathern shows how lexical distinctions in one domain may connect with, or connote, distinctions in other domains. However, her assertion that there is no geographic boundary between wild and domestic areas may be too strong. The landscape experienced by most Hageners is one of scattered gardens, fallow, grasslands and woodlands. The two last categories are not regularly used but they *are* anthropogenic. They are dispersed across the landscape and forest spirits are located there. Further, at the extremities of the known land, 'certain sparsely populated areas . . . which yield a concentration of [wild] things, come near to being considered a wild domain' [ibid.].

Among Etolo and Kubo I elicited no terms that were similarly contrastive. The Etolo word *sege*, meaning 'place', might connote 'somewhere else, not here' [Dwyer 1990: 12] and conjure up 'a distinction between non-domesticated and domesticated realms' [ibid.: 12]. But at other times, as in **ne sege** 'my place', the word 'embodied the sentiment of belonging – to the village, the gardens, the pandanus orchards, the sago groves, the forest, the streams and the mountains' [ibid.: 12–13]. **Sege** was more than these things, for it was also the Etolo word for 'rain' and for 'time'. There was no contrastive term.

The Kubo word **do** is translated by some people as 'bush' but it too lacks lexical contrast. Its primary referent seems to be those portions of the forest that, at a given moment, are not in use. If the unforeseen happens and people are obliged to camp overnight before reaching a house then the temporary shelter they make is a **do muson** or 'bush house'. They do not generalize this label to more permanent houses or to those associated, albeit briefly, with specific subsistence activities. Nor to my knowledge are **o wai** and **o fia**, which I translate as domestic and wild pig, implicated as metaphors in the social sphere. It is true that for Kubo 'wild pigs' wander in the 'bush' – they are sometimes called **do o** – but this does not connote peripheral places. Far from it: the connotation is of potentially abundant production because wild pigs provide Kubo with most of their protein.

My point remains. The potential exists to contrast periphery and centre where these have distinctive use values. The Kubo and Etolo languages reveal no such contrasts; at the least, the languages of many Highlanders move in this direction.

Finally, among Kubo, Etolo and Siane the emergence of gradients in the use of land and in gender-based associations with land is matched by differences in the form and distribution of invisible beings. Gradients of these sorts are few among Kubo. *Their* invisible beings permeate the landscape, routinely appear in the form of animals and, in invisible form, interact freely with the spiritual essences of people. The visible and invisible worlds of Kubo are co-extensive. By contrast, gradients in the visible world characterize the landscape of Siane and, in parallel, *their* invisible beings are often located at or beyond the peripheries of local territories, assume nebulous forms, and interact only indirectly with either the material or spiritual manifestations of people. They may be propitiated, or appealed to for favours or blessings, but they can rarely be controlled. The visible and invisible worlds of Siane are neither reciprocal as with Etolo nor co-extensive as with Kubo.

If the correlations summarized here are not fortuitous, and I do not think they are, then important theoretical implications follow. Intensification of production will simultaneously alter people's perceptions of both the visible and invisible world. At the outset a landscape in which use values are generalized, extensive and ungraded and in which invisible beings are all-pervading must be understood in totality as a landscape of human action and interaction. Hence, it is 'cultural'; there is no 'nature' and no contrast. But a landscape in which use values are particularized, intensive and graded and in which invisible beings are of nebulous form and peripherally located provides different opportunities. Here, through a process of cultural accretion, is the potential to invent 'nature'. Intensification and its correlates combine to categorize the visible world. Culture is internalized; it implodes. The created periphery of the visible world, increasingly divorced from human contact and understanding, emerges as 'nature'. It is imaginary and, in a materialist world, may stand for the 'other'. Differences I have read into the landscapes of Kubo, Etolo and Siane inform my grasp of this theoretical position. None of these people, however, should be taken to exemplify either a beginning or an end on what is a divaricating continuum of possibilities. Each is unique. Each offers to us a different appreciation of the relation between people and the landscapes within which they live.

There are places where 'culture' is the sum of the visible and invisible worlds and where it is these latter that provide sources of metaphor as people seek to comprehend 'self' by opposing it to 'other'. There are other places where perceptions have altered, where the invisible world has moved beyond the periphery of the visible and has lost relevance to the daily affairs of women and men or has been put aside. In these places the visible world may be conceived as the sum of culture and nature and these latter may, through metaphor, inform understanding of 'self' and 'other'. Little wonder, therefore, that for many Westerners the domain of the sacred has been relocated from the invisible world to the imaginary place that we call 'nature' – to the 'wilderness'.

An evolutionary rider may reinforce the argument. It may also, perhaps, help fill the gap between the ecological and cognitive fields with which I deal. The gap, of course, is social relations. The origin of self-aware, purposive and intentional beings – that is, I take it, of humans who are necessarily social – can only have been located in a landscape of familiar relationships. The heightened perceptual awareness that marked this transition can only have been of *those* relationships – in space and in time and with all classes of interactive beings. Thus, it can only have been *that* constellation of relationships that was represented by self-awareness and permanently patterned purpose and intention. There was nothing else to grasp. The genesis of 'self' was itself an inverted metaphor because, at the outset, relationship was construed as 'self'. That which we, as Western thinkers, regard as 'other' was, in the beginning, the essence of 'self'; there was no 'other' as we understand it because the self-aware, purposive and intentional being was necessarily part of those relations. The domain of 'nature', which we have abstracted from that ever-changing pool of relationships, did not exist.

Other Lands, Other Lives

Western thought has mistaken the periphery for the primal. To a large extent the conservation movement has compounded the error by sanctifying the perceived primal.[10] The consequence is alienation because, at base, it was the domain of action and interaction – the substantive centre (which was total), not the imagined periphery – that was both primal and sacred.

The mistakes are ontological. They impact on cognition. In acting upon, or seeking to understand, *their* landscape, which is assumed to be global, Western thinkers are informed by *their* cognitive categories. That is unavoidable. And, were it not that *other* people occupy that land, it might be harmless or merely self-destructive. After all, there is no a priori basis for claiming any one intellectual tradition as 'correct'; each may be judged, only in context, as apt or inapt. All such traditions have legitimacy as means of comprehending self and the world. Again, inasmuch as the conservation movement acts out its nostalgia for the 'vanishing past' – that is, the periphery – there may be no cause for alarm. No matter what changes occur to the earth there always will be a periphery, a place of attachment and a focus of action, for nostalgic longing. The rare and endangered building may assume the same value and satisfy the same need as the rare and endangered plant. Even individual people might attain heritage status. 'Nostalgia' is not a threatened species. Within the bounds of Western thought the voices of dissatisfaction and protest will need more powerful arguments as they seek to change the world. Otherwise they will merely change with it.

Other people *do* occupy the land and, I have argued, their perceptions differ from those of Westerners and from one another. I refer particularly to third- and fourth-world peoples who, so often, are underprivileged or rapidly becoming so. Victims of global hegemony, obliged to participate in global economies, they are dispossessed and effectively disenfranchised. These people, of course, have been and remain the 'objects' of anthropological inquiry. Increasingly *their* lands have become 'objects' of interest to conservationists. How should *we* perceive them or act responsibly toward them? To date our failures have been profound. I am not encouraged to think that things are improving.

The methodology of ecological science asserts a separation of environment and organism. This conditions the questions asked, the analyses performed and the answers proposed. But if the 'objects' of inquiry are people who know no such separation, then ecological science is unable to reveal those 'objects' either to us or to the people themselves. My geographical analyses attempted to avoid the 'environment-organism' dichotomy.

The now blossoming field of evolutionary ecology, which seeks 'explanations for cross-cultural and historical variation in human social behaviour' [Smith 1988: 222], fares no better. The rigour is impressive as is the disciplined approach to the formulation and testing of hypotheses. But the commitment to selection – 'natural' or otherwise – is the problem. It assumes a primal state and, thus, if *we* have misapprehended the primal then *they* are not revealed. Again, despite a controversial literature, I think that Marxist anthropologists speak, ultimately, with the same voice. The domain of the primal is coincident for these two schools of thought: it is 'nature' as *we* have perceived it and I have argued that some people – all, I suspect, in the category we like to call 'hunters and gatherers' – know no such condition.

These different approaches may achieve great elegance, they may position other people within the categories of our thought and contribute to our imagined understanding. But they cannot elicit the essential and unique humanity of these people. The search for man – and woman! – in a state of 'Nature' is doomed to failure. The primal human, I have argued, was cultural with no experience of nature.

The ecologist, perhaps, may intrude upon the lives of other people, seek answers to his or her questions, go away, and contribute to the excited dialogue of a select company of like-minded scholars. With luck he or she may be only an innocent participant in, and witness to, the hegemonic process. The conservationist cannot be let off so easily. The struggle for land is unequal. The categories of thought that inform the conservationist may have no analogue among those whose land is at stake. And, in these cases, to give precedence to *our* categories is, by fiat, to deny the land rights of others [Wright 1990]. Again, where the strategies of conservation acknowledge the proprietorial rights of others, seeking only to 'manage' wildlife in joint arrangements with those others, then often, it seems, those strategies are founded in the perception of a 'proper' trajectory that passes from nature to culture. The people shall be 'tamed', 'civilized', and made like us, for their own good and the good of the 'wilderness'. That is not good enough.

My intellectual debt to Roy Wagner is considerable. His book *The Invention of Culture* [1981] influenced the title of this chapter. Elsewhere he has written: 'the dangers of working out our own problems on the soils and in the hearts and minds of other peoples should not be overlooked' [Wagner 1977: 409]. I have argued for more intimate connections between the geographies of the visible and invisible worlds than, I think, have been previously suggested. My excursions across the lands and minds of other people reveal much diversity and challenge central tenets of *our* lives. It is our responsibility to those people to acknowledge and act on the challenges.

Notes

I thank the University of Queensland for granting periods of leave, the Papua New Guinea National Government for awarding research visas, and the University of Papua New Guinea and Papua New Guinea National Museum for affiliation. I thank Rofaifo, Etolo and Kubo people who have been my friends, hosts and teachers and Monica Minnegal for her comments.

1. Benterrak, Muecke and Roe 1984.
2. See Beek 1987 and Feld 1982 for other interpretations.
3. The interpretation of the painted design as female genitalia is from Beek [1987] and was for Bedamuni people.
4. The sexual and gender imagery of the Kubo dance costume is multivocal. To Kubo observers the dancer *is* a fully plumaged Raggiana Bird of Paradise and, therefore, despite biological reality, *is* female. Females are attracted to the dancer and yet the sexual provocation seen in his movements is directed toward other males. Themes of transvestism, homosexuality and heterosexuality are merged in a costume that has higher purposes. The spirits of the dead, attracted by the beat of the drum and the singing, see the bird of paradise and are drawn to assist the living: to cure the ill, release game to hunters or assist novices in their transformation to men. Beek 1987 provides an extended discussion of the many layers of meaning that may attach to the similar Bedamuni dance costume.

5. The historical contexts within which I worked may, of course, influence my interpretations. I record them briefly here, referring to significant impacts, not to first contacts. By 1972, Rofaifo (Siane), of the Highlands, had experienced three decades of influential contact. They grew coffee as a cash crop, paid taxes, participated actively in national elections and understood the requirements of modern law-enforcement agencies. Their familiarity with Christianity was of long standing; it was accommodated within, but did not dominate, people's everyday lives. Indeed, there were signs of cynicism in response to unfulfilled promises and an expectation that desired change might be best facilitated through involvement in more secular arenas. By 1979, Etolo of the Southern Highlands Province had been hosts to Christian pastors for ten years and the impact of the church was pervasive. Government patrols, though intermittent, were accepted as sources of authority and as channels through which people might relay hopes and aspirations about future change. The commitment to change facilitated by missionization was strong and traditional ritual expression in abeyance. The experience of government help was one of frustration and abandonment. Through these ten years of contact Etolo communities had consolidated as larger villages. Though this had quantitative effects upon subsistence patterns there was little qualitative shift in the subsistence base. It was not until 1984 that people in the northern parts of Kubo territory (Western Province) enjoyed prolonged contact with the outside world. In that year a mission station (the Evangelical Church of Papua New Guinea) was established and this was followed, in 1985, by government-sponsored health facilities and, in 1988, by a primary school. Earlier contacts with pastors from the Seventh Day Adventist Church had not been sustained and, at best, the influence of Government was indirect. Gwaimasi village, where I lived, was distant from these impacts. In 1986–87 people travelled for two days to seek medical attention and for three or more days to communicate with Government officials. Subsistence practices and ritual expression were little altered but people enthusiastically awaited profound, though barely understood, changes to their lives.

6. For details of Kubo subsistence, see Dwyer and Minnegal [1991, 1992].
7. There are many similarities between Kubo and Gebusi in the beings that inhabit their invisible worlds; for details of the latter, see Knauft 1985a. See also Shaw 1990 on the related Samo.
8. For details of Etolo subsistence, see Dwyer 1990 and Kelly 1988.
9. My anthropological study among Rofaifo was limited to a minor investigation of hunting and more extended work in ethnoclassification [Dwyer 1974, 1976]. I have drawn on Salisbury [1962], who lived with Siane speakers a few kilometres south of Leu village, on an extensive literature concerning Highland agricultural systems, and particularly on Hide's [1981] major work among Sinasina, about 25 km WNW from Leu. My own information on the invisible world of Siane is meagre and I have relied on Salisbury 1965.
10. The biblical 'wilderness' was peripheral and essentially unused. It presented challenges from, or permitted communion with, the invisible world. The rise of Western materialism, entrenched in the values of the Judaeo-Christian tradition, has progressively shifted this perception. Through several centuries peripheral domains were tamed, conquered and colonized and their diverse physical forms collected, displayed and subjugated. In these ways the geographic extent of the perceived periphery was diminished and fear of the 'wilderness' abated. This paved the way for the current shift to attitudes of nostalgic longing and sanctification that have adhered to the supposed 'wilderness'. See Wright 1990.

Bibliography

Barth, F., *Cosmologies in the making: a generative approach to cultural variation in inner New Guinea*, Cambridge: Cambridge University Press, 1987

Beek, A. van, 'The way of all flesh: hunting and ideology of the Bedamuni of the Great Papuan Plateau (Papua New Guinea)', Unpublished Ph.D. thesis, University of Leiden, 1987

Benterrak, K., S. Muecke and P. Roe, *Reading the country,*

Fremantle: Fremantle Arts Centre Press, 1984

Douglas, M., *Natural symbols*, Middlesex: Penguin Books, 1970

Dwyer, P.D., 'The price of protein: five hundred hours of hunting in the New Guinea highlands', *Oceania* 44, 1974, pp.278–93

——, 'An analysis of Rofaifo mammal taxonomy', *American Ethnologist* 3, 1976, pp.425–45

——, 'Animal metaphors: an evolutionary model', *Mankind* 12, 1979, pp.13–27

——, 'The perception and reconstruction of the apt-organism', *Journal of Social and Biological Structures* 11, 1988, pp.221–33

——, *The pigs that ate the garden: a human ecology from Papua New Guinea*, University of Michigan Press: Ann Arbor, 1990

Dwyer, P. D. and M. Minnegal, 'Hunting in lowland tropical rain-forest: towards a model of non-agricultural subsistence', *Human Ecology* 19, 1991, pp.187–212

——, 'Ecology and community dynamics of Kubo people in the tropical lowlands of Papua New Guinea', *Human Ecology* 20, 1992, pp. 21–55

Feil, D., *The evolution of highland Papua New Guinea societies*, Cambridge: Cambridge University Press, 1987

Feld, S., *Sound and sentiment: birds, weeping, poetics and song in Kaluli expression*, Philadelphia, University of Pennsylvania Press, 1982

Hide, R.L., 'Aspects of pig production and use in colonial Sinasina, Papua New Guinea', Unpublished Ph.D. thesis, Columbia University, 1981

Ingold, T., *The appropriation of nature: essays on human ecology and social relations*, Manchester: Manchester University Press, 1986

——, 'Notes on the foraging mode of production', in T.Ingold, D. Riches and J. Woodburn (eds), *Hunters and gatherers 1: history, evolution and social change*, Oxford: Berg, 1988

Kelly, R.C., *Etoro social structure: a study in structural contradiction*, Ann Arbor: University of Michigan Press, 1977

——, 'Etoro suidology: a reassessment of the pig's role in the prehistory and comparative ethnology of New Guinea', in J. Weiner (ed.), *Mountain Papuans: historical and comparative perspectives from New Guinea fringe highlands societies*, Ann Arbor: University of Michigan Press, 1988

Knauft, B.M., *Good company and violence: sorcery and social action in a lowland New Guinea society*, Berkeley: University of California Press, 1985a

——, 'Ritual form and permutation in New Guinea: implications of symbolic process for socio-political evolution', *American Ethnologist* 12, 1985b, pp.321–40

Lawrence, P. and M. Meggitt (eds), *Gods, ghosts and men in Melanesia: some religions of Australian New Guinea and the New Hebrides*, London: Oxford University Press, 1965

Lévi-Strauss, C., *The savage mind*, London, Weidenfeld and Nicholson, 1966

Ohtsuka, R., *Oriomo Papuans: ecology of sago eaters in lowland Papua*, Tokyo: University of Tokyo Press, 1983

Salisbury, R., *From stone to steel: economic consequences of a technological change in New Guinea*, Melbourne: Melbourne University Press, 1962

——, 'The Siane of the Eastern Highlands', in P. Lawrence and M.J. Meggitt (eds), *Gods, ghosts and men in Melanesia: some religions of Australian New Guinea and the New Hebrides*, London: Oxford University Press, 1965

Shaw, R.D., 'The Bosavi language family', *Pacific Linguistics* (Series A), 70, 1986, pp.45–76

——, *Kandila: Samo ceremonialism and interpersonal relationships*, Ann Arbor: University of Michigan Press, 1990

Smith, E.A., 'Risk and uncertainty in the "original affluent society": evolutionary ecology of resource-sharing and land tenure', in T. Ingold, D. Riches and J. Woodburn (eds), *Hunters and gatherers 1: history, evolution and social change*, Oxford: Berg, 1988

Strathern, M., 'No nature, no culture: the Hagen case', in C. MacCormack and M. Strathern (eds), *Nature, culture and gender*, Cambridge: Cambridge University Press, 1980

Turner, V.W., *The ritual process*, Harmondsworth, Middlesex: Penguin Books, 1974

Wagner, R., 'Scientific and indigenous Papuan conceptualizations of the innate: a semiotic critique of the ecological perspective', in T. Bayliss-Smith and R. Feachem (eds), *Subsistence and survival: rural ecology in the Pacific*, London: Academic Press, 1977

——, *The invention of culture*, Chicago: The University of Chicago Press, 1981

Wilden, A., *System and structure: essays in communication and exchange*, London: Tavistock Press, 1972

Wright, J., 'Wilderness and wasteland', *Island* 42, 1990, pp.3–7

Chapter 7

The Concept of Vital Energy Among Andean Pastoralists

Hiroyasu Tomoeda

Andean Pastoralism

More than 4,000 metres above sea level in the Central Andes (Peru and Bolivia) is an extensive plateau (**puna**, 4,000–4,500 m) where agriculture is almost impossible, and where live about 200,000 pastoralists herding llamas (*Lama glama*) and alpacas (*Lama pacos*). This does not mean that the people do not have access to cultivable land. On the contrary, the herding communities in general have small plots of communal land in the lower zone (**suni**, 4,000–3,500 m), where they cultivate potatoes or other Andean tubers. But even with this the pastoralists depend for almost all of their food on the farmers who are more adept in the cultivation of maize and other cereals in the temperate zone (**quichua**, 3,500–2,500 m) of the valleys.

The pastoralists travel down to the farmers' villages several times a year with their male llamas carrying jerky, wool, and other products of the **puna**, which are exchanged for agricultural products. This economic complementarity, long established between the herders and farmers, does not imply equal social status. The farmers tend to regard the herders as inferior, and marriage contracts rarely occur between them. On the other hand, the herders seem to be content to take advantage of the economic benefits, reconciling themselves to behaving as **waqcha** (poor, socially as well as economically) and letting the farmers feel generous on occasions when they exchange products.

In this paper I shall discuss the ritual performance and song of the Andean pastoralists. I shall focus my attention on (1) the perception and conception of vital energy, and (2) the function of poetry in the ritual context. I say 'poetry,' but what I treat here is a series of phrases sung only in a specific ritual context, rooted in a pastoralist way of life. Thus the poetry may lose its validity if we take it out of the total contexts (ritual and daily life). Reading those song phrases presented in the section 'songs of the **ayllusqa** ritual', we can obtain some impression of a rather simple, naive lyricism.[1]

Ritual of August

First, I shall describe the details of a ritual I observed in August 1981 on the **estancia** of Chicurumi, Province of Aymaraes, Department of Apurimac, Peru. It was called **ayllusqa** or **agustukuy** and makes reference to camelid fertility. The first sequence of the ritual, which is called **vispera**, consists mainly of the preparation of an offering which will be made the next day. It took place at three in the afternoon and lasted approximately three hours.

Vispera

3.00 p.m. In a corner of the house of a herder, a man puts a bundle (**mesaqepi**) on an improvised table, which when opened reveals its contents: **illa** (lithic zoomorphic figurines); **kiqmi** (braided ropes of llama wool); **pichuwira** (male llama fat); sea shells, three **waiqa** (small bags) which contain coca leaves, **llampu** (maize flour) and incense (rosin powder); several **chuwa** (wooden receptacles); **taku** (a piece of fine clay); mineral objects; three **qero** (wooden cups), etc. On the wall are hung bronze bells and woollen llama ropes. In addition, the owner places on the **mesaqepi** ears of maize of different colours, a small drum (**tinya**) and a bottle of rum (**aguardiente**).

3.05 p.m. The man brings burning embers in a broken ceramic pot which he places under the **mesaqepi** and adds incense and coca leaves which produce smoke that is offered to the **mesaqepi**. The man takes five ears of maize and places them one by one in the **chuwas**. He shucks one ear of yellow **muruchu** maize and places the kernels in one **chuwa**, which he then gives to his wife; she turns the **chuwa** over into a mortar (**muray**) and proceeds to grind the kernels with a pestle (**kutana**). The man, seated to the right of the woman, chooses three coca leaves (**kintu**) and adds them to the ground corn while the woman pauses in her task. She then places **chapchu** (roughly ground maize) into the **chuwa**, to which he adds a piece of **pichuwira**, three leaves of coca **kintu** and a bit of incense.

3.10 p.m. The man puts the **chapchu** into the fire as an initial offering; then he offers a shot of rum to the **illa** figurine, spilling some drops on the **mesaqepi** and drinking the rest (the act of **tinka**). The woman and the other observers carry out a **tinka** in the same manner. The man then brings one **chuwa** from the **mesaqepi** takes an ear of yellow **muruchu** maize, which he then shells into the **chuwa** and gives to his wife, who proceeds to prepare **llampu** (fine maize flour, lit. 'soft'). The process of grinding is interrupted briefly so that the husband can add three **kintu** and a little bit of pulverized **taku** clay, and the white flour turns into a reddish brown powder.

The first **llampu** prepared by the woman is placed back into the **chuwa**, together with its cob. The husband returns it to the **mesaqepi** and, taking another **chuwa** which contains **granada** maize, proceeds to remove the kernels and hands them to his wife.

3.28 p.m. The woman begins once more to prepare a second **llampu** for the **kanta** (figure of a female alpaca). The previously prepared **llampu** was for the **kanta** of female llama. The new **llampu** is finished and is returned to the **chuwa**, which the man takes back to the **mesaqepi**. He then gives his wife a new **chuwa** which contains red **chumpi** maize which he has already shelled. The wife begins to grind the corn in order to prepare the third **llampu**. One kernel is spilled from the mortar. She picks it up and, taking a cup of rum, spills some drops on the mortar and pestle while she mumbles **'ay Chicurumi, Pachamama'** and drinks the rest, pausing so that the husband can add three **kintu** and **taku**. This **llampu** will be used for the **kanta** of male llama.

3.48 p.m. The third **llampu** is ready, the man hands a **chuwa** of white **almidon** maize to his wife, who begins once again to grind the kernels while the man takes the last **chuwa** from the **mesaqepi** and the three ears of **muruchu** corn that had been left there, loosening the kernels. While doing this, the bottle of aguardiente, which was at his side, tips when he moves his arm. Taking a cup he spills some drops of liquor on the floor and says, '**Pachamama** is asking for **aguardiente**.'

3.58 p.m. The preparation of the fourth **llampu** for the **kanta** of male alpaca is finished and the man takes it in its **chuwa** to the **mesaqepi**. The woman takes the kernels of the white corn and begins to grind them, while the husband adds the ground clay. This time he does not add the **kintu**, although he has selected them. The **llampu** for the baby animal is ready and the man places it on the **mesaqepi**. Having concluded the grinding, the woman receives from her husband three **kintu** and chews them, and then drinks the liquor from the cup.

4.11 p.m. The man, holding a copper knife, proceeds to scrape the following materials in this order: **illa** figurine, shell, and three kinds of mineral objects. These scrapings are added to the five **chuwas** that contain **llampu**.

The wife grinds three ears of **muruchu** corn one by one to prepare three **chapchus**. The husband adds some **pichuwira** fat, three coca leaves and incense. The three **chuwas** with **chupchu** in them are also placed on the **mesaqepi**. The owner performs a **tinka** in front of the **mesaqepi** and bids the others do the same.

4.25 p.m. When the **tinka** is concluded, the man places the five **chuwas** which contain the **llampu** on the floor, and the couple proceed to perform a **tinka** to them with the rum. Then the woman plays the **tinya** drum for some minutes and begins to sing:

Sumaqllata mamallay	Beautifully, mother
Muyurinki mamallay	You will go around, mother
Hermanay paisanaqa	My sister, townswoman
Mamallay allpaqaqa	My mother alpaca

4.29 p.m. The man takes a piece of llama fat (**pichuwira**), divides it in two, gives one half to the woman and keeps the other. Both vigorously knead the fat into balls, which they mix with **llampu** until they obtain the form of a male llama and a female llama (the figures are made by the man and his wife respectively). In the process they use all the **llampu** without leaving any residue. While this is going on, a chicken enters the room. The owner does not permit the fowl to eat the maize left on the floor because, if it did, 'in the same manner the condor would eat one of our little alpaca babies'. Neither chicken nor dogs are allowed to come near the place of the ritual.

The daughter, who has remained behind in the kitchen, brings a pot of hot **chicha**, and the mother interrupts her activity to fill the two **qeros** with the maize beer, which she hands to her husband. He then spills some of it on the floor and drinks the rest; the woman does the same, and then the spectators are invited to drink the **chicha**.

4.46 p.m. The head, neck and body of the male llama are already distinguishable in the dough kneaded by the husband. With one finger he makes a small hole in the part that corresponds to the back of the llama, and, placing in it pieces of incense, he closes the opening. When the figurine is finished, it is placed in the **chuwa**. The man places a leaf of **kintu** next to the head of the figurine and two **kintu** on each side of the back.

The man begins to knead another piece of **pichuwira** for the **kanta** of the male alpaca. The woman, finishing the **kanta** of the female llama, opens its back, places incense inside and closes the opening. She places the figurine in a **chuwa** and sprinkles a handful of coca leaves over it.

4.56 p.m. The woman asks her husband to give her some **pichuwira** to make the figure of the young animal. Both are vigorously kneading the fat and from time to time add some **llampu**. The man notices that the woman has made a mistake: she has made only one figurine instead of three. The man begins to give shape to the figurine of the male alpaca; the front part of the neck is retouched so that the shape of a **piskayoq** becomes more visible. He perforates the back, places the incense inside, and deposits the figurine in the **chuwa**.

The man takes the **kanta** of the female llama that his wife has made and kneads it again. He divides the ball into three equal parts and with one of them begins to shape a small figurine.

5.08 p.m. The man finishes the three small figurines. He places them on a **chuwa** and spreads incense over their backs. He again kneads **pichuwira** to make the figure of the female alpaca. The woman gives kneaded dough to her husband so that he may shape the **kanta** of an offspring which is also called **musoqkancha** (lit. 'renewed corral'). The woman now kneads the dough left by her husband, while he opens the back of the offspring **kanta** to introduce the incense and places the figurine in a **chuwa**. The woman gives the dough to her husband, who breaks it into three pieces, shapes the three figurines of the female alpacas, places them in a **chuwa** and spreads incense on their backs.

5.23 p.m. In front of the owners are five **chuwas** in which one can see nine figurines (or **kanta**) of camelids made of **pichuwira**, maize flour (**llampu**), coca **kintu** leaves, incense and mineral scrapings. In the first **chuwa** there are three **maqesana** (female llamas); in the second, three **allpacha** (female alpacas); in the third, a **piara** or **carguero** (male transportation llama); in the fourth, a **wanso** (male alpaca); and in the fifth, **musoqkancha** or offspring. Together with the cobs of corn that have been used, the figures are placed in their **chuwas** in such a way that their posteriors face the couple.

5.26 p.m. The man proceeds to carry out the first **tinka** of the **piara** with liquor, invoking these phrases (this act is called **samay**):

> Wanso . . . piara samaykamusaq hampuchkanqa, pionnimpas hampuchkanqa, maqesanakuna hamuchkanqa, clasin clasin riatillumpas hamuchkanqa, todo completo hamuchkanqa, waskampas kimsa docena sartalla hamuchkanqa, gastumpas waranqa mediollalla hamuchkanqa, pichuwirampas Chalhuanca uraylla-manta hamuchkanqa, chuyallampas Chalhuanca urayllamanta hamuchkanqa.
>
> (**Wanso, piara** I desire to come, its **peon** should come, **maqesana** should come, her ear ribbons of all colours as well, it should all come complete, three dozen rope bundles should come too, its expenses of one thousand and a half should come too, its **chuya** [**aguardiente**, lit. 'transparent'] from far below Chalhuanca.)

He allows some drops of liquor to fall on the **piara** figure and drinks the rest. Then the wife does a **tinka** with the **aguardiente** provided by the husband and also makes invocations. The man asks the observers likewise to make **tinka**. The **wanso** is also invoked in a **tinka** in the same order: by the man, his wife, and the spectators. Then the **tinka** of the **maqesana** and the **allpacha** are invoked, and while one person recites invocations, the rest shout '**wajoo, wajoo**'. The wife begins to play the **tinya** and to sing.

5.56 p.m. The man proceeds to wrap all the **kanta** in a shawl with their **chuwas**, taking care not to damage them, and places the bundle on the **mesaqepi**. One of the three **chapchus** is burned as the ritual of vispera is concluded, while the others are kept for the next day.

Dia

I will now narrate the second sequence of the ritual, **dia**, which, together with the 'preceding evening' (**vispera**), forms a whole unit known as **ayllusqa**. Because of snow, the **dia** I describe had to be postponed for one week, and during that week the owner suffered continuous nightmares due to the delay in offering the **kanta** to the mountain spirits. The ritual took place on 21 August in one corral of Chicurumi.

8.20 a.m. Men and women invited by the owners begin to gather and converse inside the house. They are offered some rum while the owners ask them to help in the ritual. Those who accept are called **peones**. Then the owner addresses an old man and asks him to take the role of **masa** (lit. 'son-in-law') since he knows the rituals well, and because the owner is ignorant of them, as he is not from these parts.

8.42 a.m. All go to the **tinkana** corral. The owner carries the **mesaqepi** on his back and in one hand the **tinya** drum, while his wife carries a full huge earthenware **chicha** jar. One peon carries **taya** (a bush), firewood and embers, others carry diverse packages, while from a nearby corral, the old **masa** drives the animals towards the **tinkana** corral.

8.53 a.m. Together with the herd about 80 head, mostly alpacas, the owners arrive at the **tinkana** corral where they are welcomed by two men and a woman. The corral is shaped like a figure eight and located some 400 m from the owners' house. One of the circles measures 8 m, the other 5 m. At the junction there is an opening 1.50 m wide, and the wall constructed with stones is 1.30 m high. On the eastern part of the larger circle there is an entrance which was later closed off with a blanket.

9.08 a.m. The **mesaqepi** is left in a corner of the corral. The owner unties the bundle, removing coca leaves and selecting **kintu**. The **masa** prepares a fireplace by digging to a certain depth and, after making the sign of the cross, places firewood in it. The **masa** receives from the owner 25 leaves of coca **kintu**, lights the fire with the embers and adds **kintu** and **llampu** while the others converse and chew coca. The owner, taking off his hat and facing east, kneels in front of the **mesaqepi** and unties the bundle.

9.27 a.m. Following the instructions of the **masa**, the owner approaches the fire pit, takes a **chuwa**, and drinks **aguardiente** from it, sprinkling some drops over the fire. The wife does the same and the old **masa** closes his act. The old man places incense in three **chuwas**. The woman feeds it to the fire slowly in order to create smoke and not flames.

9.38 a.m. The wife offers two **qeros** of **chicha** to her husband. He proceeds to make a **tinka** in front of the fire, first with the vessel in his right hand in honour of Pachamama, and then with the left hand vessel in honour of the Apu or mountains.

The owner carries out **tinka** to the **chapchu** which is placed in a **chuwa** on the **mesaqepi** with two **qeros** of **chicha**, while invoking **samay**. He drinks the **chicha** that he has in his right hand, and once finished, he switches the vessel to his left hand, and the one in that hand to his right hand, and drinking that, returns the empty **qeros** to his wife. When this **tinka** is completed, the man picks up the **chuwa** in both hands, lets his breath fall on the **chapchu** (this act is called **samay** also, lit. to respire or to take a rest), and sprinkling it over the fire lets some drops of **chicha** fall on it.

The owner's wife asks an old woman to play the **tinya** during the ritual. She is offered a small bottle of rum in payment. The old woman plays the **tinya** and presently begins to sing. When she is finished, only the men shout '**wajoo, wajoo**'. The owner distributes coca while the old lady continues with the singing, and discussion arises about the order in which the **canta** figurines should be burned.

10.00 a.m. The first **samay** begins. The owner standing in front of the **maqesana** (female llama figurines) recites the following:

> Samaykamusaq, hamuchkanqa maqesana hamuchkanqa, hamuchkanqa, Huayunca pasaqmanta, Parionakunapa, don Fabian Huarcapa hamuchkanqa, Huamanikunapa hamuchkanqa, Sallcatamanta kimsa corralla hamuchkanqa, wasi gastumpas Chalhuanca uraymanta hamuchkanqa, tragullampas pichuwira, todo completo hamuchkanqa.
>
> (I am going to ask that the **maqesana** come, that she come from the sector of Huayunca, from the Parionas, from that of Don Fabian Huarcapa, that she come from the Huamanis, from the area of the Sallcata three corrals, that the food for the house come from far

below Chalhuanca, the drink, the **pichuwira**, everything complete, that everything complete may come.)

Then the wife performs a **tinka** with two **qeros** of **chicha**, while the owner does it with a **chuwa** full of rum. The other people also do their **tinka** one by one, while the **masa**, kneeling in front of the **mesaqepi**, shakes the **waiqa** (the small bag that contained coca leaves) over the figurines of the **maqesana**. When each invited person has finished his invocation, the **masa**, shaking his bag, reiterates, '**hampuchun, hampuchun**' (let them come, let them come!).

10.06 a.m. The **maqesana** have received their **tinka**, and the old woman passes the drum to the patron, who plays it. When he concludes, all the men shout '**wajoo, wajoo**'. The wife takes the **tinya** and sings, and once the song is finished, returns the instrument to the old woman, who continues with the recital.

The congregation has finished the third **tinka** to the **maqesana** (the first two with **chicha** and the third one with rum) imploring the **samay**. The **masa** takes the figurines to the fire pit, spreads coca, grease and incense on them and puts them in the fire with the heads facing east. The fire is moderate so that the offering burns slowly and smokes sufficiently. The old woman continues singing. One of the guests places a pot of **chicha** on the fire in which the figurines have been placed moments before.

10.15 a.m. The second **samay** directed to the female alpacas (**allpacha**) is begun. The ritual repeats itself, and once it is over, the wife of the owner sings while playing the **tinya**. The **masa**, making invocations, carries out the **tinka**, and at the end, sings accompanying himself on the **tinya**. The old woman continues with the songs. During the second **samay**, a couple arrives at the corral, and the old **masa** burns the figures while the old woman sings, and we hear some invocations.

10.38 a.m. The third **samay** to the male llama (**piara**) is carried out in the same way as those preceding. At the completion of three consecutive **tinka**, each participant receives coca leaves from the owner or from the **masa**. The old woman continues singing while the old **masa** completes a **tinka** and implores:

Ay inka piaralla hamuchkanqa, machu Cucuchimanta vecinokunapas hamuchkanqa pachaq pionnimpas allin yuyayniyoq hamunqa.

(Oh! that the inca **piara** from the great Cucuchi [a mountain] might come; that his neighbours might come; that one hundred **peons** with good memory might come.)

11:06 a.m. The old man and a helper bring some braided ropes (**kiqmi**) and woollen ribbons (pink and carnation red). The ribbons are called **waita** or **tika** (flower), and the **samay** of the **piara** continues.

One **peon** selects 25 coca leaves, braids them together with a skein of pink wool, and leaves them in one **chuwa**. He then begins to make another bundle.

11.20 a.m. The **samay** to the **piara** is finished. On the **mesaqepi** there is now one **chuwa** with the figure of the **piara** and another with that of the **wanso** (male alpaca) whose **tinka** now begins. The ceremony takes place as before.

11.30 a.m. The **masa** picks up the braided ropes and hands them to the wife of the owner, who hangs them around her neck. Three male helpers go to the centre of the corral and bring three male llamas that have been previously selected by the herder. They are young animals. They face east. The wife hands over one of the ropes, and the herder ties the hind legs of the llamas, forcing them to sit. The seated llamas are tended by the three men who place their hands on the necks and backs of the animals. The owner sprays water called **ñawin yaku** (water from the fountain's eye) over the heads and backs of the llamas, and then he repeats it with **ñawin aqa** (**chicha**).

The herder then gets a piece of fine **taku** clay and with it draws three branching lines over the backs of the llamas. His wife then hands him his three **kintu** leaves and **llampu** which he buries between the shoulder blades of the animals.

11.40 a.m. The **masa** brings the 25 bundles of coca leaves that have been selected previously. He digs a small hole of about 5 cm with a knife in which he places the leaves and the **llampu** and covers them. Meanwhile, the owner perforates the ears of the llamas and adorns them with the pink ribbons. The owner sprinkles coca leaves over the backs of the animals and places a coin on the middle llama's back. All those attending surround the llamas and take the coca leaves which they chew.

11.48 a.m. The wife takes off her shawl (**lliqlla**) and covers the llamas' heads with it. The **masa** burns incense in front of the animals' noses, and the owner, carrying the **illa** figurine, shakes it over the covered heads of the animals. The owner breaks a handful of **wayllaichu** grass into three portions and places each bundle along the backs of the llamas, tying the grass with the animal's own wool.

11.53 a.m. The **masa** burns the figures of the male alpaca and llama, and they begin the **tinka** to the bound llamas. The owner places himself behind the llamas. A man places some bells around the neck of the central llama and causes the bell to sound while the **tinka** is continuing. As the bells sound, the **masa** distributes coca leaves from a shawl spread out in front of the llamas to those people who have finished the **tinka**. The owner's wife also performs the **tinka** with two **qeros** of **chicha** and one **chuwa** of liquor and recites an invocation. When she is finished, she too receives coca leaves.

12.20 p.m. The **tinka** of the young llamas is ended, and the owner frees the animals, while '**wajoo, wajoo**' is shouted in unison. The old man advises that now is a good moment to give names to the llamas and the central one is named Rompecalle (street-buster).

12.35 p.m. The three helpers bring three young male alpacas to the centre of the corral and place them close together facing east. The owner forces them to sit using the **kiqmi** rope which is handed to him by his wife. The owner sprays the alpacas with **nawin yaku** and **nawin aqa**, and with the **taku** clay draws the branching lines on the wet parts. The **masa** hands him three **kintu** leaves and with **llampu** given him by the wife the herder buries the leaves in the thick wool between the shoulder blades of the alpacas. Then they tie the three bundles of **wayllaichu** grass on their backs and sprinkle coca leaves on them. As before, a coin is placed on the back of the middle animal.

The congregation surrounds the alpacas once more and takes coca leaves from their backs, which they chew or place in their coca bags. The owner adorns the ears of each alpaca with a pair of ribbons and they are also named: Nueve Decimos (a valued silver coin), Wamanpusaq (eight condors), and Cerro Blanco. The owner's wife covers the head of the alpacas with her shawl and the **masa** gives them incense. The congregation carries out one by one the **tinka** and **samay** with two **qeros** of **chicha** and one **chuwa** of rum while the owner shakes his **kipuna** bag which contains coca leaves. Before beginning the **tinka** some of the attending parties place coins on the animals, which are called

voluntad or goodwill.

The **masa** brings the figurine of the **musoqkancha** from the **mesaqepi** and repeatedly passes them under the noses of the alpacas. Then he burns them in the pit. The **tinka** and **samay** are ended, and the owner frees the animals.

1.31 p.m. The herder and his wife, carrying in their right hands **qeros** full of **chicha**, go around the inside of the corral, driving the animals towards the exit, and as the animals leave the corral, the herder and his wife spill the contents of their **qeros**. During this phase one hears repeatedly **'wajoo, wajoo'**. In the middle of the empty corral the owner's wife and a female guest dance to the **tinya** and the songs they sing.

1.35 p.m. The owner takes the last **chapchu** to the fire pit and burns it as the final offering.

The **masa** counts the coins given by the guests during the **tinka** of the animals and, handing them over to the owner, informs him of the amount of money donated. The man deposits the money in a separate **waiqa**, or **samay** bag. The old man puts all the things on the **mesaqepi** in order while informing the owner.

2.12 p.m. Helped by the **masa**, the owner ties the **mesaqepi** on his back. Everybody returns to the house. The camelids driven out of the corral are seen dispersed in the distance.

Songs of the Ayllusqa Ritual

The **ayullusqa** is carried out with songs from which springs the whole world of the herders and which help us understand the deep meanings of the ritual. The rite is repetitive, cyclical, and becomes monotonous. The prime objective for the person performing the ritual is without doubt, the increase of the camelid herd. The initial phases of the ritual are expressed in song as if man and domestic animal have met again after a year has passed (this ritual is an annual event), and as if both have lost their strength and vitality. This is expressed in the following song:

> Chaychallaraqmi yuyarimuyki / Pisikallpalla wawallaykiqa / Perdonawanki licenciawanki / Chicurumillay perdonawanki / Chayllaraqmi yuyarimuyki / Pisikallpalla patronayki / Pisikallpalla patronchallayki / Ñam hermana hermanachayqa.

(Only now do I begin to remember you / Your son with little strength / You will forgive me, give me licence / Chicurumi, forgive me / Only now do I remember you / Your **patrona** with little strength / Your little patron with little strength / Yes, sister, my little sister.)

The I-you address in the song refers not only to the pair man-beast, but also to the man-spirit world (Pachamama, Apu and other tutelary gods). The meeting (**tinkoq**) of these three beings (man, beast and divinity) is a necessary precondition for animal fertility which the herder so ardently desires, and his desire is satisfied only by the offering of the gifts in the ritual to the gods. Should he neglect them, the gods will become angry and any kind of ill luck might befall him, such as the diminution of his herd through sickness or theft, or even his own death (human death is sometimes interpreted as a result of the mountain having devoured a man's heart). Thus the day of the meeting is crucially important and is expressed in the following song:

Kunan punchaoqa qollqe kurralpi / Lunes punchaoqa qori kurralpi / Imayllamantam duenollay ruwawanki / Diallaykiqa, santuykiqa, mamallay.

(Today in the silver corral / Monday in the golden corral / Out of what are you making me, owner mine? / It is your day, it is your saint's day, mother.)

As I mentioned before, the three beings, who meet in a special place (**uiwa tinkana** corral) and at a determined time, are all in a state of near exhaustion and tiredness, and it is necessary to replenish the lost forces in order to achieve an increase in fertility:

Patronchallay dueñochallay / Kamaqchallayman igualaykuway / Kamaqchallayman cabalaykuway / Inciensollaykiwan chuyachay-kiwan / Wirallaykiwan llampullaykiwan / Kukachaykiwan kintuchaykiwan / Qollqe platupi, qori platupi.

(My patron, my owner / Please even out my forces / Please replenish my forces / With your incense and your liquor / With your fat and your **llampu** / With your coca and your **kintu** / On the plate of silver, on the plate of gold.)

Kamaq can, perhaps, be translated as soul or spiritual force. From the context of the song we can derive that incense, **chuya**

(liquor), **wira** (llama grease), **llampu** (maize flour) and coca **kintu** are material representations of the concept of **kamaq**. **Kamaq**, once materialized, becomes tangible to the senses in that incense is smelt, the **llampu** touched, and the coca tasted, and therefore perceived by man.

It seems that it is the llamas who sing the song asking these things to be given (**alcanzarles**); yet, these are no more or less than offerings given to the gods. Of all these offerings, the most important is coca, as we shall see. The herder who made the **kanta** (animal figurines) explains the force that coca has, and why he added it to the ground corn. Only when **kamaq** has been **alcanzado** (succoured) and has revitalized itself in the exhausted beings is it possible to increase the herd, which is what men desire:

> Kamaqchallayman alacanzaykuptikiqa / Kikichallaysi chauchurichkasaq / Kikichallaysi mallkirichkasaq / Mallkichahina wayllachahina / Chiri wayrapi chauchurichkasaq.
>
> (If you would succour me, my force / I would be sprouting roots by myself / I would be shooting branches / Like the tree, like the **waylla** grass / In the cold wind I would be taking root.)

Many of the songs collected need to be studied in depth. Here I will only deal with one aspect of them. The songs that sing of the desire to increase the herd, with their bucolic character, inherently contain the force of green nature. Herders use a series of metaphors when they refer to the increase of fertility of the herd, for example 'I would be shooting branches / Like the tree, like the **waylla** grass', with its obvious allusions.

According to the man who was shaping the animals from the **llampu**, the ground corn represents the body of the animal while the coca represents its force (**kamaq**). Coca plays a symbolic role of great importance. For example, it is the 'shawl' that envelopes the animal. (It is said that the wool grows rapidly if the animal is wrapped in the shawl.) It is the 'pasture' that the animal eats, while the rum is water (the **chuwa** full of liquor is called **qocha** [lake]). The participants chew the coca leaves spread on the backs of the animals. The ears of the figurine **piara** and the loads (food) of the llamas are represented with coca leaves. And, finally, when the animals receive the **tinka**, three **kintu** leaves are mingled with the animal's wool so that it may grow rapidly and abundantly (in the case of the alpaca) or so that the animal may have great strength (in the case of the llama).

The metaphor used to express the increase of the herd through the **waylla** grass has already been mentioned. This plant, a kind of **ichu** (bunch grass; possibly *Calamagrostis*), grows in the **puna** and is eaten by the camelids. The three branching lines drawn on the foreheads and backs of the animals are associated with **wayllariy** (branching like the **waylla**). A handful of **waylla** grass is also tied along the spine of the animal with its own wool. Coca and **waylla** are used as symbols of increase and abundance of wool. **Waylla** is the clothing and finery of the camelids according to the following song:

> Qollqe kanchapi sayallachkani / Qori wayllalla coronayoqlla / Qori wayllalla dia pachayoqlla / Ima kuyayllam bajaykamuni / Qayka kuyayllam hermana purisaq / Dueñullay qollqe wayllawan tipawanki.
>
> (I am standing in the silver corral / With my crown of golden **waylla** / With my clothing of golden **waylla** / Oh! how much admired have I come down / How greatly admired, my sister, I will walk / With silver **waylla** you will sit me down.)

Clothing is made out of spun and woven wool. It is very interesting that in a figurative way, wool derives from another vegetable material. The woollen ribbons which adorn the animals' ears are also figuratively plants, since they are called **tika** or **waita** in the following song:

> Amapola tikawan churawanki / Arequipachallay multicolores / Tupawasqachu? Luciasqachu? / Ima kuyayta qayka kuyayta / Amapola tikari lucillawasqapaschu? / Sumaqllata mamallay waitarinki.
>
> (You will adorn me with the poppy flower / Many colours from Arequipa / Do I look good? Do I look fine? / How loved and how much admired / Doesn't the poppy flower look good on me? / Beautifully you will bloom, my mother.)

Once the vitality (**kamaq**) of the camelids has been regenerated through the ritual process, the animals, figuratively dressed in **waylla** clothes and decorated with flowers, express their happiness and contentment, and demonstrate their latent power in the next song:

Chicurumillay pampachapiqa / Ima kuyayllam patronay kani / Qayka kuyayllam dueñollay kani / Acerollas kallpayqa / Acerollas tulluyqa.

(On the pampas of Chicurmi / How admired I am, my **patrona** / How admired I am, my owner / My force is like steel / My bones are like steel.)

The vital force thus replenished becomes usable by the herders. They prefer to express this by relating it to their customary treks to the valleys. During the phase when they carry out the **tinka** and **samay** for the male llamas, they sing as follows:

Kaylluchallayta churakapuway / Sillwichallayta churakapuway / Viajero kachkani, pasajero kachkani / Wasichallanchi pasallasaqña / Ama pañallay waqapuwankichu / Hermanullayki kallachkaniraq / Qari qariraq kapullawan / Chicurumillay jovenchallaqa / Reqsisqalla kachkani / Llaqtan llaqtan purimunaypaq / Sara mamaqa sasa kayllanallas / Sara mamaqa sasa sillwiñalla / Apamuchkani karguchalla ruwananchipaq / Obligacionllata rurachkanchik / Santo santata servichkanchik / Maykamaraqcha purillasaqpas / Maykamaraqcha samarisaqpas / Pero maqtaqa qari qarilla.

(Put on my chest adornment / Put on my saddle cloth / I am a traveller, I am a passer-by / I am already leaving our house / Don't you cry for me, my little sister / I am still your brother / Very brave very brave / Is the young man of Chicurumi / I am well known / For going from town to town / It is difficult to bring mother maize / It is difficult to carry mother maize / I am bringing it to pass the cargo / So that we discharge our obligation / To serve the saints / How far need I go / and where will I rest? / But I am a strong young man.)

Poetics of the Ayllusqa Ritual

After describing the ritual performances and songs, what we must confirm first of all are the following characteristics or tendencies of the ritual song-discourse when compared with ordinary verbal communication.

Prohibition of Certain Everyday Terms

For example, the terms 'llama' and 'alpaca' are not always used in the ritual process, but are replaced by other specifically ritual terms: **maqesana** (female llama), **piara** (male llama), **allpacha** (female alpaca) and **wanso** (male alpaca). **Trago**, a common word which means rum made of sugar cane, must be called **chuya** (an adjective of any pure or transparent liquid), and **chicha** or **aqa** (maize beer) is **qocha** (lake). Coca leaf is **pastu** (from Spanish pasture) or **kintu** (a ritual term). Those who use the everyday terms are sanctioned ritually.

Abuse of Spanish

Though it is true that many Spanish words are nowadays employed in Quechua, their use is very prominent in the ritual song phrases. Both **aiparimuway** (Quechua) and **alcanzamuway** (quechuanized form of the Spanish verb 'alcanzar') mean here 'deliver me' (imperative mode). But the employment of the non-customary wording may cause some poetic effect on the part of a native speaker of Quechua. Thus, foreign vocabularies are adopted to give a poetic impression.

Abuse of Limitative and Diminutive Suffixes

The Quechuan dialects spoken in the Apurimac region are known as 'sweet' or 'tender' compared with other dialects (for example, the Cuzquenian dialect spoken in the southern part of Peru). One of the reasons it is so qualified is the frequent use of limitative and diminutive suffixes. In the case of the ritual song, this tendency is more conspicuous, and the emotional mood of affection or intimacy a speaker (singer) has toward the referents is over-emphasized. For example, if a man says '**patronchallay, patroncha**' he is demonstrating much affection for his patron, using **-cha** (diminutive) repeated twice and **-lla** (limitative), to a degree that seems excessive in any ordinary discourse.

Restricted Use of the Person of Verbal Inflexion

Quechua distinguishes the singular and plural of the first, second and third persons. But in the sung discourse of the ritual the singular of the first and the second persons is used almost exclusively, and use of the inclusive plural of the first person (**ñoqanchis**) and of the singular of the third person (**pay**) are very rare.

Sung Discourse as Lyric Poetry

Taking into account the six functions of language differentiated by Roman Jakobson [1981: 27] we can say the following about the ritual song as poem.

(1) The referential function which is important in any daily communication retreats in the ritual song, and the poetic function becomes more dominant.
(2) At the same time, in relation to the above-mentioned, the emotional function and connative function are also prominent.

As a result the song of the ritual is characterized as (1) lyric expression linked with the emotive function using the first person (I), and (2) supplicatory expression imbued with the connative function using the second person (you). Logically, the song lacks the characteristics of epic poetry which, focused on the third person, strongly involves the referential function of language.

Using the language in a different way from usual communication, all the things and situations, whether real or imagined, are represented poetically and aesthetically. In this poetic discourse, it may happen that a denotative articulation which one term has in ordinary use is lost and its meaning becomes ambiguous, but it is possible to produce another one. As an example here we have the expression **huerta** (Spanish 'kitchen garden'). The ritual performer treats a kind of portable altar as a sacred object, which is called **mesaqepi** (lit. 'sacred bundle') in daily use, or even in the ritual when one indicates it as a mere object. But if he is prompted to refer to the spiritual world it symbolizes in relation to himself, he may call it **huerta** as a metaphor. Once referred to as kitchen garden, the referent (sacred bundle=spiritual world) becomes very intimate to him and makes him conscious of a necessary attendant respect that a kitchen garden should receive. Unlike other large plots of cultivable land, the kitchen garden is very near his residence, well protected by stone or adobe walls, and constantly supplied with water. These characteristics of a kitchen garden are attributed also to the spiritual world when one uses **huerta** with reference to the **mesaqepi**. The performer must then treat the sacred bundle as he does his kitchen garden to rear such plants as flowers, onions, carrots and other kinds of vegetables which will not grow without receiving man's constant attention. Moreover, it is worth mentioning here that **huerta** is a female noun in Spanish, and the spiritual world is often personified as Pachamama (the earth goddess).

Now let us consider the contents of the ritual song. I shall choose some examples out of many sung phrases I collected, and present them in seven categories. I shall arrange them so as to clarify the meanings of the song and not necessarily in accordance with the time and sequence in which they were sung.

(1) Scenes from the pastoralists' life	
Kaylluchallayta churakapuway	Put on my chest adornment
Sillwichallayta churakapuway	Put on my saddle cloth
Viajero kachkani	I am a traveller
Pasajero kachkani	I am a passer-by
Acerollas tulluyqa	My bones are like steel
Acerollas kallpayqa	My power is like steel
(2) Debilitation of vital energy	
Kamaqchallayman igualaykuway	Please even out my force
Kamaqchallayman cabalaykuway	Please replenish my force
Inciensollaykiwan chuyachaykiwan	With your incense, with your **chuya**
Qollqe platopi qori platopi	On the plate of silver, on the plate of gold

(3) Merging of categories

Tanto tantolla tukuykuchkani	So much I appear to be
Achka Achkalla tukuykuchkani	So many I appear to be
Ayway patowan mezclarikuspa	Getting mixed with wild ducks
Ayway wachwawan yaparikuspa	Counting myself with wild **wachwas**

(4) Restoration of vital energy

Kamaqchallayman	If you would succour
Alcanzaykuptikiqa	My force
Kikichallaysi chauchurichkasaq	I would be sprouting roots by myself
Mallkichahina	Like the tree
Wayllachahina	Like the **waylla** grass
Chiriwayrapi	In the cold wind
Chauchurichkasaq	I would be taking roots

(5) Renovation of marking

Arequipachallay multicolores	Many colours of Arequipa
Tupawasqachu	Do I look good?
Luciawasqachu	Do I look fine?
Amapola tikawan churawanki	You adorn me with poppy flower
Ima kuyayta hayka kuyayta	How loved and how much admired

(6) Reestablished Categories

Sutireqsillam hermanachayqa	Well known my sister
Iscahuacallay llaqtachallampi	In Iscahuaca village
Sutireqsillas Lopezchallay	Well known
Warmiqa	The wife of Lopez
Sutireqsillas Hilariollay	Well known
Warmiqa	The wife of Hilario

(7) Terms of Address

Ñam hermanachallay hermana	Now my sister, sister
Patronchay dueñochay	My patron, my owner
Mamallay mama	My mother, mother
Ñañallay ñaña	My sister, sister
Wauqellay wauqe	My brother, brother
Paisanallay paisano	My townsman, townsman

The examples in the first group are a part of many song phrases related to experiences of the pastoralists' daily life. From them we can see that travelling to ensure food supplies is one of their important activities. This part of the song, if detached from the total context of the ritual, seems only to present a scene of accustomed travel to which a pastoralist is conveying his strong emotional feeling under the guise of a male llama. But the ordinary experience receiving poetic expression assumes another appearance in relation to the ritual. Such experiences as preparation for travel, leave-taking, or a laborious trek which in general are depicted in the final sequence of the ritual, signify the disposal of vital energy, in contrast to its restoration, found in the first part of the ritual process.

Accordingly, at the beginning of the ritual all the beings are presumed to be in a state of lost force, which is expressed with phrases from the second group, or similar. The force is referred to as **kamaq** (which can be translated as vital energy) in the sung discourse, whose debilitation or weakening receives different but related expressions, contrasting with the forceless/forceful state. The **kamaq** (vital energy) implies not only physical power (**kallpa** in Quechua) but also a spiritual one. For example, such contrasts as forget/remind or harsh/gentle are employed to qualify the actual condition of vital energy. The state of **kamaq** is also perceived by human senses contrasting cold/warm, hard/soft, dry/wet or dusty/cleaned. Phrases like 'not come into mind', 'forget', and 'lack of gentleness' are social and emotional expressions of lost **kamaq**. Thus the weakening or loss of **kamaq** is transmitted by physiological, emotional and social codes. Here we must emphasise that human sensitivity is very important in Andean religion and ritual. Andean people perceive rather than understand their religious world, while lacking instrumental devices or special apparatus which will make them sense transcendental religious experience (e.g. masks, elaborate paraphernalia, temple or shrine-like constructions).

The weakening of **kamaq** perceived by human senses is manifested in various ways, including visually. Resulting necessarily from the lack of **kamaq**, the idea of 'weak to eyes' is also expressed in the song, as in the phrases of the third group. I have said that sung discourse refers to pastoralists' daily experiences. So with these phrases. A shepherd – a woman or a child – is sitting on some place a little above and far from where her herd is, contemplating the llamas and alpacas dispersed on the **puna** and grazing on the pasture. A flock of wild ducks or other aquatic birds floating on a little marsh are also in her panoramic visual field. The llamas and alpacas far from her and the aquatic birds near her are seen as if they were a part of the same flock or group. This beautiful scene we can appreciate in the afternoon of the **puna**, as the animals huddling together in the morning gradually become separated as the sun rises. But the shepherd presents this beautiful scene perceived by herself as 'though I am only one or two, yet I appear to be many', which sounds as if a llama were bluffing in counting himself with others (birds) which are not of his category.

I have already mentioned that weakened **kamaq** can be represented in various dimensions. The same is true of its restoration, which constitutes a part of the ritual performance to which the phrases of the fourth group correspond. Furthermore, singing 'I'm remembering you,' 'you are turning back to me', or 'you are going around', one versifies that the relationship between two beings is to be recovered as it was. The restoration is also perceived by the ritual performer himself when he is warming, cleaning or softening any object. Only when the restoration of **kamaq** is perceived is the rejoicing 'shooting like **waylla** grass or tree' recited. But the restoration is the beginning of the weakening process of **kamaq** which is to be disposed and dispersed in the time and space in which ordinary life is carried on. The conception of **kamaq** in disposal is, so to speak, entropic (with centrifugal aspects in the ritual performances and discourses), and the ritual may be understood as a periodic attempt to repress the inevitable increasing process of entropy (with its centripetal aspects).

Those phrases categorized in the fifth group are related to the conception just mentioned. The coloured ribbons (**arete**) with which the ears of animals are adorned are called 'flower' (**waita** or **tika**), as a metaphor referring to the beauty of adornment in the ordinary life context. But the 'flower' being put at the turning point of the **kamaq** process conveys a meaning in the ritual context, for it implies the cycle of life and death (blooming and withering). So the animals are demonstrating a somewhat reluctant attitude toward the acceptance of the adornment. Another phrase, 'blood river takes me off', which refers to a little bleeding at the moment of ear-marking, confirms the implication of 'flower=ribbon'. Ear-marking is usually done at the same time as ribbon adornment, though this was not the case for the ritual I described in the section 'ritual of August'.

Ear-marking, as well as ribbon adornment, is a sign of ownership, like a cattle brand, and allows the differentiation of one group or category from another. That is, with a new or renewed sign the world is demarcated, rearticulated and recategorized, and accordingly the daily life is carried on. In this sense, earmarking, ritually cutting off a piece (or pieces) of the ears, comes to have a practical value in distinguishing the individual owners of domestic animals. To this correspond the song phrases of the sixth group. These are sung in the final phase of the ritual and express the pleasant and comfortable union of the same category (**layamasi**), of the same language (**parlaqmasi**), or of companions. Only by shouting, and shouting the given names, does the ritual performer/singer confirm the recategorized and reestablished order of the world.

What we must discuss finally is the problem of the beings in the poetic and symbolic world which we have examined. In the time (one's day) and space (qualified as a golden corral) of the llama ritual there are three kinds of beings: humans, animals and deities. The gods are not visible but the performances and things of the ritual are expressed as if they were rejoined there with the other two kinds of being. The three beings present in the conception and consciousness of the ritual performers are never referred to directly in the sung discourse, though they may also be present in the case of the invocatory discourse or the performance. The message 'With your **chuya** and incense, on the plates of gold and silver, replenish my force', although poetic, refers to a part of the ongoing ritual. The concordance between sung discourse and performance is as follows: 'golden and silver plates': a small wooden plate (**chuwa**); 'with your **chuya** and incense': rum and rosin powder; 'even out of my **kamaq**': to do offering. But the relationship of the agent to the receiver of action (in sung discourse) is man to animal when the offering is directed to animals, or man to god when it is put into fire. In the latter case he who implores is god, because he is in a state of weakened **kamaq**. But if we see the whole of the ritual the subject who implores must be taken as man.

The apparent purpose or motive of the ritual is to implore the gods for increase of animals, but in the ritual song, at least, there never appears an expression such as 'I (man) ask you (god) increase of him (animal)'. In fact, neither the gods nor the animals are referred to directly in the song at all. The three beings (man, god, animal) which are distinguished from each other in everyday experience, and even in the ritual sequences, lose their specific reality in the poetic and symbolic presentation of song. This world, on the contrary, is constituted only by 'I' (the first person) and 'you' (the second person). That is, the beings who appear in the song are only 'I' and 'you'. So the tripartite relationship of man/god/animal is presented with the bifurcate relationship of 'I'/'you', without the 'he'. Now, to discuss what 'you' and 'I' represent or mean we must examine the actions and states to which each of the two beings is subject. Then the following is confirmed.

(1) The actions and states of the subject 'I' are, so to speak, intransitive in relation to 'you'. For example, 'I myself take roots', 'I am, stand, walk, or go down admired,' or 'I am thirsty', 'I am only one', 'I appear to be many yet' (ironical expression).
(2) The actions and states of 'you' are common in part with those of 'I'. So we find many phrases in which the subject 'I' is replaced by 'you'. But we never find the case of 'You are thirsty,' or 'You appear to be the only one'. From the view point of **kamaq**full/**kamaq**less, 'I' is to be the subject of any negative state or action, but 'you' never. The state with **kamaq** is common between 'I' and 'you', but the state without **kamaq** is only proper to 'I'.
(3) The action of the subject 'you' is transitive directly to 'I'. For example, 'you' forgive, do something favourable, replenish **kamaq**, make sit, put on adornment for 'I', or say something bad, forget or abandon (negation of these actions is asked for 'you' by 'I'). Only after receiving the direct action of 'you' are the actions or states of 'I' established. So 'if you even out my **kamaq** for me, I will take root by myself in a harsh environment' is basic interaction between 'I' and 'you'. It is worthwhile mentioning here that Andean pastoralists employ in this sung discourse only true persons from a linguistic viewpoint. Thus, Benveniste [1966: 225–36] has distinguished true persons (the first and second person) from the non-personal third person.
(4) Only one action of 'I' transitive to 'you' is imploration (taking the imperative mode grammatically). The agent of implor-

ation is exclusively 'I', and never 'you'. Basically 'I' is a being who asks or implores any favour or pardon from 'you', who conversely takes the position of patronage.

Having confirmed the two beings' relationship in the sung discourse, we will discuss the seventh group. Here are relative terms which 'I' employs frequently to address himself to 'you'. From them it is very clear that the concrete relation between 'I' and 'you' is confined to mother/child, brother/sister, and patron/protege (curiously, never father/son), which are all basic, simple and intimate relations in Andean society. The relationship of 'I' and 'you' applies whether identical, fraternal or patronal. That is, the relationship of the three beings is understood as the direct product of the three sets: {man, animal, god}, {I, you}, {mother/child, brother/sister, patron/client}. Moreover, the intimate and interiorized relation of 'I/you' is strengthened by the excessive use of limitative and diminutive suffixes.

Only the first and the second persons are employed in the sung discourse, so the interaction of 'I' and 'you' constitutes the discourse universe [Ducrot and Todorov 1972: 317–24] which refers to the transcendental or religious experience going on in ritual time and space. In this situation, we must confirm that 'I' is the singer himself. One assumes that the phrasing of the song is prescribed to a large extent by tradition, and how to employ the two persons is also preconditioned. But which phrases to choose and thread, which relative term to use to address oneself to 'you', or what emotion to express, are all a matter for the singer. Practically, this song cannot be sung in unison by two or more. As a result the singer 'I' is involved with his flesh and blood in the poetic and symbolic world he himself is referring to. Thus, we can now say that 'I' might be animal, man, or even god when the singer addresses himself to 'you' with 'my patron'.

The three beings of man/animal/god denotated in the ordinariness of human experiences are reduced to the two beings of 'I' and 'you' in the ritual experience, which is nothing less than a poetic world. This situation, in which the three beings differentiated in ordinariness come into a close and intimate relation as two beings, exists only in the ritual. In ordinary time and space, in turn, the relationship grows weaker and the intimacy cannot be maintained eternally. Through daily activities **kamaq** diffuses, the multi-colours of the ear-ribbons fade out, the

mesaqepi becomes dusty, and the intimate relation becomes unconcerned and cold. Having thus conceived, a pastoralist invites his relatives and friends, gathers his llamas and alpacas and conjures up the gods, realizing his **ayllusqa** in the sacred corral. The Quechua verb **aylluy** means 'to unite'.

Note

1. The description of the ritual process and song (the sections 'Ritual of August' and 'Songs of the **ayllusqa** ritual') is based on a previous article [Tomoeda 1985: 277–99]. I owe much to Dr Enrique Mayer, who translated my original paper from Spanish. A more detailed discussion of the sung discourse (the section on 'Poetics of the **ayllusqa** ritual') is developed in Tomoeda 1988: 110–32.

Bibliography

Benveniste, E., 'Structure de relations de personne dans le verbe', in *Problémes de linguistique générale*, Paris: Editions Gallimard, 1966

Ducrot, O. and T. Todorov, 'Référence', in *Dictionnaire encyclopédique de sciences du langage*, Paris: Editions du Seuil, 1972

Jakobson, R., 'Linguistics and poetics', in S. Rudy (ed.), *Roman Jakobson: selected writings III*, The Hague: Mouton, 1981

Tomoeda, H., 'The llama is my chacra: metaphor of Andean pastoralists', in S. Masuda, I. Shimada and C. Morris (eds), *Andean ecology and civilization*, Tokyo: University of Tokyo Press, 1985

——, 'Songs from llama and alpaca fertility rituals', in J. Kawada and J. Nomura (eds), *Studies in oral traditions IV*, Tokyo: Kobundo, 1988 (in Japanese)

Part II

Relations Between Specific Domesticates and Human Populations

Chapter 8

Glutinous-Endosperm Starch Food Culture Specific to Eastern and Southeastern Asia

Sadao Sakamoto

In the traditional or ritual life of Japan, particularly at the end of the year, a special festive cake is prepared from glutinous rice for the coming New Year. The rice is first steamed after immersion in water for twenty-four hours, after which it is pounded in a wood or stone mortar with a wooden pestle (Figure 8.1). The product is called **mochi**, is decorated with the fruit of *Citrus aurantium* L. and with the leaves of a fern, *Gleichenia japonica* Spr, and put in the alcove in the main room, **tokonoma**, or in other ritual places such as the **kamidana**, the shelf for the Gods, and the **kudo**, the kitchen stove. Japanese also eat glutinous rice in soup with vegetables or meat, called **zoni**, at the first breakfast of the New Year. **Mochi** is additionally offered to Shinto shrines and Buddhist temples. For celebrating family birthdays, the entrance of children into school, weddings, or other festive occasions, steamed glutinous (or waxy) red rice is prepared, stained with the red seeds of the Azuki bean, *Vigna angularis* (Willd) Ohwi et Ohashi, with which it is cooked. Thus, for ritual occasions Japanese prefer to prepare and eat glutinous rice, while boiled, non-glutinous rice is eaten at regular meals.

In some parts of Japan similar **mochi** cakes are prepared from glutinous foxtail millet or common millet. Indeed, the ritual use of glutinous endosperm cereals can be seen not only in Japan but also in other East Asian countries.

In Japan, glutinous rice is also important in non-ritual activities. Rice straw of glutinous form is used for making **shimenawa**, a decorative rope hung on the **torii** gate of Shinto

Figure 8.1. Pounding of Japanese **mochi** cake
a) Steaming glutinous rice in basket steamer
b) Pounding **mochi** in a stone mortar with a wooden pestle

Figure 8.2. The **Torii** gate of a Shinto shrine decorated with **shimenawa**

shrines (Figure 8.2), and the making of **shimekazari** (a hanging decoration at the entrance door) or **takarabune** (a treasure boat signifying prosperity) for a prosperous New Year. In the past, glutinous rice straw was used even more widely to make rope, mats and shoes than the straw from non-glutinous rice because of the softness of the fibres [Sakamoto 1989].

In this paper I will establish the kinds of cereals in which glutinous-endosperm cultivars are found, their geographical distribution, and the nature of cultural practices specifically based on starch use within the distributional range. The study illustrates some aspects of the close relations which may obtain between specific plant domesticates and a people's way of life.

The Characteristics of Non-glutinous versus Glutinous Starch

Non-glutinous (or non-waxy) and glutinous (or waxy) starches are differentiated by the relative amount of two types of starch components, amylose and amylopectin. Non-glutinous starch consists of about 20 per cent amylose and 80 per cent amylopectin, while glutinous starch contains only amylopectin. The two can be distinguished clearly and easily by staining with an iodine iodine-potassium (I-KI) solution. Non-glutinous starch stains blue, while glutinous starch stains brownish red.

Glutinous reserve starch can be found only in glutinous (waxy) cereal varieties, and can be observed only in the endosperm (triploid) or in pollen grains (haploid). Starches detected in other plant tissues or organs, such as in leaves, stems, embryos or other diploid parts, are always non-glutinous, even in the glutinous varieties. The reserve starch deposited in vegetatively propagated organs, such as root tubers, bulbs or corms, is exclusively non-glutinous. There is no report of wild grain species containing glutinous starch. They have only non-glutinous starch.

It is well known that the glutinous character is controlled by a single recessive gene. This means that the glutinous character must have resulted from a recessive mutation occurring at the non-glutinous locus. This gene has been located on particular chromosomes, for example, chromosome I in rice and barley and chromosome IX in maize. The spontaneous mutation rate of this locus has been observed in maize grain and pollen grain and barley pollen grain. It occurs at the level of 10^{-4} – 10^{-6}, similar to the frequencies of other gene loci.

Both non-glutinous and glutinous varieties are known to be found in seven cereals of the family Gramineae, namely rice (*Oryza sativa* L.), foxtail millet (*Setaria italica* P. Beauv), common millet (*Panicum miliaceumn* L.), sorghum (*Sorghum bicolor* Moench), Job's tears (*Coix lacryma-jobi* L. var. *ma-yuen* Stapf), barley (*Hordeum vulgare* L.) and maize (*Zea mays* L.). The occurrence of the glutinous character is not related to phylogenetic groups in Gramineae but occurs as a parallel variation in different grass groups. For example, glutinous varieties are found in Asian rice but not in African rice (*Oryza glaberrima* Steud). Glutinous forms are observed in barley but not in bread wheat (*Triticum aestivum* L.): both dispersed from the Middle East to East Asia in prehistoric times and have been cultivated extensively as winter crops. Glutinous forms are common in East Asian landraces of foxtail millet and common millet, but not in Japanese barnyard millet (*Echinochloa utilis* Ohwi et Yabuno), even though these three millets have played quite an important role in the nutritional life of East Asian countries. Cereal crops, including the glutinous form, are almost exclusively diploid (2x) species, with the exception of common millet, which is tetraploid (4x). A recessive glutinous mutation at the non-glutinous locus can be expected and the phenotypic expression of the glutinous character can be found rather easily in diploid cereal crops. On the other hand, its expression is almost nil in tetraploid or polyploid plants. This is the reason why we cannot find glutinous varieties in bread wheat (6x), Japanese barnyard millet (6x) and finger millet (*Eleusine coracana* Gaertn), which is tetraploid (4x).

The earliest record of glutinous cereal varieties can be found in the Li Ki (Record of Rites), a Chinese classical book of the Han Dynasty (206 BC–AD 220). The terms **tu** and **shu** indicate glutinous rice and glutinous common millet respectively. In Pen Tsao Kang Mu, a famous Chinese Herbal and Materia Medica written by Shi-Chen Li [1578], glutinous varieties of rice, foxtail millet, common millet, sorghum, Job's tears and barley are described, indicating their popularity in Chinese society of the sixteenth century. Glutinous starch was first stained brownish red using the iodostarch reaction for rice by Gris [1860], and then repeated for common millet, sorghum, maize, Job's tears and barley, in that order [Eriksson 1969]. Very recently non-glutinous and glutinous perisperm types have been also reported in *Amaranthus hypochondriacus* L. [Okuno and Sakaguchi 1981; Sakamoto 1989, 1993; Sugimoto et al. 1981].

Geographical Distribution of Glutinous-Endosperm Varieties in Seven Cereal Crops

Mainly based on data recently compiled by Sakamoto [1982, 1989], the geographical distribution of glutinous endosperm varieties in seven cereal crops can be summarized briefly.

Rice (*Oryza sativa*). Rice is traditionally cultivated extensively in East Asia and glutinous forms have been well known in China for 2,000 years (see above). After examining 1,042 indigenous landraces from nine geographical regions from India to Japan, Nakagahra [1985] concluded that: (1) in India, Sri Lanka and Nepal there were almost no glutinous forms; (2) waxy varieties were abundant in samples from Indochina, Thailand, and Myanmar (Burma); while (3) rather high frequencies for non-

Table 8.1. Geographical distribution of non-glutinous and glutinous landraces of foxtail millet [Sakamoto 1982]

Regions	No. of strains	non-glutinous	glutinous	glutinous[5] (%)
Japan (1972–9)				
Tohoku	2	0	2	(100)
Kanto	2	1	1	(50)
Chubu	19	8	11	57.9
Kinki	42	1	41	97.6
Shikoku	33	3	30	90.9
Kyushu	19	8	11	57.9
Okinawa	22	4	18	81.8
subtotal	139	25	114	82.0
Korea (1977–8)	90	30	60	66.7
Cheju Island (1973–7)	12	7	5	41.7
subtotal	102	37	65	63.7
China (1975)[1]	25	22	3	12.0
Taiwan (1972–7)	28	3	25	89.3
Lan Yü (1977–8)	19	8	11	57.9
Batan Island (1973–7)	22	22	0	0
Luzon Island (1977)	4	2	2	(50)
Halmahera (1976)	5	5	0	0
Thailand (1974)	1	0	1	(100)
Nepal (1975–6)	5	5	0	0
India (1977)[2]	32	32	0	0
Afghanistan (1977–8)	13	13	0	0
Central Asia (1974)[3]	7	7	0	0
Europe (1975–9)[4]	4	4	0	0
total	406	185	221	54.4

1. from National Institute of Agricultural Research, Japan
2. from two Institutes in India
3. from All-Union Institute of Plant Industry, Leningrad
4. from three botanical gardens in Europe
5. Figures in parentheses indicate percentages obtained from very few strains

glutinous and low frequencies for glutinous ones were found from Japan and northern China. Watabe [1967] has indicated that glutinous rice cultivation in Southeast Asia appears to focus on northern and northeastern Thailand and Laos, spreading to the surrounding regions of Myanmar, Yunnan, Vietnam and Cambodia.

Foxtail millet (*Setaria italica*). Foxtail millet has been cultivated very widely in most parts of Eurasia since ancient times. As shown in Table 8.1, of the 406 samples collected by the present author at various locations in 1972–79 from Europe to East Asia, 221 had glutinous endosperm (54.4 per cent) and 185 had non-glutinous endosperm (45.6 per cent) [Sakamoto 1982]. The former was found in Japan, Korea, China, Taiwan, Luzon Island in the Philippines and Thailand, but only the non-glutinous form was found in samples from the Batan Islands of the Philippines, Halmahera Island in Indonesia, Nepal, India, Afghanistan, Central Asia and Europe. Recent additional data indicate that glutinous forms were found in three samples from southern Yunnan and Guizo Provinces of China and in sixteen from northern Myanmar. The obvious gaps in the geographical distribution of the glutinous form have been recognized in two places. One is the Bashi Channel between Taiwan, including Lan Yü Islands, and the Batan Islands. The other seems to lie between Nepal-India and Assam. It is assumed that the Bashi Channel gap is related to the importance of the glutinous varieties in the ritual life of the native people inhabiting the Taiwanese mountains and the Lan Yü Islands. The other gap can be explained by the clear difference in food preferences between Nepal-India and Assam.

Common millet (*Panicum miliaceum*). Common millet has also been very popularly and extensively grown in Eurasia since ancient times. In spite of the wide occurrence of this crop in Eurasia, studies of the geographical distribution of glutinous varieties have not been made. To the extent that this could be checked by the author, the glutinous form was detected only in specimens from Japan, Korea and China. The samples from India, northern Pakistan, northeastern Afghanistan, Central Asia and Europe included only non-glutinous forms.

Sorghum (*Sorghum bicolor*). Sorghum is a typical African domesticate and the most important millet in Africa. This crop was brought to the Indian Subcontinent about 4,000 years ago and from there it has spread widely in Asia. The glutinous form of this crop has been reported from Java, India, the Philippines, China and Japan. According to the author's survey, the glutinous form is very common in Korea and Japan. Of ten samples from northern Myanmar, seven were non-glutinous, while the remaining three were glutinous. However, very few glutinous samples have been identified from Taiwan, Halmahera and Nepal. All the samples collected from Ethiopia by the author were non-glutinous.

Job's tears (*Coix lacryma-jobi* var. *ma-yuen*). Job's tears is widely grown as a grain crop in East and Southeast Asia, and it is also cultivated as a medicinal plant in China, Korea and Japan. Kempton [1921] observed both non-glutinous and glutinous forms in samples from Burma, India, the Philippines and China. According to a recent collection from northern Myanmar made by the author, of twenty-four samples, twenty-two were glutinous and two were a mixture of both forms. However, in East Asia, mostly the glutinous form is grown.

Barley (*Hordeum vulgare*). The glutinous form of barley is found in China, Korea and Japan. In Japan it is found on the islands of the Seto Inland Sea, areas facing the Seto Inland Sea and in northern Kyushu. According to Nakao [1950], this form can be classified into three morphological varieties, which have naked and purple-coloured grain in common: *violaceum* Kcke, *sikangense* E. Åberg and *purpureum* Nakao.

Maize (*Zea mays*). Glutinous maize was first found by Collins [1909] in a collection of specimens from China. This type of maize has generally been reported from northern Myanmar, Assam, China, Korea and Japan. Maize was domesticated in the New World, but the glutinous variety originated in East Asia, having developed after the introduction of maize.

The geographical distribution of the seven cereal crop species with glutinous-endosperm varieties is summarized in Table 8.2. It can be seen that the glutinous varieties of rice, foxtail millet, sorghum, Job's tears and maize are distributed over wide areas of eastern and southeastern Asia from the mountainous region of Assam to Japan and Indonesia. The glutinous form of common millet and barley is confined, so far, to China, Korea and Japan, while the glutinous form of these seven crops is not found in the western half of Eurasia, especially in the wide areas between India and Europe, and Africa and the Americas.

Table 8.2. Geographical distribution of glutinous landraces in seven cereal crops

Areas	rice	foxtail millet	common millet	sorghum	Job's tears	barley	maize
India							
Nepal	±			±			
Assam (India)	+				+		+
Myanmar	+	+		+	+		+
Thailand	+	+					
Laos	+						
Vietnam	+						
Philippines	+	+		+			+
Indonesia	+	+		+	+		
Taiwan	+	+		+	+		
China	+	+	+	+	+	+	+
Korea	+	+	+	+	+	+	+
Japan	+	+	+	+	+	+	+
Primorskii							+

Glutinous-Endosperm Starch Food Culture in Eastern and Southeastern Asia

In this paper the use of glutinous-endosperm cereal crops for ritual or daily food preparation is called 'glutinous-endosperm starch food culture'. Several typical examples of this culture in eastern and southeastern Asia are introduced [Sakamoto 1989].

Japan

A classification of randomly selected foods made from rice in Japan in terms of starch type (non-glutinous versus glutinous) and type of ingredient (grain versus flour) is illustrated in Figure 8.3, in which each food name is written in Japanese. It can be seen that glutinous rice starch plays as important and wide a role as non-glutinous rice starch. Techniques for the production, processing and preparation of local foods made from glutinous rice are complex.

In the mountain villages where paddy rice is difficult to grow, several kinds of millets have been used as traditional staple food

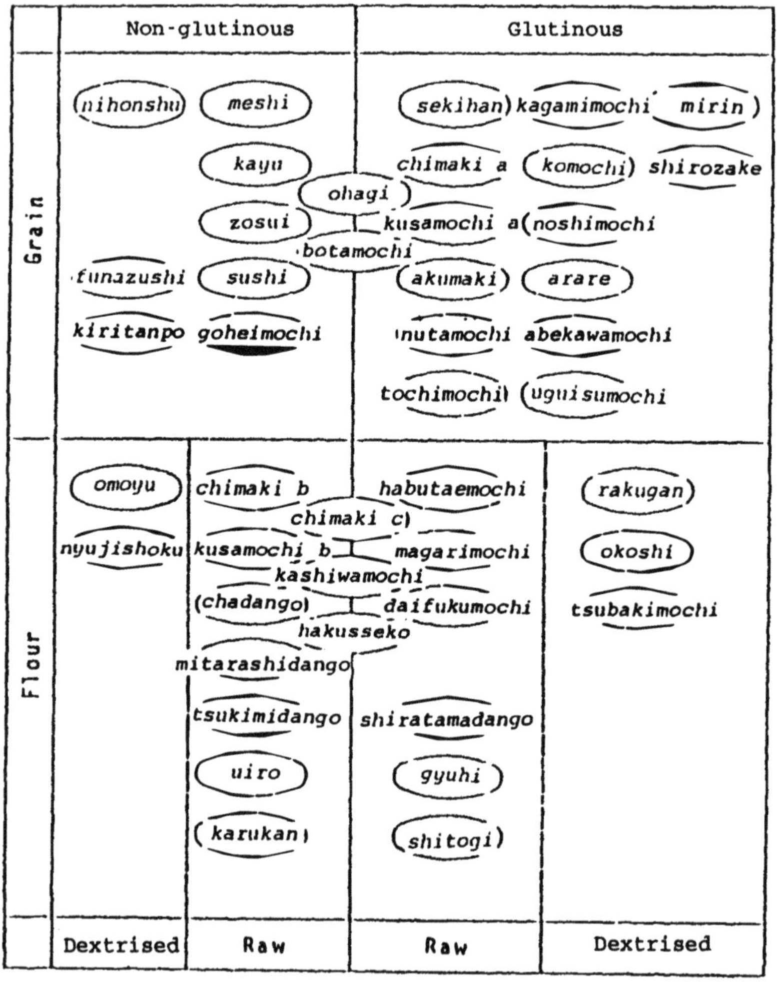

Figure 8.3. Classification of foods made from rice in Japan in terms of starch type and form of ingredients [Sakamoto 1989]

materials until recent times. For examples, in Outoson, Nara Prefecture, in the central part of the Kii Peninsula, foxtail millet, common millet, Japanese barnyard millet, sorghum and finger millet have been grown on a large scale. According to Takei et al. [1981], glutinous grains of foxtail millet and common millet are steamed and then pounded to make **mochi**. A dumpling is also prepared from the flour of glutinous common millet or from half-and-half mixed flour of non-glutinous and glutinous foxtail millet. Glutinous sorghum is ground into flour after removing its harsh taste by immersion in running water, then steamed or boiled for the preparation of dumplings.

Extensive utilization of glutinous forms of grain crops could be found until about forty years ago in Simo-Hoya, in the suburbs of Tokyo, where upland field cultivation has traditionally been practised. Upland rice, wheat, barley, foxtail millet, common millet, Japanese barnyard millet and sorghum have been grown there. The glutinous type of upland rice was frequently used not only to prepare **mochi** cake, but also to cook red-coloured steamed rice at various ritual occasions during the year. **Mochi** cakes were eaten to celebrate the New Year and then often preserved in clean salt water kept in a cool barn for up to about ten months for use as a ready food during the busy farming seasons. Steamed glutinous rice cooked with red Azuki beans or red cowpeas (*Vigna unguiculata* Walp) was served at rites of passage or other ceremonies. Glutinous-type grains of foxtail millet, common millet and sorghum were used frequently to prepare **mochi** cakes [Editorial Committee of Hoya City 1986].

Taiwan

In mountain villages on Taiwan the subsistence economy is based on swidden agriculture. For example, according to Sasaki and Fukano [1976], at Kinuran, a Rukai village in southern Taiwan, glutinous foxtail millet is regarded as the most important crop. Polished grains of glutinous foxtail millet are pounded in a wooden mortar with a wooden pestle or ground with a stone mortar, and then water is added to make a kind of wet dough. This is further kneaded by hand and made into a loaf 15–20 cm in length. It is then wrapped in banana or taro leaves and boiled in hot water. The product has a smooth surface and is called **abai**, a kind of dumpling, which is an indispensable food for guests. In Kinuran, local beer is also prepared from glutinous foxtail millet. The dumpling, **abai**, is kneaded by hand, then put in a clay pot with seeds of a chenopod (most probably a domesticated type of *Chenopodium album* L.) and water. Natural fermentation occurs in the sealed pot after two or three days and a sour-tasting beer called **kabawan** is produced. This plays an important role not only at the harvest festival, but also at other festivals and ritual ceremonies in Rukai society, and it is also served to guests.

Southwestern China

The intensive utilization of glutinous rice and diverse ways of cooking it are known from southwestern China. Zhou [1984] reports various glutinous rice foods and their preparation methods observed among Yao and Zhuang people of Gaungxi Province, among Dong and Miao people of Guizho Province, and among Dai people living in the southern part of Yunnan Province, as summarized in Table 8.3. In particular, Dong, Miao and Dai grow glutinous rice extensively and, interestingly, they eat glutinous rice at every meal, cooked with the steamer. At the same time they frequently pound steamed glutinous rice with a wooden mortar and pestle to make glutinous rice cakes, identical to Japanese **mochi**. Steamed glutinous rice dried in the shade and popped in frying oil is locally called **mihua**. Glutinous rice dumplings are made from rice flour, boiled in water, and then rolled in sugar or soybean powder. Glutinous rice also plays a significant role in ritual life. At a frame-raising ceremony for a newly built house, a red cloth and a panicle of glutinous rice decorate the ridge of the house, and the owner of the house scatters round-shaped glutinous rice cake from the ridge to the people participating in the ceremony. He then serves steamed glutinous rice, a soup with glutinous rice cakes and glutinous rice beer.

Table 8.3. Glutinous-endosperm rice culture in southwestern China (compiled from Zhou 1984)

glutinous rice food type: people	steamed glutinous rice	**mochi**[1]	**chimaki**[2]	**narezushi**[3]	local wine
Yao	ritual	+	+	+	
Zhuang	ritual		+		
Dong	daily	+		+	+
Miao	daily			+	
Dai	daily	+	+	+	

1. glutinous rice cake
2. steamed glutinous rice-flour cake
3. fermented glutinous rice with fish

Continental Southeast Asia

From southwestern China to Assam through Laos, northern Thailand and northern Myanmar, intensive cultivation of glutinous rice and its utilization in daily meals and/or on ritual occasions have been reported by Watabe [1967], who called this area the glutinous rice zone. Marshall [1922] reported that the Karen of Myanmar eat non-glutinous rice for ordinary meals, but they also grow many local landraces of glutinous rice that are boiled or steamed for an early morning meal or for special feasts. Steamed glutinous rice is sometimes mixed with sesame seeds and pounded in a mortar to make a sticky paste, locally called **tometopi**.

Insular Southeast Asia

Glutinous rice is grown widely, if in relatively small quantities, in many parts of insular Southeast Asia, where it is linked either to specific agricultural rituals or to more general festive occasions, such as the end of the Muslim month of fasting. Here, I mention just three geographically diverse instances.

According to Conklin [1980], in Ifugao Province of northern Luzon, Philippines, rice beer made from glutinous rice is used for festivals and served to neighbours, friends and visiting celebrants from other places. All field owners brew beer year-round for this purpose. The overall local rice-production ratio of glutinous to non-glutinous varieties is about 1:5. Dove [1985] reports that the Kalimantan Kantu' grow both non-glutinous and glutinous rice varieties, planting a minimum of one non-glutinous and one glutinous variety in each swidden. They prepare a special dish consisting either of cakes or a gruel made from glutinous rice. Finally, in Gamomeng village on the west coast of Halmahera, eastern Indonesia, bamboo rice is occasionally prepared for special events, such as a child's birthday or a visit by friends or relatives [Poniman and Takaya 1988]. First, a piece of banana leaf is put along the inner surface of a green bamboo stem 60–120 cm in length and 3 cm in diameter. Then, glutinous rice is put inside it. Coconut milk is poured over the contents and the bamboos roasted at an angle over the fire. When the outside of the bamboo is burnt, the rice is done; the bamboo is split into two halves and the sticky rice is cut and served. A very similar cooking method for glutinous rice was reported from northern Thailand by Iwata [1966]. The cooking method for glutinous rice using bamboo stems found in Halmahera may have been dispersed from continental Southeast Asia together with glutinous rice cultivation.

Origin and Dispersal of Glutinous-Endosperm Starch Food Culture

Glutinous-endosperm starch food culture based on the glutinous varieties of seven domesticated cereals (rice, foxtail millet, common millet, sorghum, Job's tears, barley and maize) is specific to eastern and southeastern Asia, along a broad arc traceable from Assam-Myanmar to Japan. I have noted here that this culture is particularly extensive and diverse in the mountainous areas running from Assam to southwestern China, through northern Myanmar, northern Thailand, Laos and Vietnam. I have elsewhere proposed that the centre of origin of this culture might be traced to this area [Sakamoto 1989], schematically indicated in Figure 8.4 by hatching. The main distribution of glutinous landraces of the seven cereal crops involved is also indicated in this figure, which clearly shows that the distribution of glutinous varieties and their utilization for food is confined only to eastern and southeastern Asia. From the centre this culture has dispersed eastward to Japan and southeastward to insular Southeast Asia.

Figure 8.4 shows a very obvious distributional break between Assam and peninsular India. In other words, a preference for glutinous-endosperm starch for certain foods has never spread westward from the hypothesized distributional centre to other parts of the south Asian subcontinent and the Himalayan mountain foothills, including Bhutan, Sikkim, Nepal, India and Sri Lanka. This apparent distribution cannot be explained easily, other than to suggest that people in those areas prefer non-sticky foods.

The preference for eating glutinous-endosperm starch or sticky foods must have been established fairly early in the evolution of cereal farming. In continental Southeast Asia it is

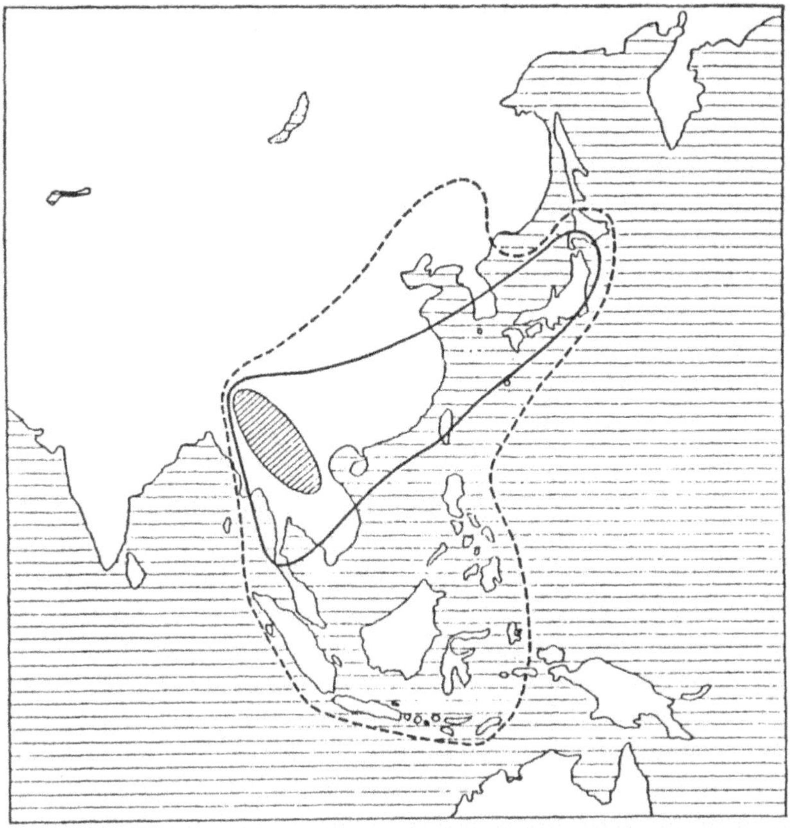

Figure 8.4. The centre of origin of glutinous-endosperm starch food culture (hatched), main distribution of the culture (solid line) and geographical distribution of glutinous landraces found in seven cereal crops (dotted line) [Sakamoto 1989]

widely assumed that people intensively collected and ate tuberous corms of wild taro and yam during the period of food-gathering and proto-cultivation, which would have provided a diet rich in sticky starch. It is also assumed that during the incipient agricultural stage they domesticated and developed many vegetatively propagated crops, including taro, yam and banana. This particular preference for eating sticky foods must be the effective basis for the selection of glutinous-endosperm mutants from normal non-glutinous cereal crops, which must have occurred when people in this area first introduced upland rice, different kinds of millet and finally lowland rice [Sakamoto 1989; Sasaki 1982]. At the time of the introduction of maize from the New World to East Asia in the fifteenth or sixteenth century, glutinous varieties of six cereal crops had already been established and were being widely utilized. Consequently, it might be hypothesized that people preferred the glutinous to the non-glutinous maize. This would indicate the existence of a food preference which also exerted selective pressure in favour of glutinous-endosperm forms of cereal crops in general.

Bibliography

Collins, G.N., 'A new type of Indian corn from China', *USDA Bureau of Plant Industry Bulletin* 161, 1909, pp.1–30

Conklin, H.C., *Ethnographic atlas of Ifugao: a study of environment, culture and society in northern Luzon*, New Haven and London: Yale University Press, 1980

Dove, M.R., *Swidden agriculture in Indonesia: the subsistence strategies of the Kalimantan Kantu'*, Berlin: Mouton, 1985

Editorial Committee of Hoya City (ed.), *Documentary Report on Simo Hoya's Folklore*, Hoya, 1986 (in Japanese)

Eriksson, G., 'The waxy character', *Hereditas* 63, 1969, pp.180–204

Gris, M.A., 'Note sur la Fécule du Riz', *Bulletin Societe Botanique de France* 7, 1860, pp.876–8

Iwata, K., *Birthplace of Japanese culture: ethnological expeditions to rice cultivating people in Southeast Asia*, Tokyo: Kadokawa Shoten, 1966 (in Japanese)

Kempton, J.H., 'Waxy endosperm in Coix and Sorghum', *Journal of Heredity* 12, 1921, pp.396–400

Marshall, H.I., *The Karen people of Burma: a study in anthropology and ethnology*, Columbus: Ohio State University Press, 1922

Nakagahra, K., The birthplace of rice and its cultivation, Tokyo: Kokinshoin, 1985 (in Japanese)

Nakao, S., 'On the waxy barley in Japan', *Seiken Ziho* 4, 1950, pp.111–13

Okuno, K. and S. Sakaguchi, 'Glutinous and non-glutinous starches in perisperm of grain amaranths', *Cereal Research Communication* 9, 1981, p.305

Poniman, A. and Y. Takaya, *Field research notes on traditional agriculture: I, Sumatra, Lombok, Flores, Timor and Maluku*, Kyoto: Nokobunka Kenkyu Shinkokai, 1988 (in Japanese)

Sakamoto, S., 'Waxy endosperm and perisperm of cereals and grain amaranth and their geographical distribution', *Journal Japanese Society of Starch Science* 29, 1982, p.41

——, Glutinous-endosperm and -perisperm starch food culture, Tokyo: Chuokoronsha, 1989, (in Japanese)

——, 'Characteristics and ethnobotany of waxy perisperm starch found in a species of grain amaranths, *Amaranthus hypochondriacus L.*', *Food and Foodways* 5, 1993 (in press)

Sasaki, K., *Shoyojurin culture complex; from Bhutan, Yunnan to Japan*, Tokyo: Nihon Hoso Shuppan Kyokai, 1982 (in Japanese)

——, and Y. Fukano, 'Swidden cultivation and agricultural rituals in a Rukai village', *Research Bulletin of the National Musuem of Ethnology*, Japan 1, 1976, pp.33–125

Sugimoto, Y. et al., 'Some properties of normal- and waxy-type starches of *Amaranthus hypochondriacus* L.', *Stärke* 33, 1981, pp.112–16

Takei, E. et al., 'Kii Sanchi ni okeru Zakkoku no Saibai to Riyo narabini Awa no Tokusei (Cultivation and utilization of millets in the mountainous area of Kii Peninsula, and characteristics of foxtail millet)', *Kikan Jinruigaku*, 4, 1981, pp.115–54

Watabe, T., *Glutinous rice in northern Thailand*, Kyoto: The Center for Southeast Asian Studies, Kyoto University, 1967

Zhou, D.S., Utilization of glutinous rice: sticky foods found in minority tribes, in K. Sasaki (ed.), *Under Shoyojurin forests in Yunnan*, Tokyo: Nihoo Hoso Shuppan Kyokai, 1984 (in Japanese)

Chapter 9

Creating Landrace Diversity: The Case of the Ari People and Ensete (*Ensete ventricosum*) in Ethiopia

Masayoshi Shigeta

Introduction

This paper proposes an analytical framework for the ethnobotanical study of the human-plant relationship as a case study of domestication, understanding domestication not as an event, but as a process of person-plant interaction [Anderson 1960]. Most previous discussions of plant domestication have supposed that humans have intentionally selected useful plants from among wild ones. This presumption is also reflected in our general attitude towards the classification of the taxonomic 'varieties' of cultivated plants. It has long been believed that human 'intention' and 'utility' are crucial to the creation of a new 'variety' of cultivated plant and for maintaining its diversity.

In the last two decades there has been a new trend in the study of plant domestication, which places the emphasis on human-plant interaction [Bye 1979; Odum 1971; Rindos 1984; Shigeta 1988]. The term 'domestication' is deliberately employed here as a neutral word to avoid implying that there is any 'intention' on the part of humans to transform plants into 'cultivated plants.' In Japanese, there is no proper translation or alternative term for 'domestication'.

'Domestication' is one of the key words in this study, because I believe that the essence of the subsistence activities of agriculturalists can be best understood as a process of interaction between humans and plants. It must have been in the early ('pre-modern') agricultural societies that human-plant interactions played the most important role in the subsistence of both humans and plants. In his comprehensive analysis of human-animal interactions based on socio-cultural research amongst Eurasian pastoral nomads, Tani [1976: 3] has pointed out that the subsistence activities of agricultural people can also be described as a kind of interaction-oriented behavior. However, he excludes gathering and the utilization of semi-cultivated plants from consideration. I have discussed human-plant relationships in relation to the wild plant utilization of agricultural peoples elsewhere [Shigeta 1987a].

The ecology of individual crop species and the differentiation of varieties have been dealt with to some extent by agronomists, though rarely examined in relation to specific human societies. Nor has the way in which cultivated plants survive in response to particular human cultures been given much attention in anthropological studies. One pioneer work is Bye [1979], on the interaction between Tarahumara Indians in South America and the useful plants of the genus *Brassica* [but see also Jackson, Hawkes and Rowe 1980 and Brush, Carney and Huaman 1981].

Cultivated plants entrust their reproduction almost entirely to human beings. As a result, cultivated plants have come to hold many convenient characteristics for us, such as the loss of seed dormancy and shattering and the gigantism of useful parts. At the same time, however, depending on humans is the cultivated plants' best means of survival, and we might almost say that cultivated plants have 'chosen' humans as their agent of reproduction. Cultivated plants rely on humans for their reproduction and, at the same time, the characteristics of the plants themselves determine a part of a people's culture and society.

The unintentional actions of the Ari that have led to the maintenance of genetic variation in wild ensete populations, and which I call folk in-situ conservation, is a typical example of such human-plant relationships [Shigeta 1990].[1] I would like to propose here that the two aspects of plant domestication, i.e. how people utilize a plant and how the plants rely on people for reproduction, must be studied together in one society. Only with such first-hand data will we be able to trace historical changes in human-plant relationships and examine the typology of the relationships. This study constitutes a prelude to an overview of human-plant interactions and to the examination of human-plant relationships as a conceptual framework in the study of domestication. I believe that the study of the ensete-Ari relationship with regard to the creation and maintenance of 'variety' provides a good counter-example to the long-standing conventional view on plant domestication.

First, the Ari way of classifying ensete, based on their recognition of the characteristics of landraces, is analysed. I contend that Ari activities towards the landraces play an indispensable role in maintaining and increasing their diversity. Secondly, I contend that the diversity in ensete plants, which serves as a major predicate for Ari cognition, is neither generated by human effort alone nor solely by the plants themselves, but rather, generated by the interaction of both. I demonstrate that the ability to recognize diversity in a plant population is a major dynamic force for 'diversified selection' in interaction between people and plants. I propose to call this type of diversified selection 'cognitive selection'.

Ari Recognition of Ensete Landraces

Vernacular Names

Ensete, *Ensete ventricosum,* is a giant monocarpic evergreen perennial related to the banana, the root, leaf-stems and inner-bark pseudostem of which provide food. It flowers and seeds at the end of a life-cycle of between eight and fourteen years, depending on altitude. Ari call the cultivated ensete **agemi**, while **gela** denotes the wild one, as well as the seedlings spontaneously growing in the garden. They recognize not only the difference between the wild and cultivated ensete but are also keenly aware of distinctions between varieties of the latter. These varieties are classified by the Ari, and each has its own name.

In order to distinguish Ari categories from 'varieties' in the conventional Western sense, I prefer to use the term 'landraces,' which is a unit for grouping cultivated ensete plants by their common characteristics reflected in specific vernacular names. I have recorded more than one hundred vernacular names for cultivated ensete landraces. Of these, seventy-eight names can be identified as labelling distinguishable landraces, after excluding synonyms, dialectal and geographical variants.[2]

Table 9.1. Frequency distribution and the order of named ensete landraces recalled by 39 informants

Landrace names	Informant Number 31	32	29	35	33	28	21	38	6	34	37	39	3	4	10	22	23	42	15
1. gena	1	3	1	2	10	2	3	4	10		7	5			2	1		4	2
2. joolak	11	9	7	16				8	2				1	1	1		1		
3. karta	3	13	3	11	14	1	2	3	6		5		6					3	
4. kaksa	13			9								2		3	3		3	2	
5. mooset	14	14	11	10	4	5		5	3	5			2		6	6			
6. garacha			13						8		2		7	4				1	3
7. salta			15	6					9		4	1						7	
8. daakai	8	7				7			1		9		3				2		
9. maga	20	15		12			5	6			1			2					
10. oosade	6		10	7		9							5		5		4		
11. alaka				1	3	6			5			3							
12. arpa	21	12		4	11				7							2			
13. gufalak			9	8	7					10				5		3			
14. gaya			14					1						6	7				
15. intada				5	13									7		7			
16. tsala	5	4	4	13	2														
17. babsul, babsuloo			17				7				10								
18. buguni	2				1	3				1									
19. mana, monna, moonet						4			4								5		
20. zinka	17		12					2											
21. dampa			6									8							
22. dibla				15						7									
23. gola, goolet			5				1												6
24. ila				3									4				6		
25. komcha, kumcha		18				11				6									
26. segenda, siigenda			8													5			4
27. shupalak	18		16													4			
28. sigi, sire				17	12	14													
29. washinga	4		2																
30. worija mocha	7	5				8													
31. zelgina											3	7					7		
32. a sarat, asarett					6					2									
33. anetsa, antsa		2																	
34. arbi								11		8									
35. barga												9						6	
36. boobalakk												4						5	
37. dama						13					8								

	16	20	27	36	41	2	9	19	25	26	30	40	8	18	1	17	5	11	24	12	A	B	C
	3		1	1	2		2		1	3	3	1		3		3		2			27	82	3.0
	2	6	6	4	1	1	3	4	5	1											20	90	4.5
	6		2	2	4				2		2	2				2					20	92	4.6
		5			3	3	1	3			1	4	1	4	2		1	1			19	64	3.3
	1								4	4	4				3						17	101	5.9
		4				4				5		3		1	1	1	2			1	16	60	3.8
4					5	5							2						1		11	59	5.3
			5	3					3												10	48	4.8
5				6						2											10	74	7.4
							5				5								2		10	58	5.8
							4	1				5									8	28	3.5
								2													7	59	8.4
																					6	42	7.0
						2															5	30	6.0
					6																5	38	7.6
																					5	28	5.6
														2							4	36	9.0
																					4	7	1.8
		2																			4	15	3.8
		1																			4	32	8.0
				5																	3	19	6.3
								5													3	27	9.0
																					3	17	4.0
																					3	13	4.3
																					3	35	11.0
																					3	17	5.6
																					3	38	12.0
																					3	43	14.0
			3																		3	9	3.0
																					3	20	6.6
																					3	17	5.6
																					2	8	4.0
		3																			2	5	2.5
																					2	19	9.5
																					2	15	7.5
																					2	9	4.5
																					2	21	10.5

Table 9.1. (Continued)

Landrace names	Informant Number 31	32	29	35	33	28	21	38	6	34	37	39	3	4	10	22	23	42	15
38. damnett																			5
39. gofa, gopo												6							1
40. kaago, kaakett					15	12													
41. lusakk						10		9											
42. maza	19						4												
43. shooka		10								4									
44. tapna				14	9														
45. alfakomtso																			
46. anka											6								
47. ankuma							6												
48. baisametocchi																			
49. bargedd							10												
50. buttomosu		6																	
51. chanli	10																		
52. daatsakan					5														
53. gaste chanli	9																		
54. harbi													8						
55. irsuna								10											
56. katsumi		11																	
57. nodikutu							9												
58. nodopuls					8														
59. salkaaraakett										3									
60. salpri		17																	
61. satsa										9									
62. sekaar							11												
63. shoolaka		16																	
64. sippa		1																	
65. tsaami genna		8																	
66. tsisi	12																		
67. tsooka	15																		
68. tsoparakk								7											
69. uba joolakk															4				
70. ursinda							8												
71. zoota	16																		
	21	18	17	17	15	14	11	11	10	10	10	9	8	7	7	7	7	7	6

Total number of landraces referred to

A: Number of informants referring to each landrace.

B: Sum of the numbers of recalled order. Total references made to each landrace.

C: Average of B.

16	20	27	36	41	2	9	19	25	26	30	40	8	18	1	17	5	11	24	12	A	B	C
		4																		2	9	4.5
																				2	7	3.5
																				2	27	13.5
																				2	19	9.5
																				2	23	11.5
																				2	14	7.0
																				2	23	11.5
											4									1	4	
																				1	6	
																				1	6	
											3									1	3	
																				1	10	
																				1	6	
																				1	10	
																				1	5	
																				1	9	
																				1	8	
																				1	10	
																				1	11	
																				1	9	
																				1	8	
																				1	3	
																				1	17	
																				1	9	
																				1	11	
																				1	16	
																				1	1	
																				1	8	
																				1	12	
																				1	15	
																				1	7	
																				1	4	
																				1	8	
																				1	16	
6	6	6	6	6	5	5	5	5	5	5	5	4	4	3	3	2	2	2	1	298	1729	

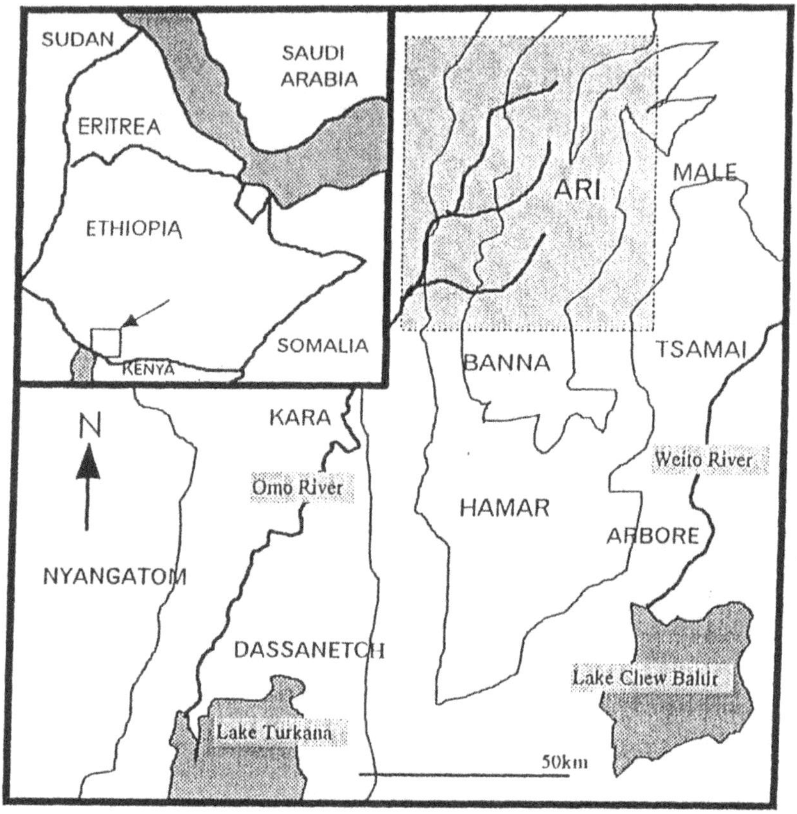

Figure 9.1. The Ari and their neighbours

Table 9.1 shows the frequency distribution and ranking of ensete landraces recognized by thirty-nine informants.[3] They refer to 71 different landrace names, and 298 names in total. The smallest number of recounted names was one, by Informant No. 12, while the most was 21, by Informant No. 31. The average number of landraces referred to by informants was 7.6. The mode was between five and six landraces.

The highest frequency of reference was recorded for **gena** landrace, which 27 out of 39 informants referred to. The second was **joolak,** followed by **karta, kaksa,** and **mooset** in descending order. Interestingly, only ten landraces were referred to by more than ten informants, and only 16 landraces by more than five informants. The remaining 55 landraces were mentioned by only one to four informants. Twenty-seven landraces (38 per cent of the total number) were referred to by only one informant. It is noteworthy that among the Ari, popular landraces amount to ten, the others being known only to a few people.

Geographical Differences in Ensete Landrace Recognition

In the lowlands (**dawla**), data were obtained from three traditional chiefly territories.[4] Table 9.2 shows the six landrace names that were referred to with the highest frequency by the people residing in these geographical areas. In the lowland, the landrace names that were referred to frequently were common in all three territories, i.e. **kaksa**, **gena**, and **mooset**. But the number of landrace names which were referred to by only one informant was also uniformly high in all three areas.

Table 9.2. Most frequently provided landrace names: a comparison of Sida, Biyo and Baka chiefdoms (lowland) with the highland area

Chiefdom:		Lowland		Highland
	Sida	Biya	Baka	
Order				
1	kaksa	gena	gufalakk	gena
2	garacaha	kaksa	mooset	karta
3	gena	alaka	asarat	daakai
4	joolak	karta	arpa	joolak
5	karta	salta	buguni	maga
6	mooset			mooset

In the highland area called Dizi, approximately between 1,600 and 1,800 m, ten informants gave a total of 113 landrace names. On average, each informant gave 11.3 landrace names, a number comparatively higher than those of the lowlands.

The results can be summarized as follows:

(1) Five to six landrace names were widely known to the people of all the territories.

Table 9.3. Frequency distribution and order of named landraces recalled by 22 informants in the Sida area indicating age and sex

Landrace names	Informant Number 38	(17)	12	(20)	18	11	(23)	8	9	(42)	10
kaksa				5	4	1	3	1	1	2	3
garacha		1	1	4	1					1	
gena	4	3			3	2			2	4	2
joolak	8			6			1		3		1
karta	3	2								3	
mooset	5										6
oosade							4		5		5
salta								2		7	
gaya	1										7
alaka									4		
daakai							2				
maga	6										
mana				2			5				
arpa											
ila							6				
zinka	2			1							
alfakomtso								4			
anetsa				3							
arbi	11										
babsul					2						
baisametocchi								3			
barga										6	
boobalakk										5	
damnett											
dibla											
gofa, gopo											
gola, goolett											
gufalakk											
harbi											
intada											
irsuna	10										
kusakk	9										
segenda											
tssoparakk	7										
uba joolakk											4
zelgina							7				
Total number of landraces referred to	11	(3)	1	(6)	4	2	(7)	4	5	(7)	7
Age and sex	7	(11)	13	(14)	15	17	(18)	20	20	(20)	25

Numbers in parentheses are those of women informants.

A: Number of informants referring to each landrace.

16	24	30	1	2	3	4	6	19	5	15	A	B	C
		1	2	3		3		3	1		14	33	2.3
			1	4	7	4	8		2	3	12	37	3.0
3		3					10			2	11	38	3.4
2				1	1	1	2	4			11	30	2.7
6		2			6		6				7	28	4.0
1		4	3		2		3				7	24	3.4
	2	5			5						6	26	4.3
4	1			5			9				6	28	4.6
				2		6					4	16	4.0
							5	1			3	10	3.3
					3		1				3	6	2.0
5						2					3	13	4.3
							4				3	11	3.6
							7	2			2	9	4.5
					4						2	10	5.0
											2	3	1.5
											1	4	
											1	3	
											1	11	
											1	2	
											1	3	
											1	6	
											1	5	
										5	1	5	
								5			1	5	
										1	1	1	
										6	1	6	
						5					1	5	
					8						1	8	
						7					1	7	
											1	10	
											1	9	
										4	1	4	
											1	7	
											1	4	
											1	7	
6	2	5	3	5	8	7	10	5	2	6	116	434	
25	25	25	25	30	30	30	30	30	40	40			

B: Total references made to each landrace.
C: Average of B.

(2) The majority of the landraces were only known to a limited group of persons or to an individual.

(3) Clear differences between the highlands and the lowlands with regard to landrace names could be found, although many landrace names were common in both areas.

Differences in Landrace Recognition According to Age and Sex

Data on age differences in landrace recognition obtained from 22 informants of the *Sida* chief's territory are shown in Table 9.3. The names of the landraces are arranged in descending order of frequency of reference.

In terms of the number of landrace names referred to, there seems to be no significant difference between different age groups. Informant No. 38, who was seven years old, gave an exceptional number of names (n=11). However, there was no significant correlation between age and the number of landrace names mentioned (r=-0.018). Therefore, it can be concluded that regardless of age, people know the popular landrace names as well as a few rare ones.

Since only a few data were obtained from females, it might not be appropriate to judge sexual differences in landrace recognition of ensete. However, as far as the present data show, there was no bias for women to know certain landrace names (informants Nos. 20, 23 and 42 were women and are in parentheses in Table 9.3). Among the Ari, propagation and transplanting are mainly done by men. Women mainly engage in the harvesting, processing and cooking of ensete. However, both women and men keep close contact with the important aspects of the life cycle of ensete. Agricultural activities such as weeding and thinning leaves are undertaken by both sexes. Thus, the results suggest that there is not much difference in Ari landrace recognition in terms either age or sex. This also implies that knowledge of ensete cultivation, so critical to the Ari way of life, is widely and rather evenly distributed, regardless of age and sex.

The Process of Learning Landrace Names

Girls of five to six years old are quite often seen helping or imitating their mothers' work in the ensete field, engaged in activities such as cutting, processing and food preparation. Both boys and girls help their parents, and even engage in agricultural activities themselves by the age of five to six. It is likely that young Ari children acquire various kinds of knowledge about ensete, such as landrace names, propagation, differential characteristics and methods of processing, through the observation of daily household activities. I asked eleven pupils in the sixth grade of Metsar Primary School to list all the landrace names they knew, to determine when children started acquiring ensete knowledge (Table 9.4). The age of the students ranged from twelve to eighteen years. There was only one girl student (Informant No. 6).

The students gave 49 names in total, and about 4.5 names per person on average. The minimum number of landrace names recounted per person was two, while the maximum was eight. Table 9.4 indicates that the overall tendencies do not show much difference from the results presented in Table 9.1. When the children were asked where they learned the names of ensete landraces listed, most of them acknowledged that they learned them at home, where ensete plants were actually planted in their home garden. Usually, Ari men of between fifteen to eighteen are married, and independent of their parents, building their own houses. Girls are married between thirteen and sixteen. Sixth-grade primary school students, who sometimes drop-out of school to marry, appear to have as complete a knowledge of ensete landraces as older Ari people.

The Meaning of Landrace Names, and their Etymologies

The etymologies of the many landrace names are presently not known to the Ari. Of the 78 names recorded, only 18 were revealed to have certain meanings. These are listed in Table 9.5. It is noteworthy that the meanings of most of the well-known landrace names are lost. Five names are derived from wild plant names, while six indicate kinds or tastes of food. For example, **tsala** was originally the local name of *Arisaema* sp., whose way of

Table 9.4. The knowledge of ensete landrace names among sixth-year primary school children in Metsar village

Landrace names	Informant Number											Frequency of landraces referred to
	9	14	5	8	10	4	6	1	2	13	15	
kaksa	5			1	4		1	3	2	1		7
karta	2	7	5			3		1				5
gena	1	1	1		1	2						5
joolak	3		4	3	3					3		5
osade				2	5				1		2	4
gufalak	8					4				2		3
mooset			3	4							1	3
garacha	4	2					2					3
daakai	7	3					3					3
salta					2			2				2
baka sulay		6					4					2
maga	6	4										2
mana									3			1
gaya						1						1
gel			2									1
tilagiees				5								1
intada		5										1
Number of recalled landraces	9	7	5	5	5	4	4	3	3	3	2	49
Age	18	17	15	16	14	14	13	12	14	17	14	

processing very much resembles that of ensete. **Zergi** means 'wheat,' and people explain that this name came from the similarity of tastes between wheat and ensete. In other cases, the names of locations, ways of propagation, personal names and so on have become attached to the landrace.

The folk etymologies provided for ensete landrace names were not consistent between informants. Some explained that the landrace name **oisi** meant bamboo, because it grows as quickly as bamboo. However, many others were not so sure. Similar ambiguity can be seen in ensete landrace names that are the same as, or related to, food. I have witnessed several cases where people easily changed the meaning of landrace names, according to the context of the discussion. Given such limited and inconsistent data it may be safest to treat vernacular landrace names as arbitrary labels to distinguish groups of plants.

Table 9.5. Meaning of vernacular landrace names

Landrace name	Gloss	Explanation
1 **barga**	finger millet	Whole appearance looks like finger millet
2 **tsala**	*Arisaema* sp.	Taste and colour of the fermented starch like that of *Arisaema* sp.
3 **arbi**	a weed, Compositae	?
4 **oisi**	bamboo	Grows fast like bamboo
5 **zergina**	**zergi** = wheat	Taste and colour like that of wheat
6 **ila**	flour	Tastes like cereal flour when made into fermented starch
7 **washinga**	**washi** = fermented starch	? Suitable for making **washi**
8 **daatsakan**	**daatsa** = porridge	Suitable for making porridge
9 **goolet**	**gola** = a kind of alcoholic beverage	Suitable for making beer
10 **mooset**	**mosa** = corm, or name of recipe	Suitable for cooking as **mosa**
11 **sekaar**	sugar	Tastes sweet
12 **gofa**	place-name	Brought from Gofa area
13 **intada**	by itself	Grows suckers by itself
14 **katsumi**	spear	Appearance is like a spear
15 **mana**	caste name	?
16 **worja mocha**	personal name	?
17 **ubajoolak**	**uba** = place-name	Mainly cultivated in Uba area. Appearance differs from **joolak**
18 **tsaami gena**	**tssami** = white	More whitish than **gena**

Differential Characteristics of Landraces as Criteria of Recognition

The importance of Ari landrace names as identifying labels cannot be discussed apart from how people give such names by recognizing the characteristics of each landrace. In other words, it is important to find out what kinds of characteristics are used in order to identify the plant groups by landrace names.

In the case of cultivated ensete, the various characteristics of ensete plants *per se* are the focus of the identification. In this sense, to regard landrace names as 'mere labels' might be misleading. Naming a discrete population of cultivated ensete must be more complex than the mere labelling of the colour categories, which are essentially continuous in nature. As the naming of colour categories is a secondary process of cognition [Lenneberg 1967], it is necessary to categorize the colour spectrum before naming plants. However, a person's recognition of plants begins with the ability to discriminate the varied discontinuous characteristics of the plant population.

In the case of ensete, there are occasions when the recognition of ensete by people occurs almost simultaneously with the recognition of a group of plants. Of course, there might be cases of recognition without recalling landrace names, but it should be remembered that this is exceptional in Ari-ensete relationships.

In such circumstances, Ari-ensete identification can best be explained by the analogy of human face-to-face relationships. People identify individual ensete plants in their field just as they remember an individual person by his or her face. Thus in this case the landrace name is a kind of proper noun.

Identification of individual ensete plants is also made easier by the fact that Ari are surrounded by ensete plants in their garden from the beginning to the end of the ensete life-cycle. Through such an intimate relationship with the ensete, people can remember many landrace names. It is plausible to assume that rare landrace names are preserved through this manner of identification.

Identification of Landraces Through Analogy

The other and more common role of landrace names is, of course, to serve as a kind of label to facilitate identification in terms of shared classificatory knowledge. People recognize variation in the ensete population according to their classificatory rules and judge or select the necessary observable (perceptual) variations from overall variation. As a result of this conceptualization of variation (or the process of recognition) a specific landrace name is chosen from the repertory of existing landrace names. Compared to face-to-face identification, this mode of landrace identification is achieved through analogy, i.e. the identification of similarity among several indicators (characteristics).

Table 9.6 lists the fourteen ensete landrace names with the characteristics that are popular among the people of Dumtseter village. Eight characteristics employed in their classification are listed in Table 9.7.

Pigmentation of the plant body is always among the first that people recognize as an important characteristic for the identification of landraces. Characteristics of secondary importance vary, depending on the landrace. These characteristics can be grouped into three:

(1) Morphological characteristics in outer appearance.

(2) Characteristics which cannot usually be directly observed but which can be recognized through a long period of observation, i.e. those related to the life-cycle and means of propagation.

(3) Characteristics that are directly related to the utilization of ensete, such as usage and taste.

Ari can pick out a landrace easily, even in an unfamiliar garden, because they identify a landrace name with outward perceptual criteria (1). With information acquired through long-term familiarity (2) and utilization (3), the identification becomes more correct.

I do not deny that this form of identification might also possibly be employed in the case of face-to-face identification. Even if an ensete landrace name is deemed self-evident by

Table 9.6. Some ensete landraces with their characteristics

Vernacular name of landrace	Characteristics
aasha	Little pigmentation on the midribs and petioles. Roundish at the basal part. Compared with **daakai**, it has wider leaves and is blackish at the base of the pseudostems.
alaka	Little pigmentation on the midribs and petioles. Corm quality is similar to **daakai**, but **alaka** boils more quickly. Narrow leaves with no creases.
daakai	Thin red stripe on the midribs and petioles. Blackish at the base of pseudostems. Leaves with creases along with veins.
garecha	Very whitish midribs and petioles. Plant height is low and slow growing. Takes six years to reach the height of a two-year **daakai**. Short petioles. The corm and leaves are large.
gena	Gigantic. Reddish brown to black midribs and petioles. Used only for the fermented food stuff called **washi**. The corm cannot be eaten. Takes about five years to mature.
intada	Peculiar landrace which lost its apical dominance. Has voluntary suckers like those of bananas. The corm and basal part of the pseudostems do not grow large. Leaves are utilized for various material purposes. Not utilized for food.
joolak	Red pigmentation on the midribs and petioles. Red spots on the basal part of the pseudostems. Narrower leaves than **shuupalak**.
kaksa	Red pigmentation on the midribs and petioles. Leaves with creases and easily torn. Not suitable for wrapping purposes. Corm has a bitter taste. Takes three years to flower.
mooset	Red stripe on the midrib and petioles. The corm is boiled easily and gives floury starch.
oosade	Dark red to black colour on the midrib and petioles.
salta	Red pigmentation on the midrib and petioles. Narrower leaves and longer petioles than those of **shupalak** and **joolak**. Compared with **tsala**, it grows as quickly as bamboo.
shupalak	Red pigmention on the midrib and petioles.
tsala	Even pigmentation from the basal part of the petioles to the top. Slower growing than **salta**.
zinka	Red pigmentation on the midrib and petioles. Long and wide petioles. Narrow and tough leaves. Petioles twist upwards.

someone, a person would often list reasons that would enable categorization in terms of characteristics in the order (1)-(2)-(3). Figure 9.2 illustrates a case of Ari identification using these

Table 9.7. Criteria for classification of ensete by the Ari people

1	Pseudostem: colour, pigmentation, length
2	Basal part of pseudostem: colour, pigmentation, shape (swelling)
3	Leaves: width, shrinkability, thickness, toughness
4	Plant configuration: height, openness of the leaves
5	Growth speed: maturity duration, flowering period
6	Use: good for fermented products, for boiling and for wrapping
7	Taste: sweetness, fluorines
8	Propagation: by artificially induced bud, by self-induced bud

classifactory rules. This scheme is deliberately constructed by myself as a bifurcate key-table. The order of the characteristics reflects, as much as possible, those mentioned by Ari farmers. However, for some readers, Table 9.6 may contain far more information than Figure 9.2, and not every Ari follows completely the same method of identification. Therefore, although Figure 9.2 is complete and correct for the Ari at each branching point, individual informants may sometimes fail to identify certain landraces.

Most probably, Ari reach the right answer by using the first group of characteristics. Their means of recognition may be similar to pattern recognition: in other words, they identify the specific characteristics and the stratification is apparent to them at a glance.

Landrace Groups

Popular landraces are usually identified by the second and third types of characteristic discussed in the previous section on analogy. This is particularly true in the case of landraces that have specific uses. These specialized landraces can be categorized into three groups:

(1) **Washi**: specialized for use in **washi** (fermented ensete) processing (**gena** is a representative landrace of this group).

(2) **Mosa**: specialized for use in **mosa** cooking, that is, the boiled corm and basal parts of pseudostem (**mooset** and **maga** are two representatives).

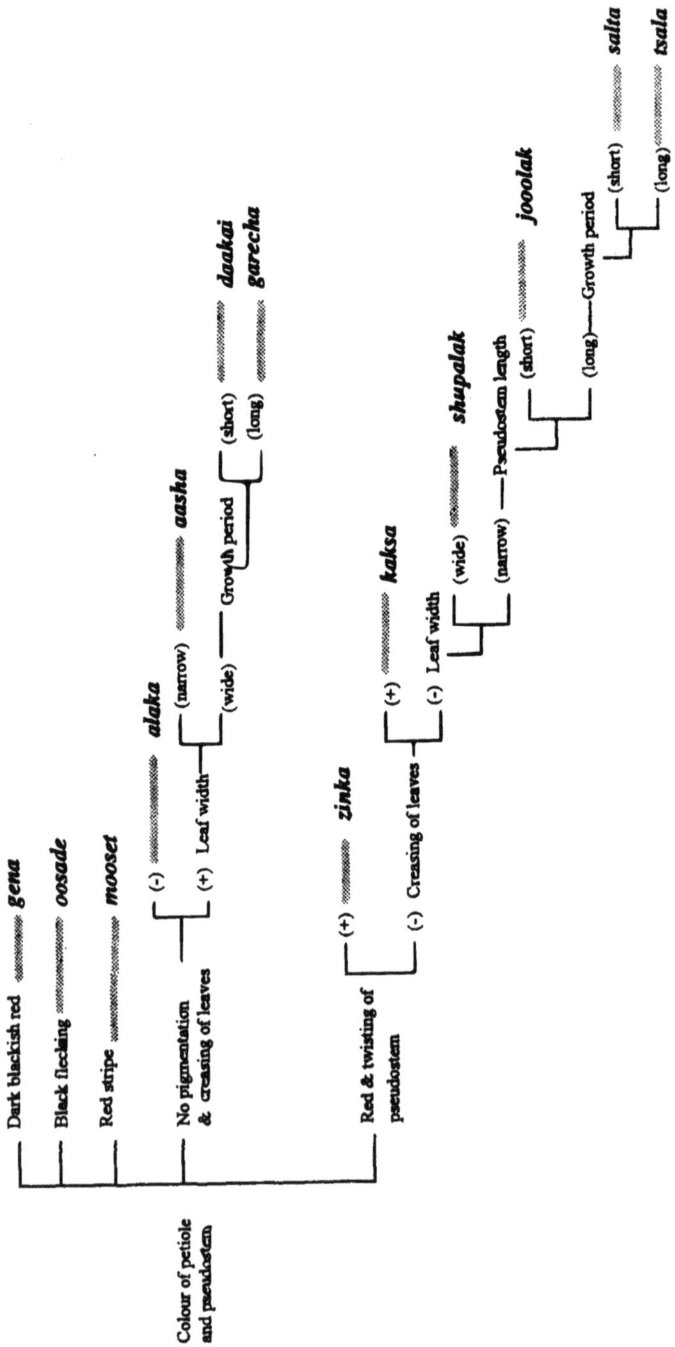

Figure 9.2. Classificatory dendrogram of thirteen ensete landraces

(3) **Kala**: specialized for use of **kala** (leaves) for wrapping and other purposes (**intada** is the best example of this group).

This landrace grouping is my own and Ari do not volunteer the existence of three different groups of this kind. Indeed, all the representative landraces listed above can be utilized for all the purposes given, without any kind of biological or cultural restriction, and none of the three groups have special collective names. For example, it is quite possible to eat the small corms of **intada**, a landrace usually cultivated for its leaves, although Ari never do so.

However, people clearly recognize these three distinctive characteristics as important and remember them in relation to landrace names. Thus information on 'tasty' ensete is transmitted through human memory of landrace names. Maintenance of landraces by the Ari through the recognition of three groups of distinctive characteristics must play a crucial role as an important selective force in forming the real landrace groups. For example, **gena** is believed to be the most suitable ensete landrace for processing **washi**. The basal parts of the pseudostem are large enough to increase the total yield of fermented **washi**. On the other hand, the corm is relatively small and fibrous, and therefore is not suitable for consumption as food.

In landraces of the **mosa** group, such as **maga**, **mooset** and **alaka**, the taste of boiled corm is the first priority, since the starch yield from the pseudostem is rather low. The landrace called **kaksa** is well known for its easily torn and therefore useless leaves. This landrace can be seen as a case of negative specialization of the third (**kala**) group. **Intada** may be the most typical landrace of the **kala** group. As mentioned earlier, ensete is a monocarpic crop that only provides flowers and seeds at the end of its very long life-cycle. Ordinary ensete does not grow suckers, though *Musa* banana, to which it is related, does. However, **intada** is an exceptional landrace that produces suckers before it flowers. **Intada** literally means 'grows by itself'. It is a very interesting landrace which has lost its apical dominance. **Intada** has neither been described taxonomically nor hitherto reported in the literature.

We should not overlook the fact that a landrace such as **intada** is firmly maintained by the people even though its main use is for its leaves. The existence of **intada** is perpetuated by the fact that the ensete is a multi-purpose crop most parts of which are utilized not only as food but also for material artefacts.

Diversity of Ensete as a Plant Species

Genetic Identity of Ensete Landraces

It has been tacitly presupposed that the ensete population has enough genetic variation to be recognized and classified into 'landraces' by the Ari people. Before considering this point further, it is necessary to clarify whether a so-called landrace is really a botanical, or genetically uniform, entity. It might be claimed that Ari vernacular names are a poor guide in such a matter, and natural to presume so because most landrace names are only shared by a few people.

At an early stage in my investigations, I thought that there might be some inconsistencies between ensete genetic identity and landrace. However, this doubt has now been completely removed for the following reasons:

(1) Vegetative propagation of ensete always ensures a correspondence between genetic identity of parent and offspring. More than two hundred clonal seedlings can be obtained by this method.

(2) People inform each other thoroughly of the characteristics and the vernacular names of a landrace when they process the corm of ensete in order to produce seedlings for propagation.

(3) Throughout the process of propagation, transplanting, and other husbanding activities, as well as in utilization, the vernacular names are always used.

(4) When people exchange seedlings or make gifts to relatives or friends they always mention the name and often the place of origin.

Furthermore, since ensete is consumed after the three to eight years of the growing period, individual plants are exposed to view for a long period of time. The ensete garden, **tika haami**, surrounds the house where people spend their daily life. People not only have many opportunities to learn the names of individual ensete landraces through conversation, but also live in close physical proximity to the plants. They often know the vernacular name as well as the life history of individual ensete plants.

Such evidence demonstrates why the Ari conserve both the landrace and its name together. I have called this means of maintaining ensete landraces 'folk conservation' [Shigeta 1987b].

However, there have been very few studies of the genetics of ensete, mainly because its life-cycle is too long to obtain useful data. So far, only one Ethiopian agronomist has tried to analyse quantitatively the difference of yield between two ensete varieties [Bezuneh 1984: 113–14].

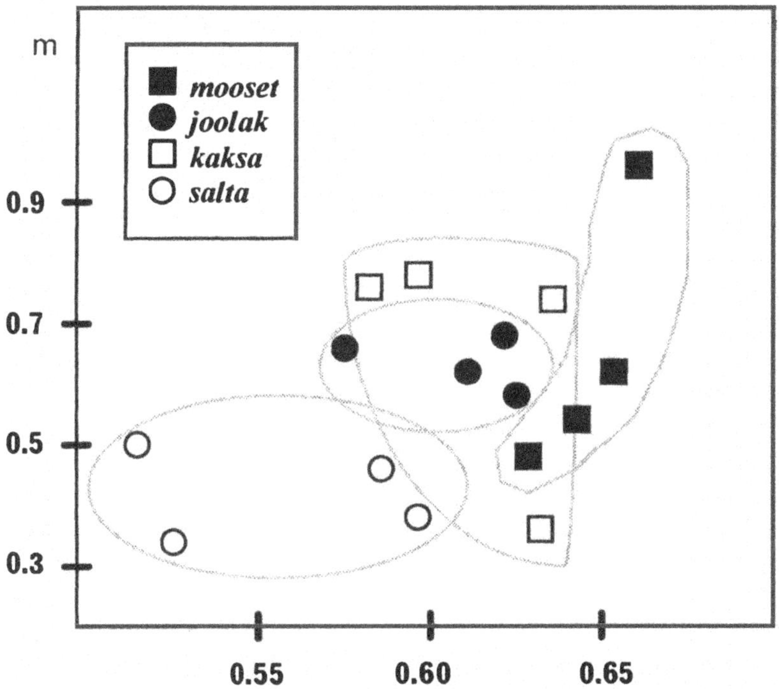

Figure 9.3. Differentiation of four ensete landraces by quantitative characteristics. The vertical axis is leaf width; the horizontal axis is the leaf length to pseudostem length ratio.

Figure 9.3 shows the relationship between the leaf length: pseudostem length ratio (abscissa) and the leaf width (ordinate) of four different landraces, **mooset**, **joolak**, **kaksa**, and **salta**. For the measurement of such characteristics, four plants of each landrace at the same growing stage were selected, and three pseudostems with leaves were removed from individual plants. The length of each pseudostem and leaf, and the maximum leaf width, were measured.

Three landraces, except **mooset**, were rather difficult to distinguish from one another in terms of the colour and pattern of the midrib. Both **joolak** and **kaksa** looked very similar in their colour pattern, leaf width and overall appearance. However, **kaksa** was clearly distinguishable from the others because the leaves were more prone to tear than the other landraces. **Salta** looked very different overall. Only **mooset** had fine red stripes on the pseudostems.

All these diagnostic characteristics of ensete landraces may contribute to their easy identification by Ari. However, the peculiarity of the landrace cannot necessarily explain those where different landrace names are given to what appears to be genetically identical populations. Also, I suspect that sometimes the same name is given to two genetically different populations.

The former case most probably occurs with rare landrace names which are known to a very few people. In some cases, people themselves were aware of possibly misleading landrace names. For example, some people explain that the names **lusakk**, **oosade**, and **maga** are synonymous, but currently used in three different areas, Laida, Sida, and Wuba respectively.

At present, there is no effective way of solving the problems of genetic identity. Although it does not mean that such an attempt is impossible, it is extremely difficult, costly and time-consuming to examine rigorously the characteristics of ensete under present circumstances using conventional methods. It is practically impossible to identify the synonymity of two individual plants growing separately in distant locations and doubtful whether one can judge the synonymity by a cursory examination of morphological characteristics. Therefore, for the time being, we can only estimate the genetic identity of ensete landraces, first through the comparison of very rough but measurable morphological characteristics, and secondly from the information provided by people who have a long experience of cultivating ensete. Taking all the available evidence into account, however circumstantial, there is no reason why we should not follow the Ari identifications.

Diversity in the Cultivated Population of Ensete

Ari farmers usually know twenty or more vernacular ensete names. However, the number of landraces which farmers know is usually more than that of the landraces actually planted in their own **tika haami**. This may have some connection with the fact that having many kinds of ensete, as well as large numbers of them, is prestigious.

The list of landraces shown in Table 9.8 indicates that two informants knew more names of ensete landraces than they cultivated in their own fields. For example, Informant A planted only five landraces in his field but was acquainted with seven additional names of ensete. Figure 9.4 illustrates the positions of twenty-seven ensete plants including ten landraces in Informant B's field. The spatial arrangement of the individual ensete plants in the **tika haami** follows no particular rules. Rather, they are planted at random. This is a typical example of the mixed cropping pattern of this area. Ensete plants are rarely planted in a single line, and the same landraces, not more than three, are

Table 9.8. Knowledge of names of ensete not grown in own garden

	Landraces cultivated in own garden	Landraces not cultivated but known to informant
Informant A	**gena, intada, joolak, kaksa, salta** (5)	**alaka, arpa, gaya, gufalak, karta, maga, shupalak** (7)
Informant B	**aasha, daakai, garecha, gena, joolak, kaksa, mooset, oosade, salta, shupalak, zinka** (11)	**gaya, intada, gufal Raka, karta, maga** (6)

Notes

1. Interview conducted in the ensete garden of each informant
2. Numbers in parentheses indicate the total number of landraces mentioned

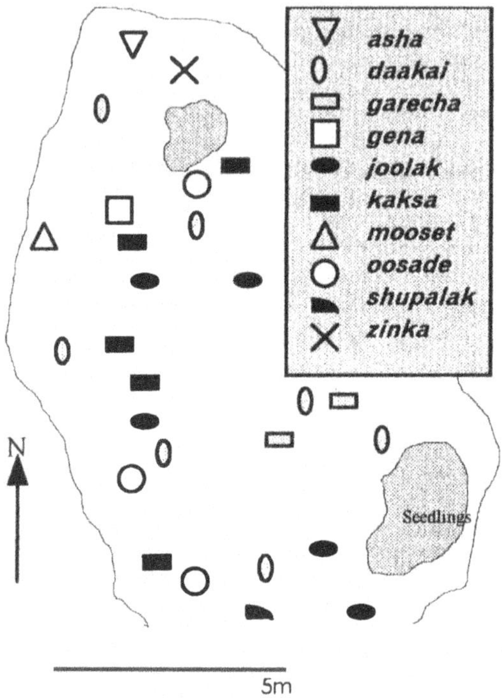

Figure 9.4. An example of an ensete planting pattern in a garden

seldom grouped in one place.

All this attests to the general tendency of the Ari to seek increased diversity in ensete landraces. It is their hope that the landrace names increase as an indicator of diversity. Their sophisticated method for vegetative propagation, which can produce more than two hundred offspring at one time, ensures the process of diversification, or at least prevents the loss of landraces. In order to analyze local variation in landrace diversity, as well as that of individual gardens, further data collection and more detailed analysis are needed.

Diversity in Wild Ensete

It is more difficult to assess the degree of variation in wild ensete than in the cultivated population. There is no practical way of examining observable variation and it may be meaningless to measure the morphological characteristics for quantitative comparison, since growth stages are not known for the wild population. However, it is possible to identify considerable variation in qualitative characteristics in wild populations.

I have observed 48 wild ensete plants selected from 300 to 600 plants growing in natural habitats located 2 km south of Dumtseter village. Of the 48 plants, 22 had red pigmentation on the pseudostem, while eight had yellow to green stems without the red pigment. The leaves of three plants were easily torn, showing **kaksa**-like characteristics. Among the plants observed, none showed the **intada**-like way of propagation with suckers.

It should also be noted that wild ensete is entirely propagated by seeds. Therefore, it may be plausible to assume that there are some genetic variations within wild populations.

The Ari call wild ensete **gela** and protect specific populations ritually. This conservation effort also involves attention being paid to variation among the **gela**. The Ari are well aware of several different 'landraces' in wild ensete. They call them wild **kaksa**, wild **gena** and so forth. More than ten landrace names were mentioned as those found among wild ensete populations. All these names were the same as those of the cultivated ensete. Thus the recognition of variation in the wild population is mostly based on observation of the cultivated populations.

Although **gela**, wild ensete, is not identical with cultivated ensete, **agemi**, it is labelled with the same landrace names. In some instances the wild ensete has been transformed into a cultivated landrace. This process of transformation will be discussed in the next section.

Human-Plant Relationships in Creating New Landraces

Origins of Diversity in Ensete Landraces

The genetic diversity of cultivated ensete arises directly from either bud mutations occurring in cultivated populations or through the introduction of new variations from outside, i.e. from wild populations, or from outside Ari land. Certain landraces, such as **gofa** and **wbajoolak** are, in fact, known by the Ari to be of foreign origin.

I have not come across any case of new landraces originating from a bud mutation. However, I have obtained four circumstantial but clear facts indicating gene flow between wild and cultivated ensete populations (Figure 9.5). These are: (1) that wild populations maintain sexual propagation with a high frequency of flowering individuals; (2) that there are, though rarely, flowering ensete among cultivated populations; (3) that there are ritually protected areas of wild ensete near the village; and (4) that wild ensete is a predominantly out-crossed plant, pollinated by bats.

I have one case of seed-propagated ensete being introduced into cultivated populations. The mother plant was cultivated in a **tika haami**. It happened to flower, and consequently fruit, and finally some of its seeds germinated.

In this case, and in a few similar ones for which I indirectly collected the data, the seedling was first called **gela**. The Ari never remove seedlings from a **tika haami** but foster them with great care. After a year or so, when morphological characteristics become sufficiently evident to allow identification, the ensete plant is named according to an already known landrace or is said to be given a new name. (Those landrace names with very low frequencies in Table 9.7 were sometimes explained as such cases, though there was no direct evidence to prove this.) At this point, the **gela** is transformed into a landrace in the **agemi**.

In the case I observed, the naming had not yet taken place, but the owner clearly identified its mother plant by the vernacular landrace name. Around this **gela** plant were many seeds in the soil. Since the ensete seeds have a hard seed coat, they are often buried for a long period of time without germinating. Buried seeds may act as a kind of seed bank, ensuring the genetic diversity of the cultivated ensete population. Once a group of plants is established as a landrace, it is spread vegetatively in the same way as other landraces. This makes it quicker for the landrace to spread and become known to the people. This new ensete is called **afi agemi**, literally 'seed ensete'. The 'seed ensete' is valued for those occasions when people exchange seedlings with friends, or when they sell ensete products in the barter markets.

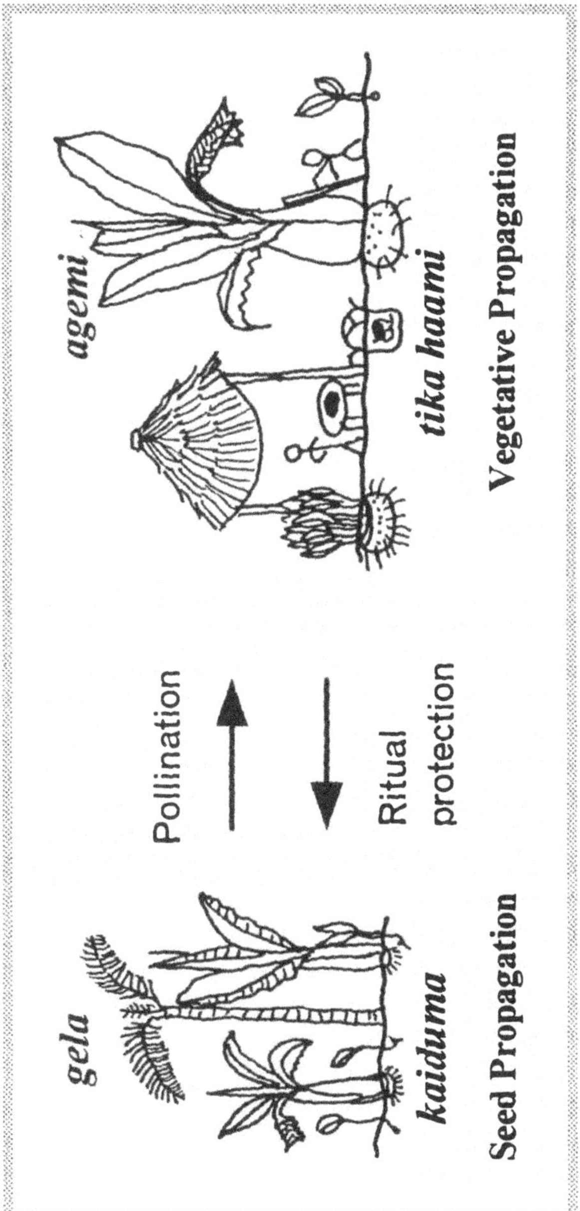

Figure 9.5. Schematized relationship between cultivated and wild population of ensete

The Dynamics of Creating Diversity in Landraces

We often endow variety names to cultivated plants and their use is in many cases significant. This is obvious in the case of those rice varieties which Japanese know very well. When we 'select' rice varieties, some means of evaluation is always needed, such as yield and tolerance to damage.

The connotation of the word 'selection' is that the better ones are kept, while the worse ones are eliminated. The word also implies a choice of superior varieties from a mixture of variants. However, it seems that the attitude of the Ari towards ensete landraces does not necessarily fit this concept of selection. This is, first, because the Ari place primary importance on the characteristics which are not directly related to practical use. These characteristics are mainly morphological, related to the outer appearance of ensete plants. Boster [1984] refers to those characteristics relating to outer appearance as 'perceptual distinctiveness'.

In describing and classifying ensete, the characteristics related to practical use, such as taste and usage, are only mentioned to a very limited extent in the case of a few popular landraces. To the contrary, most of the other landraces are referred to by their outer appearance. Such landraces are known to relatively few people but each has its own vernacular landrace name.

If humans only recognized the practical use factors in identifying landrace variation, these landraces would be reduced to a few superior ones. The majority of the inferior landraces would have perished. If such a means of recognition is the only and best one, those landraces with no apparent practical use might lose their reason for existing. As a result, all of the vegetatively propagated cultivated plants such as ensete would narrow the range of their genetic variation.

But that is not the case. There exists a diversity of landraces, and the people preserve even minor landraces with no apparent practical use, giving each of them specific vernacular landrace names. By doing so, the Ari, though unintentionally, prevent ensete from decreasing in either genetic diversity or number of landraces. Their means of recognition also ensures the introduction into the cultivated population of new variation found in the wild population. Although I did not come across such a case, new landraces created by bud mutation are also likely to be brought into the cultivated population and given a proper name. Thus, genetic diversity in the population of cultivated ensete can be maintained.

There appear to be two co-existing and conflicting selective forces operating on the ensete population. One force pushes towards the increased diversity of ensete landraces and the other toward their reduced diversity. The cultivated population of ensete maintains its diversity first because people value not the usage but the characteristics of outer appearance as a means of distinguishing landraces, and secondly because there are far more such characteristics than utilitarian traits. Such human practices have the effect of 'diversifying selection.' The main driving force for diversifying selection is human observation, discrimination of various plant characterics and naming. I would like to propose here the concept of 'cognitive selection' to describe such behaviour. By contrast, 'utilitarian selection' reflects human behaviour which decreases variation in the long run. We may presume that cognitive and utilitarian selection combine in reality, producing the effect generally called 'artificial selection'. However, artificial selection has usually been deemed synonymous with utilitarian selection. I will emphasize this fact by using the appellation 'artificial selection in a narrow sense'.

We can state the main characteristics of cognitive selection concisely. Cognitive selection is a universal phenomenon found in human attitudes towards plants. The first process in cognitive selection is that people distinguish perceptible differences in a plant population and designate specific names. These perceptible differences are the characteristics of outer appearance such as shape, colour, size and surface characteristics. The second process in cognitive selection is the human behaviour that sustains the diversity of plants. Cognitive selection does not stratify individual plants nor bring about the extinction of inferior individuals. Cognitive selection, as a result, increases diversity in a given population. In this sense, it is synonymous to 'diversifying selection'. Cognitive selection can co-exist with utilitarian selection (artificial selection in a narrow sense) which results in less diversity and more uniformity. Cognitive selection is a prerequisite of utilitarian selection. Utilitarian selection takes place just after cognitive selection. When the criteria of cognitive selection coincide with those of utilitarian selection, cognitive selection will cease to function, and diversity will decrease.

A Mechanism for Creating Diversity in Landraces

The mechanisms through which cultivated populations of ensete maintain their diversity can be summed up as follows. Genetic exchange occurs between the cultivated and wild populations of ensete. Bud mutations can take place in the cultivated populations which are maintained by vegetative propagation. Some of the wild populations are preserved by the Ari in ritual sanctuaries. The people protect ensete seedlings grown in **tika haami**. Furthermore, the method of vegetative propagation encourages the discovery of bud mutation, the establishment of new landraces and their popularity. People engage in cognitive selection, that is they recognize landraces, classifying and naming them on the basis of external distinctiveness. Selection pressures may be placed on ensete landraces which favour their utilitarian value. All these factors interact to maintain the diversity of ensete landraces.

Conclusion: Diversity of Landraces and Ari Cosmology

I have emphasised the role of human-plant relationships for creating and maintaining the diversity of ensete landraces. However, I have not contended that the Ari intentionally try to diversify their ensete landraces. Ari rarely speak of the merits or significance of the diversity of ensete landraces, nor do they try to interpret diversity in terms of insurance and risk management.

So we need to ask what, if any, utilitarian benefit might be gained by maintaining the diversity of ensete landraces. I repeatedly posed this question in the field, asking why Ari planted so many kinds of landraces in their gardens ('**Arekane agemi aynet bedmi kordee?**'). Their answers were all negative, not apparently realizing the utilitarian importance or practical usefulness of maintaining diversity in ensete landraces. Most people said, 'To have as many kinds of landraces as possible is better (**wanna**; good, fine, beautiful, tasty)'. Or, they insisted that the practice was merely a well-established custom.

If I had asked them why they kept such a custom for so long and why they considered it good to have many landraces, I expect their answer would have become purposive: 'because various kinds of landraces are desirable for different ways of cooking'; 'they have different tastes'; 'the strength of the leaves and growing periods differ'. This assumption has been confirmed by subsequent research. But as repeated in previous sections, selection based on such criteria of practical usefulness would lead ultimately towards a decrease in variation.

The diversity of ensete landraces is a reality. Ari attitudes towards this reality are apparently affirmative. They appreciate the diversity of landraces in their fields. However, their explanations for it are not necessarily a precondition of that diversity. The diversity can best be understood in terms of a kind of non-purposive causality born from the interactions between human beings and plants. The Ari appreciation of diversity is firmly sustained by behaviour through which they perceive and even create different ensete landraces, that is through cognitive selection, not as a result of practical usefulness, which is merely one of the consequences of such diversity.

Notes

The main Ethiopian field research on which the present study is based was conducted between November 1986 and February 1987, and from October to December 1989. I stayed in Metsar village in the former Bako-Gazer Woreda (District), Geleb na Hamer-Bako Awaraja (Province) in Gemu-Gofa Kifla-hager (Region), Ethiopia. The area has now been integrated into Bako-Gazer Awaraja, South Omo Province. The research area was approximately 35 km north of Jinka, the present capital of South Omo Province, and in the Sida domain, the territory of one of the traditional local chiefs of Ari. The research was made possible by a grant-in-aid for the Overseas Scientific Research Department from the Japanese Ministry of Education, Science and Culture. I would like to express my deepest gratitude to the project leaders, Professors Keiichi Sakamoto and Katsuyoshi Fukui.

1. The Ari people, linguistically classified just a decade ago, belong to the Omotic stock of the Afro-Asiatic group [Fleming 1976] and are one of the least studied ethnic groups of southwestern Ethiopia, and of Africa in general. The contemp-

orary Ari are, uniquely, sedentary agriculturalists in tropical Africa. Their sophisticated techniques of environmental manipulation, including the management of ensete fields, need much further investigation.

2. I asked informants to list all the names of ensete landraces which they knew. Interviews were made individually to avoid influences from other informants. I made sure that a name was a single entry in my records. During the interview, I did not proffer any information, unless asked, as to whether the informants had mentioned a landrace name more than twice. Also, it should not be immediately concluded that an informant was ignorant of a particular landrace even if he did not name it. Using this method, the number of landraces identified by informants may be an underestimate of actual landrace recognition in their daily life. Informants tended to recount more landrace names when they were asked in their own gardens than was the case in an isolated room. A separate survey was also made using a check list of ensete vernacular names compiled by myself. However, using this test, people always seemed to register more names than they would ordinarily recognize, it being prestigious to have many kinds of landraces in a garden.
3. In the first column of Table 9.1 the names of landraces are shown in descending order of frequency mentioned by 39 informants. The numbers in the middle column show the order of respective landraces mentioned by an informant. For example, informant No.31 mentioned landrace **gena** first, **bubuni** second, **karta** third, and so on. The total number of landraces mentioned by each informant is listed in the bottom row of the table. The final three columns indicate the number of informants referring to each landrace, total references made to each landrace and its average.
4. In the territory of the Sida chief, Sida baabi, 22 informants gave 36 different landrace names, for a total of 116 references, 5.3 names per person on average. The minimum number of landraces named was one, and the maximum was eleven. Among the landrace names recounted, **kaksa, garecha, gena,** and **joolak** were the five most frequently named, by more than a half of the people asked. Four landrace names were named by more than five persons, while eight were mentioned by less than five but more than two persons. Of the

36 landraces, twenty names were only given by one person. In the territory of the Biya chief, Biya baabi, three informants mentioned thirteen names, for a total of twenty references, of which **gena** and **kaksa** were mentioned by all three informants. However, eight names were only mentioned by one informant. In the territory of the Baka chief, Baka baabi, the situation was almost the same as in Biya. Three informants gave 32 names in total, half of which were only referred to by one informant. In contrast, only two names, **gufalak** and **mooset**, were mentioned by all three informants.

Bibliography

Anderson, E., 'The evolution of domestication', in Sol Tax (ed.), *Evolution after Darwin*, Vol.II: *The evolution and man, culture and society*, Chicago: University of Chicago Press, 1960

Bezuneh, T., 'Evaluation of some *Ensete ventricosum* clones for food yield with emphasis on the effect of length of fermentation on carbohydrate and calcium content', *Tropical Agriculture*, Trinidad 61(2), 1984, pp.111–16

Boster, J.S., 'Classification, cultivation, and selection of Aguaruna cultivars of Manihot esculenta (Euphorbiaceae)', *Advances in Economic Botany* 1, 1984, pp.34–47

Brush, S. B., H.J. Carney and Z. Huaman, 'Dynamics of Andean potato agriculture', *Economic Botany* 35(1), 1981, pp.70–88

Bye, R. A. Jr, 'Incipient domestication of mustards in Northwest Mexico', *The Kiva* 44(2–3), 1979, pp.237–56

Fleming C. H., 'Omotic overview', Part 3, 13, in M. Lionel Bender (ed.), *The non-Semitic languages of Ethiopia*, Monograph No.5, African Studies Center, Michigan State University, 1976

Jackson, M. T., J.G. Hawkes and P.R. Rowe, 'An ethnobotanical field study of primitive potato varieties in Peru', *Euphytica* 29, 1980, pp.107–13

Lenneberg, E. H., *Biological foundations of language*, New York: John Wiley, 1967

Odum, E. P., *Fundamentals of ecology*, London, Philadelphia: Saunders, 3rd edn, 1971

Rindos, D., *The origins of agriculture: an evolutionary perspective*, Orlando, Florida: Academic Press, 1984

Shigeta, M., 'Cognition, utilization and weediness of plants: man-plant relationships in the Acholiland, southern Sudan', *Ahurika-kenkyu* 31, 1987a, pp.27–62 (in Japanese with English summary)

——, 'Variation and folk conservation of ensete (*Ensete ventricosum* (Welw.) E. E. Cheesman) landraces in southwestern Ethiopia', *Japan Journal of Breeding* 37(2), 1987b, pp.278–9 (in Japanese)

——, 'A case of man-plant relationships: ensete cultivation and utilization of the Omotic Ari in southwestern Ethiopia', *Kikan-Jinruigaku* 19(1), 1988, pp.191–271 (in Japanese)

——, 'Folk in-situ conservation of ensete (*Ensete ventricosum* (Welw.) E. E. Cheesman): towards the interpretation of indigenous agricultural science of the Ari, southwestern Ethiopia', *African Study Monographs* 10(3), 1990, pp.93–107

Tani, Y., 'On pastoral culture', *Jinbungakuhou* 42, 1976, pp.1–58 (in Japanese)

Chapter 10

Human Cognition as a Product and Agent of Evolution

James Boster

This paper explores two complementary ways of examining the relationship between evolution and cognition: the first arguing that human cognition has evolved to make sense of other evolutionary products, the second describing the impact of human cognition on the selection of domesticated plants. The first section of the paper reviews research that delineates the correspondences among different biological classification systems. A series of studies establishes that: (1) Jivaroans, scientific ornithologists, and ornithologically naive United States undergraduates agree in their recognition of patterns of resemblance among a collection of South American bird specimens; (2) variation in the agreement among the diverse groups of informants seems to reflect variation in the clarity of the pattern of resemblance [Boster, Berlin, and O'Neill 1986; Boster 1987]; (3) the diverse groups agree on their categorization because they attend to the same features to differentiate the sets of specimens, apparently reflecting a pan-human perceptual strategy that selects those features of a collection of organisms that yield the most informative categorization [Boster and D'Andrade 1989]; (4) there is greater intracultural variation than cross-cultural variation in biological similarity judgments [Boster 1991]; and (5) the classificatory strategies at work here are apparently related to more general features of human categorization such as typicality judgment [Boster 1988]. These results suggest that many features of human cognition are best understood as having evolved to interpret and discriminate other products of evolution.

The second section of the paper concerns the consequences of human strategies of discriminating and categorizing natural kinds, focusing on the relationship between classification, cultivation and selection of Aguaruna cultivars of manioc (*Manihot esculenta* Crantz). Here I review research that argues that the procedures cultivators use to distinguish crop varieties have the effect of selecting for increased variability in the features they use to distinguish the crops: cultivators change the world in the process of understanding it in such a way as to make their procedures for understanding more appropriate [Boster 1984a; 1984b; 1985]. This section completes the argument that human cognition has co-evolved with the natural world; it both shapes and has been shaped by other living things.

My objectives in writing this paper are self-indulgent: to summarize some of my own contributions to the study of ethnobiology and to describe the broader context of that work and its implications. I do not report any new experimental findings; my effort instead is directed to sketching out the overall coherence of a series of publications over some years. The most important propositions will be numbered for future reference.

Cognition as a Product of Evolution

I will begin by reviewing research that describes and explains the correspondences among different systems of biological classification. I will first summarize the evidence for various sorts of universals in folk biological classification. I will then argue that the source of these commonalities in classification lie both in the natural order (the objective organization of biological diversity) and in the character of human cognition.

A number of researchers have described strong correspondences among the systems of biological classification of different cultural groups. Four types of universals have been described: universals in the linguistic form of classification, in the mapping of categories on to their referents, in the judgment of similarity among organisms, and in the choice of characteristics of organisms to attend to.

Berlin and his collaborators [Berlin 1972, 1976; Berlin, Breedlove and Raven 1973, 1974] have documented strong universals in the linguistic form of biological classification. They have shown that (I) most folk biological classification systems are shallow taxonomies. Most named categories are at the folk generic rank (e.g. oak, robin, otter), which is the most psychologically salient rank of the folk taxonomy. Some of these, especially culturally important species, are further subdivided into folk specific categories (e.g. live oak, pin oak), while all but the most anomalous are included in superordinate 'life-form' categories (e.g. tree, bird, mammal).

Various researchers have also noted profound regularities in the ways that folk biological categories map on to scientific species [Berlin 1973; Bulmer 1970; Diamond 1966; Hunn 1975]. This work has argued that (II) folk biological classifications tend to 'carve nature at the joints', naming the same objective discontinuities in nature as those recognized in scientific classification.

For each of these first two universals, the evidence and argument is linguistic; both depend on examining people's names for organisms and observing either how the names relate to each other (as when one examines the inclusion relationships among categories in a folk taxonomy), or how they relate to their referents (as when one examines the mapping of folk categories on to their biological ranges). However, there are a number of problems inherent in depending on the linguistic form of folk taxonomies to argue for universals. One is convincing fellow investigators of the naturalness and flexibility of your analytic framework: this appears to be the principal criticism of Berlin's theory of folk taxonomic rank [Ellen 1979; Healey 1978–9; Hunn 1982; Randall 1976]. Secondly, this approach to universals presumes that informants all agree what organisms should be called. In fact, informants typically do not universally agree on the identification of plants and animals. Finally, depending on named category as a clue to categorization limits one's choice of cultural groups to investigate: one can't elicit information from people who have no prior experience of and hence no names for the organisms.

Given these difficulties in documenting the first two types of universals in biological classification, I have chosen to focus my own research on documenting possible universals in people's discrimination of biological diversity using methods that do not depend on people's names for organisms. This can be termed a psychological (as opposed to a linguistic) approach to universals in biological classification; it exploits the fact that when people confront biological diversity, they are able to make a whole host of discriminations among organisms, only a small fraction of which are actually marked by named categories. For example, we can readily discriminate between different instances of 'Eucalyptus' even though we may only have one name that captures the whole category.

The central finding of this work has been that (III) culturally diverse groups (Jivaroans, scientific ornithologists, and ornithologically naive United States undergraduates) agree in their recognition of patterns of resemblance among a collection of South American bird specimens[1] [Boster Berlin and O'Neill 1986; Boster 1987].

Boster et al. [1986] confronted the problem that earlier arguments for a strong correspondence between folk and scientific biological classification systems depended on methods that falsely assumed that native informants agree in classification. Instead, we used the pattern of disagreement in identification to infer the informants' implicit categories. We showed that folk and scientist alike were responding to the same objective pattern of resemblance among birds by demonstrating that the folk confuse species that the scientists classify as related.[2]

Boster [1987] tested the generality of these findings. Because both the Jivaro and the scientists are expert ornithologists, it is conceivable that the high agreement between them is a consequence of the fact that all have had ample opportunity to study the organisms and have represented their understandings in explicit nomenclatural systems. The question that arises is whether informants who have no prior knowledge of the birds recognize the same underlying pattern of similarity and difference as that recognized by the Jivaro and the scientists. To test this, I asked University undergraduates who had never seen these South American birds before to sort them according to their overall similarity.[3] Despite the cultural differences among the three informant groups and the differences in the cognitive tasks each performed, all groups recognized substantially similar patterns of resemblance among the bird specimens. It appeared that recognition of the pattern of resemblance among organisms does not depend on formal training in taxonomy, intimate knowledge of the organisms, or possession of named categories for the specimens.

The second major result of this research was that all groups of informants agreed more strongly on the pattern of resemblance among non-passerine birds (e.g. woodpeckers, toucans) than on the pattern of resemblance among passerines (e.g. wrens, tanagers). We interpreted this difference in people's categorization of these two subsets of birds as evidence that (IV) the objective pattern of similarity among organisms varies in its clarity and that alternative biological classification systems correspond only to the extent that the objective pattern is clear [Boster et al. 1986; Boster 1987].

The next paper in this series, Boster and D'Andrade 1989, addressed a question suggested by this work: is the source of cross-cultural similarities in biological classification in the world alone or is it in the mind as well? Given the strong agreement among diverse informant groups and the apparent dependence of the strength of agreement on the clarity of the objective pattern of resemblance, it was conceivable that the objective pattern of resemblance of specimens alone was responsible for the agreement among culturally distinct groups. In other words, it was conceivable that the correlational structure of the attributes of the bird specimens was sufficiently strong that, no matter what attributes were attended to, the same classification would have resulted. Boster and D'Andrade used nine size and seven colour measurements of the passerine and non-passerine bird specimens used in the earlier experiments to help assess why the diverse groups of informants agree on their categorization. We articulated two possibilities: the structured-world and the structured-mind hypotheses. The structured-world hypothesis held that diverse groups of informants may attend to radically different characters of organisms but, because the characters themselves are intercorrelated, the same structure is discerned no matter which characters are chosen. The structured-mind hypothesis held that in addition to assuming the intercorrelation of characters, diverse groups of informants choose the same salient attributes of the specimens to differentiate them.

The results supported the structured-mind hypothesis. Not only do diverse cultural groups categorize the birds in similar ways but they appear to pay attention to the same characteristics. We found that all informant groups, whether they were South American Indians, United States undergraduates, or ornithologists, based their categorization of the passerines on the same set of characters and based their categorization of the non-passerines on a different suite of characters. (V) Agreement among informant groups appears not to be simply due to the correlational structure of the stimuli; the different groups apparently share a perceptual/cognitive strategy of selecting those features that will best discriminate the collection of specimens.

These findings, suggestive of a pan-human (or possibly pan-mammalian) perceptual strategy for making sense of biological diversity, prompted further inquiry into the psychological basis of cross-cultural agreement in folk biological similarity judgment. Accordingly, Boster [1991] investigated the relationship between the intracultural variation and cross-cultural universals in biological similarity judgment. There were three major results from this study. First, there was a high degree of consensus among the United States undergraduate informants on the sorting of both subsets of birds; the pattern of agreement among informants fit Romney, Weller and Batchelder's [1986] cultural consensus model. Secondly, agreement with the aggregate of one's own group is correlated with agreement with the aggregates of responses of other groups of informants. Thirdly, aggregates are closer than individuals are, and there is more intra-cultural than cross-cultural variation in judging the similarity of the birds. I interpreted this result as providing further indication that (VI) cross-cultural universals in biological similarity judgment are ultimately based in the characteristics of individual cognition: groups agree because individuals do.

Boster [1988] also attempts to relate the classificatory strategies at work in these experiments to more general features of human categorization such as typicality judgment. 'Typicality' refers to the fact that often some members of categories are regarded as more representative of the category than other members. For example, chairs are regarded as more typical examples of furniture than rugs, robins are regarded as more typical examples of birds than penguins. The objectives of the paper were to show that typicality judgments (at least of birds) could be explained by the structure of the scientific taxonomy and that diverse informant groups (United States undergraduates, Jivaroans and scientific ornithologists) all responded to the birds in ways that reflected the birds' relative structural typicality. The first objective was met by demonstrating that (VII) typicality ratings of birds [Rosch 1975] are more strongly correlated with the number of related species than with the frequency of the birds in the observers' immediate environment or with the frequency of mention of the birds in written materials [cf. Mervis, Caitlin and Rosch 1976]. In other words, typical birds are those that are members of large orders and families, not those whose names are read most often or that are seen most frequently. This confirms the interpretation of Rosch and Mervis [1975] that resemblance to other category members rather than familiarity best predicts informants' typicality judgments.

The second objective was met by showing that, due to the structural basis of typicality effects, informants respond to passerines in other ways that reflect the passerines' structural position as most similar to other birds. (VIII) Informants make finer discriminations of the most typical birds (passerines) than among less typical birds (non-passerines) yet disagree more often in identifying them, and report greater difficulty in judging the similarities among them. Specifically, in spite of the fact that the United States undergraduates had not seen these South American bird species before and did not have names for any of them (with the occasional exception of woodpeckers and toucans), most informants regarded the passerines as more like their idea of typical birds. If typicality judgments were only a result of familiarity, the United States informants would not consistently pick one or another of the novel specimens as more typical examples of birds. United States undergraduates also said they had more difficulty sorting the passerines than the non-passerines. Similarly, the Jivaroan informants agreed with each other much more often on the names of the non-passerine specimens than they did on the passerines. Paradoxically, the Jivaroan informants appear to have tried to discriminate the passerine specimens more finely than the non-passerines, thereby tackling a more difficult task with this group of specimens: they merge genera and species of non-passerines while distinguishing species of passerines. This last finding suggests that the way to explain some aspects of human thought is to look at the structure of what they are thinking about: here, there is evidence that the internal structure of categories reflects the objective structure of the domain.

The Evolution of Cognition and of Natural Kinds

I would now like to argue that these results, taken together with those of other ethnobiological and cognitive researchers, suggest that many features of human cognition are best understood as having evolved to interpret and discriminate other products of evolution. I will begin by summarizing where we have come so far in the argument.

First, the natural world is well structured. Evolution can be regarded as a mechanism that generates natural structure, creating a pattern of similarity and difference among organisms. Organisms may appear similar either through diverging little from a common ancestor (phylogeny) or through convergent adaptation to particular niches (convergence). The resulting pattern of similarity and difference is robust and is recognized in similar ways by members of diverse cultural groups (findings II, III). The universals in the form of folk biological classification systems as shallow taxonomies reflects the universal experience of biological diversity as evolutionary products with a branching pattern of similarity and difference (finding I).

Secondly, human cognition is highly responsive to the natural structure. The clarity of the objective pattern of similarity among organisms appears to be responsible for variation in the degree of correspondence among alternative biological classification systems (finding IV). The objective pattern of similarity also appears to determine which organisms humans will regard as most prototypical (finding VII).

Thirdly, it appears that human response to natural structure is governed by a coherent cognitive faculty that affects perception and memory and other aspects of mental performance. Humans actively construct an understanding of natural structure, extracting those characteristics of a diverse assemblage of organisms that will best discriminate them (finding V). It appears to be the sharing of perceptual/cognitive strategies at the individual level that is responsible for the consensus among culturally diverse groups in classification: cross-cultural universals in folk biological classification are based in individual cognition (finding VI). The objective structure coherently influences many different aspects of cognition at once, affecting typicality judgments, discrimination, errors in identification and the perceived difficulty of the task (finding VIII).

Many other characteristics of human cognition can be understood as having been forged to make sense of biological diversity. Natural kind categories appear to have a privileged place in human category formation. It is as though a strategy of categorization that works best for organisms were adapted to make sense of novel objects such as human artefacts.

For example, Rosch's notion of basic level object (a generalization of Berlin's concept of the folk generic) was seen as capturing the correlational structure of the features of concrete objects [Rosch, Mervis, Gray, Johnson and Boyes-Braem 1976]. The basic level objects maximize similarity of instances within categories and maximize difference between instances of different categories. It is the most abstract level that can be captured by a single visuo-spatial representation. The correlational structure reflects the organic coherence of living things and the function or purpose of artefacts. Most organisms are characterized by very high feature intercorrelations because many of the features of organisms reflect different aspects of an adaptation for a particular function. For example, the feathers, hollow bones and wings of birds all reflect an adaptation for flight. These features do not freely vary in the biological world; they come as a cluster because they are all parts of an adaptive package for flightedness. Artefacts tend to have weaker correlational structures than living things because their purposes are more limited: a chair minimally need only serve to be sat upon, it need not breathe, feed and reproduce as well [cf. Waddington 1966].

Similarly, the formation of categories on the basis of family resemblance to a prototype makes the most sense when dealing with natural kinds (as opposed to artefact or abstract categories). Evolution often produces assemblages of related organisms which literally share a family resemblance to one another but do not share necessary and sufficient diagnostic features (much to the dismay of systematists!). This is part of what makes the prototype-surround tactic in recognizing natural kinds such a powerful one. It allows humans to recognize the gestalt coherence of features even when confronted with instances that are missing one or another of the prototype's suite of intercorrelated features. Typicality and correlational structure are linked: typical members of categories capture the correlational structure of features better than atypical members and share more features with the prototype for the category [Malt and Smith 1984]. Children early on grasp the greater coherence of natural kinds and make more inductive inferences about natural-kind categories than about artificial categories [Gelman and O'Reilly 1988].

There is also substantial evidence for strong neural localization of the capacity to discriminate living things. Several investigators have shown that individuals may suffer specific deficits in their identification or recognition of animals or other living things as a result of various kinds of insults to the brain, including herpes simplex encephalitis [Sartori, Job, Miozzo, Zago and Marchiori 1993; Sheridan and Humpreys 1993; Silveri and Gainotti 1988], Alzheimer's dementia [Silveri, Daniele, Giustolisi and Gainotti 1991], infarct [Hart, Berndt and Caramazza 1985; Hart and Gordon 1992], or other forms of brain injury [Basso, Capitani and Laiacona 1988; Farah, Hammond, Mehta and Ratcliff 1989; Farah, McMullen and Meyer 1991; McCarthy and Warrington 1988, 1990; Young, Newcombe, Hellawell and de Haan 1989]. The same subjects are usually reported to have much better recognition of artefacts and other non-living things. While some authors have argued that these apparently category-specific deficits are best explained as a consequence of the intrinsic differences in the visual discriminability [Gaffan and Heywood 1993] or familiarity [Funnell and Sheridan 1992; Stewart, Parkin and Hunkin 1992] of living and non-living things, on balance the results suggest a striking difference in the perceptual processing of natural and unnatural kinds.

If these aspects of human cognition have been shaped by evolution to make sense of biological diversity, we are probably not alone. These perceptual skills are probably homologous with those of other higher vertebrates. In other words, if one could train chimpanzees, rats, or even pigeons to do the bird-similarity judgment tasks described here, I expect that the results would be similar to the human responses. Among the most important tasks that animals face in adapting to an environment is recognizing the similarity and difference among other organisms and forming appropriate categories of them. There is an enormous fitness advantage to being able to recognize that a novel plant species is related to a species already known to be either edible or poisonous. It would be a similarly fatal error to mistake a potential predator for a prey species. In fact, not only is categorization an essential survival skill, but there is some evidence that animals (and plants) exploit the characteristics of other species' capacity for categorization to their own advantage. One can interpret the widespread phenomenon of mimicry in this light as a response to the category systems of potential predators: one can avoid being eaten by masquerading as something inedible. There would be no point in the masquerade if the predator did not form categories of edible and inedible [Cott 1966].

One need not assume that animals have been selected to be systematists directly; the capacity to recognize the similarity and difference of different species is probably part and parcel of the perceptual apparatus that allows animals to recognize either the constancy of an object from different perspectives or in different settings, or to recognize the similarity of individuals from the same species [Lorenz 1966; Herrnstein 1984].

The capacity for categorizing natural kinds has probably been more completely researched in pigeons than in any other non-human species. Pigeons appear to be much more adept at learning natural categories than at learning artefacts, readily discriminating photographs containing people [Herrnstein and Loveland 1964], trees, bodies of water or individual humans [Herrnstein, Loveland and Cable 1976] and even fish [Herrnstein and de Villiers 1980] but not more artificial categories such as cubes [Cerella 1977] or cartoon characters [Cerella 1980, 1982]. Furthermore, pigeons can readily learn to discriminate oak leaves from non-oak leaves but had much more difficulty discriminating an individual oak leaf from other oak leaves [Cerella 1979]. It is likely that pigeons are not narrowly specialized on recognizing natural objects *per se*, but rather at recognizing 'polymorphous categories' that have the correlational structure of natural kinds. For example, pigeons do quite well at discriminating letters, making a similar pattern of mistakes to the one that humans make [Blough 1982].

The evidence on the acquisition of natural-kind categories by pigeons is suggestive of the sort of data one would want on the categorization skills of higher vertebrates generally. One might expect evidence of neural modules specialized for the recognition of gestalts and family resemblances among instances in biological domains and other domains with prototype-surround category structures. The numerous reports of category-specific deficits in the discrimination of living things described above suggest the existence of some such neural sub-system in humans [Hart and Gordon 1992], most likely involving temporo-limbic structures [Sartori et al. 1993; Silveri et al. 1991].

Cognition as an Agent of Selection

To review, in the first section of this paper I summarized evidence of a number of universals in folk biological classification. I argued that these universals have their origin both in the fact that the objective pattern of similarity and difference among organisms is clearly structured and that human beings share a domain-specific cognitive strategy for selectively attending to those features of a collection of organisms that will best discriminate them. I also speculated that other aspects of human cognition and categorization could be understood as having evolved (and hence were best suited) to discriminate biological diversity. In this second section of the paper, I argue that when humans use these same cognitive strategies to discriminate domesticated plants and animals, they exercise sufficient control over the reproduction of the domesticates to select for perceptual distinctiveness. I argue that the process of differentiating the cultivars on the basis of the most diagnostic characters has the effect of increasing the range, continuity and independence of variation of those characters. The end result of this process is to increase the distinctiveness of the cultivars and thus increase the number of cultivars that can be distinguished and maintained in cultivation. I have developed this argument in detail elsewhere [Boster 1984a, 1984b, 1985] in describing the selection of Aguaruna cultivars of manioc (*Manihot esculenta* Crantz); here I simply summarize the model and evidence and frame it in the broader context of human biological classification.

The Aguaruna Jivaro of the tropical forest of northern Peru maintain more than a hundred cultivars of manioc, their most important crop. I studied their classification, cultivation and selection of manioc cultivars in an effort to understand how they could tell the difference among so many different varieties and why they would bother. It appeared that manioc cultivars are selected not just for their ability to survive against natural pests and disasters (natural selection), or for their culturally desirable flavours or yields (cultural selection), but for being distinctive (perceptual selection). Although small inventories of cultivars could conceivably be maintained without good discriminating features by simply remembering where one planted each one, the cultivator's ability to observe and remember perceptible distinctions is critical to the maintenance of an inventory of a hundred varieties. Before a new cultivar can be selected for yield or flavour or other non-directly observable characters, it must be distinguishable on the basis of its combination of characters from those already in cultivation. A cultivar that fills a gap in the morphological continuum of existing cultivars would be more likely to be accepted than one that is extremely similar to cultivars already maintained. Cultivators identify cultivars on the basis of characters that show the greatest range of variation and perceptual salience. In Aguaruna cultivars, these are leaf-shape, petiole-colour, and stem-colour. Cultivars that are similar to each other in these 'good' characters are more likely to be confused with one another than those that are similar in other characters. Thus perceptual selection is not random, but is directed toward increasing variation in those characters that are already regarded as valuable for identification. They thereby enhance the value of the characters they choose in distinguishing the varieties. If the cultivators could not distinguish among cultivars, their cultivar inventories would shrink to those clones that produce the most planting material, due to the random loss of rarer cultivars.

This process of selective acceptance of cultivars on the basis of perceptual distinctiveness has eight important consequences that are borne out [Boster 1985]: (1) The overall range of variation in important taxonomic characters is increased over wild relatives. (2) Large perceptual gaps between cultivars on these taxonomic characters are gradually filled; at saturation, all cultivars are separated by just-noticeable differences on a morphological continuum. (3) Important taxonomic characters of the cultivars species-wide tend to vary independently of one another. (4) Geographic races defined by taxonomic characters are absent; the cultivar inventory maintained by the Aguaruna represents a large proportion of the total range of variation in taxonomic characters of the cultivated species as a whole. (5) The important taxonomic characters that distinguish cultivars in the inventory maintained by the Aguaruna vary independently from one another just as the characters of the species-wide collection of cultivars do. (6) Aguaruna cultivators use the same salient characters to distinguish cultivars as do the botanists. (7) Aguaruna cultivators treat the continuously varying cultivars as approximately equally difficult to distinguish perceptually. (8) Aguaruna cultivators confuse those cultivars they regard as most similar to one another; similarity of cultivars on the most important taxonomic characters (petiole-colour, stem-colour, and leaf-shape) independently contribute to the confusion of cultivars.

This model helps explain why cultivated plants show such extreme variation in features that have little to do with the utility or survival of the plant [Dodds 1965; Harlan 1975; Yen 1968]. It is precisely those taxonomic characters that have very few functional consequences that are most valuable perceptually as characters to discriminate the cultivars; a cultivator could not accept taxonomic characters that hurt flavour or yield, or that made cultivation more difficult. Pigmentation characters are ideal because they have relatively little adaptive significance for the plants (compared with stem habit, for example). Although occasionally native cultivators may deliberately select for the strange or bizarre [Harlan 1975: 110, 138], it is likely that most selection of perceptual distinctiveness is unconscious.

There is a procrustean aspect to this process: in effect, the cultivators force the cultivars to fit their perceptual strategies. While I have argued that the perceptual/cognitive strategy of discriminating organisms evolved to make sense of biological diversity, when it is applied to discriminate the 'captive audience' of domesticated species, it has the paradoxical effect of making them less like wild species. As indicated earlier, biological form is generally characterized by very high feature intercorrelations. It is this aspect of organisms that humans exploit in forming prototype-surround categories of them. However, because cultivators discriminate cultivars on the basis of combinations of characters (not simply on each character independently), they exert a selection pressure to break down the intercorrelations among characters by retaining cultivars with novel combinations of characters. For example, if pigmentation in different parts of the plant co-varied (to either all light or all dark pigmentation), an unusual plant that had lightly pigmented petioles but darkly pigmented stems would be much more likely to be retained in cultivation. Humans are not the only species who apparently change the world in the process of understanding it. Human perceptual selection of manioc varieties is analogous to other processes of perceptual co-evolution (e.g. moths and bats [Roeder 1963]; pollinators and plants [Barth 1985; Jones and Little 1983]; predators and prey [Cott 1966]). The co-evolution of the visual and sensory systems of pollinators and the floral morphology of the plants they pollinate is the most similar to the case of human-crop interactions. The plants have an 'interest' in making themselves attractive to potential pollinators and must advertise their incentives of nectar and pollen in ways that match the sense modalities of the pollinators. Hence, the predominant pigmentation of the flowers matches the visual system (or relative lack of it) of the pollinator: bird-pollinated flowers tend to be red, bee-pollinated flowers blue and yellow, bat- and moth-pollinated flowers white. As in the process of selection for perceptual distinctiveness described above, the perceptual strategy used by the pollinator in finding nectar and pollen sources exerts a selection pressure on the plants; plants that do not advertise in the appropriate sense modality are less likely to be pollinated and set seed. For insects as well as humans, the use of the perceptual strategy has the effect of enhancing its value for discriminating the plants.

In sum, as the biological world has radiated, the capacity to recognize the order in that radiation has co-evolved. The evolution of human cognition to understand the natural world is part of a more general process among living things. Mind has evolved both to understand and to shape nature.

Notes

I would like to thank Stephen Beckerman, Peter Dwyer, Steven Gaulin, Cornell Guion, Ted Lowe, Kim Romney, Margaret Rubega, Sadao Sakamoto and Art Weiss for their helpful comments and suggestions, Roy Ellen and Katsuyoshi Fukui for their work on this volume, and our Japanese hosts for arranging the conference.

1. It is important to be clear that the experiments used to demonstrate this agreement among alternative classification schemes either implicitly or explicitly constrained the informants to make use of morphological (as opposed to functional or utilitarian) criteria to discriminate the specimens. If one elicits judgments about biological similarity using methods that allow the informants to make use of other knowledge they have about the organisms, they very well may use it. For example, when Boster and Johnson [1989] asked informants to judge the similarity of fish using an unconstrained pile sort task, the expert fishermen made use of their extensive knowledge of the behaviour and utility of the fish to sort them, while novices (who were limited to basing their similarity judgments on the pictured shapes of the fish) sorted the fish more closely according to the scientific classification.
2. In the Jivaroan bird experiments, adult Aguaruna and Huambisa informants were asked to identify prepared bird specimens laid out on long tables. All the birds were native to the surrounding tropical forest. We constructed two measures: one a measure of how each pair of bird species were confused with one another by Jivaroan informants (specimen overlap) and the second a measure of how close each pair of bird specimens were in the scientific classification (taxonomic distance). Our measure of the correspondence between the folk and scientific systems of classification was simply the correlation between these two measures.
3. In the United States research, University of Kentucky students were asked to participate in two specimen-sorting experiments, a free pile sort and a successive pile sort. Subjects were screened to ensure that they had no formal training in zoology nor any familiarity with these South American birds. In the free pile sort, subjects were asked to arrange specimens into groups on the basis of their overall similarity. Each subject performed this task separately on subsets of forty passerine specimens and forty non-passerine specimens. The judged similarity of each pair of specimens was measured by counting how often the pair was placed in the same group. In the successive pile sort, subjects were first asked to place the specimens into groups on the basis of their overall similarity. Subjects were then asked to divide whichever group they thought most heterogeneous into two coherent subgroups.

The subject continued to divide groups until all specimens were separated. Next, the subject's initial groups were restored and the subject asked to merge the most similar pair of groups. The subject continued to merge groups until all specimens were merged. Each subject performed the task separately on subsets of fifteen passerine and fifteen non-passerine specimens. These subsets of specimens were chosen such that the passerine and non-passerine subsets had the same underlying scientific taxonomic structure. For this purpose, the Piciformes (including woodpeckers, toucans, galbulas and jacamars) were chosen as a representative non-passerine order. The judged similarity of each pair of specimens was measured by counting the rank order in which they were split apart. The ornithologist who identified the specimens also performed the dividing experiment, using species names on index cards as stimuli rather than as specimens. These judgements broke ties in the taxonomic distances, disambiguating, for example, which species in a genus are most similar. Again, the judged similarity of species was measured by counting the rank order in which they were split apart. The data analysis involved comparing the judged similarity of each pair of specimens with the measure of confusion by the Jivaro in the bird identification experiments and with proximity in scientific classification.

Bibliography

Barth, F., *Insects and flowers: the biology of a partnership*, Princeton: Princeton University Press, 1985

Basso, A.E.C., Capitani and M. Laiacona, 'Progressive language impairment without dementia: a case with isolated category specific semantic defect', *Journal of Neurology, Neurosurgery and Psychiatry* 51, 1988, pp.1201–7

Berlin, B., 'Speculations on the growth of ethnobotanical nomenclature', *Language in Society* 1, 1972, pp.51–86

——, 'Folk systematics in relation to biological classification and nomenclature', *Annual Review of Ecology and Systematics* 4, 1973, pp.259–71

——, 'The concept of rank in ethnobiological classification: some evidence from Aguaruna folk botany', *American Ethnologist* 3, 1976, pp.381–99

——, D. Breedlove and P. Raven, 'General principles of classification and nomenclature in folk biology', *American Anthropologist* 75, 1973, pp.214–42

——, *Principles of Tzeltal plant classification: an introduction to the botanical ethnography of a Mayan-speaking people of Highland Chiapas*, New York: Academic Press, 1974

Blough, D.S., 'Pigeon perception of letters of the alphabet', *Science* 218, 1982, pp.397–8

Boster, J., 'Classification, cultivation, and selection in Aguaruna cultivars of Manihot esculenta (Euphorbiaceae)', in G. Prance and J. Kallunki (eds), *Ethnobotany in the neotropics: Advances in economic botany 1*, 1984a

——, 'Inferring decision making from preferences and behavior: an analysis of Aguaruna Jivaro manioc selection', *Human Ecology* 12, 1984b, pp.343–58

——, 'Selection for perceptual distinctiveness: evidence from Aguaruna cultivars of Manihot esculenta', *Economic Botany* 39, 1985, pp.310–25

——, 'Agreement between biological classification systems is not dependent on cultural transmission', *American Anthropologist* 89, 1987, pp.914–20

——, 'Natural sources of internal category structure: typicality, familiarity, and similarity of birds', *Memory and Cognition* 16, 1988, pp.258–70

——, 'The information economy model applied to biological similarity judgment', in J. Levine, L. Resnick and S. Teasley (eds), *Socially Shared Cognition*, American Psychological Association, 1991

——, B. Berlin and J.P. O'Neill, 'The correspondence of Jivaroan to scientific ornithology', *American Anthropologist* 88, 1986, pp.569–83

——, and R. D'Andrade, 'Natural and human sources of cross-cultural agreement in ornithological classification', *American Anthropologist* 91, 1989, pp.132–42

——, and J. Johnson, 'Form or function: a comparison of expert and novice judgments of similarity among fish', *American Anthropologist* 9, 1989, pp.866–89

Bulmer, R., 'Which came first, the chicken or the egg-head?', in J. Pouillon and P. Maranda (eds), *Échanges et communications:*

Mélanges offerts à Claude Lévi-Strauss, Vol. 2, The Hague and Paris: Mouton, 1970

Cerella, J., 'Absence of perspective processing in the pigeon', *Pattern Recognition* 9, 1977, pp.65–8.

——, 'Visual classes and natural categories in the pigeon', *Journal of Experimental Psychology: Human Perception and Performance* 5, 1979, pp. 68–77

——, 'The pigeon's analysis of pictures', *Pattern Recognition* 12, 1980, 1–6

——, 'Mechanisms of concept formation in the pigeon', in M.A. Goodale and R.J.W. Mansfield (eds), *Analysis of Visual Behavior*, Cambridge, MA: MIT Press, 1982

Cott H.B., 'Animal form in relation to appearance', in L.L. Whyte (ed.), *Aspects of form*, Bloomington: Indiana University Press, 1966 (1951)

Diamond, J.M., 'Classification system of primitive people', *Science* 151, 1966, pp.1102–4

Dodds, K., 'The history and relationships of cultivated potatoes', in J. Hutchinson (ed.), *Essays on crop plant evolution*, Cambridge: Cambridge University Press, 1965

Ellen, R.F., Introductory essay, in R.F. Ellen and D.A. Reason (eds), *Classifications in their social context*, London: Academic Press, 1979

Farah, M., Hammond, Mehta and Ratcliffe, 'Category-specificity and modality-specificity in semantic memory', *Neuropsychologia* 27, 1989, pp.193–200

——, P. McMullen and M. Meyer, 'Can recognition of living things be selectively impaired?', *Neuropsychologia* 29, 1991, pp.185–93

Funnell, E. and J. Sheridan, 'Categories of knowledge? Unfamiliar aspects of living and nonliving things', *Cognitive Neuropsychology* 9, 1992, pp.135–53

Gaffan, D. and C. Heywood, 'A spurious category-specific visual agnosia for living things in normal human and nonhuman primates', *Journal of Cognitive Neuroscience* 5, 1993, pp.118–28

Gelman, S. and A.W. O'Reilly, 'Categories and induction in young children', *Cognition* 23, 1988, pp.183–209

Harlan, J., *Crops and man*, Madison, WI: American Society of Agronomy, 1975

Hart, J., R.S. Berndt and A. Caramazza, 'Category-specific naming deficit following cerebral infarction', *Nature* 316, 1985, pp.439–40

—— and B. Gordon, 'Neural subsystems for object knowledge', *Nature* 359, 1992, pp.60–4

Healey, C., 'Taxonomic rigidity in folk biological classification: some examples from the Maring of New Guinea', *Ethnomedizin* 5, 1978–9, pp.361–84

Herrnstein, R.J., 'Objects, categories, and discriminative stimuli', in H.L. Roitblat et al. (eds), *Animal cognition*, Hillsdale, NJ: Lawrence Erlbaum Associates, 1984

—— and D.H. Loveland, 'Complex visual concept in the pigeon', *Science* 146, 1964, pp.549–51

——, D.H. Loveland and C. Cable, 'Natural concepts in pigeons', *Journal of Experimental Psychology: Animal Behavior Processes* 2, 1976, pp.285–302

—— and P.A. de Villiers, 'Fish as a natural category for people and pigeons', in G.H. Bower (ed.), *The psychology of learning and motivation*, Vol.14, New York: Academic Press, 1980

Hunn, E., 'A measure of the degree of correspondence of folk to scientific biological classification', *American Ethnologist* 2, 1975, pp.309–27

——, 'The utilitarian factor in folk biological classification', *American Anthropologist* 84, 1982, pp.830–47

Jones, C.E. and R.J. Little, *Handbook of experimental pollination biology*, New York: Scientific and Academic Editions, 1983

Lorenz, K., 'The role of gestalt perception in animal and human behavior', in L.L. Whyte (ed.), *Aspects of form*, Bloomington: Indiana University Press, 1966 (1951)

Malt, B. and E.E. Smith, 'Correlated properties in natural categories', *Journal of Verbal Learning and Verbal Behavior* 23, 1984

McCarthy, R. and E. Warrington, 'Evidence for modality-specific meaning systems in the brain', *Nature* 334, 1988, pp.428–30

——, 'The dissolution of semantics', *Nature* 343, 1990, p.599

Mervis, C., Caitlin and Rosch, 'Relationships among goodness of example, category norms, and word frequency', *Bulletin of the Psychonomic Society* 7, 1976, pp.283–4

Randall, R., 'How tall is a taxonomic tree? Some evidence for dwarfism', *American Ethnologist* 3, 1976, pp.543–53

Roeder, K., *Nerve cells and insect behavior*, Cambridge: Harvard University Press, 1963

Romney, A.K., S.C. Weller and W.H. Batchelder, 'Culture as consensus: a theory of culture and informant accuracy',

American Anthropologist 88, 1986, pp.313–38

Rosch, E., 'Cognitive representations of semantic categories', *Journal of Experimental Psychology: General* 104, 1975, pp.192–233(b)

—— and C. Mervis, 'Family resemblances: studies in the internal structure of categories', *Cognitive Psychology* 7, 1975, pp.573–605

——, Rosch, C.B. Mervis, W.D. Gray, D.M. Johnson and P. Boyes-Braem, 'Basic objects in natural categories', *Cognitive Psychology* 3, 1976, pp.382–439

Sartori, G., Job, Miozzo, Zago and Marchiori, 'Category-specific form-knowledge deficit in a patient with herpes simplex virus encephalitis', *Journal of Clinical and Experimental Neuropsychology* 15, 1993, pp.280–99

Sheridan, J. and G.W. Humphreys, 'A verbal-semantic category-specific recognition impairment', *Cognitive Neuropsychology* 10, 1993, pp.143–84

Silveri, M. and G. Gainotti, 'Interaction between vision and language in category-specific semantic impairment', *Cognitive Neuropsychology* 5, 1988, pp.677–709

——, A. Daniele, Daniele, Giustolisi and Gainotti, 'Dissociation between knowledge of living and nonliving things in dementia of the Alzheimer type', *Neurology* 41, 1991, pp.545–6

Stewart, F., A.J. Parkin and N.M. Hunkin, 'Naming impairments following recovery from herpes simplex encephalitis: Category-specific?', *Quarterly Journal of Experimental Psychology: Human Experimental Psychology* 44A, 1992, pp.261–84

Waddington, C.H., 'The character of biological form', in L.L. Whyte (ed.), *Aspects of form*, Bloomington: Indiana University Press, 1966 (1951)

Yen, D.E., 'Natural and human selection in the Pacific sweet potato', in E.T. Drake (ed.), *Evolution and environment*, New Haven: Yale University Press, 1968

Young, A., Newcombe, Hellawell and de Haan, 'Implicit access to semantic information', *Brain and Cognition* 11, 1989, pp.186–209

Chapter 11

Agrarian Creolization: The Ethnobiology, History, Culture and Politics of West African Rice

Paul Richards

Introduction

In linguistics, creolization is the process whereby two languages converge to form a third language that then develops independently of either parent. Nineteenth-century linguistics thought largely or exclusively in terms of language formation by divergence and descent. Pidgins and creoles were seen as marginal to the main processes of language formation, pathological products of cultural miscegenation. But as the modern world has increasingly come to recognize 'globalization' as a key element in contemporary culture change, so creole linguistics has moved centre stage in debates about the origin and development of language. Now, in the work of, for example Bickerton [e.g. 1975], creolization is seen simply as one set of universal possibilities for language formation (a bio-programme) activated by certain kinds of social conditions. Ocean-borne trade from the fifteenth century onwards vastly enlarged the demand for communication between groups with little or no prior contact. Many pidgins (reduced trade-contact languages) emerged where the mercantile networks of the Portuguese, Dutch, British etc. touched the shores of the Atlantic and Pacific. In a number of such places these nodes of world trade gave rise to new but permanent communities of merchants, brokers and labourers with heterogenous linguistic and cultural backgrounds. It was typically in these circumstances that pidgins began to develop into fully fledged creole mother tongues, making a future full community life possible without linguistic dependence upon past incoherence and diversity of origins.

In this chapter I wish to address, in a speculative way, the possibility that some kinds of agrarian development might be based on an analogous process of 'cultural creolization', which I would wish to define (provisionally) as a characteristic and self-conscious commitment to exploit synergy between external and indigenous elements in the elaboration of farming technologies in parts of the world where the diverse backgrounds of the participants in an agricultural system erect barriers to the elaboration of evolutionary models of agrarian change. I need to make two points clear at this juncture. First, by 'models of agrarian change' I mean those internal images through which members of society guide and negotiate the realization of common futures. Secondly, any fully worked-out theory of agrarian creolization will require extensive engagement with current debates in semiotics, culture theory and material culture studies. This is an agenda for a book rather than a chapter, and all I hope to achieve in the following pages is the ground plan of a suggestive case study, to be quarried in future theoretical analysis. My aim is to hint, by example, at some of the ways in which a concept of agrarian creolization might be useful in progressing 'beyond nature and culture' in anthropological debates about technological change in agriculture.

My case study concerns the old rice-cultivation zone of the Atlantic coast of the western half of West Africa. Here (uniquely) cultivation of Asiatic rice has been grafted on to farming systems and technologies formed for the cultivation of indigenous African rice, to create distinctive 'creolized' cultivation processes and (agri)cultural values with an independent character and life of their own. It is perhaps more than coincidental that this part of the Atlantic basin is also known for its contributions to linguistic creolization from an early date.

Cultivated Rice in West Africa

There are two cultivated species of rice in West Africa, *Oryza glaberrima* and *O. sativa*. According to recent research in rice genetics the introduced species *O. sativa*, now responsible for most of the region's rice output, has two distinct centres of domestication, one in India, the other in east Asia [Glaszmann 1987, 1988]. When *O. sativa* first reached West Africa from Asia is not known. *O. glaberrima* is found only in Africa, and was most probably domesticated by West African peoples in ancient times. A likely ancestor, the annual wild rice *O. barthii* (*O. breviligulata*), is widespread in wetlands from the Gambia to Lake Chad. The ancestry of *O. glaberrima*, complicated by gene transfer between wild and domesticated types in the region, is not yet fully understood. These two cultivated rice species are not easily distinguished in the field. Typically, *O. glaberrima* types have small pear-shaped grains, with red pericarp and olive to black seedcoat colour, simple-branched erect panicles, and short rounded ligules. But small pear-shaped grains are a feature of some *sativa* types, and some *glaberrima* types have pointed ligules.

African rice is sometimes referred to loosely as 'red rice' and Asian rice as 'white rice'. But red pigmentation in the grain pericarp is a dominant character in both species. Plant breeders select for the white-skinned types preferred in international trade. By contrast, some ethnic groups – e.g. the Mende of Sierra Leone – prefer red-skinned types. Red pigmentation from the pericarp, some of which is transferred in cooking to the grain of partially milled ('rough') rice, is seen as a guarantee of quality, since local rices are thought to have a firmer texture and better flavour than imported white-skinned types. Red colouring also has a symbolic significance. Among the Mende 'red rice' (rice soaked in palm oil) is an important feature of sacrifices to the ancestors. But colour, it should be reiterated, is no guide to species: it serves only to distinguish long-established 'traditional' cultivars from 'trade rice' and modern varieties acquired through development projects.

Rice is a self-pollinating crop. Inter-specific hybrids are generally infertile, and although some have been developed by researchers few if any have proven viable. West African farmers are very knowledgeable about rice types and sometimes plant both species side by side in one field. Some local rice researchers suspect that this may, in time, have given rise to natural inter-specific crosses. But so far as is known the two species remain distinct.

Zones of Rice Cultivation in West Africa

Rice is the main staple food of about ten to fifteen million West Africans living in the coastal region of the western half of West Africa (from Casamance in Senegal to the Bandama river in central Ivory Coast). It is convenient to refer to this area as the West African Rice-subsistence Zone (or WARZ for short) to distinguish it from two other West African settings in which rice is an important crop, but not the sole or major staple: riverine cultivation in the drier savannas from the Senegal River to Lake Chad, and the zone of recent commercial rice cultivation in the forests and wetter savannas of Ghana and Nigeria.

The zone of riverine rice cultivation is important historically, since the Inland Delta of the Niger in Mali has been proposed as the cradle for the domestication of *O. glaberrima*. However, rice is not now the main staple in this more arid part of the West African savanna, because cultivation is limited to the major river valley basins. Other staples – pearl millet, *Digitaria exilis* (so-called 'hungry rice') and sorghum – are important in flood recession and dryland conditions over much of the savanna zone in West Africa. A Mande creation myth [Dieterlin 1957] assigns the origin of these West African grain staples (including rice) to the valley of the Upper Niger – the Mande heartland. This is a point of view some independent observers are inclined partially to endorse on ethnobotanical and environmental grounds, though others [notably Wigboldus 1986, 1991] are inclined to argue that sorghum and pearl millet have a less ancient pedigree as West African crops.

The zone of modern commercial expansion of rice covers, principally, Nigeria, Ghana and eastern Ivory Coast. Today, Nigeria is the largest producer of the crop in West Africa. Some dry-season riverine cultivation of *O. glaberrima* is still found in the far north of the country, on the edge of the Sahel, but the bulk of Nigerian rice comes from the wetter Guinea Savanna and Forest zones to the south, where white yam (*Dioscorea rotundata*) is, historically, the more important staple. Rice in these areas is mainly a commercial interloper associated with colonialism [but cf. Brydon 1981]. In western Nigeria it seems first to have been introduced in the mid-nineteenth century by ex-slaves raised in Sierra Leone. Some improved cultivars promoted by the National

Agricultural Cereals Research Institute still contain Sierra Leonean germplasm (e.g. the Mende short-duration upland type *jete*) in their make-up.

Outside the WARZ, rice is a modern crop, associated with urbanization and the expansion of commerce in the colonial period: yam is the traditional rural staple, a fact apparent in its continued ritual salience (in annual New Yam ceremonies, for example) throughout a zone stretching from eastern Ivory Coast to eastern Nigeria [Arua 1981; van der Breemer 1990]. In Ghana during the 1970s business entrepreneurs (some from the military) invested in machines and fertiliser for the rapid expansion of rice cultivation in the extensive grassy bottom lands north of the forest to meet urban market demand. The state has promoted rice at the expense of yam cultivation in the Ivory Coast [Chauveau, Dozon and Richard 1981]. Van den Breemer [1990] shows that in the eastern part of the country, groups related linguistically and politically to the yam cultivators of the Ghanaian forest zone are still highly ambivalent about those of their number who cultivate rice, fearing that the commercialism that accompanies rice-farming will undermine the social values upon which the rural community is based and deplete the fertility of the forest.

West of the Bandama river in the Ivory Coast, however, among groups speaking the Mande languages of the Upper Niger basin or the West Atlantic languages of the Upper Guinea coast, material cultures and moral economy have long been shaped by the experience of cultivation of rice in forested environments. Of present-day countries comprising the WARZ the nucleus is formed by Sierra Leone and Liberia. To this nucleus should be added the south-western quadrant of the Ivory Coast, most of the Republic of Guinea (at least as far north as Futa Jallon), Guinea Bissau and the valley of the Casamance in southern Senegal. The distinctive feature of the societies of the WARZ is that the greater part of rural society (men and women, young and old) is directly involved in rice cultivation. This contrasts with the situation in the savanna zone to the north, where rice is a speciality crop formerly sometimes typically associated with slave settlements and, more recently, with cultivation by women. An important irony of political economy follows from this: in the West African savannas the cultivation of rice tends to be correlated with a marked degree of peasant stratification, whereas in the WARZ it is associated with the survival of egalitarian values [Linares 1981, 1991; Watts and Carney 1990].

Among West African regions where rice is today an important crop, the WARZ is unique in the extent to which social institutions and cultural values have been shaped by rice cultivation. Only here, I would suggest, is it possible to seek valid parallels and comparisons with those Asian societies similarly dependent on rice as a staple [cf. Bray 1986]. The possibility of making such comparisons would seem to be especially intriguing in view of the fact that the origins of West African rice agriculture are independent of events in Asia (being based on *O. glaberrima*). But unlike Madagascar – the other source of potential comparison in Africa – the long-term interaction between crop and society in the WARZ is not complicated by direct cultural transmission. Further, rice cultivation in the WARZ in the modern period has been 'politicized' in ways that differ significantly from circumstances that have prevailed in, say, Japan or Sri Lanka (and for that matter, in other parts of West Africa). It will be suggested below that in some respects the continued involvement of peasant producers in the WARZ in rice is subversive of modern state power (a fact that may have some bearing on the emergence of, for example, Casamance regional separatism in Senegal). I will outline ethnobotanical reasons for finding the inventive elaboration of indigenous rice-farming practices the last resort of an 'uncaptured peasantry'. As Tacitus conceded in the case of the matrilineal, semi-nomadic, shifting-cultivating, cattle-minded Germans 2,000 years ago, poverty-with-freedom is not to be despised: it may contain the seeds of future developments [Richards 1992].

Early Foci of African Rice Cultivation

It is a reasonable assumption that the domestication of *O. glaberrima* must have occurred within the zone of distribution of the putative ancestor, *O. barthii*. For want of firm historical evidence, Portères [1976] attempts the reconstruction of a likely scenario for the emergence of African rice cultivation by drawing upon ethnobotanical evidence, principally rice germplasm collected in the francophone countries of the region, especially

Guinea-Conakry [also Portères 1966]. (It is to be regretted that Portères seems never to have collected material in Liberia and Sierra Leone, since both countries are of focal importance in the WARZ.)

Portères [1976] identifies three areas of rice germplasm diversity in West Africa: the Inland Delta of the Upper Niger in present-day Mali, the Gambia-Casamance-Guinea Bissau coastal-riverine zone, and the Guinea dorsal (the watershed between the Upper Niger basin and the rivers draining to the Atlantic coast through the forests of present-day Sierra Leone, Liberia and western Ivory Coast). He considers the first of these three areas the likely focus of domestication (on genetic grounds), and treats the other two as centres of secondary dispersal. Today, these three *foci* are associated with three distinct techniques for growing rice [Dresch 1949]: flood advance and recession techniques in the Inland Delta [Harlan and Pasquereau 1969], estuarine wetland systems along the coast [cf. Linares 1981; van der Drift 1992], and dryland-wetland catenary cultivation systems on the Guinea dorsal [Richards 1986].

Pending fuller examination of the region's rice germplasm resources (including systematic scrutiny of representative collections from Sierra Leone and Liberia) it might be wise to suspend judgement on Portères' interpretation. An alternative, a-centric, model might prove preferable [cf. Harlan, de Wet and Stemler 1976]. Early human communities probably gathered wild rice throughout the region. *O. barthii* is widely distributed in deep-flooding wetlands and grassy clearings throughout the forest and savanna zones West Africa, and even today the practice of gathering wild rice in times of hunger has not entirely lapsed in parts of Sierra Leone. Gathering would have exerted its own pressures upon *O. barthii* (selection for types less prone to shattering, for example) and early steps towards planting might have taken place over a wide range of sites occupied by early gatherer-hunters (there is archaeological evidence for stone-age occupation of sites within the Upper Guinean forest-savanna ecotone, e.g. at Yengema in Sierra Leone). Porteres' [1976] suggested date for these first steps towards a rice agriculture based on *O. glaberrima* – c. 2000–3000 BP – appear reasonably conservative and consistent with other fragmentary data concerning West African agricultural origins [Harlan et al. 1976; Wigboldus 1986].

The issue of when and where varieties of *O. sativa* were first introduced into West Africa is almost as unclear as the question of the origins of *O. glaberrima.* It is generally and probably rightly assumed that the repertoire of rices available to farmers in the WARZ (especially those in coastal districts) was greatly expanded from the early sixteenth century by introductions direct from Asia through the agency of the Portuguese. However, it seems to me not necessary to suppose that the Portuguese were the first or sole agents of the spread of *O. sativa* in the region.

Specimens of *O. glaberrima* were collected in central America in the 1950s. Evidently, the plant crossed the Atlantic with the slave trade (perhaps accidentally) and may have been cultivated subsequently by slaves or ex-slaves for their own subsistence (cf. the situation regarding sorghum in the New World. There are reports in Arab sources of rice cultivation in West Africa prior to the arrival of the Portuguese in the late fifteenth century. It is impossible to say whether these early reports refer to *O. sativa* or *O. glaberrima.* Lewicki [1974] speculates that an apparent reference to rice in North Africa in the classical period (prior to the first records of *O. sativa* in Egypt) might be evidence that *O. glaberrima* crossed the Sahara northwards. Today, West African farmers regularly acquire varieties through accidental introductions. A recent example is a range of rices in Sierra Leone now referred to as 'OAU', allegedly derived from unhusked grains found in sacks of milled rice imported for the conference of the Organization of African Unity in Freetown in 1980. If *O. glaberrima* crossed the desert northwards in classical times and travelled across the Atlantic during the period of the slave trade, and if farmers today acquire new varieties from sacks and packing, there is no good reason to suppose that *O. sativa* could not have reached West Africa via the trans-Saharan trade, perhaps accidentally. Nor is it necessary to suppose that the Portuguese (or any other merchant-adventurers in the West African coastal trade) had a definite policy for the introduction of *O. sativa*. Indeed, it may be doubted that they were aware of any significant difference between local rice and the rice species they had encountered in Asia. A few unhusked grains swept out of a ship-board food store would be all that might be needed for local farmers to add significantly to their germplasm stock.

Recent work on *O. sativa, O. glaberrima* and presumed

ancestors, using techniques of isozyme and DNA analysis, confirms the distinctiveness of *O. glaberrima* and the plausibility of a relationship with *O. barthii*. In future it may be possible, using such techniques [cf. Dally and Second 1990; Glaszmann 1987, 1988], to map out more explicitly the ancestry of different varieties of rice in the region, and so throw more light on the issue of domestication and the spread of early cultivation.

Rice Types and Farming Systems

Throughout the WARZ most small-scale producers cultivate a mix of dryland and wetland rice cultivars – typically from 50:50 to 80:20 in favour of dryland types – often in catenary sequences linking free-draining upland soils, river-terrace sandy or silty moisture retentive soils, and hydromorphic soils in inland valley swamps. Management of this upland-wetland continuum is still largely by rotational fallowing of forest secondary successions and wooded savannas, though locally important specialized wetland rice farming systems are to be found in certain coastal and riverine environments (principally, the salt-water exclusion farming systems in estuaries from the Gambia to Conakry, the tidal-pumped mangrove farming systems of the Great and Little Scarcies, and the flood-retreat/flood-advance systems of the riverine grasslands behind Turner's Peninsula in southeast Sierra Leone).

In recent years there has been a tendency throughout the region to compensate for a rise in population pressure by increased use of the wetland end of the catenary spectrum. Historically, this tendency has been most marked in the wetter savannas and forest-savanna transition, where dryland rice cultivation is at its most marginal because of lower rainfall and a shorter wet season (<2000mm per annum, five or more dry months). Semi-permanent cultivation of valley swamps (alternating wet-season rice cultivation with dry-season cultivation of crops such as cassava, sweet potatoes and groundnuts) is now the norm in some of the more heavily populated parts of northern Sierra Leone and Guinea. Elsewhere (in eastern Sierra Leone, Liberia and western Ivory Coast) forest fallowing remains viable, and planting dryland rice on the upper part of the catenary continuum still attracts the greater emphasis.

Both cultivated species of rice in the WARZ contain strongly rooting types adapted to rain-fed uplands, types capable of rapid elongation of the first internode to match variable flood depths in river flood plains, and salt-tolerant types adapted to brackish coastal wetlands (e.g. mangrove swamps), as well as regular wetland types suited to planting in paddies. Some glaberrima types are especially quick-ripening, but in comparative trials tend to yield less than equivalent sativa types. They suffer from a number of agronomic disadvantages (shattering of the ripe grains) and processing defects (being hard to husk and clean). On the credit side of the balance, however, they typically exhibit evidence of durable resistance to local pathogens, e.g. blast, and continue to grow strongly even in depleted and low-fertility soils (in these conditions they appear capable of out-competing weeds that would overwhelm sativa types).

As a generalization it can be stated that wherever *O. sativa* can be planted in the WARZ there is an equivalent *O. glaberrima* type. With the more recent introduction of sativa types, perhaps especially since the beginning of the colonial period, it seems probable that some glaberrimas will have been squeezed out of the better favoured environments. *O. glaberrima* survives, however, in specialized niches (many farmers in Sierra Leone reserve short-duration glaberrimas for use as a hunger-breaker crop in advance of the main harvest) and in areas of especial environmental difficulty. For example, it would seem that increased land pressure and rainfall uncertainties in northwest Sierra Leone in recent years have caused farmers to reserve high-yielding improved cultivars of *O. sativa* for better land and to turn increasingly to the use of hardy glaberrima types on poorer soils.

The current relative importance of *O. sativa* and *O. glaberrima* in the WARZ, and their distribution over different soils and farming systems, can be judged from a national sample survey of rice germplasm usage undertaken in Sierra Leone during the 1987 harvest season [Lipton, Pain and Richards 1993]. Sierra Leone lies at the heart of the WARZ and is more or less equally divided between forest and wetter savannas. The country contains examples of all the major rice-farming systems in the WARZ with the exception of the coastal salt-exclusionary systems (these are found further north along the West African coast). Rice germplasm was inventoried and collected in just over 600 household farms in three major rice cultivation settings in Sierra Leone: the *boli* wetlands of north-central Sierra Leone, the estuarine mangrove swamps of the northwest of the country and the upland-wetland catenary continuum in five localities along the savanna-forest gradient.

Some 480 samples of about 180–200 rice types (as named and distinguished by farmers) were planted out in uniform conditions (according to whether they were dryland or wetland types) for observation at Rokupr Rice Research Station. The number of days to 50 per cent flowering, culm height and other growth characteristics were recorded. From these data it has been possible to surmise that a number of samples collected under different names belong to a single type (this situation arises where research-station releases have been named after the person first introducing the seed: informal seed distribution from farmer to farmer is the most important channel of diffusion in Sierra Leone). More commonly, however, samples collected under a single name often proved to be quite varied in their performance when grown under uniform conditions at Rokupr. In some cases the explanation seems to be that the name used by the farmer is a category label, referring to the kind of performance to be expected of the seed type in question, or reflecting the circumstances under which the rice was acquired.

In Mende (the main language of the south and east of Sierra Leone), for example, **yaka** refers to any tall, long-duration, flood-tolerant variety suitable for cultivation in inland valley swamps under low management; **bongo** to short-duration rices suitable for planting at the bottom of the catenary profile as hunger breakers; **ndogbogbi** to general-purpose types suited to a range of upland sites with different fallow histories; and **helekpo** to varieties acquired from elephant dung (of which more below).

Sierra Leonean farmers do not recognize *O. glaberrima* types as a group. This is not surprising since, even where entries included in the observation collection were concerned, it was not possible for rice scientists always to judge from the phenotype which samples belonged to which species. Farmers do recognize, however, that glaberrima types sometimes survive as weedy associates of *O. sativa*. Where these weedy associates are encountered in a field of *O. sativa* Mende farmers refer to the glaberrima off-types as **sanganya**.

Sanganya is sometimes used as a category name for a range of rices with glaberrima-type grain characteristics (small, pear-shaped, readily-shattering grains with stubborn red skins), but in most cases other *O. glaberrima* types have specific names (such as the quick-growing – 100–110-day – types the Mende call **mala** and **pende**). Farmers sometimes rogue **sanganya** and plant it separately. It is said to have been a major upland type in Mende country in former times. It still survives as a crop for two reasons: it is considered especially 'sweet', and it will tolerate poor soils and skimpy weeding.

Disregarding caveats about species recognition and the diversity masked by category names, it appears that the observation collection comprised about 160 (minimum) distinct *O. sativa* types and about twenty (minimum) distinct *O. glaberrima* types (from about 600 farms). From the farm germplasm inventories it is possible to estimate that *O. glaberrima* would account for about 5 per cent of the national rice crop by area in 1987 (and slightly less than 5 per cent – due to typically lower average yields, compared to *O. sativa* – of total rice output in the country). *O. glaberrima* was hardly encountered in the two specialized wetland environments, but specimens were found on all parts of the soil catena in dryland-wetland continuum sites in drier savanna locations. In some villages in northwest Sierra Leone glaberrimas accounted for up to forty per cent of all rice planted. In wetter, more forested districts in the south and east of the country *O. glaberrima* seems now to be reserved for specialist use, either as a hunger-breaker crop on moisture-retentive soils or as a catch-crop where labour for weeding is limited. Hunger-breaker rices are sometimes referred to as a group as 'three-month rices' (i.e. varieties ripening within 90 to 120 days). A substantial proportion of *O. glaberrima* samples in the observation collection fell into this category.

Viewed from a historical standpoint there can be little doubt that the cultivation of glaberrimas declined in the face of further dissemination of *O. sativa* from and through research centres such as Rokupr Rice Research Station (established 1934) during the colonial period. There is some possibility, however, that *O. glaberrima* types are now experiencing a relative renaissance in parts of Sierra Leone in recent years due to adverse and deteriorating soil and rainfall conditions. Evidence on patterns of adoption and abandonment of rice varieties among the 600 farm households in the survey just cited suggests that acquisition and rejection rates for *O. glaberrima* types are not significantly different from those for research station releases of *O. sativa*. In fact, in the five years up to 1987 as many farmers reported adopting *O. glaberrima* types *new to them* as had adopted a modern release of *O. sativa*, an unexpected and remarkable finding. Although *O. glaberrima* today occupies only a modest place in the rice economy of Sierra Leone it is more than a historic relic.

That glaberrima rices take their place on merit (rather than as mere historical survivals) is suggested by the following incident. A Mende farmer interviewed in Kogbotuma (Moyamba District) in October 1987 described to us an unintended experiment with a glaberrima type recently introduced from the northern part of the country. His seed stocks of a preferred sativa type were insufficient to cover the area he had cleared ready for farming early in 1987. Unwilling to see his labour wasted, he accepted, for want of anything better, a loan of seed of the unfamiliar glaberrima, and planted this alongside the favoured sativa. Half way through the season the household found itself in the middle of an unplanned experiment. Half the *O. sativa* plot and all the *O. glaberrima* plot were abandoned to weeds when the women of the household were called away to deal with a bereavement. The informant read 'treatment and control' into this pattern of unintended neglect. The section of unweeded *O. sativa* yielded nothing, but the *O. glaberrima* alongside it was far less seriously affected by weed competition. In fact observers were hard pushed to detect where the weeding had stopped. As a result, the farmer proposed to pay much more attention to this unfamiliar but tolerant variety in subsequent years.

Selection and Experimentation by Farmers

The idea of experimentation to explore the potential of unfamiliar rice germplasm seems to be widespread among the societies and cultures of the WARZ. Rice is largely self-pollinating and farmers can maintain varietal distinctness to a considerable extent by roguing at harvest. Some natural out-crossing takes place at the margins of plots, and spontaneous intra-specific crosses are encountered from time to time.

It is not uncommon to come across farmers with an apparently intuitive understanding of hybridization. Squire [1943, cited in Richards 1985] reported that Mende farmers sometimes carefully selected their seed rice only from the centre of a stand to avoid accidental out-crossing. In other cases, farmers seek consciously to isolate new strains from this mixed material [Richards 1986].

Edge-reaped seed is known to contain a higher proportion of mixed types (including natural crosses). This was explained to me by one informant in Lalehun (Gola Forest, eastern Sierra Leone) in terms of a human analogy – seed rice from the centre of the field was likened to a 'mother' surrounded by her 'children'. 'Mother' rice from the centre of the plot is expected to breed true to type, whereas identical phenotypes from the farm edge are recognized as sources of some of the variation one might normally expect to find among numerous offspring. (Family resemblances are a frequent topic of conversation among Mende villagers.)

Off-types subsequently isolated from edge-reaped material are likely to be carefully conserved and planted in small trial plots to assess their potential. In Mende rice 'extracted' from other rice through roguing and subsequent experimentation in this way is known by the category name **mbeimbeihun** ('rice in [from within] rice'). A trial is referred to by the Mende word **hugo** (literally 'to investigate within'). The material, if promising, may then be planted out more extensively and subjected to mass selection (selection and reservation of panicles at harvest according to the characteristics of the desired phenotype). Over time, this selection process on a strongly in-breeding annual such as rice will tend to reduce heterozygosity in the planting stock.

Some types subject to such selection may be regarded as approximating to true varieties. Other local rice types in the WARZ are best regarded as members of landraces (a pool of germplasm from which a number of characteristic phenotypes are recurrently 'extracted').

In short, then, the available evidence suggests that farmers in the WARZ are especially interested in, and appreciative of, variation in rice germplasm, and that they attempt to manipulate it in their favour. It is appropriate to ask why these farmers place emphasis on such manipulations. What do they hope to gain thereby?

The main aim always seems to be to ensure better exploitation of the range of opportunities on the dryland-wetland continuum. In Mogbuama, a village in central Mende land studied in detail in two seasons (1983 and 1987), about a hundred farm households planted about fifty distinct rice types, divided into three broad categories: about thirty 4-month rices suited to free-draining soils on the upland-wetland continuum, about ten 3-month rices suited to lower-slope moisture-retentive soils, and about ten long-duration flood-tolerant types for water courses and swamps. There was a significant turnover of rice types between 1983 and 1987. Overall, about 18 per cent of rice area in 1983 was planted to rices no longer being grown in Mogbuama in 1987. In 1987 new types accounted for about 23 per cent of the rice area planted. Varieties new to the farmer concerned since 1983 accounted for more than 50 per cent of all rice planted by fifty-two farmers active in both years. Although some of these changes were forced by accident or indigence, others were adaptive changes directed towards a more versatile use of soil resources, especially at the lower end of the soil catena. Short-duration types suited to low-lying land and planted as an insurance against pre-harvest hunger were the types most eagerly sought: in interviews, farmers reported two or three distinct phases in the acquisition of such rices in Mogbuama. Formerly, the quickest rices of this kind ripened in about 125 to 130 days but the main current varieties take only 110 to 115 days. All were acquired through informal seed-distribution channels. Several were selections by farmers.

This regular sifting and selection of a wide range of rice types appears to be a general feature of rice farming throughout the WARZ. The underlying logic is as follows. Lacking labour, and cultivating low-fertility land in hazardous and unstable environments, the region's farmers have chosen to match their planting materials to the land as they find it rather than undertake expensive soil-quality and land-shaping improvements, as in many parts of Asia. In effect – to draw upon a computer analogy – West African rice farmers have learnt that it is more cost-effective to manipulate the software than rebuild the hardware of the cultivation system.

Rice, Elephants, Rogues and Witches

Although it may prove that African rice cultivation first developed in specialized riverine and coastal wetland environments, there seems little doubt that the forest-savanna ecotone – which today approximately bisects the WARZ east-west – is the historical axis around which rice farming systems of the dryland-wetland continuum type developed. Portères' [1976] Guinea dorsal centre of germplasm diversity and secondary dispersal lies along this line in a region otherwise identified as a possible focus for the emergence and differentiation of languages classified by Greenberg [1966] as southwest Mande. Mende belongs to this group. It might seem reasonable therefore to hazard a link between the subsequent spread of these languages and the opening up of a significant forest-edge frontier of agricultural settlement that has moved progressively deeper into the forests of Liberia and Sierra Leone over the past half millennium [cf. Murphy and Bledsoe 1987].

Apparently consistent with this notion of a rice frontier moving south and east through the western half of the Upper Guinean forest block, it has been hypothesized [Livingstone 1958] that the distribution of the human sickle-cell gene reflects increased exposure to malaria with agricultural intensification. Sickle-cell gene confers some degree of immunity to malaria, and heterozygotes for the sickle-cell gene survive and become more numerous as a percentage of the population in regions of high malaria risk. Populations closest to the Guinea dorsal (where rice agriculture has a long history) have some of the highest rates of sickle-cell gene in the region, and groups closest to the forest frontier (in southeast Liberia) – where it is suggested that hunting, gathering and low-intensity cultivation of roots and tubers gave way to rice agriculture only in the mid-nineteenth century [d'Azevedo 1962] – have some of the lowest.

This association between rice and increased risk of malaria implies that the forest was settled only at considerable psychic cost. Even today groups like the Mende suffer some of the highest (malaria-related) infant mortality rates in the world. Historically, they see themselves as having engaged in a struggle to tame the forest, both to harness its practical resources by physical efforts and to neutralize its mystical dangers through intellectual agility.

When asked about the origins of rice cultivation in their communities, Mende farmers sometimes begin with reference to the forest elephant. With restricted labour capacity to clear the high trees of the upland rain forest, the first rice farmers probably exploited natural gaps and clearings in the canopy, especially grassy swamp basins. Even today such swamp basins are sought out by farmers contemplating a move into high forest along the Sierra Leone-Liberia border [Davies and Richards 1991]. These swampy clearings are a favoured dry-season habitat of elephants. In fact, such areas may have been first opened up, or at least significantly enlarged, by these 'bulldozer herbivores' [Kortlandt 1984]. Having thinned the forest, let in light, puddled swampy soils and encouraged the growth of grassy plants and broad-leafed creepers, elephants also supplied the seed.

Elephant droppings sometimes contain large amounts of undigested rice; during the hungry season these droppings were once regularly collected for food. The gatherers would notice from time to time that 'elephant dung' rice (**helekpo** in Mende; cf. Temne **pa rank**) contained unfamiliar types. These were rices grazed in farms belonging to distant villages, perhaps beyond the scope of regular human contact through high forest. Here, then, was one important early source of potential novelties upon which diversified planting strategies could be elaborated as farmers began to press up-slope on to the high forest interfluves from swamp basins first puddled by their elephant rivals.

Mende historiography often traces the formation of forest-edge communities to spots where a hunter killed an elephant [Hill 1984]. Eye-witness accounts of elephant kills in the Gola Forest dwell on the small army of helpers that would quickly assemble, its first task being to erect a camp of temporary shelters for the time – often several weeks – that it might take to butcher the carcase, smoke the meat and share out the spoils. A hunter with the skill regularly to engage the wider elephant 'society' in a struggle for habitable terrain was likely to have the regular material resources to hand, as well as the mystical powers and leadership potential, to shape a more permanent human community and abate its anti-social tendencies. Fittingly, such a leader would claim the ivory tusks and have them carved into trumpets as insignia of chieftancy, to symbolize this transfer of the forest from elephant to human domain. To his followers –

working at the other end of the elephant – would fall the gift of rice germplasm with which the forest soils could be put to work to secure the subsistence of the new community.

As links developed with European traders on the coast ivory became a significant export from the region. The story circulated that between Sierra Leone and Cape Mount rice farms were to be found fenced with elephant tusks [Rodney 1970]. However fantastic, the tale reflects a grain of truth, since in the seventeenth century this area was noted for its greater emphasis on direct trade with Europe (England especially) in ivory, camwood and other forest products, and a correspondingly smaller emphasis on the slave trade, than elsewhere on the West African coast [ibid.]. Many of the famous Afro-Portuguese ivories were in all probability carved at this period in communities first won from the forest in the manner just described. But seventeenth-century coastal chiefs knew where their priorities lay; Rodney records that trade with Europeans might be suspended when the local people were busy making their rice farms.

The mercantilism of this creolizing Atlantic rice coast was not restricted to commodities alone, however, for along with the ivory trade went the beginnings of an import-export economy in ideas and artefacts concerning subsistence rice cultivation.

It is from a Dutchman engaged in trade at Cape Mount in the mid-seventeenth century that we have one of the earliest documented accounts of a catenary rice-farming system in the Upper Guinean forest region. As reported to the Amsterdam geographer Dapper (1668 [Jones 1983]), the system was based on three rice crops: a short-duration crop planted in low-lying muddy places in January or February, a second crop on middle slopes a month or two later, and a 'great' crop of longer duration rice on uplands when the rains were well set. The Dutch observer makes it plain that by planting three crops of different duration on different soils farmers were seeking to minimize labour constraints and also ensure a crop in the hungry season [cf. Richards 1986]. Even at this early date (c.1640) some farmers at Cape Mount were reported to have insufficient rice of their own to cover annual requirements, perhaps due to a diversion of labour into activities associated with overseas trade, and to make up shortages they relied on imports from Gola country to the north. Thus even at this early date rice cultivation in the inland parts of the WARZ was beginning to articulate with the Atlantic economy.

A second early account from an area further west along the coast of Sierra Leone confirms the logic and efficacy of the catenary system as a framework for the flexible management of difficult environments with restricted labour, but it also illustrates (perhaps more unexpectedly) that local agricultural techniques were of interest across the Atlantic!

In the 1780s, Thomas Jefferson (at the time US envoy to France) became interested in the topic of rain-fed rice cultivation. His primary concern was to find a substitute for wetland rice in South Carolina, and so ease the rigours of slavery and disease in that state. Discovering that dryland rice grew in abundance in West Africa, he commissioned an American sea-captain, Nathaniel Cutting, to collect seed samples and information on the correct mode of cultivation. On the coast of Sierra Leone Cutting sought the advice of an Anglo-African trader, William Cleveland, resident in the Banana Islands (close to Freetown). Cleveland's account [Cutting 1790] explains the clear distinctions drawn by local farmers between short-duration early-season varieties (suited to cultivation on low-lying moisture-retentive soils), medium-duration varieties intended for uplands, and long-duration varieties grown in inland valley swamps (these were considered poor eating and unhealthy, a view still shared by farmers in the region today). Cutting then fulfilled the second part of his commission from Jefferson by purchasing a keg of dryland rice on the Dembia River (in Susu country, north of present-day Conakry) which Jefferson distributed to planters throughout South Carolina and Georgia, where (according to entries in Jefferson's diaries over a number of years) it proved a considerable success.

A second eighteenth-century account of subsistence rice-farming practices along the Atlantic coast of the WARZ illustrates the opposite phenomenon, the attempted introduction of an English innovation (the sickle) into West African rice-harvesting, and the mobilization of cultural defences by local rice farmers to fend off a clear threat to the varietal distinctiveness upon which effective management of the dryland-wetland continuum depended.

The English reformer Henry Smeathman [1783] reports that although, to him, the sickle was an obviously superior labour-saving innovation, it was rejected by farmers in Sherbro country for fear of being accused of witchcraft. To Smeathman this was an instance of blatant African irrationality. But now it seems more likely that Smeathman had failed to appreciate what William Cleveland had made clear to Thomas Jefferson: that local methods of rice farming depend upon the segregation of planting material into distinctive types suited to the three main catenary positions. Harvesting each head individually allows rejection of rogues, and so facilitates variety maintenance. Harvesting by bunch, with a sickle, leads to greater mixing of the different types. This would potentially affect the capacity to manage catenary rice-farming systems in a flexible way.

It was rather as if Smeathman had tried to insist on the use of a new word from a foreign language without paying attention to the locally forged 'grammatical' context in which it would be expected to function. In suspecting that they would be charged with witchcraft, Smeathman's laggard adopters may have had a clear understanding of the adverse social consequences of his proposed innovation, the confusion in the following year likely to result from farmers having to sow what they had (so efficiently, but so carelessly) reaped. It is interesting to note that a modern development project in Mende country is attempting to re-introduce local rice farmers to the benefits of panicle selection at harvest, although there is no clear evidence that such knowledge has ever been lost. Farmers cheerfully accept this 'innovation' in the absorptive spirit of all good creolizers. It matters not from where an idea originates. It matters only that it works in context. Many WARZ farmers have handed me 'traditional' rice types for which I can trace a clear pedigree to colonial or post-colonial research. That is not the point. To them, if it works, it belongs to the culture, and therefore has an ageless endogenous pedigree. It is a valid utterance in the 'language' of agrarian creolization.

Agrarian Creolization: Innovation and Rice Politics in the WARZ

Today, rice continues to exercise a powerful influence over the moral economy of the societies of the Upper Guinea Coast. In Sierra Leone and Liberia – the two central countries of the WARZ

– rice accounts for about 50 per cent of total per capita food-energy intake. The ability to offer one's clients and supporters rice (in effect a synonym for 'food') is still the first test of political legitimacy in these countries.

In former times, political leaders were rural-based and commanded the labour and land to secure large stocks of rice *in situ* using local cultivars and production techniques. From the colonial period onwards, power and authority moved to the cities. Rural folk still preserve a modicum of political independence by cultivating their own rice using traditional methods. However, introduced methods – first based on mechanization, and later centred on the Southeast Asian Green Revolution (GR) bio-technology package for intensive cultivation of wetlands – have not proved economically (or environmentally) sustainable. Soils are too poor and yields too low to sustain the costs of mechanization. The GR package is too labour-intensive, in a setting where labour is as yet a greater constraint on production than shortage of land and lacks any tendency to spread spontaneously. Some degree of spontaneity in the diffusion process is essential, since the costs of constantly promoting and 'pushing' the package are too high for government or donors to sustain indefinitely.

Not all elements in the GR package are equally inappropriate to local circumstances, however. Farmers 'cream off' the bits they like, most notably some of the introduced improved varieties. The ones they take are those that enhance their repertoire of planting choices, and which work well without fertiliser. This selection and integration of external elements within an indigenous contextual framework appears to involve cognitive and communicative processes not entirely dissimilar to those required for linguistic creolization. Creolization draws upon elements from diverse linguistic sources to create a new language according to (arguably) general principles of language reduction and elaboration [Bickerton 1975]. Similarly, I envisage that it may be the case that agricultural 'creolization' draws freely from external elements but integrates them into a local system with a productive logic ('grammar'?) of its own. In the process the external elements quickly lose much of their alien distinctiveness and re-emerge with local significance.

The theoretical basis for this proposition will require further elaboration on another occasion. All I wish to note here is that if the argument can be sustained, then the historical basis for the fact that farming systems in the WARZ are especially prone to a process of 'creolization' will not be hard to find. Here we have a situation involving the geographical convergence of two non-interbreeding species of a single crop – a crop, furthermore, in which farmer selective pressure tends to maintain varietal distinctiveness and a coastal socio-economic context deeply transformed by mercantilism from the time of the Portuguese. The scene is thus set for the emergence of a system of rice agriculture combining elements of both the indigenous *O. glaberrima* and introduced *O. sativa* systems. Even if the concept of 'agarian creolization' turns out to be no more than a loose linguistic analogy, it nevertheless helps focus upon the point that it would be wrong to suppose that *O. sativa* types were ever seen as replacements for *O. glaberrima* types. Rather, farmers sought enhancement of their repertoire of planting choices, both species being incorporated into the new enlarged system in a manner analogous to the way words from different, apparently incongruent languages are welded into a creole tongue through relexification.

We lack direct evidence concerning species and types grown at earlier periods. But the Dutch evidence that farmers were using a repertoire of rices segregated into distinct duration classes at a period (in the mid-seventeenth century) when *O. sativa* was first perhaps beginning to spread widely suggests that repertoire management is the long-established key to farmers' germplasm evaluation strategies. It seems unlikely, given this kind of framework, that the introduction of *O. sativa* types had a revolutionary impact on West African rice-farming systems. This still seems to be the case today. When modern varieties are released they appear not to lead to the mass *replacement* of local cultivars (a feature sometimes associated with the GR in Asia). Rather, they are used for *repertoire enhancement*, e.g. diversification of the choices possible in management of the soil catena (typically, farmers use modern varieties on better land where their responsiveness to high fertility is put to best use and reserve local types for use on more difficult sites). In this way modern varieties improve the productivity of the total system as much by driving local varieties into niches to which they are especially well suited as by their own contribution to higher yields. Like words in a creole language, each rice, whether ancient or exotic, is open to take up a new 'meaning' – a new position within the emergent system. It is possible that a few of the 'least fit' local varieties have been squeezed into extinction at the end of this creolizing experience, but so far, in the normal course of settled evolution of farming systems, loss of rice germplasm biodiversity does not seem to be a marked feature of the WARZ. Where there is a problem, however, is in mobilization of institutional and political support for *repertoire enhancement* systems of germplasm management. Such systems have high levels of internal independence (farmers continue to make a lot of key choices for themselves). External inputs are not needed in bulk, nor necessarily on a regular, repeated basis. Once farmers have access to superior genotypes, they continue to multiply their own seed stock.

In such cases the seed delivery process has low visible impact and is seen by sponsors as being unjustifiably expensive to maintain. On the basis of certificate seed rice distributed over the years, development agencies in Sierra Leone expected to find that about 40 to 50 per cent of all farm land would currently be planted to modern varieties (the national Rice Research Station at Rokupr has been active since 1934). Field surveys [Lipton, Pain and Richards 1993] suggest a typical figure closer to 15–20 per cent. Judging by subsequent analysis of types collected during the survey and later grown under observation at Rokupr, some of this discrepancy is accounted for by those releases that have been so well absorbed into local 'creolized' farming systems that they now have local names and are used in places and ways not originally intended. Perhaps as importantly, farmers have continued to exert selective pressure on released varieties that have undergone some degree of spontaneous hybridization with local types (more research is urgently needed in this area). It is as likely, however, that a significant proportion of released materials has been lost or discarded by farmers who found that the material failed to perform up to the mark in local conditions (in such cases adopters are in effect re-screening released material for unsuspected adverse genotype-environment interactions). This on-the-job adaptation by farmers is valuable and all to the good. But in the absence of a clearly marked product and a highly visible impact sponsors tend to become discouraged.

It is to be expected, perhaps, that outside sponsors should seek to have a short-term dramatic impact. It can be argued that the real onus of support for farmer-adaptive agricultural systems lies elsewhere. Agricultural 'creolization' requires external stimulus as well as indigenous innovation: it is in the slow but steady synergy between external and internal inputs that the adaptive strength of such systems lies. This implies a strong and direct relationship between farmers and national agricultural research sustained over a long period with the expectation of incremental improvements rather than progress measured in quantum leaps. But here the WARZ has a political problem. The process of 'creolization' in agricultural systems is in large measure locally controlled, whereas states such as Liberia and Sierra Leone are dominated by urban-based mercantile interests. Investing resources in steadfast support for rice research buys little in the way of political legitimacy and influence when the rural poor are free simply to cream off elements in research packages and use them as they see fit to elaborate their own survival systems further. This is the point at which urban factions lose faith in technological improvement and turn instead to rice imports or charity as the source of patronage through which to shape the political process. Why, otherwise, in West African countries as dependent on rice as Japan or Sri Lanka, would politicians berate rice researchers for their irrelevance to, and lack of impact upon, the national economy, or step aside when a politically inspired riot threatens to wreck an internationally renowned rice research centre, thus causing the entire loss of the national rice germplasm collection?

But if the direct synergy between farmers and research is (as suggested above) an established factor in the survival strategies of the rural poor throughout the WARZ (and one resting on deep historical roots), then it is vital that both elements in the equation of agricultural 'creolization' be well-sustained. The evidence of field surveys in one country – Sierra Leone – suggests that farmers continue to innovate upon a basis provided by research, and that biodiversity tends thereby to be well conserved. There are two current threats to the system, however. The first is that *in situ* rice research, well connected to the farming community, will be neglected. The second is the curse of war and insurrection, spreading out of Liberia, but now reaching into Sierra Leone, and threatening Guinea and the Ivory Coast. Farmers faced by nothing worse than poverty can conserve and continue to exploit rice germplasm biodiversity. Farmers fleeing for their lives before murderous rebels cannot. Currently, peace would be one of the greatest contributions to conservation of rice biodiversity in the forests of this part of West Africa.

Conclusion

In this chapter I have argued that the historical trajectory of agricultural change in those parts of West Africa dependent on rice as the major staple has tended to emphasise the flexible management of germplasm resources above all other areas of farming. Seed-type repertoire management, I have suggested, has been a substitute for land shaping and water control in communities beset by poor soils and labour shortage. The effectiveness of this option is in part a reflection of the status of rice as a self-pollinating crop. Additionally, West African farmers have been stimulated to an interest in repertoire enhancement by long historical exposure to rice types from *two* distinct species. I speculate that the co-existence over five hundred years or more of *O. glaberrima* and its non-interbreeding analogue *O. sativa*, by establishing a broader than usual base for varietal experimentation, has positively encouraged a tendency towards 'agrarian creolization' (a characteristic commitment to exploit the synergy between external and indigenous elements in the elaboration of farming systems). This pioneering spirit came into its own in the opening up of a settlement frontier on the north-western flank of the West African tropical rain forest, where the elephant was a source of fruitful ideas about agricultural innovation. The forest frontier brought people and elephants into rivalry and helped feed an international market for ivory. Conservationists are now struggling to save the African elephant, and have chosen a ban on world trade in ivory as their principle weapon. Protecting and enhancing the viability of a 'creolizing' agricultural system that tends towards the flexible management of germplasm resources requires an altogether different mental orientation on the part of conservationists. Respect for the values of science (whether the science of farmers or researchers) and commitment to regional conflict resolution will here be as important as any strictly biological strategy for the conservation of African rice. Nothing could more precisely illustrate the need for the fullest cooperation between students of biology and society to find ways of thinking about the future beyond nature, beyond culture.

Bibliography

Arua, E.O. 'Yam ceremonies and the values of Ohafia culture', *Africa* 51(2), 1981, pp.694–705

Bickerton, D., *Dynamics of a creole system*, Cambridge: Cambridge University Press, 1975

Bray, F. *The rice economies*, Oxford: Blackwell, 1986

Brydon, L. 'Rice, yams and chiefs in Avtime: speculations on the development of a social order, *Africa* 51(2), 1981, pp.659–675

Chauveau, J., J.P. Dozon and J. Richard, 'Histoires de riz, histories d'igname: le cas de la moyenne Ivory Coast', *Africa* 51(2), 1981, pp.621–58

Cutting, N., 'Nathaniel Cutting to Thomas Jefferson, from St Marc, island of St. Domingue, 6th July 1790', in T. Jefferson, *The papers of Thomas Jefferson, volume 17, 6th July to 3rd November*, ed. J.P. Boyd, Princeton, N.J.: Princeton University Press, 1965

Dally, A.M. and G. Second, 'Chloroplast DNA diversity in wild and cultivated species of rice (Genus *Oryza*, section Oryza), Cladistic-mutation and genetic-distance analysis', *Theoretical and Applied Genetics* 80, 1990, pp.209–22

Dapper, O., 1668. Cited in A. Jones, *From slaves to palm kernels: a history of the Galinhas country (West Africa) 1730–1890*, Wiesbaden: Steiner Verlag, 1983

Davies, A.G. and P. Richards, *Rain forest in Mende life: resources and subsistence strategies in rural communities around the Gola north forest reserve (Sierra Leone)*, Report to the UK Overseas Development Administration, Department of Anthropology, University College London, 1991

d'Azevedo, W., 'Some historical problems in the delineation of a Central West African region', *Annals of the New York Academy of Sciences* 96, 1962, pp.512–38

Dieterlin, G., 'The Mande creation myth', *Africa* 17(2), 1957, pp.124–38

Dresch, J., 'La riziculture en Afrique Occidentale', *Annales de Geographie* 58, 1949, pp.295–312

Glaszmann, J. C., 'Isozymes and classification of Asian rice varieties', *Theoretical and Applied Genetics* 74, 1987, pp.21–30

——, 'Geographic pattern of variation among Asian native rice cultivars (*Oryza sativa* L.) based on fifteen isozyme loci', *Genome* 30, 1988, pp.782–92

Greenberg, J.H., *The languages of Africa*, Bloomington: University of Indiana Press, 1966

Harlan, J., J.M.J. de Wet and A.B.L. Stemler (eds), *The origins of African plant domestication*, The Hague: Mouton, 1976

Harlan, J. and J. Pasquereau, 'Decrue agriculture in Mali'. *Economic Botany*. 23, 1969, pp. 70–74

Hill, M.H., 'Where to begin? The place of the hunter founder in Mende histories', *Anthropos* 79, 1984, pp.653–6

Kortlandt, A., 'Vegetation research and the "bulldozer" herbivores of tropical Africa', in A. C. Chadwick and S. L. Sutton (eds), *Tropical rain forest*, Special Publication of the Leeds Philosophical and Literary Society, 1984

Lewicki, T., *West African foods in the Middle Ages*, Cambridge: Cambridge University Press, 1974

Linares, O., 'From tidal swamp to inland valley: on the social organization of wet rice cultivation among the Diola of Senegal', *Africa* 51(2), 1981, pp.557–94

——, *Power, prayer and production*, Cambridge: Cambridge University Press, 1991

Lipton, M., A. Pain and P. Richards, *Rice research in Sierra Leone & Sri Lanka: policy, spread and impact*, World Bank, Washington D.C. 1993

Livingstone, F.B., 'Anthropological implications of sickle-cell gene distribution in West Africa', *American Anthropologist* 60, 1958, pp.533–62

Murphy, W. and C. Bledsoe, 'Kinship and territory in the history of a Kpelle chiefdom (Liberia)', in I. Kopytoff (ed.), *The African frontier: the reproduction of traditional African societies*, Bloomington: Indiana University Press, 1987

Portères, R., 'Les noms des riz en Guinee' (in nine parts), *Journal d'agriculture tropicale et de botanique applique*, 13, 1966

——, 'African cereals: Eleusine, Fonio, Black Fonio, Teff, Brachiaria, Paspalum, Pennisetum, and African Rice', in J. Harlan, J.M.J. de Wet and A.B.L. Stemler (eds), *The origins of African plant domestication*, The Hague: Mouton, 1976

Richards, P., *Indigenous Agricultural Revolution: ecology and food-*

crop farming in West Africa, London: Hutchinson, 1985
——, *Coping with Hunger: hazard and experiment in an African rice farming system*, London: Allen and Unwin, 1986
——, 'The versatility of the poor: indigenous wetland management systems in Sierra Leone', in T. Bierschenk and G. Elwert (eds), *Folgen der Entwicklungshilfe*, Campus Verlag: Frankfurt, 1992
Rodney, W., *A history of the Upper Guinea Coast, 1545–1800*, Oxford: Clarendon, 1970
Smeathman, H., Appendix to C.B. Wadstrom, *An essay on colonization*, London, 1794 [1783]
van den Breemer, J.P.M., 'The diffusion of dry rice cultivation among the Aouan of Ivory Coast', in J.P.M. van den Breemer et al. (eds), *The social dynamics of economic innovation: studies in economic anthropology*, Leiden: DWSO Press, 1990
van der Drift, R., 'Arbeid en alcool', PhD thesis, Linden: University of Linden, 1992
Watts, M. and J. Carney, 'Manufacturing dissent: work, gender and the politics of meaning in a peasant society', *Africa* 60 (2), 1990, pp.207–41
Wigboldus, J., 'Trade and agriculture in coastal Benin c. 1470–1660; an examination of Manning's early-growth thesis', *Dertig Jaar Afdeling Agrarische Geschiedenis, AAG Bijdragen* 28, Landbouwhogeschool-Wageningen, 1986, pp.299–380
——, 'Pearl millet outside Northeast Africa, particularly in Northern West Africa: continuously cultivated from c. 1350 AD only?', in R.E. Leakey and L.J. Slikkerveer (eds), *Origins and development of agriculture in East Africa: the ethnosystems approach to the study of early food production in Kenya, Studies in Technology and Social Change* 19, Technology and Social Change Program, Iowa State University: Ames, Iowa, 1991

Chapter 12

Co-evolution Between Humans and Domesticates: The Cultural Selection of Animal Coat-Colour Diversity Among the Bodi

Katsuyoshi Fukui

Introduction

That the domestication of animals was accompanied by variation in the colour of animal coats and the maintenance of their polymorphism has been pointed out by Zeuner [1963: 61–2] and Nozawa [1982: 6; 1994: 189–99]. The phenomenon of coat-colour polymorphism is the main characteristic of domesticated animals and has become an important basis for determining whether the animals in prehistoric rock paintings were domesticated or not. For instance, one can clearly determine on the basis of coat-colour polymorphism that the cattle painted around 3000 BC on the rocks of Tassili in the Sahara were domesticated animals. It is thought that this phenomenon is more diagnostic of domestication than even the shattering habit that comes with the cultivation of wild plants.

As reasons for the coat-colour polymorphism that arises from the domestication of animals, Nozawa [1982: 6] gives the actualization of latent genetic variations and depletion in the pressure of natural selection, but this does not explain how latent variation becomes effective through the process of domesticating wild animals. It can be surmised that it was a result not only of increased human intervention in the sexual reproduction of animals through the process of domestication but also of human cognition of the phenotype of that latent genetic variation and its incorporation into the cultural system. In other words, the polymorphism of coat-colour that accompanies domestication can be said to be both a cultural and a genetic phenomenon.

Travelling through East Africa, one can see a great diversity of coat-colours among cattle, as well as among goats and sheep. Finch and Western [1977] relate this to differences in altitude and analyze it from the standpoint of ecology. They say that one tends to find cattle of lighter coat-colour in the lowlands where the sun rays are strong, while one sees the dark-coated cattle more in the highlands because they drink more water and their coats absorb more heat from the sun than do the light-coated varieties. Moreover, they argue that when it comes to coat-colour variation, ecological factors are more strongly at work than those of social selection.

However, anthropological studies of the Bodi [Fukui 1979b; 1988b; 1991], as well as of other pastoral societies in East Africa, have shown that coat-colour polymorphism corresponds with the pastoral cultures of these various peoples.[1] The purpose of this chapter is to clarify, using concrete examples from the pastoral Bodi of the southwestern part of Ethiopia (Figure 12.1), what kind of cultural devices underlie the phenomenon of coat-colour polymorphism in the process of the domestication of animals.

In East African pastoral societies, it has been noted that domesticated animals are individually identified on the basis of coat-colour variation and are often given names. Furthermore, these pastoral peoples maintain a symbolic relationship with domesticated animals, especially cattle, that exceeds the material level of their milk or blood as food sources. As they live their whole lives with their domesticated animals, the animals become an indispensable focal point not only for the human relations within their society but even for their entire world view.

In the present chapter, I will systematically show not only how Bodi comprehend coat-colour variation in domesticated animals, especially cattle, and how this is connected with other aspects of their social system and culture, but also how they maintain coat-colour polymorphism and how it arises from cultural devices that reflect their concept of folk genetics.

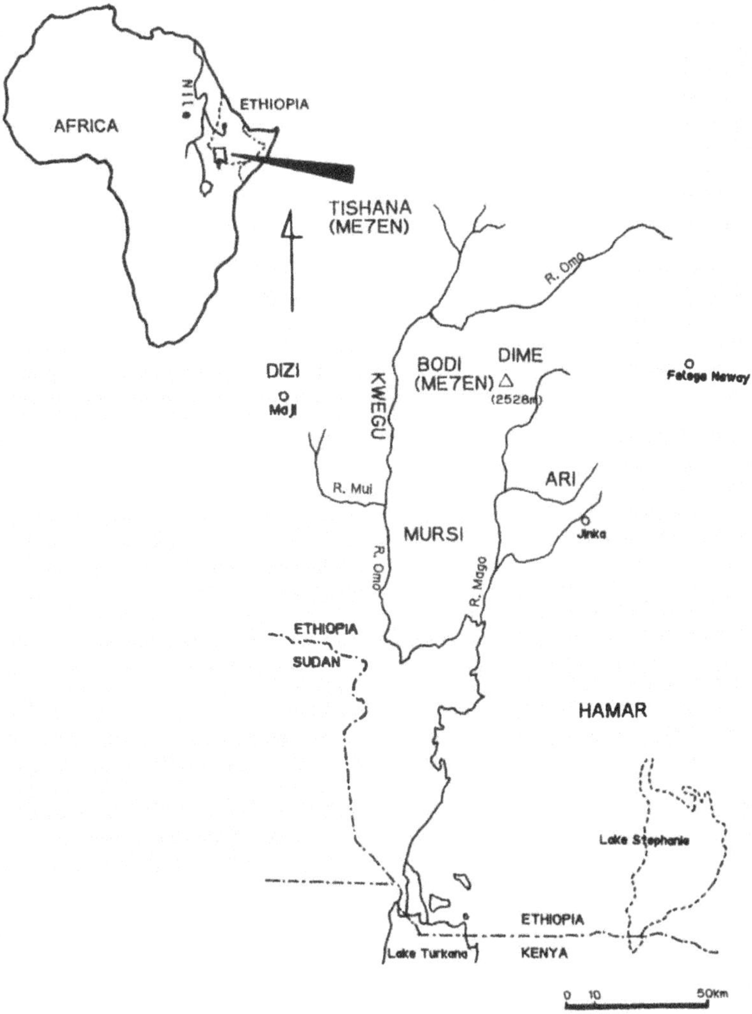

Figure 12.1. The Bodi and neighbouring peoples

Identification of Coat-Colour Polymorphism Among Cattle

The Position of Cattle within Bodi Society

The livestock that the Bodi breed includes cattle (**bi**, pl.: **bio**), goats (**tongo**, pl. **tena**), chickens (**kobut**, pl. **kobuwo**) and dogs (**roso**, pl. **rosiya**). We regard these four kinds of animals as livestock, but from the Bodi's point of view cattle and goats, and even chickens, are included under the same term **me7en** (humans) as the Bodi themselves, while dogs are called human servants (**gaima**). The concept of what we generally term livestock does not exist for the Bodi. By contrast, any animals outside these four types, such as lions and buffalo, ants, dragonflies, mites, fish, and so on, are generally called **kagenya** (sig. **kaget**). Bodi eat chicken only when they are sick, and they never eat eggs. Then again, the Bodi rarely hunt, so they do not use their dogs for anything in particular.

The only livestock Bodi depend on in their daily lives are cows and goats. These two animals are the most important ones for them, on both a material and a conceptual level. Cattle in particular are indispensable to the Bodi in virtually every aspect of their culture, so much so as to outdistance the importance of goats. As will become increasingly apparent later, cattle are not only the objects of personal identification for each member of the society, they are indispensable participants in many Bodi rituals. In short, if cattle were to disappear from Bodi society, it would mean that the Bodi would lose the foundation of their mental existence including their reason for living. Although it may be almost impossible for us to imagine their world, this is not an exaggeration by any means.

Cattle are also extremely important from an subsistence perspective. In addition to their milk, the blood taken from their jugular veins is the main staple of Bodi daily diet, along with the maize they cultivate in their swiddens. As far as goats are concerned, they use only their meat and their skins. While goats may be killed only for their meat, cattle are usually sacrificed only during a ritual. Cattle and goats are rarely kept in the same compounds, goats being bred individually in an area far away from where the cattle are grazing. There being no regularly held market in Bodi land until 1989, they are used to walking one or two days through a mountainous region to where agricultural peoples live. But the Bodi would never set out for this regularly held market with their cattle to trade them for something. At most, they might take kefir or meat from livestock that have died from sickness to a nearby agricultural people to trade it for grain.

The Cattle Compound 'Family Register'

I have thought for some time that as long as the Bodi depend on cattle, the point of departure for any understanding of their society and culture must be their system of comprehending and maintaining cattle. So, after first learning their language and identifying each person as well as grasping their overall interpersonal relationships, and after about half a year had passed since my first arrival in a herding camp, I began my research on cattle.[2]

I started by studying the ways Bodi understand the concept of livestock. At the initial stages of my survey I showed informants the coat-colour patterns that had been drawn up by Dyson-Hudson [1966], based on his research amongst the Karimojong people in Uganda. But Bodi coat-colours of cattle are so diverse that it was impossible to fathom the underlying regularities in this way. So I turned next to the more concrete task of compiling a record of 'kinship' relations (genetic relatedness) within the cattle compounds. I did this by sketching each individual animal within the livestock compound, and on the basis of these sketches asking for the animal's name, its owner and where it came from, or what its kinship relations were, and so on. The Bodi untiringly provided me with the names of each individual animal, but it required more time to ask them about the cattle's histories and owners.

In this way, through observation and questioning, I was able to acquire information on more than 421 head of cattle in 21 cattle compounds (**tui**, pl.: **ori**). Among these, there are some cattle compound 'kinship' records that I drew up twice at different times, so the actual total number of individual animals I surveyed is actually higher than this. However, I would like to leave the detailed analysis of the record of 'kinship' relations for another occasion and limit my discussion here to providing an example of one cattle compound (Table 12.1) from which I will extract the principles of identifying individual cattle in Bodi society. The kinship relations between cattle owners in this compound are shown in Figure 12.2.

Each head of cattle is given a name. As an example let us look at No.1 in Table 12.1, **oi de ludi** (usually pronounced **oi da luuto**). **Oi** refers to the bull and **ludi** refers to the coat-colouring and the fact that the head and rump are black while the rest of the body is white. Moreover, **de** (which becomes **te** or *ja* if the preceding word ends in a consonant) is in the genitive case, so **oi de ludi** means 'the bull with the black head and rump'. In the case of No.2, since **bhongai** means 'male calf,' **bhongai de ludi** means 'the male calf with the black head and rump'; No.3, **lun ja gelli**, becomes '**lun> luch** (ox) + **gelli** (black head and rump and white back with black spots); No.4, '**bi de bhileji**' becomes '**bi** (cow) + **bhileji** (yellow); and No.9, **mor te bha-bhileji** becomes '**mor** (heifer) + **bha-bhileji** (yellow-coated with a white stomach).

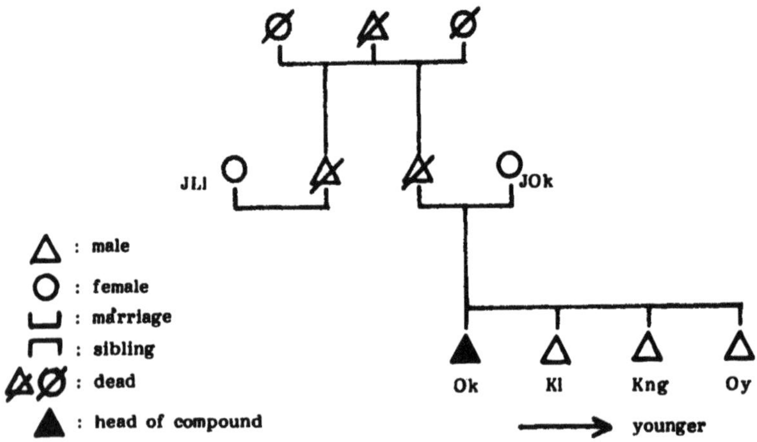

Figure 12.2. Kinship relations of cattle owners for compound inventorized in Table 12.1

What is clear from the examples given above is that Bodi names for cattle are made up of composite terms based on the cattle's gender, age and coat-colour. In addition, there are cases such as **mogut te golonyi** (cow with the horns in the shape of the crescent moon and with a red coat) where the name emphasises the shape of the horns. Even so, it can generally be stated that

Table 12.1. Cattle composition of a Bodi compound (Ok)

No.	Cattle Name	Attribute (Sex, Age . . .)	Cow-Calf Relation	Owner*
1.	*oi de ludi*	bull		Ok
2.	*bhongai de ludi*	bull calf		Ok
3.	*lun ja gelli*	ox		Ok
4.	*bi de bhileji*	dry cow		Ok
5.	*bhongai de shimaji*	bull calf	calf of No.4	Ok
6.	*bi de bhileji*	dry cow		Ok
7.	*bhongai de chobur-gidhangi*	suckling bull calf	calf of No.6	Ok
8.	*bi de bhileji*	milk cow with a dead calf		Ok
9.	*mor te bha-bhileji*	calf		Ok
10.	*mor te gelli*	pregnant heifer		Ok
11.	bi de moji	dry cow		Ok
12.	*bhongai de ludi*	suckling bull calf	calf of No.11	Ok
13.	*bi de bhaasi*	milk cow		Ok
14.	*bhongai de koro-chokaji*	suckling bull calf	calf of No.13	Ok
15.	*bi de tula-holi*	dry cow		Ok
16.	*bhongai de bila-golonyi*	suckling bull calf	calf of No.15	Ok
17.	*bi de kilindi*	dry cow		Ok
18.	*mor te kilindi*	calf		Ok
19.	*bi de eldi*	milk cow		JL1
20.	*mor te cha7i*	suckling heifer calf	calf of No.19	JL1
21.	*bi de kilindi*	milk cow with a dead calf		JL1
22.	*bi de gidhangi*	milk cow with a dead calf		Ok
23.	*bi de bhileji*	pregnant dry cow		Ok
24.	*bi de cha7i*	milk cow		JOk
25.	*mor te shimaji*	suckling heifer calf	calf of No.24	JOk
26.	*oi de kilindi*	bull		K1
27.	*bi de kilindi*	milk cow with a dead calf		K1
28.	*bi de bha-gidhangi*	milk cow		K1
29.	*bhongai de ludi*	suckling bull calf	calf of No.28	K1
30.	*bhongai de bhilasi*	bull calf		Kng
31.	*bi de golonyi*	milk cow		Kng
32.	*bhongai de kalmi*	suckling bull calf	calf of No.31	Kng
33.	*bhongai de bhilasi*	bull calf	calf of No.31	Kng
34.	*mor te el-shimaji*	heifer		Kng
35.	*mor te golonyi*	heifer		Kng
36.	*bi de golonyi*	milk cow with a dead calf		Oy
37.	*mor te bha-golonyi*	heifer		Oy

Total female 24 = milk cow (*roine*) :5 + milk cow with a dead calf (*tukan*) :5 + heifer (*mor*) :8 + dry cow (*usi*) :6
male 13 = bull (*oi*) :2 + ox (*luch*) :1 + bull calf (*bhongai*) :10

* See Figure 12.2.

cattle names in Bodi society are terms of classification based primarily on coat-colour. In other words, all Bodi cattle coat-colours are clearly recognized and placed within the framework of their classifying system.

In addition to recognition of gender, age and coat-colour, the Bodi also have an extremely acute grasp of the condition that each of their cattle is in, and they have appropriate terms to express these conditions. For instance, there are the terms **usi** (dry cow), **roine** (milk cow), **tukan** (mother cow with a stillborn calf), **mich'irit** (cow that does not want to copulate with the bull), **ilol** (a cow that copulates but does not conceive), and so on.

Cognition and Classification of Colours

What we can conclude from the example of one cattle compound family register is that the principles of identifying individual animals go beyond the characteristics of gender and age, and are based on recognition of a large variety of coat-colours (Figure 12.3, Plates 12.1–12.2). Bodi have terms corresponding not only to colours such as red and white, but also use terms to describe the pattern, such as **ludi** 'black head and rump only, everything else white'. Since this cognition of colour and pattern is directly connected to the central aim of this chapter, the Bodi's understanding of folk genetics, I would now like to turn to a discussion of the classification of the colours and patterns most important to the Bodi.

We commonly express the conditions of the natural world, including coat-colour, using two different sets of terms: colour and pattern or colour configuration. But the Bodi do not express colour and pattern separately, using instead the single term **ch'ore** or **a7engi**. **ch'ore** refers to the coats of all mammals, including humans, but it is also used to describe the diversity of colours and patterns to be found not only on furless animals but also in plants and in the natural world in general. Dissatisfied with this general term, I asked whether there might be a term that referred more directly to colour/pattern itself. As a result, I became convinced that the term that most closely approximates to our notion of colour and pattern is **a7engi**, and from then on I decided to use this term whenever discussing colour and pattern. The term **a7engi** most aptly described the content of my project,

(a) *bhilasi*

(b) *ludi*

(c) *gelli*

(d) *bhaasí*

(e) *bholigaasi*

(f) *eldi*

Figure 12.3. Some examples of black and white patterns in Bodi cattle

so it was acceptable for my purposes, but I cannot at this point deny that there exists in Bodi language a much broader concept than this for colour and pattern. Here, however, I will use **a7engi** instead of **ch'ore**, even though **ch'ore** derives originally from 'coat'.

Without limiting myself to cattle coat-colour, I conducted a survey of Bodi classification of colour, using, among other things, ninety-eight colour cards [see Fukui 1979a and 1979b]. I tried to think of ways of recording Bodi understanding of cattle coat-colour without using photographs, but decided in the end that the most important thing was first to grasp their recognition of colours in general using colour cards. This was due to the fact that, in addition to cards being easy to transport and use, whichever ones I ended up using, the colours themselves could be measured against a spectrophotometer and expressed in terms of numbers or coloured cards that could be easily compared.

Chromatic terms are the words people use to classify and schematize various colours ranging on a scale from purple to red in electromagnetic wavelengths within the range of 380–780 nm. If one defines culture as the total system of classifications of the universe pertaining to a given society [Sturtevant 1964], then the classification and cognition of Bodi colour would indeed be culture itself. Moreover, although it is not surprising that Bodi colour terms, and also their criteria of colour recognition, differ from our own [see Conklin 1955], it is possible to a certain extent to reproduce their colour terms using such subjective methods as a chromaticity diagram.

It is possible to discern the following eight basic Bodi colour terms:[3]

A) **golonyi**: corresponds to red.

B) **nyangaji**: corresponds to orange; high saturation on the chromatic scale leads to merging with **golonyi**; low saturation on the chromatic scale leads to merging with **gidhangi**.

C) **shimaji**: corresponds to purple; red-purple is recognized with the two terms **golonyi** and **shimaji**.

D) **cha7i**: includes colours ranging from yellow green to grue (green and blue).

E) **bhileji**: corresponds to yellow; low saturation of yellow on the chromatic scale is called **gidhangi**.

F) **gidhangi**: not limited to grey, colours that are low on the chromatic scale at wavelengths of around 575–600nm converge in this category; low chromatic saturation of yellow and yellow-green overlap as **gidhangi**.

G) **holi**: pure white.

H) **koro**: corresponds to black; also used to refer to what we would call deep blue purple at frequencies above 60 per cent.

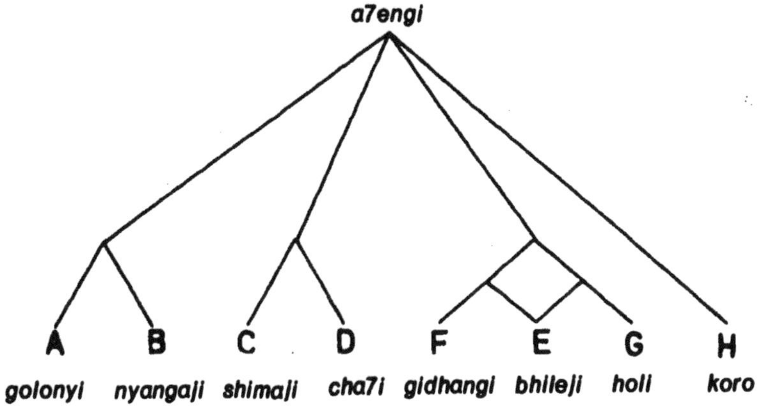

Figure 12.4. Bodi classification of colour

What I discovered of special interest was that these eight basic colour categories are seen as being related to one another in a particular way (Figure 12.4). For instance, I was told that **golonyi** and **nyangaji** are **nganiya** ('consanguinous'). If we were to say that red and orange belong to the same family, anyone could understand what we were talking about, but it would be hard for us to imagine that **cha7i** and **shimaji** are closely related. It is true that in terms of wavelength, green, blue and purple are consecutive. In addition, **gidhangi** and **bhileji** are also considered **nganiya**, as are **bhileji** and **holi**. As mentioned earlier, **gidhangi** includes not only grey, but reds or yellows that are low on the chromatic scale. Only **koro** stands independent of the various colours, but within the Bodi concept of **a7engi**, which includes pattern as well as colour, **koro** has a **nganiya** relationship with several patterns, as I will explain later. Since these kinds of close relationships among basic colour terms will become important to the thesis of this chapter – the cognition of coat-colours and the view of genetics and categorization of cattle – I would like to examine them in more detail.

The colour classification just reported was obtained by

interviewing individually a total of twenty-five informants, both men and women, using ninety-eight cards. In their responses to the colour cards, one can detect not only the basic colour terms I have outlined above but also composite terms made up of more than one of the basic colour terms or of concrete terms borrowed from other elements of the natural world, and also heterogeneous composite terms made up of a combination of colour terms and concrete words. Of these, the basic colour terms make up 83.3 per cent (the number of times basic colour terms were used: 2043/2450: 98 cards x 25 people); composites of colour terms made up 8.3 per cent (214/2450); concrete words made up 5.8 per cent (143/2450); and heterogeneous composite terms made up a mere 0.7 per cent (18/2450). Of the concrete terms that came up in this context, if one selects only those terms that derive from cattle coat-colours, one can divide them morphologically into two kinds. One would be the accumulation of simple terms such as **moji, puraji, shokaji, derdi** and **shinde**, and the other would be those that are accompanied by the suffix **-idi**, as in **kultere-idi, gulbho-idi** and **lele-idi**. It is highly likely that these last three terms are loan words that were originally names for something else, but once they came to be used to express cattle coat-colours they lost their original meanings. For instance, the concrete word **bato-idi**, which derives from **bato** or 'gnat', was explained by some informants as simply a term to describe a coat-colour, but originally it described the greyish colour of a large swarm of gnats. On the other hand, the origin of **kultere-idi**, which is used on rare occasions to describe a blue-purple or dark blue tint, as well as that of the term following it, **lele-idi**, are at this point unclear. **Lele-idi** is sometimes used to refer to a dark red or deep purple.

Next I would like to look at concrete terms deriving from single lexical elements that are said to originate from cattle coat-colours. Colour and pattern configuration terms that derive originally from cattle coat-colours are specifically referred to as **rongo de nandi. Rongo** means 'name', while Bodi say the meaning of **nandi** is derived from the cattle. Also, the original meanings of the colour terms **moji, puraji, shokaji, derdi** and **shindi** used specifically for coats have been lost. Regardless of their meanings, I would like to look at how Bodi paired those colours that derive from coat-colours with the colours on the ninety-eight colour cards.

(1) **moji**: often said to resemble a lion; used to describe dark or dull reds, oranges and yellows.

(2) **puraji**: used for yellow and red that is tinged grey.

(3) **shokaji**: used for light green, or grey-tinged green or blue, or else for grey.

(4) **derdi**: said to describe perfectly the colour associated with tincture of iodine; cards that were shades of dark red, dark yellow-orange, deep purple, or deep red-purple were placed in this category.

(5) **shindi**: used to describe a bright red like salmon pink.

Cognition and Classification of Patterns

As mentioned earlier, cattle coat-colours include not only what we refer to as colours but also all the various pattern configurations. Thus in order to elucidate the Bodi cognition of cattle coat-colour, it is important first to acquire an understanding of their cognition of pattern in combination with their cognition of colours.

With this purpose in mind, I drew eighty-two cards (ca. 27 x 100 mm) of all the various patterns I could come up with and conducted a survey of the Bodi's recognition of these various patterns by showing them the cards. These cards included not only black and white pattern configurations (sixty-five cards) but also red and white (seventeen cards). This was because I wanted to investigate what kind of naming would result if the colour on the patterned part of the card were altered. Then I showed each card, one by one, to two informants and asked them 'what colour/pattern is this?' (**da a7engi atang?**).

When I did this, they readily and easily replied with the appropriate name for almost all the cards. There was a great variety of patterns on each of the eighty-two cards, and almost all of them were ones that we would not be able to identify if we were asked. For instance, a Japanese person might be able to identify a red dot in the middle of a white background as the national flag of Japan, but we would not be able to come up with a single expression for a black dot on a white background. But the Bodi call this latter pattern **gelli** and when the black dot is changed to red, they use the term **gel-golonyi** (or red **gelli**) (see Figure 12.5). The black dot is called **toole** and the red dot **toole golonyi** (or red **toole**). However, **toole** is used not only when the red or black shape is round but as a general term for spots of all kinds, whether triangular in shape, square, or some other variety.

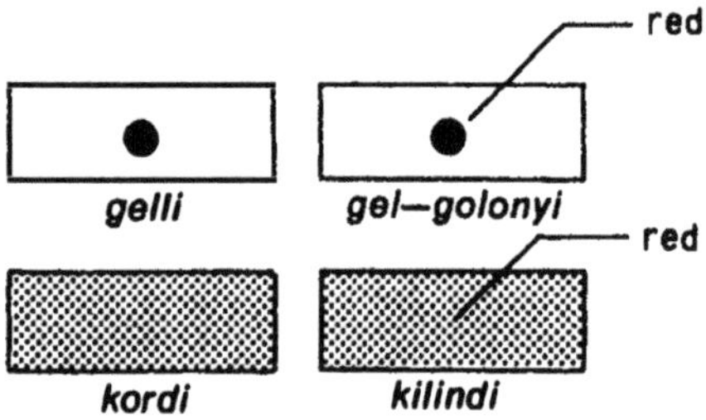

Figure 12.5. Cognition of patterns formation (1)

Bodi call small black speckles **kolset**. A white background with a lot of black **kolset** is called **kordi**, and when the black speckles are changed to red, this same pattern is called **kilindi**. The word **kilindi** originally derives from the coat-colour of the giraffe which is called **kilin**. In general, when the black part of the pattern is changed to red, a composite term is created reflecting those colour terms. The only case I discovered, other than **kilindi**, where a completely different name was used instead of a composite term was **seroji**. **Seroji** was not included in the various patterned cards I had made at random, but it refers to a pattern that contains more than three different colours.

I learned that not only were the Bodi easily able to identify almost all the various patterns on the eighty-two cards, but also that there were close relationships among the patterns and that these made up a system of classification. This is illustrated in Figure 12.6. If one looks carefully, one can discern several analogous properties in the patterns that are grouped under each roman numeral. For instance, if we look at roman numeral I in Figure 12.6, we see that broken lined or speckled patterns are called **kordi**, a series of straight or of wavy lines in a pattern is called **tulk'a**, and a pattern with a series of thick horizontal lines is called **bhilasi**. Comparing these to the other groupings of

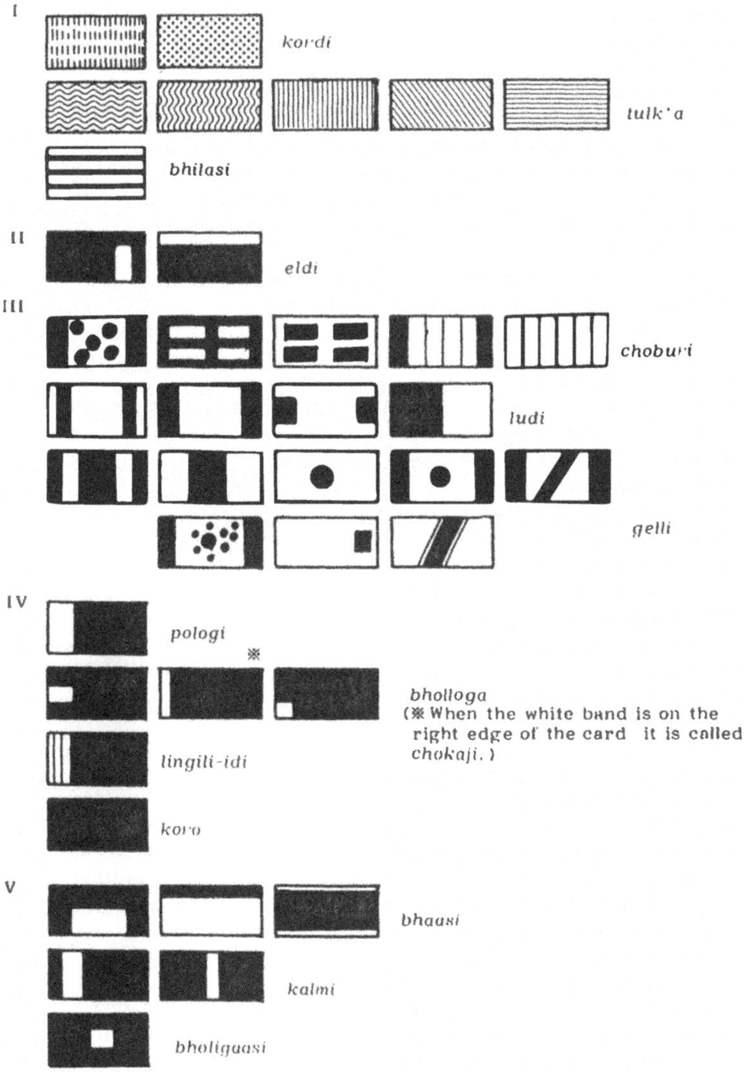

Figure 12.6. Bodi classification of typical patterns

patterns, even outsiders would agree that a unifying essential quality can definitely be discerned that makes it possible to treat these patterns as one group. As in their classification of colours, the Bodi refer to this kind of grouping as **nganiya**.

Once one has internalized the principles behind this classification system, it becomes possible to predict to a certain extent what the names of the patterns will be, even if new ones are added to the large number of patterns. For instance, several black dots on a speckled background is given the composite name **choburi-kordi**. This can be thought of as the combination of **choburi**, or a pattern that contains several black dots or spots, and **kordi**, a black-speckled pattern (see Figure 12.7). When speckled dots are concentrated on both ends of a pattern, this is called **lu-kordi**, a composite of the terms **ludi** and **kordi**. The black part on the end of the pattern is called **luset**, and when this black **luset** is on both sides of a white background it is called **ludi**. When the black part, or **luset**, is changed to a speckled pattern, or **kolset**, then the resulting pattern is **lu-kordi** (**lu** + **kordi**). With this kind of combination, the number of possible composite terms stretches out infinitely.

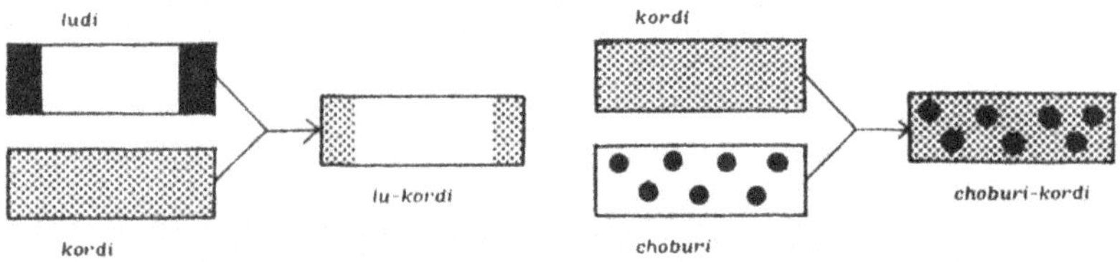

Figure 12.7. Cognition of pattern formation (2)

Although this may be intrinsically interesting, we need to go on and ask, on what this rich conceptual framework is based? In the process of asking Bodi to provide the names for the eighty-two colour cards, I learned that their cognition of these various patterns has its roots in the coat-colours of the cattle. This was made clear by the fact that they responded to certain of the patterned colour cards with 'there are no cattle with this coat-colour.' Of the eighty-two cards I showed them, the four patterns shown in Figure 12.8 were the ones for which they were finally unable to come up with terms, and in each case the reason repeatedly given was the one just mentioned, 'there are no cattle with that coat-pattern.'

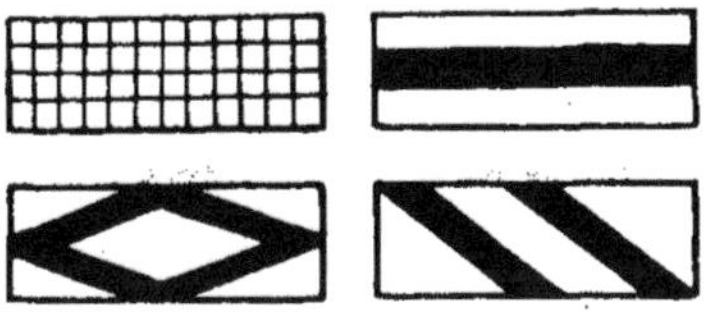

Figure 12.8. Patterns the Bodi cannot identify and name

There is further proof to support the contention that Bodi cognition of patterns is based on the coat-colour of cattle, namely the fact that when one reverses the pattern in Figure 12.6, roman numeral IV, called **bholloga**, the name is changed to **chokaji**. In other words, if the left side is white, it is called **bholloga**, but if the right side is white, it is called **chokaji**. Bodi are, of course, looking at the card as if it were a head of cattle, with the left side as the head and the right side as the tail. They call the coat-colour of the cattle with a white head **bholloga**, but cattle with black coats where the lower part of the tail is white **chokaji**.

A Cognitive and Classificatory System for Cattle

As is clear from the 'family register,' cattle are individually distinguished by their coat-colour and given names that combine such characteristics as coat-colour, gender, age, and so on. I would now like to discuss some wider aspects of the Bodi cognitive system centered on coat-colour classification.

Cattle of all kinds of coat-colour are recognized according to the colours and patterns we have examined. For instance, a bull with black speckles is called **oi de kordi**. As I have mentioned, **oi** means bull and **kordi** refers to the black speckled pattern. An ox with a vertical striped pattern is called **lun ja tulk'a (>luch de tulk'a)**, **luch** being the term for ox and **tulk'a** referring to a number of patterns involving straight and wavy lines. A milk cow with a black head and tail-tip is called **bi da luuto (>bi de ludi)** and a milk cow with black spots on its back is called **bi de gelli**. Just like the colours and geometrical patterns we have looked at so far, these cows with their various coat-colours also have close interrelationships. Even more interesting is the fact that a group of cows that falls in the 'same' coat-colour category is called by the same name as a human clan (**kabchoch**). To talk simply about the close interrelationships between colours or

Figure 12.9. Bodi classification of cows with black-white coat and their clans (*kabchoch*)

geometric patterns Bodi use the term **nganiya**, with its emphasis on consanguinity or friendship, but when it comes to categorizing the cattle themselves they use the term **kabchoch**, equivalent to a human clan. This, plus the fact that use of this term is limited to the cow, must be noted, setting aside for the moment the reasons why this might be so.

It is instructive to compare the classification of a **kabchoch** of the highly valued white and black cow (Figure 12.9) with the geometric pattern classification (Figure 12.6) described earlier. For instance, if we compare numeral I in Figure 12.6 with numeral I in Figure 12.9, we find a corresponding pattern in both the cattle grouping and in the geometric pattern grouping. The vertical striped **tulk'a** pattern, the speckled **kordi** pattern and the broad horizontal striped **bhilasi** pattern can all be said to correspond clearly in both cases.

Next I would like to compare Figure 12.6 numeral III with Figure 12.9 numeral III. A cow with a coat like a Holstein is called **choburi**, but if one looks at the geometrical patterns, one can see that the same term, **choburi**, is usually used to refer to a pattern that contains many large patches. A cow with a black head and rump is called **ludi**, but with the geometric patterns **ludi** refers either to a pattern where both ends are black or where one half is black and the other half white. I have never actually seen a cow among the Bodi that was half black and half white, but a banner that was made from cattle skin with that pattern was called **ludi** (see Figure 12.14). To continue, a cow whose head and rump are black and whose central back area is covered with black spots is called **gelli**, but if one looks at the geometric patterns referred to as **gelli**, they are all characterized by spots that tend toward the middle and not the edges. In other words, **gelli** is a characteristic where the spot tends towards the middle section, and this distinguishes it from other kinds of pattern.

I have already explained the geometrical pattern called **bholloga** (Figure 12.6, IV), but if one compares the **bholloga** coat-colour in Figure 12.9 (III) to **chokaji** one can clearly see how the names of the geometrical patterns are created from the coat-colour of cattle. Up to now, we have seen how the classification of geometrical patterns compares with the classification of cattle coat-colour, with the one based on the other. The same kind of analogy can be drawn in the case of colour classification. The close interrelationships between colours that we have looked at earlier also correspond extremely well with the classification of cattle coat-colour. I would like now to take a look at the classification of both pattern and colour and see how it relates to the names given to each **kabchoch** (Table 12.2).

Table 12.2. Classification of cows and their 'clan' names

	'Clan' name	Kinds of coat-colour pattern
I.	*bio baldonyi luibuli, bio baldonyi kangach ngoli, bio baldonyi dubare*	1) *bhileji* 2) *gidhangi* 3) *holi* 4) *moji* 5) *muge-idi* 6) *gushur-idi* 7) *bha-moji* 8) *saimani-idi* 9) *tula-holi* 10) *tula-moji* 11) *polog-bha-gidhangi*
II.	*bio baldonyi sela cha7i, bio baldonyi iranyi, bio baldonyi goga, bio baldonyi sela ngundi, bio baldonyi sinimodala*	12) *golonyi* 13) *nyangaji* 14) *kelali-idi*
III.	*bhogine*	15) *gachi-idi*
IV.	*bio baldonyi dombuli, bio baldonyi bonde hana, bio baldonyi hana koliyo*	16) *bha-golonyi* 17) *gel-golonyi* 18) *lu-golonyi* 19) *chobur-golonyi* 20) *bhila-golonyi* 21) *bha-nyangaji* 22) *gel-nyangaji* 23) *lu-nyangaji*
V.	kongogila derto, kongogila bhilechoito	24) *derdi* 25) *kulankulan*
VI.	*bio baldonyi durber, bio baldonyi binda, bio baldonyi alachuli, bio baldonyi lugi7a*	26) *shimaji* 27) *cha7i* 28) *bha-shimaji* 29) *el-shimaji* 30) *nyabha-shimaji* 31) *el-chea7i* 32) *kalmi-cha7i* 33) *seroji* 34) *gulbho-idi* 35) *gure-idi* 36) *diktach*
VII.	*koro da ganyu bikilini, malan kilin*	37) *mulen-idi* 38) *bato-idi*
VIII.	*pajjaka*	39) *shimachi-gidhangi* 40) *cha7k-holi* 41) *tiri-idi*
IX.	*koro da ganyu malanshali, bio baldonyi jingo korkori*	42) *koro* 43) *bholloga* 44) *pologi* 45) *chokaji* 46) *lingili-idi*
X.	*lugushala oinya choburi, kachkach gera bhongage waya, kachkach ger kabule, bio baldonyi bhonde waya, bio baldonyi bhonde lomuli*	47) *ludi* 48) *choburi* 49) *gelli* 50) *gel-poka-idi* 51) *nyabha-koroda* 52) *kabara-koroda*

Table 12.2. Continued.

	'Clan' name	Kinds of coat-colour pattern
XI.	*longonyi de ziga, longonyi ko7oli bor, koro da malanshali, bio baldonyi gimmako*	53) *kalmi* 54) *bhaasi* 55) *bholigaasi* 56) *mulk'a-idi*
XII.	*elameli jameliso, bio baldonyi bonde laka, chagu malisi*	57) *eldi*
XIII.	*gela, kongoboro, oliya milotola, bio baldonyi haggasa*	58) *kordi* 59) *kili-moji* 60) *bhilasi* 61) *tulk'a* 62) *kurkuri-idi*
XIV.	*bio baldonyi pajjaka, kiroshala ongo seleji.*	63) *kilindi*

What can be clearly seen in Table 12.2 is that those cows whose coats are in a close relationship in terms of colour classification are placed in the same clan. In Clan I there is **bhileji**, **gidhangi**, **holi**; in Clan II, **golonyi** and **nyangaji**; in Clan VI **shimaji** and **cha7i**; and in Clan IX **koro** and **bholloga**, included with other white and black patterns. As for geometric patterns, as we have already seen in Figure 12.9, cows whose coats exhibit patterns that are in close relationship with one another are placed in the same clan, but it is clear that Bodi colour classification, just like their classification of geometric patterns, is based on these kinds of cattle. I will not go into a detailed explanation of each of these coat-colours, but they can be determined for the most part from the terms I have given for colours and geometric patterns. For instance, **bha-golonyi** in Clan IV, a composite term made up of **bhaasi** and **golonyi**, means 'a red coat with white stomach area.' However, it should be noted that the sixty-three kinds of cattle coat-colour given in Table 12.2 do not represent all the coat-colours that are known to the Bodi. This is because terms that appeared in my discussion of colour terminology, such as **kultere-idi**, **puraji**, **shokaji**, **gulbho-idi**, **lele-idi**, **shindi** and so on, are not to be found here. This is an inevitable result of the kind of survey I conducted where I gave the informants the names of cow clans and asked them to come up with as many coat-colours as they could that would fit under each clan name. It can be assumed that the variations of coat-colour are much greater in number. But, looking at the examples given in related surveys, it can at least be assumed that this chart includes almost all the coat-colours that are most important to the Bodi.

The next question which arises is what the fourteen clan names to be found in Table 12.2 might mean. There are some for which the Bodi themselves can no longer provide meanings, but there are enough where they can to see how these might relate to the classification of cattle.

First let us look at clan I. The name **bio baldonyi**, which is used for many clans, comes from the phrase **bio barit te anyi**, which means 'the cow that is my virility.' **Luibuli** is a composite name made up of **lui** (white feathers used in dance) plus **buli** (a bird's name). The meaning of the following term, **kangach ngoli**, is unknown. The third term, **dubare**, is the clan name of the sitatunga (*Tragelaphus spekei*). Almost all animals are divided by clan, and these classifications are connected with their system of cattle classification through coat-colour. One can find this kind of derivation of cattle clan names from animal clans in the word **pajjaka** found under numerals VIII and XIV in Table 12.2. **Pajjaka** is the clan name for the giraffe (**kilin**). The coat-colour term **kilindi** actually refers to a coat with reddish patches like that of a giraffe and thus they are connected with cattle in terms of coat-colour.

If we look at the coat-colours that fall under clan I, the first three (1, 2, and 3) correspond, as I explained earlier, to the basic colour terms yellow, grey and reds low on the chromatic scale, and white. **Moji** is a dark yellow or even dull reddish colour that is close to **gushur-idi** (6), the colour of a lion's coat. **Muge-idi** and **saimani-idi** (8) are just 'cow names' and their meanings have been lost. **Saimani-idi** corresponds to such shades of colour as dark yellow-orange or dark yellow-green, as I mentioned above. **Bha-moji** (7) refers to a '**moji**-coloured cow with white stomach area'. **Tula-holi** (9) and **tula-moji** (10) refer to the 'white cow with vertical stripes' and 'the **moji**-coloured cow with vertical stripes' respectively. Finally, **polog-bha-gidhangi** (11) means a '**gidhangi**-coloured cow with a white face and stomach area.'

In clan II **sela cha7i** means feathers of **cha7i** (grue) and **iranyi** refers to a leather coat of goat skin called **jun**, but the rest are unknown. Included here are **golonyi** (red), **nyangaji** (orange) and **kelali-idi**. **Kelali-idi** is very close to the colour orange and derives from the name of the **kelela** insect, which makes its nest in trees.

In clan III, **bhogine** derives from the agricultural term **bhoginedo** (to clear the land by cutting fallow in the swidden field). Moreover, the coat-colour listed under this classification as **gachi-idi** derives from **gachit** (*Eragrostis abyssinica,* a grain cultivated only in Ethiopia). As a colour it is close to 'bright orange.'

Dombuli, which is found in the next clan (IV), derives from **dombuloch**, the name of a clan of the Bodi themselves. **Bonde hana** refers to 'the upper reaches of the Hana River (which flows from east to west through the heart of Bodi country),' and they say that there is a very large rock called **bha-golonyi** (lower half white; upper half red) in the upper reaches of that river. The **hana** in **hana koliyo** also refers to the Hana River, but the derivation of **koliyo** is unknown.

Clan VI contains cows of the following coat-colours: (28) **bha-shimaji** (a purple coated cow with white stomach area), (29) **el-shimaji** (purple coated cow whose left front foot is white), (30) **nyabha-shimaji** (cow with purple ears), (31) **el-cha7i** (grue cow with white left front foot), (32) **kalmi-cha7i** (grue bull with a belt of white around its middle), (33) **seroji** (cow with a coat pattern that combines more than three colours), (34) **gulbho-idi** (a coat-colour, the etymology of which is unknown), (35) **gure-idi** (a coat-colour derived from a particular bird) and (36) **diktach** (meaning unknown).

Kongogila in clan V refers to the round piece of wood that married women of the Mursi people of Southern Bodi wear on their lower lips, but when the words **derto** (**derdi** colour) or **bhilechoito** (a yellowish colour) are added to this the overall meaning becomes unclear. The coat-colours listed under this clan are **derdi** and **kulankulan**. As I explained earlier, **derdi** is a dark red or dark yellow-orange, a colour that resembles tincture of iodine.

The derivation of **durber** in clan VI is unknown, but **binda** and **lugi7a** are place names in Bodi. The derivation of **alachuli** is unknown but it refers to **cha7i** (grue). The coat-colours listed under this clan include **shimaji** (purple) and **cha7i** (grue), but even the Bodi themselves could not explain what these colours have to do with the place names. However, judging from the fact that when the Bodi move to **Lugi7a** in the Southern part of Bodi, for example, they first ritually sacrifice a **bhongai de shimaji** (a male calf with a purple coat) and then build a new settlement before they can begin to drink the water there, suggests that clan names undoubtedly have a deep connection to historical oral tradition. They call cattle linked through sacrifice to the new land to which they have moved **bi de gerebai**.

The clans **koro da ganyu bikilini** (our black cow with the giraffe pattern) and **malan kilin** (giraffe that leads the herd), listed under numeral VII, include the coat-colours (37) **mule-idi** (a colour like a rhinoceros: colour in the dark reds or dark oranges) and (38) **bato-idi** (derived from 'swarm of gnats', referring to greyish orange or greyish blue) but the reasons given for this are difficult to explain.

As mentioned earlier, the clan name **pajjaka** (VIII) is connected to the name of a giraffe clan, and includes such coat-colours as (39) **shimachi-gidhangi** ('greyish blue-purple', 'greyish purple', 'greyish red-purple') and (40) **cha7k-holi** ('bright blue-purple,' 'dull blue-purple). The clans under numeral IX called **koro da ganyu malanshali** (our black, buffalo) and **bio baldonyi jingo korkori** (meaning unknown) include **koro** (black), and coat-colours in Table 12.2 that are deemed to be close to **koro**. **Malanshali** is the name for a buffalo clan.

All the clan names under numeral X are of unknown origin, but include some coat-colours of particular significance to the Bodi: (47) **ludi**, (48) **choburi** and (49) **gelli**. Also included in these clans are cows with the coat-colours (50) **gel-poka-idi** (cow with **poka-idi**-coloured (unknown) spots on its back), (51) **nyabha-koroda** (black ears) and (52) **kabara-koroda** (black eyes).

The meanings of the clan names under numeral XI are unknown. In addition to the important coat-colours listed under these clans, (53) **kalmi**, (54) **bhaasi** and (55) **bholigaasi** (for reference see Figure 12.9), there is also (56) **mulk'a-idi** (meaning unknown). The meanings of the clan names listed under numeral XII are also unknown but they only include the coat-colour (57) **eldi** (Figure 12.9).

The meanings of the clan names under numeral XIII are also unknown, but in addition to the important coat-colours (58) **kordi**, (60) **bhilasi**, and (61) **tulk'a** (Figure 12.9), they also include (59) **kili-moji** (giraffe-pattern with **moji**-coloured patches) and (62) **kurkuri-idi** (meaning unknown). The last clan listed under numeral XIV, called **bio baldonyi pajjaka** and meaning 'my virility, clan of giraffe' is linked with the pattern of a giraffe's coat-colour.

As we can see from the above, there are fourteen cow clans or **kabchoch**, and cows of all the various kinds of coat-colour to be found in Bodi can be placed as a rule under one of these clan designations. It is not certain what kind of criterion Bodi use to decide which coat-colours fall under which clan, but it appears that there are extremely important implications here for understanding their pragmatic relationship with cattle. Inferring from the few clan names for which meanings can be traced, the explanations found in their connections with the animal world and in oral histories may be of importance, if only because they allow us to explore the Bodi's symbolic world. However, in addition to these kinds of symbolic explanations of why a certain kind of coat-colour falls under a particular clan, it is necessary to follow a yet different thread of inquiry. To be specific, I wish to argue that the kind of cattle classification that relies on coat-colour is connected to Bodi understanding of the genetics of coat-colour.

To demonstrate this will require, in addition to Bodi knowledge of cattle coat-colour genetics, an investigation of their understanding of genetics treated as a natural scientific method. Probing this with limited data, I hope to show that this knowledge is the basis for selection of coat-colour polymorphisms.

Tracing Cattle Genealogies and the Cognition of Cross-Breeding

Tracing Cattle Genealogies as far back as Sixteen Generations

As mentioned in the section on the cattle 'family register' Bodi keep track of the mother/calf relationships among their cattle. There is little doubt that these mother/calf relationships have been understood ever since herders or their forebears first domesticated animals from the wild. However, the Bodi know not only the mother/calf relations for two generations but several generations of these relationships among the cattle in the compounds nearest them.

When I started my survey of cattle in Bodi society, I had no idea that the Bodi had this deep level of knowledge concerning their cattle. In the compound of one of my informants was a milk cow that he and his wife were particularly fond of. It was called **mogut te golonyi** (red-coated cow with horns the shape of the crescent moon) and it was prized as the smartest cow leading the herd in the pasture. One day, when I asked my informant the 'matrilineal' background of this cow, he explained to me her 'lineage' back sixteen generations (Figure 12.10).

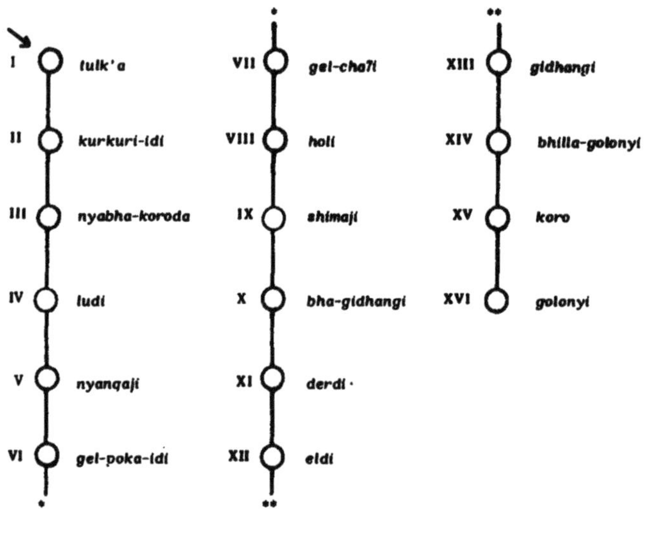

Figure 12.10. Genealogy of the cow called *bheliyach*

However, there are two things to note here. One is that while the 'matrilineal' relationships, which could also be called a genealogy mediated by coat-colour, are remembered by way of the system of classification of coat-colours discussed earlier, how closely this reflects the genetics of coat-colour is questionable. It may be that we should keep the two separate for the purposes of analysis. At the very least, however, it can be assumed that the genetics of Bodi cattle breeding and their classifications of cattle are indivisibly connected. Bodi (or their genetic precursors) have observed the heredity of their cattle coat-colours for several hundred or more generations, and have transmitted these results to their descendants.

The second point to note is that we are relying on Bodi

informants' memories and not on their confirmation of the facts based on having actual cows with clear, undeniable 'matrilineal' relationships. Whether the source of our information comes from one person or many, it is extremely difficult to reconstruct what happened in the past using this kind of information. We know only too well how such information can become altered through livestock profits or losses or the social situation at a particular time [see Fukui 1986]. Of course, the information I present here is that of the Bodi at a particular point in time, although I believe I was generally able to determine whether or not an informant was purposely lying.

Let us return now to the genealogy of the cow **mogut to golonyi** that I referred to earlier. The mother of **golonyi** (red) was **koro** (black), **koro's** mother was called **bhilla-golonyi** (thick red belt around the middle), the mother of **bhilla-golonyi** was **gidhangi** (grey or low saturation of red), **gidhangi's** mother was **eldi** (left front foot white, the rest black), **eldi's** mother was **derdi** (coat the colour of tincture of iodine), and so on. The mother cow of the first generation came from a location far from Bodi country so it would be difficult at this point to trace her. What we can say for sure is that the Bodi can trace back the genealogical relationships of their cattle a long way, at least through the female line. However, what can we say about bulls?

Examples of Cattle Cross-Breeding

In the case of the cow **mogut te golonyi**, I was so impressed by the depth of my informants' knowledge of her genealogy that I missed an opportunity of asking how the bulls might fit into the equation. However, from that point on I made sure to ask not only for the 'matrilineal' relationships of the cattle but also for their 'patriline'. As long as the Bodi continue to raise cattle in close proximity to themselves, they will maintain a clear cognizance of their 'patrilineal' relationships as well. Let us look now at the results of their knowledge of cross-breeding of particular coat-colours, citing specific examples of actual calf/ parent (both maternal and paternal) relationships that were based on their classification of coat-colour (Table 12.3).

The information in this table was collected from five informants and is limited to definitely verifiable cross-breeding

Table 12.3. Bodi cognition of cattle coat crossing

No.	Cow coat	Bull coat	Calf coat
1G*	*bha-golonyi*	*gel-golonyi*	*el-golonyi*(M)**
2G	*bha-golonyi*(C1)***	*gel-golonyi*(B1)	*golonyi*(F)
3G	*bha-golonyi*(C1)	*koro*	*bha-golonyi*(M)
4G	*bha-golonyi*(C1)	*ludi*	*golonyi*(F)
5G	*lu-golonyi*	*bhila-golonyi*	*lu-golonyi*(M)
6G	*kordi*	*gidhangi*	*koro*(F)
7G	*golonyi*	*nyangaji*	*nyangaji*(M)
8G	*holi*	*moji*	*moji*(M)
9G	*golonyi*	*golonyi*	*gel-golonyi*(M)
10G	*koro*	*ludi*	*kalmi*(M)
11G	*nyangaji*	*golonyi*	*nyangaji*(M)
12G	*lu-golonyi*	*golonyi*	*el-golonyi*(F)
13G	*lu-golonyi*(C12)	*golonyi*(B12)	*el-golonyi*(F)
14G	*bhaasi*	*el-bhaasi*	*gel-golonyi*(M)
15G	*gel-golonyi*	*bholloga*	*bha-nyangaji*(M)
16G	*gel-golonyi*(CfC15)	*nyangaji*	*bha-nyangaji*(M)
17G	*gel-golonyi*(C16)	*gel-golonyi*	*golonyi*(M)
18G	*lu-golonyi*	*el-golonyi*	*gel-golonyi*(M)
19G	*polo-kordi*	*eldi*	*bhaasi*(F)
20G	*bhaasi*(Cf19)	*koro*	*bhaasi*(F)
21G	*bhaasi*(Cf19)	*koro*(B20)	*koro*(F)
22G	*bhaasi*(Cf19)	*nyangaji*	*bha-golonyi*(M)
23G	*bhaasi*(Cf19)	*cha7i*	*golonyi*(M)
24G	*gel-golonyi*	*bholloga*	*bha-nyangaji*(M)
25G	*kalmi*	*bha-nyangaji*(Cf24)	*lu-golonyi*(F)
26G	*el-nyangaji*	*lu-golonyi*	*el-nyangaji*(F)
27G	*el-nyangaji*(C26)	*golonyi*	*golonyi*(M)
28G	*polo-bhaasi*	*bhaasi*	*bholloga*(F)
29G	*bha-golonyi*(C29)	*shimaji*	*holi*(F)
30G	*bha-golonyi*(C29)	*eldi*	*choburi*(F)
31G	*bha-golonyi*(C29)	*moji*	*bha-golonyi*(F)
32G	*bha-golonyi*(C29)	*koro*	*holi*(M)
33G	*bha-golonyi*(C29)	*koro*(B32)	*koro*(M)
34G	*bha-golonyi*(C29)	*nyangaji*	*nyangaji*(M)
35G	*bha-golonyi*(C29)	*moji*(B31)	*bhileji*(M)
36G	*bha-golonyi*(C29)	*gel-golonyi*	*bhileji*(M)
37G	*bha-golonyi*(C29)	*el-golonyi*	*bha-golonyi*(M)
38G	*bha-golonyi*(C29)	*gel-golonyi*	*moji*(F)
39G	*holi*(Cf29)	*choburi*	*koro*(M)
40G	*holi*(Cf29)	*nyangaji*(B24)	*nyangaji*(M)

Table 12.3. Continued

No.	Cow coat	Bull coat	Calf coat
41G	*holi*(Cf29)	*gel-golonyi*(B36)	*moji*(F)
42G	*bha-golonyi*(Cf31)	*gel-golonyi*(B36)	*kilindi*(M)
43G	*bha-golonyi*(Cf31)	*el-golonyi*(B37)	*bha-golonyi*(M)
44G	*lu-golonyi*	*el-golonyi*(B37)	*gel-golonyi*(B36)
45G	*gel-golonyi*	*bholloga*	*nyangaji*(B34)
46A	*bhaasi*	*bhileji*	*cha7i*(M)
47A	*bhaasi*(C46)	*koro*	*bhaasi*(F)
48A	*bhaasi*(C46)	*gidhangi*	*bha-golonyi*(M)
49A	*bhaasi*(C46)	*cha7i*	*golonyi*(M)
50A	*nyangaji*	*choburi*	*golonyi*(F)
51A	*seroji*	*bhilasi*	*golonyi*(F)
52A	*golonyi*(Cf51)	*koro*	*gidhangi*(F)
53A	*seroji*	*ludi*	*seroji*(F)
54A	*seroji*(Cf53)	*gidhangi*(B48)	*koro*(F)
55A	*shimaji*	*eldi*	*el-golonyi*(F)
56A	*shimaji*(C55)	*eldi*(B55)	*el-golonyi*(M)
57A	*shimaji*(C55)	*choburi*	*choburi*(M)
58C	*kilindi*	*kilin-gidhangi*	*gel-gidhangi*(M)
59C	*golonyi*	*holi*	*kilin-gidhangi*(B5)
60C	*kilindi*(C58)	*bhilasi*	*bhila-gidhangi*(M)
61C	*kilindi*(C58)	*shimaji*	*bhila-kilindi*(M)
62C	*kilindi*(C58)	*holi*	*bhila-kordi*(F)
63C	*kilindi*(C58)	*holi*(B62)	*golonyi*(M)
64C	*kilindi*(C58)	*golonyi*	*gidhangi*(M)
65C	*kilindi*(C58)	*bhilasi*(B60)	*golonyi*(F)
66C	*golonyi*(Cf65)	*gidhang-golonyi*	*golonyi*(F)
67C	*holi*	*holi*(B62)	*cha7i*(M)
68C	*holi*(C67)	*holi*(B62)	*holi(saba-golonyi)*(F)
69C	*holi*(C67)	*bha-cha7i*	*eldi*(M)
70C	*kilin-bhileji*	*golonyi*	*holi*(B62)
71C	*kilin-bhileji*(C70)	*holi*	*bhila-cha7i*(M)
72C	*kilin-bhileji*(C70)	*holi*(B71)	*bhaasi*(F)
73C	*kilin-bhileji*(C70)	*holi*(B71)	*koro*(F)
74C	*kilin-bhileji*(C70)	*holi*(B71)	*el-bhaasi*(F)
75C	*kilin-bhileji*(C70)	*holi*(B71)	*holi(saba-bhileji)*(F)
76C	*kilin-bhileji*(C70)	*kilin-gidhangi*	*kilin-bhileji*(F)
77C	*bholighaasi*	*holi*(B62)	*bha-gidhangi*(M)
78C	*nyabha-cha7i*	*gidhang-golonyi*	*kilin-gidhangi*(M)
79B	*gel-shimaji*	*bhaasi*	*gelli*(M)
80B	*cha7i*	*cha7i*	*gel-shimaji*(C79)

Table 12.3. Continued

No.	Cow coat	Bull coat	Calf coat
81B	*eldi*	*kordi*	*ludi*(M)
82L	*koro*	*ludi*	*ludi*(F)
83L	*kordi*	*bhaasi*	*ludi*(M)

* Capital letter following numeral indicates abbreviation of informant's name.

** (M): male (F): female

*** (C1) means the same cow as No. 1. B and C is the abbreviation for bull and cow respectively, while (Cf) is calf and (C79) is the calf of the cow indicated in No.79.

combinations of cattle (cow, bull, and calf). Since the parents of cows that were stolen from other peoples or received as bridewealth or as sacrifices for specific rituals, or that were given in exchange with other peoples for whatever reason, are for the most part unknown, they are omitted from this table. There are a large number of examples where only the 'matriline' can be traced, but since these do not provide us with complete data for a study of the results of coat-colour cross-breeding, they too have been omitted. I have listed parent/calf relations that were definitely known to the informants, but have ranked the reliability of the information elicited as follows: G as the most reliable, A the second most reliable, and C the least reliable. This reliability ranking comes from having asked informants, at the stage of gathering information, the same questions at different times; the answers where there was no wavering of opinion ranked the highest and those where there was some obfuscation ranked the lowest. However, if I detected some uncertainty I asked them repeatedly until I felt sure that their memories were not mistaken, and these are the answers I have listed in the table. Informant G had a distinct recollection of as many as forty-five actual examples of cross-breeding, and when one collates the various combinations involved one can see that the cattle parent/ calf relationships are really well consolidated in his mind almost as if they were human kinship relationships (Figure 12.11).

I would like now to take a look at Table 12.3 to examine on the basis of some of the concrete examples listed there what kind of results arise out of the cross-breeding of which colour-coat combinations.

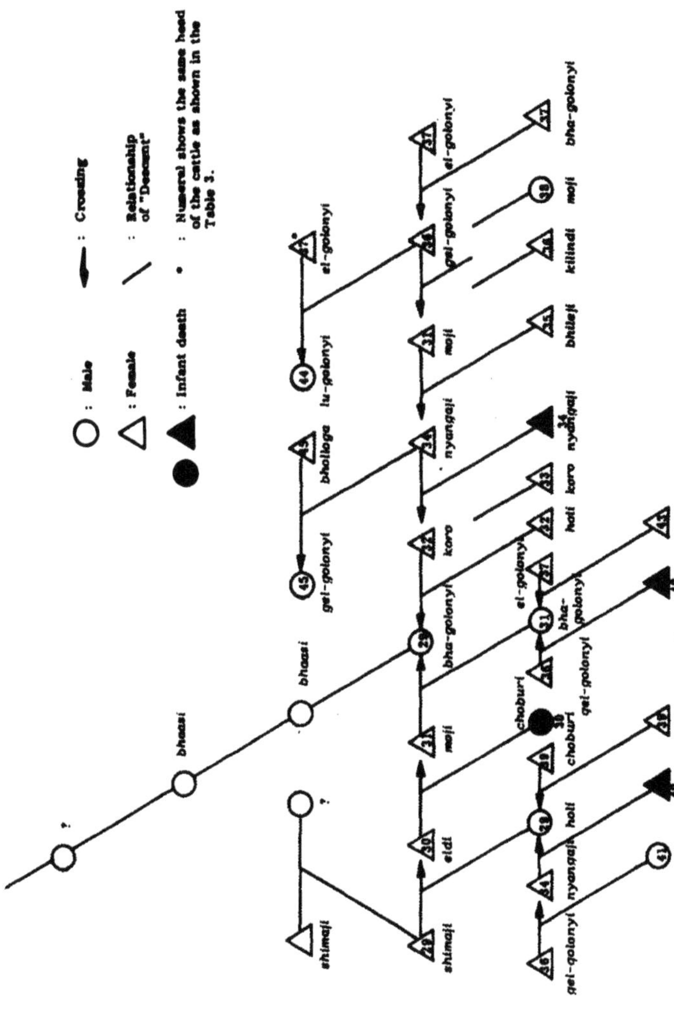

Figure 12.11. Genealogy of the cow called *bheliyach* and cognition of cross-breeding among the Bodi

First, let us look at **bha-golonyi** (red coat with white stomach area) under 1G. At the present time this cow is under the ownership of G's primary wife, but her mother (**golonyi**: red coat) was given to them by a paternal cousin five generations ago. And her mother before her is said to have been called **lu-cha7i** (white with bluish head and rump), but as far as the father bull goes, it was in the distant compound of its original owner so both of these details were unverifiable. The **bha-golonyi** in 1G had given birth to four calves. By one of G's bulls (**gel-golonyi**: white with red head and rump and red patches on back), she had given birth to a male calf, **el-golonyi** (red with white left front foot), and a female calf, **golonyi** (red cow) but both had died early on. After that she gave birth to a calf of her own coat-colour (**bha-golonyi**) by a bull named **koro** (black) in her same herd. Her fourth calf, **golonyi** was conceived with a bull, **ludi** (white with black head and rump), that was the property of G's half-brother, who lived in the same settlement. What can be concluded from these four examples of cross-breeding is that in all cases the red of the parent animals is dominant in the coat-colour of the offspring. The only calf that exhibited exactly the same colour as its mother was the one conceived when she was bred with a black bull.

Secondly, if we look at the cross-breeding lineage that has the cow **holi** (white) at its centre (8G, 39G, 40G, 41G, 59C, 62C, 63C, 67C, 68C, 69C, 71C, 72C, 73C, 74C, 75C, 77C), we can see that all the offspring that resulted were of coat-colours different from **holi**. 68C and 75C are generally referred to as **holi**, but one has a red head (**saba-golonyi**) and the head of the other is yellow (**saba-bhileji**), so in all cases the calves exhibit some colour.

Genealogy of the Cow bheliyach who Carries Bodi Historical Tradition

Next, referring to Figure 12.11 and Table 12.3, I would like to look at the examples of the cross-breeding of the cow called **bheliyach**. **Bheliyach** is a cow that is brought from the homeland where the ancestors of the regional chiefs, which included informant G, was located ten generations earlier [see Fukui 1988a; 1994]. In Bodi society, when a new regional chief is installed, a group which includes the new chief goes to the homeland, and a ritual is conducted with the chief there who is of the same genealogical line. Their homeland is over two day's walking distance from the centre of Bodi country, and there is ordinarily little traffic between the two places. Following this ritual, the cow that is received from the regional chief of the homeland is called **bheliyach**. This **bheliyach** is brought back to Bodi country, and rituals are performed anew. After the milk is put in a ritual black gourd and has turned into kefir, which then has specific uses, it is sprayed onto the cows in the herd in the morning and in the evening, as well as on to the members of the settlement in the event that the settlement is moved to a new location. Unlike regular milk or kefir, it is generally prohibited for anyone to drink it. What should be noted is that this title of **bheliyach** is passed down 'matrilineally' from cow to calf. Moreover, as long as it can be traced through its mother to the **bheliyach** line, that cow's milk has the same historical function. The genealogy of the cow that could be traced back sixteen generations, mentioned earlier in relation to Figure 12.10, was actually that of a **bheliyach** cow that had been brought from the homeland sixteen generations earlier. This is why its lineage was more distinctly known than would normally be the case.

When G's father's father was inaugurated as regional chief, the cow that was given to G's father when they returned from the homeland was **bi de gelli** (white cow with black head and rump and with black patches on its back). G had heard that it had given birth to **bi de ludi**, **bi de gelli**, **luch de gelli** and **luch de gelli**, but several generations of the lineage were missing. G was able to trace the line back five generations from the cow called **bi de bhaasi** (black cow with white stomach). The calf it produced was also **bi de bhaasi**, but since they all belonged to G's father, he had no recollection of the bull involved. G's knowledge of the **bheliyach** 'lineage', including the bulls, began from the time that he received from his father **bi de bha-golonyi**, the calf of **bi de bhaasi**.

This **bha-golonyi** (C29) gave birth to a total of 10 calves. The first (29) was **mor te holi** (white female calf) and was conceived by a bull, **shimaji** (purple), that was the property of his father's half-brother. The second (30) was **mor te choburi** (female calf with several black patches) and was conceived by a bull, **eldi** (black with white left front foot), that belonged to the same owner. The third (31) was conceived by a bull, **moji** (lion-coloured), that belonged to a different owner (Gd) and was called **mor te bha-golonyi** (black cow with white stomach). The fourth (32) was conceived with a bull, **koro** (black), that belonged to G's half-brother and was called **bhongai de holi** (white male). The fifth (33) had the same father, **koro**, as 32 and was called **bhongai de koro** (black male). The sixth (34) was conceived with G's own bull, **nyangaji**, and was called **bhongai de nyangaji** (orange male). The seventh (35) had the same father, **moji**, as 31 and was called **bhongai de bhileji** (yellow male). The eighth (36) was a bull owned by G himself called **gel-golonyi** (white with red head and rump and red patches on its back). The ninth (37) was called **bhongai de bha-golonyi** (red male with white stomach) and was conceived with the bull, **el-golonyi** (red with white left front foot), that was owned by someone else (Bt). And finally the tenth (38) was conceived by the same bull, **gel-golonyi**, as 36 and was called **mor te moji** (lion-coloured female).

When we look at the 10 examples of cross-breeding that centre around the mother cow **bha-golonyi**, in all cases except 31, a single colour becomes dominant in the colours exhibited by her calves. Also, in the three examples (39–41) of cross-breeding that have the cow (29) **holi** (white) as their mother, all exhibit white in their coat-colours. The two examples 36 and 43 that are a result of the cross-breeding of a **bha-** and an **el-** also exhibit **bha-** (white stomach) in their colour. In the four examples (46–49) that have the mother cow **bhaasi** at their centre, **bhaasi** or **bha-** colour shows up in half of them. The same can be said for the examples 20–23. Among these, in the cross-breeding of a **bhaasi** and **koro** two (20, 47) out of the three examples exhibit **bhaasi** colour, while one example (21) is **koro**.

It may be that the data provided in Table 12.3 are too scant to allow us to reconstruct the genetic principles of coat-colour cross-breeding. However, as anticipated in the examples mentioned earlier that centred around the colours of **holi** and **bhaasi**, it does seem possible to explore some of the tendencies of coat-colour genetics even from these few examples. What is important is to acknowledge that the Bodi came up with their own view of coat-colour genetics from the vast quantities of this kind of information accumulated as a result of observing hundreds and thousands of examples of cross-breeding and passing this information down to their children and grandchildren.

The Folk Genetics of Coat-Colour

As I gradually uncovered the cognitive system underlying Bodi coat-colour classification and their knowledge of their cattle cross-breeding genealogies, I began to think that they must have evolved their own view of coat-colour genetics. And so it turned out to be.

My attention was first drawn to this possibility when I was investigating the close interrelationships between colours. When I asked informants why **bhileji** (yellow) and **gidhangi** (grey or low saturation of red) are said to be **ngonige** (brothers), I was told that it is because if a cow is **bhileji**, it gives birth to a **gidhangi** calf. In the same way a **nangaji** calf is said to be born from a **golonyi** cow. What is important here is the idea that the interrelationships between colours correspond to the coat-colour genetics of cattle.

From that time on I made a point of asking people questions relating to their views of cattle coat-colour genetics. What happens if you change the colour of the bull that you cross-breed with a **holi** (white) cow? I was told that if the bull were **ludi**, the calf would be **ludi**; if **eldi**, the calf would be **eldi**; if **holi**, then **holi**; if **golonyi**, then **golonyi**; if **gelli**, then **gelli**; if **koro**, then **koro**; if **shimaji**, then either a **shimaji**, **holi** or **bhileji** calf would be born. They added that if you were to cross-breed a **holi** bull with this same list of coloured cows the results would be the same as the previous case.

What can be positively concluded from this example is that Bodi have a clear view of genetics based on principles of dominance and recessiveness, not unlike Mendel's laws. For example, in the various cross-breeding combinations that involved **holi** listed above, one can see that the Bodi are aware that **holi** as a phenotype is difficult to reproduce. That is, a white phenotype occurs only when homogeneous alleles are involved, whereas a roan is produced in the case of heterogeneous alleles. If we return for a moment to the earlier example of cross-breeding, a **holi** coloured calf can be produced only if the white coat allele is homogeneous. If a **holi** carries a heterogeneous allele, it is very difficult to produce a **holi** calf. Principally, for one to be born, a **holi** has to be bred with another **holi**, and only when the alleles are homogeneous. The reason why a **shimaji**, **holi** or **bhileji** calf is produced in the case of a **holi** cross-bred with a **shimaji** is that a **shimaji** is equivalent to what we would call a roan.

Cha7i (grue) is a coat-colour that also corresponds to a roan. It is said that if one cross-breeds a **cha7i** cow with a **golonyi** bull, a **golonyi** calf is produced; if with a **bhileji** bull, then a **bhileji** calf; in the case of a **holi** bull, then either a **cha7i** or a **holi** calf; and in the case of a **koro**, then either a **cha7i** or **koro** calf. Then again, if a **golonyi** cow is the object of cross-breeding the following coat-colour combinations are produced: if it is bred with a **koro**, a **nyangaji** is produced; if with a **nyangaji**, then a **golonyi** is produced; if with a **gidhangi**, then a **gidhangi**; if with a **bhileji** then a **bhileji**; if with a **shimaji**, then a **shimaji**; and if with a **cha7i**, then a **cha7i**. Especially in the last two cases, the results are limited to **shimaji** and **cha7i** respectively. As examples of coat cross-breeding combinations that always produce a certain coat-colour in the calves, they gave the following: if one cross-breeds a **gelli** with a **ludi**, one always obtains either a **gelli** or a **ludi** calf as a result, but in rare cases one might get a **nyabha-koroda** (white with black ears). And again, in the case of cross-breeding a **kordi** with a **bhilasi**, one always obtains either a **kordi** or a **bhilasi** calf. I have summarized these various examples of cross-breeding combinations in Figure 12.12.

The Bodi folk genetics that has as its basis this law of dominance has presumably evolved, as I pointed out earlier, from their observation of countless instances of coat-colour cross-breeding and their passing down of the results of these various combinations through the generations. It goes without saying that their folk genetics developed entirely independently of Mendel's Law, which forms the foundation of modern genetics, and that it has instead emerged out of the Bodi way of life itself. Recent studies by Nozawa [1995], based on an examination of the eighty-three cases of cross-breeding mentioned above (Table 12.3), have shown that only twenty cases cannot be explained in terms of modern genetics. Indeed, he is of the opinion that cognition of coat-colour among the Bodi is so fine as to go beyond the capacity of modern genetics.

Selection of Bulls Based on Coat-Colour

Based on this kind of genetics, I began wondering how Bodi go about selecting a bull. In the case of the cow, the two major

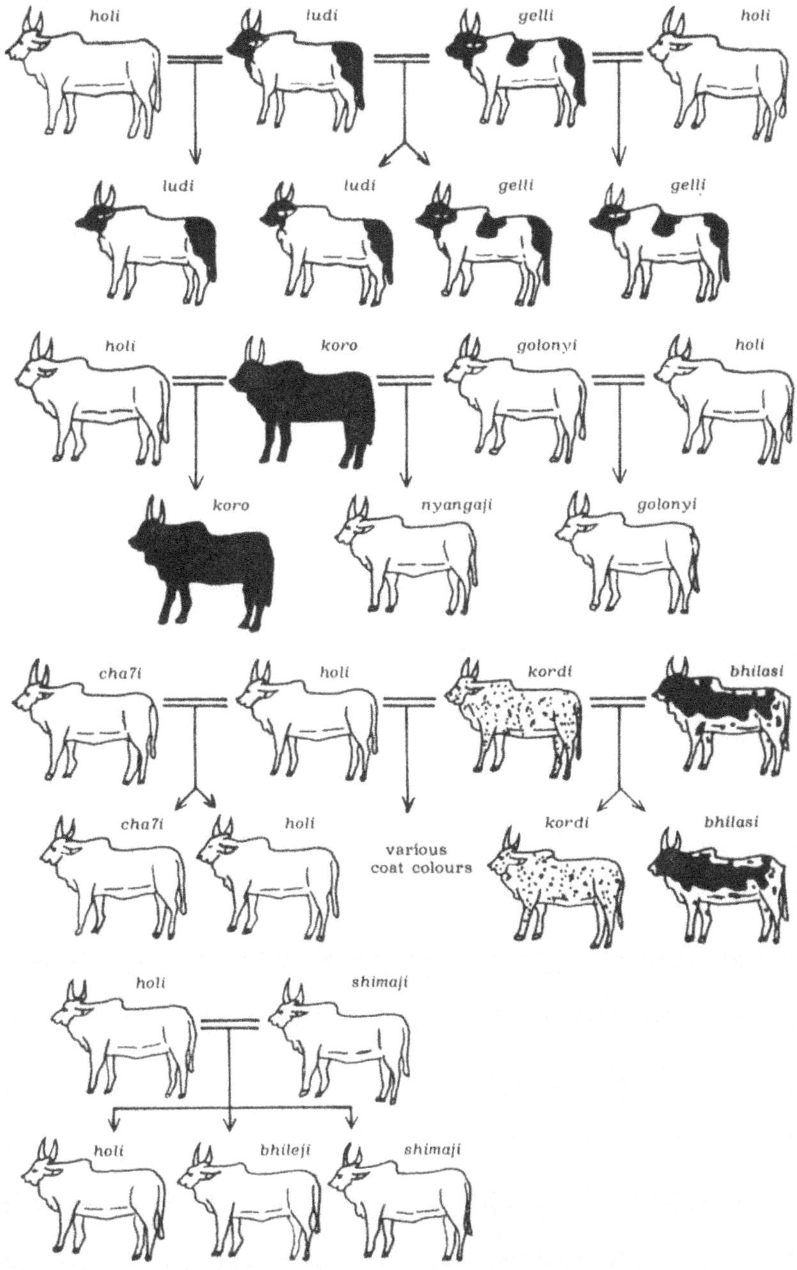

Figure 12.12. Folk genetics of Bodi cattle

Table 12.4. Cattle types and numbers in one Bodi compound

Coat-colour		Male	Bull calf	Ox	Bull	Female	Heifer	Milk cow	Dry cow
	1. *golonyi*					3	1	2	
	2. *shimaji*	1	1			1	1		
A*	3. *cha7i*					2	1	1	
	4. *bhileji*					4		1	3
	5. *gidhangi*					1		1	
	6. *moji*					1			1
	7. *ludi*	4	3		1				
	8. *gelli*	1		1		1	1		
	9. *kalmi*	1	1						
B*	10. *bhaasi*					1		1	
	11. *chokaji*	1	1						
	12. *eldi*					1		1	
	13. *tulk'a*					1			1
	14. *bhilasi*	2	2						
	15. *bha-golonyi*					1	1		
	16. *bha-bhileji*					1	1		
	17. *bha-gidangi*					1		1	

Table 12.4. Continued

Coat-colour	Male	Bull calf	Ox	Bull	Female	Heifer	Milk cow	Dry cow
C* 18. *el-shimaji*					1	1		
19. *bhila-golonyi*	1	1						
20. *choburi-gidhangi*	1	1						
21. *kilindi*	1			1	4	1	2	1
Total	13	10	1	2	24	8	10	6

* A : Colour, B : Pattern, C : Colour + Pattern

requirements are that she be healthy and a good milk producer. From among the male calves produced by this kind of cow they choose the ones that attempt to mate most aggressively. With this prerequisite satisfied they next select on the basis of coat-colour. And that becomes the colour configuration that that owner will identify himself with throughout his life. As I will explain later, this is referred to as **morare** or **bhoyoch**, and is related to the personal identification that occurs between humans and cattle based on coat-colour. Bodi say that having many bulls in one compound leads to them fighting one another so it is best to have only two. I can think of two possible reasons for choosing two. One is that in Bodi culture, even numbers are considered auspicious while odd, or uneven, numbers are avoided because they are viewed as inauspicious. The other reason may be their desire to have enough cows in their herd to make two bulls necessary, and I will give an example. When I had lived with the Bodi for about six months, the regional chief gave me the Bodi-style name **oinyahenne,** which means 'bull again.' Bodi carefully observe the mating of their bulls. For instance, they might say, 'The reason that cow is pregnant is that she mated (**moguto**) with **oi de nyangaji** (orange-coated bull) in the compound (**tui**).' They also say that once a cow has mated with a bull and conceived, it does not mate again until after it gives birth.

Next I would like to examine how the selection of bulls based on coat-colour is carried out by using a concrete example. Table 12.4 is a product of reorganizing the 'family register' (Table 12.1) according to categories of sex, age and coat-colour of the cattle. The two bulls in the compound are **ludi** (black head and rump) and **kilindi** (covered with reddish spots like a giraffe). In Figure 12.2 the owner of **ludi** appears as Ok, and the owner of **kilindi** as K1. This is because these are the coat-colours assigned to these respective owners, or in other words, that **ludi** and **kilindi** are their **morare**, or the cattle which they identify with. The goal is to acquire as many fine examples of their **morare** as possible. If **ludi** is the **morare** in question, then in order to increase the number of **ludi** cattle as much as possible, one tries to make a **ludi** coloured male into the bull of the compound. It may have been for this reason that **ludi** males were most numerous in Ok's compound.

It is clear from the above example that the Bodi select their bulls according to specific coat-colours. If that is so, how are these particular coat-colours used in their society? In the next section I would like to provide an outline of the cultural device used in selecting coat-colour polymorphisms.[4]

The Cultural Uses of Coat-Colour Diversity

We now need to examine how coat-colour is connected with various other aspects of Bodi culture. I have attempted to organize these in Table 12.5, and have divided them very generally into the following eight subjects: (1) identity of the individual or group as seen in terms of coat-colour; (2) particular coat-colour of the cattle sacrificed during ritual installation of a chief; (3) livestock sacrificed at the outbreak of war, type varying depending on enemy; (4) livestock sacrificed to mark occasions in the human life-cycle, such as marriage, illness and death; (5) livestock sacrificed at rituals pertaining to agriculture; (6) coat-colour of cattle sacrificed in the event of moving to certain locations; (7) other: cattle from which blood is drawn to rub on the body of one who has killed a lion; cattle sacrificed in the event of seeing a monster; (8) coat-colour of cattle normally not allowed to be sacrificed. I plan to analyze and develop each of these topics in more detailed papers at some later date; for now I shall mention each one in general terms.

Identifying with Cattle of Particular Coat-Colours

Morare: *Colour Configuration that Becomes the Object of Self-Identity* In Bodi society every person is given a social name one year after birth which are all associated with particular colours and patterns [Fukui 1979b]. After this is done, the child is made to wear beads of that colour and pattern around their necks and they are brought up hearing songs associated with them, much as we would hear lullabies. Eventually, the child becomes intrinsically linked to the colour and pattern it bears and begins to show signs of personally identifying with them in an almost obsessive way. For example, a girl of about thirteen named Lilinta (dragonfly), who is associated with the colour red, becomes extremely excited whenever she sees the red colour in the highest saturation of 98 colour cards which I showed to her.

Table 12.5. Cultural uses of particular coat-colours among the Bodi

	Usage	Main objects	Main kinds of coat-colour
I.	Identity		
	1. Individual	living ox	*ludi, bhaasi, gelli, ...* (see Table 12.6)
	2. Age-Set	living ox	*shimaji, bhaasi, gelli, ...*
	3. Generation-Set	b.p*	*gidhangi, moji*
	4. Area (Banner)	coat	*ludi, gelli*
II.	Installation of Chief	b.p	*golonyi, koro, holi, ...*
III.	Enemy		
	1. Chirim	b.p	*ludi, gelli, choburi*
	2. Mursi	b.p	*derdi, kalmi, bhaasi, bha-nyangaji*
	3. Hamar	b.p	*muge-idi*
	4. Highlanders	goat	*nyangaji*
IV.	Life		
	1. Marriage (first)	b.p	except *cha7i, seroji* and *derdi*
	(second)	goat	*koro, gidhangi, bhaasi, golonyi*
	2. Disease (normal)	b.p	*holi* and *idho-idi* are the best.
	(madness)**	b.p	*bhaasi, koro, gidhangi*
	3. Death	b.p, coat	different according to clan (see Table 12.8)
V.	Agriculture		
	1. Clearing	b.p	*koro* is the best; *bholigaasi, bhaasi*
	2. Sowing	b.p	*gidhang-nyangaji, gidhangi*
	3. Rain-making	b.p	*koro, bhaasi, bholigaasi*
	4. Ridding of insects	blc.***	*seroji, bhaasi*
	5. Ridding of birds	milk	*shimaji*
		blc	*cha7i*
VI.	Migration	b.p	*bhaasi* (to Udurum), *shimaji* (to Lugi7a)
VII.	Others		
	1. After killing lion	blc.	*moji, bhileji, gushuri-idi*
	2. After seeing monster	b.p	*bhaasi, koro, ludi, gelli*
VIII.	Taboo (not used for normal ritual)		*tulk'a, tula-golonyi, kor-cha7i, ludi, derdi, seroji*

* : b and p means the blood and peritoneum of the victim respectively.

** : It also includes epilepsy.

*** : blc. means the blood taken from the jugular of cattle with a particular coat-colour.

Figure 12.13. A man hums his personal poems about his favourite ox

This is not simply because it is red, but because it is her colour.

In this way, all members of Bodi society from the age of one are given a name and at the same time assume a personal identification for life with a colour and a pattern. This pattern/colour is called **morare** (Figure 12.13). **Morare** refers not only to one's own individual colour/pattern identification but also to the same coat-colour of the cow one will receive upon reaching adulthood. Since **morare** are decided according to a system of naming [Fukui 1979b], the colours/ patterns that are designated **morare** are somewhat limited. Table 12.6 is a compilation of the results of a survey of 163 (male and female) **morare**. Among monochromatic colours, **cha7i**, **shimaji**, **golonyi** and **koro** show up frequently. In black and white patterns, **ludi**, **bhaasi** and **gelli** predominate, while with coloured patterns **bha-golonyi**, **lu-golonyi**, **gel-golonyi** and **el-golonyi** are predominant. For the Bodi, obtaining cattle, especially males, of these particular coat-colours and raising them becomes their reason for living when they are young. When their bulls are castrated, they regard it as the highest honour to be called by the name of their oxen.[5] As long as this system of inheriting **morare** exists, and they continue to identify themselves personally with them, there will be no lack of cattle with these kinds of coat-colours.

Symbolic Coat-Colours of Age-Sets It is well known that age organizations flourish in many societies of East Africa. In Bodi the age-grade is not that conspicuous, but age-sets and generation-sets, which will be discussed later, form an important part in political organization. There are seven age-sets, or **lukur**, in the **ngolit** generation to which the current chief belongs. Each of these age-sets have an ox with a particular coat-colour which they identify with. The **morare** of the most trusted person in each age-group is chosen as the symbol, and the coat-colour of that **morare** becomes the object of identification for the entire age-set. They call this **shali de lukur**. Looking at these **shali de lukur** for the age-sets of the older generation **ngolit**, they appear as follows: (1) **am-uren** (name of age-set): **polo-eldi** (coat-colour that becomes object of identification), (2) **moku-su: gelli**, (3) **am-kuis: gelli** (4) **bhog-gabsa: shimaji**, (5) **dulwasa: ludi**, (6) **chuch-gel: gel golonyi**, (7) **mok-dhogol: kalmi**. If we go back one further generation, **mojit**, and look at the age-sets of the members of this old generation who are still living, they appear

(a) *derdi*

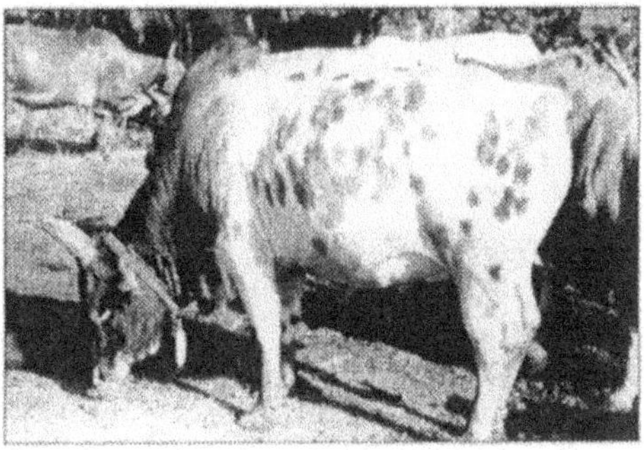

(b) *kilindi*

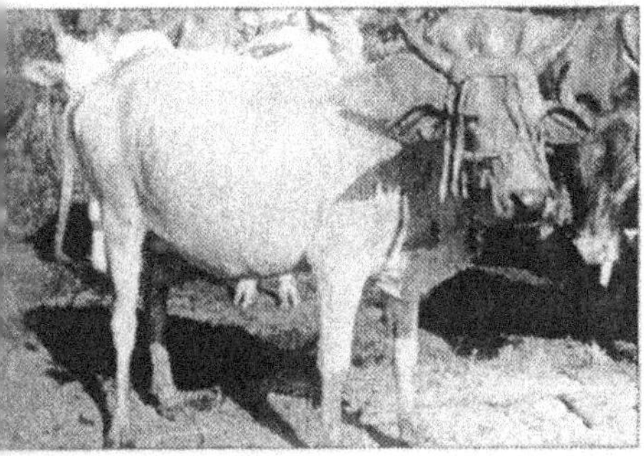

(c) *chai*

(d) *bhileji*

(e) *bha-tula-golonyi*

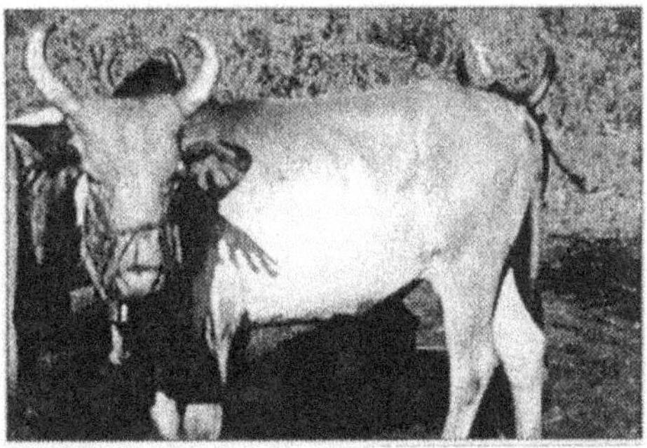

(f) *bha-shimaji*

(g) cow: *bhaasi*, its calf: *bha-golonyi*

(h) *bha-gidhangi*

Plate 12.1. Some examples of Bodi cattle coat-colour

Plate 12.2. Various coat-colours in Bodi Cattle. Following Nozawa [1994: 94] the genotypes represented here are as follows: ① C_ww ee ss; ② C_Ww eeS_; ③ C_ww E^dS_; ④ C_ww E^d_ss; ⑤ C_Ww ee ss; ⑥ C_ww E^d_ss. This notation follows Searle [1968], where WW = pigmentation almost absent, Ww = roan, ww = coloured: E^d_ = dominant extension of black, E_ = normal extension of black, e^{br}_ = brindled, ee = yellow or red; SS = self-coloured, Ss = slightly spotted, ss = spotted

Table 12.6. The main colour-patterns of the *morare*

Colour	Male	Female	Black and White Patterns	Male	Female	Colour + Patterns	Male	Female
cha7i	5	0	*ludi*	19	15	*bha-golonyi*	10	10
shimaji	3	12	*bhaasi*	12	8	*lu-golonyi*	6	0
golonyi	3	5	*gelli*	10	15	*gel-golonyi*	8	2
koro	1	8	*eldi*	8	3	*el-golonyi*	2	7
gidhangi	1	3	*kordi*	4	3	*bha-gidhangi*	1	0
bhileji	1	2	*choburi*	2	3	*bha-shimaji*	1	0
			kalmi	2	2	*kilindi*	1	3
Total*	14**	30***		57**	49***		29**	22***

* : Total number of men and women surveyed : 115 men and 48 women
The sample was repeated more than twice for both men and women shown here.
** : Total number of the *morare* sample in men surveyed : 157
*** : Total number of the *morare* sample in women surveyed : 61

as follows: (1) **am-goido: kalmi**, (2) **jariya-suuli: lu-golonyi**, (3) **bulch'om-mulo: ludi**, (4) **am-dubur: bha-shimaji**.

In the eleven age-sets and their respective objects of identification, or **shali de lukur**, listed above, the coat-colours **ludi**, **gelli** and **kalmi** each appear twice and are more frequent than any other coat-colours.

Symbolic Coat-Colours of Generation-Sets As mentioned earlier, in Bodi society generation-sets (**kek**) encompass several age-sets (see Table 12.7). For each generation there are attendant animals, and the coat-colour of these animals becomes the symbolic colour configuration for that generation.

For instance, the animal attendant on the generation of the current chief is the elephant (**ngolit**), and it follows that the symbolic colour for his generation is called **ghidangi**. The animal attendant on the previous chief Basha's generation was the defassa waterbuck (**dun**), and so their symbolic colour was the same as the coat of the **dun**, or **moji**-coloured. The animal for the generation before that was **ngolit** again, being repeated in alternate generations, and the one attendant of the generation of Galamogut, chief from four generations earlier, was the Patas (**elkumit**), which has a different coat-colour from the animals that have been mentioned up to now. The reasons for this are not clear, but the present generation of Bodi interpret it as possibly being attributable to the preference of that particular chief.

As I have already explained, the representative of a generation is the chief of that generation, but when a new chief is installed, a calf or bull with the coat-colour that symbolizes that generation is sacrificed. In this ritual the blood is painted on to the body of the chief, and further, a belt cut from the bull's peritoneum is placed on his neck. In this way a new generation is initiated.

Coat-Colours Used on Banners: Symbols of Regional Societies Bodi is a term used by the agricultural peoples of the mountainous region in the east to refer to the people who live in the eastern plains near the Omo River and who speak the **Me7en (Meken)** language, but the Bodi refer to themselves as **Me7en**. The **Me7en** in this eastern section are divided up into Chirim to the North and Mela to the South. In Mela there are three chieftains who each make up a regional group. In those regional groups that have a chief at their centre can be found banners or **longon**,

Table 12.7. Generation and cattle coat-colour in the Hana area

Generation	Chief's name	Animal	Cattle coat-colour symbolizing the generation
Present generation	Oikabur	*ngolit* (elephant)	*gidhangi*
Previous generation:			
1.	Basha	*dun* (defassa waterbuck)	*moji*
2.	Lugolonyi	*ngolit*	*gidhangi*
3.	Kangadibhoga	*dun*	*moji*
4.	Galamogut	*elkumit* (patas)*	*el-golonyi*
5.	Dhalichbhagolonyi	*ngolit*	*gidhangi*
6.	Bulasela	?	?

* Some informants gave the name of an imaginary animal (*olk'odit*), which is said to herd cattle and which stands for the colour *nyangaji*.

which symbolize that regional identity. They are of the same colour configuration of the **morare** of the chief in that particular region. While these banners are symbols of the prosperity of a particular region, they are also carried into battle with neighbouring enemies and placed in the centre of the field where councils to discuss battle strategy are conducted. These banners are made from long strips of the skin of bulls or male calves that are assembled into the colour configuration of the chief's **morare** and attached vertically to wooden poles (Figure 12.14).

In the Hana region, introduced in my earlier discussion of generation-sets, the **morare** of the present chief is **ludi** and **gelli**. The **morare** of the previous chief, Basha, was **ludi**, that of Chief Lugolonyi before him was **ludi** and Chief Kangadibhoga's before him was **gelli**, so the banners of this people have been made from bulls with coat-colours corresponding to these. In the generation of the present chief, Oikabur, a **bhongai de gelli** (male calf with **gelli** coat) was sacrificed shortly after his installation as chief and a banner was made. This calf's coat was **ludi** on the right side but **gelli** on the left so the right side of the banner was made to correspond with the **morare** of the chief and the left side to symbolize another moiety[6] different from the chief's but of the same generation. This right-hand banner is stored in chief's residence. Accordingly, since the Bodi do not have cattle of either **ludi** nor **gelli** coat-colours, they cannot make the banner that is the symbol of the regional group.

Coat-Colour Specific to Cattle Sacrificed at the Installation Ritual of a Chief

As noted earlier, in Bodi society there is the inherited title of **komorut** [Fukui 1988a, 1994]. The **komorut** is in charge of several rituals centered on rain-making and is regarded as a political authority. Although most political decisions are made by a select few in a council, mainly of elders, the name of the acting **komorut** is used through the generations to represent a regional group. Here I will refer to the **komorut** as 'chief' for the sake of convenience.

This chief formally assumes the position of **komorut** upon the performance of a series of set installation rituals. During the course of these rituals as many as ten head of cattle may be

Figure 12.14. A banner made from the hide of a particular bull

sacrificed, among which a certain number will be the cattle of a particular coat-colour. One could include in this series of installation rituals the rituals connected with the generation-set and the banners we examined earlier, but here I would like to focus on those cattle of a particular coat-colour.

First, there is the ritual called **bi de bena** ('stone' cattle). A **bhongai de golonyi** (red-coated male calf) becomes the object of sacrifice in this case, because the jewel the chief wears around his neck is red. The blood of this bull is painted on the chief and a strip of its peritoneum is placed around his neck. Then the jewel on the necklace is dipped in the blood before it is placed around the chief's neck.

In addition, **bhongai de koro** (male calf with black coat) and **luch de holi** (white ox) are sacrificed on this occasion. The **bhongai de koro** is also referred to as **bi de oonyun** (blood painting cattle) and its blood is painted on the chief's body and a strip of its peritoneum placed around his neck. It can be presumed that the reason **koro** is the coat-colour used is that the chief is associated with the gods and rain (both **koro**). Moreover, they kill a white ox, referred to as **bi de samuwonyi** (cattle whose coat is wrapped around the head). Only the chief has the blood of this ox painted on his body and a strip of its peritoneum placed on his neck, but all the people of his generation, in this case **ngolit** (elephant), wrap a long thin strip (**kudere**) of its skin around their heads. The reason why a white coat is chosen is unclear, but since, as I will explain later, this same white-coated cattle is sacrificed at times of illness, it may be because white is a symbol of 'virility.'[7]

Coat-Colour Specific to Livestock Sacrificed at the Outbreak of War

The Bodi[8] have often engaged in conflict with neighbouring peoples [Fukui 1988a, 1994], but when it comes to an all-out organized war campaign, they sacrifice a head of livestock of a certain coat-colour before initiating an attack. At these times what is significant is the fact that the coat-colour is decided according to the enemy in question.

In the case of a war being waged with the neighbouring northern Chirim people, who are also called Bodi and speak the same Me7en language, then bulls of **ludi**, **gelli** or **choburi** are sacrificed. The Bodi say that when you kill a head of **ludi**-coated cattle, a war will break out, but there may be more to this than the fact that **ludi** is the **morare** of their chief. However, they say that when war broke out with the Chirim during the time of the previous generation's chief, the two bulls **oi de choburi** and **oi de shindi** were sacrificed. In the case of a war breaking out with the pastoral Mursi, who are their southern neighbours and who are of the same Surma language group as the Bodi, it is said that **derdi**, **kalmi**, **bhaasi** or **bha-nyangaji** bulls are sacrificed. But in a recent war with the Mursi, an **oi de kordi** was sacrificed. Finally, in the case of a war breaking out with the Hamar people, who speak an entirely different Omotic language from the Bodi and live in an area three or four days' walk away, the bull to be sacrificed is said to be **muge-idi**-coloured, but in a recent war with the Hamar, an **oi de kolkoli-idi** (bull with **derdi** coat but **koro** back) was killed.

The three peoples mentioned above, Chirim, Mursi and Hamar, are all placed by the Bodi (here, the Mela people) in the category **baragara** (enemy), but the agricultural peoples living in the highlands, although they might be the object of attacks, are not included in this category. To borrow their expression, these agricultural people are called **su**, or like women and children, so they are not worth considering **baragara**. However, in March 1975, while I was living there, they waged a large-scale attack on the northeastern agricultural people (Male), killing several hundred of them, and seizing many head of cattle. At the time of this attack, they brought with them a **nyangaji** goat and held a ritual the day before the attack in which they buried it alive. When I asked why they chose a **nyangaji** goat, they responded that it was because it was mule-coloured. Actually, there is a significant metaphor at work in this response. Mules cannot survive in the lowland regions where the Bodi live, but because they can be found in the highlands of the agricultural people, the Bodi draw an analogy between the mule and the latter. By sacrificing a goat of the same colour as a mule, they enact a ritual one-sided declaration of war.

Coat-Colour Specific to Livestock Sacrificed at Rituals Associated with Marriage, Illness and Death

Marriage In Bodi society a marriage is sanctioned by the performance of two rituals, in both of which domesticated animals are sacrificed. The first, which corresponds to the announcement of an engagement, is the stage where the marriage partner is chosen. Before this ritual, the woman is called **ch'uba** (lover) but following it, she is referred to as **gangilit**. The cattle sacrificed at this time are called **bi de t'obidach**, which can be a male or female of any coat-colour except **cha7i**, **seroji** or **derdi**. In the ritual I witnessed, a **bi de golonyi** (red cow) and **luch de golonyi** (red ox) were sacrificed (Figure 12.15), but I heard that in addition to this they had also sacrificed, among others, a **bhongai de nyangaji** (orange-coloured male calf), **bhongai de golonyi** and **bhongai de bhile-nyangaji** (yellow-orange coloured male calf). From these examples, one can see that red- or orange-coloured animals tend to be selected.

After this, two to three or sometimes several months later, a marriage ritual is performed. Through the performance of this ritual the woman comes to be called **mokach** (wife) and formally moves into her husband's house to begin living with him. On this occasion goats called **tongo de salwach** are sacrificed. The coat-colours that are appropriate for this ritual are **koro**, **gidhangi**, **bhaasi** and **golonyi**, while **cha7i**, **tulk'a** and **derdi** are to be avoided.

Illness The Bodi sacrifice cattle of a specific coat-colour, not only to pray for their own recovery from an illness, but also when one of their favourite **morare** cattle falls ill. The best coat-colours for sacrifice in times of common illnesses are **holi** or **idho-idi** (colour like a white cloud), while those to be avoided are **gidhangi** or **koro**.

However, in cases of mental illness or epilepsy, cattle with coat-colours different from those sacrificed in times of common ailments are sacrificed. At such times, **bhaasi**, **koro** and **gidhangi** coloured cattle are selected. Whether it is male or female does not matter, but which coat-colour is selected follows consultation with a diviner. For this ritual in particular, the cattle are sacrificed in a thicket or near a river that is removed from the settlement.

Figure 12.15. A red cow sacrificed at the time of betrothal

The contents of the primary stomach and cuts of meat from several parts of its body are washed down the river.

Death When an older person with children dies, three kinds of cattle are generally killed. First, a head of cattle referred to as **bi de chaptach** is sacrificed, and after the blood is painted on the body, the skin is used to wrap the latter. What is of interest here is the fact that the coat-colour of this animal is different for each clan (Table 12.8). At the burial ritual, cows of coat-colours that are generally not sacrificed for other kinds of rituals become involved, such as **cha7i, seroji** and **tulk'a**. For the bereaved, the blood is thrown away and not used.

After a period of time has passed following the burial, a head of cattle called **bi de jendach** is killed. For this ritual, the

Table 12.8. Coat-colour of sacrificial cattle used for Bodi death rituals

Clan (Mela)	Coat-colour	Clan (Chirim)	Coat-colour
1) Ajit (Elma)	*cha7i*	101) Dombuloch	?
2) Artamat	*seroji*	102) Gingo	*seroji*
3) Bongito	*koro*	103) Gulach	*shindi*
4) Bosh (Limich)	*tulk'a*		*bholla-golonyi*
5) Bholei	*bhaasi, koro*	104) Kebeloch	?
6) Degit	*bhileji, moji, derdi nyabha-golonyi*	105) Kobukama	*ludi, choburi*
		106) Milroch (Ulai)	*ludi, choburi*
7) Dombuloch	*lu-golonyi,* gel-golonyi, bha-golonyi	107) Woino	*koro*
8) Gerf (Gongol)	*kilindi, seroji*		
9) Gulach	*nyangaji, derdi, shindi, moji*		
10) Irsach	*tulk'a*		
11) Jamer	*cha7i, tulk'a*		
12) Kilijach (Boka)	*tulk'a*		
13) Kolma (Gelem)	*cha7i, gidhangi*		
14) Kuriya (Saka)	*tulk'a*		
15) Marka	*koro*		
16) Moko (Bharga)	*koro*		
17) Melach (Munyach)	*kordi*		
18) Minegwach (Mis)	*seroji*		
19) Minshal	*cha7i*		
20) Ulai (Lugumaro)	*moji*		
21) Ulkui (Saigesi)	*tulk'a, mani-idi, nyabha-koroda, ludi*		

bereaved of the deceased choose whatever cattle they wish and present it as an offering to the spirit of the deceased. The more children one has the larger the number of cattle that will be offered. The coat-colours of these cattle are not specified. The blood is placed in the cattle's horn and the bereaved take turns pouring it over the stone near the head of the body; it is not used to paint the bodies of the bereaved.

The third animal to be sacrificed after someone has died is called **bi de onnyedo**. The blood and peritoneum are used just for the bereaved, but the coat-colour is not specified.

Coat-Colour Specific to Livestock Sacrificed During Rituals Associated with Agriculture

The Bodi rely on their livestock both economically and as an idiom to speak of other things, but they also cultivate sorghum and maize in swidden fields. In the rituals associated with the work in these fields and in rain-making rituals, livestock of specific coat-colours are sacrificed. These rituals include (1) the ritual of fallow-cutting, (2) the ritual of seed-sowing and (3) the ritual of rain-making. The first two together are called **bi de liba** (sorghum cow). Then, when the planting is finished and two months have passed, it becomes time for weeding the fields and (4) the ritual of insect-ridding is performed. At this time, they do not kill any livestock, but rather use blood drawn from the jugular vein of a head of cattle of a particular coat-colour. Then, when harvest time nears, they perform (5) the ritual to drive out birds and wild animals, and here too they use blood taken from the jugular vein as well as milk from the cow.

Ritual of Fallow-Cutting The Bodi use river-bank forest or secondary forests in the mountanious regions as fallow for swidden. When they cut them, they sacrifice a head of cattle of a certain coat-colour. This animal is called **bi de lim kena** (cattle for cutting trees). The most appropriate coat-colour for this sacrifice is **koro**, next in line comes **gidhangi**, but they also use **shimachi-gidhangi**, **bholigaasi**, **bhaasi** and **gelli**. It does not matter if the animal is male or female. Actual examples I collected when in the field included **bhongai de shimachi-gidhangi**, **mor de bhaasi** and **luch de golonyi**.

Ritual of Seed-Sowing When the dry season draws to its close (about March), the ritual of seed-sowing is performed in readiness for the rainy season. Livestock of specific coat-colours are sacrificed at this time. Although cattle are generally used, a he-goat of a similar coat-colour may be sacrificed. They call this animal **bi de duwon** (cow for seed-sowing). On this occasion, only male animals are sacrificed, and they say that only the two colours **gidhangi** nd **gidhangi-nyangaji** are used. These correspond to what they would think of as earth tones, but from what I heard, there are actually also cases where **kolai de golonyi** (red-coated he-goat) have been sacrificed. In the example I witnessed personally a **gidhangi** goat was killed (Figure 12.16), but the augury based on the entrails was not auspicious, so thereafter a **luch de gihangi** was killed and the blood sprinkled over a digging stick and sorghum seed before the planting began (Figure 12.17). Another three other examples I heard about all involved the sacrifice of a **bhongai de gidhang-nyangaji**.

Ritual of Rain-Making Later, when the grains begin to form and even when they mature well from start to finish, there are times when they are in danger of dying if not enough rain falls. At such times, the Bodi sacrifice cattle of certain coat-colours in order to pray for rain, and they call these cows **bi de tumonyi** (cow for the rain). In this sacrifice, it does not matter whether the animal is male or female, but the only colours permissible are said to be **koro, bhaasi** and **bholigaasi**. According to their understanding rain is regarded as **koro**, so they believe that by sacrificing a head of cattle of the same colour as rain, you can call the rain. However, when a head of **koro** or **bhaasi** cattle has been sacrificed and the rain still does not fall, they kill a female **bholigaasi**. This is related to the fact that lightening is seen as **bholigaasi**, so that by sacrificing a cow of **bholigaasi** coat-colour, they believe you can call the lightening. The reason they choose a female is based on their knowledge that female animals urinate more abundantly than males. In Bodi society there is a myth of the rain being the moon's urine, and it may be that this is related. In actual examples I heard about, **bhongai de bhaasi, mor de koro** and **luch de bhaasi** were sacrificed.

Figure 12.16. Goat killed during a seed-sowing ritual

Figure 12.17. Ox blood being sprinkled over digging stick and sorghum of the first seed sowing

Ritual of Insect-Ridding In mid-May, about forty days after planting, the insect-ridding ritual takes place. It is referred to as **mosit te libaun** (sorghum ritual) and involves taking insects out of the stems of the sorghum plants and praying that harmful insects will be driven out. After a ritual fire made from certain trees collected in the nearby thicket is set alight, they put blood taken from the jugular veins of a **seroji** cow and a **bhaasi**-coloured heifer into a gourd container. The two priests performing the ritual take branches and the blood-filled container and enter the fields. Then they dip the end of a small branch into the blood and walk around the entire field sprinkling the blood on to the sorghum and maize. However, why the coats of the cows used in this ritual for blood-drawing are **seroji** and **bhaasi** I do not know. It may be that there is no direct connection with coat-colouring, but what is interesting from the perspective of colour symbolism is that all the varieties of trees used in these **mosit** rituals, including (5), mostly have white blossoms.

Ritual to Drive Out Birds and Wild Animals Finally, when harvest time approaches, the ritual for driving out birds and wild animals is performed. This ritual is called **mosit te kangajun** (ritual for the monkey), for which blood and milk from cows of specific coat-colours is used. Though it is called 'monkey ritual,' while they are disturbing the fields, they perform another ritual at the same time to drive out the birds, especially the bird they call **woli**. They draw blood from the jugular vein of a living cow with **cha7i** (grue coat-colour) and sprinkle it over sorghum leaves they have brought from their own fields. They bring the fire lit during this ritual back to their own fields and burn the blood-sprinkled leaves. The smoke is recognized as **cha7i** (grue). It is believed that sickness escapes from the smoke of these fires.

Next, the ritual to drive out the **woli** bird is performed. At this time, milk drawn from a **shimaji**-coloured cow is placed in a cow's hoof and taken to a specific location. Once there, they cry 'Get out! Get out!' and leave the milk-filled hoof there. In this case, the reason they use a **shimaji**-coloured cow is that it corresponds in colour to the **woli** bird that is the object of their prayers.

Coat-Colour of Cattle Sacrificed During Rituals Accompanying a Move to a Specific Location

The Bodi move frequently and establish a herding camp in each new location. On average, they move about once every three months. Each location they move to must meet certain requirements before they choose it, and when they move to the two locations I will be discussing next, they perform a ritual in which they sacrifice cattle of specific coat-colours. Only after this ritual are they able to drink the water and put up the gate for their new herding camp.

When they move to **Lugi7a** in southern Bodi, they sacrifice a male **shimaji** calf. Then, when they move to Udurum, which is centrally located, they sacrifice a male **bhaasi** calf. The coat-colours of the cattle sacrificed are specific to each location and this is based on oral tradition, but it would be difficult at the present moment to provide any further explanation. The cattle that are killed when they move to certain locations are referred to as **bi de gerebai**.

Other Rituals Involving Specific Coat-Colours

We have looked at several rituals that involve specific coat-colours of livestock. Before concluding, I would like to look briefly at two further examples.

Ritual on the Occasion of Killing a Lion Bodi cattle are sometimes attacked by lions. At such times, they follow the lion's footprints, however long it takes them to track it down and kill it. After the lion has been killed, they skin it, and return home singing. As they near the settlement, the married women of the settlement come out to meet the young men and sprinkle ash over them as they return. Then, when they reach home, they draw blood from the jugular vein of the cattle whose coat resembles that of a lion and paint their bodies in the blood. Whether it is male or female does not matter, but the coat-colour is limited to **moji**, **bhileji** and **gushuri-idi** (lion-colour).

Coat-Colour Specific to Cattle Sacrificed After Seeing a Monster The Bodi say that there exists a large monster called **kilinkabur**.

Kilinkabur is the name of a clan whose people were wiped out when the ancestors of the current chief subjugated the land. The **kilinkabur** is an imaginary creature but there are people who claim to have seen it. In 1973, in a southern part of Bodi called Ginimaro, four children said they saw the **kilinkabur,** so as a result four head of cattle of a specific coat-colour (**bhaasi, koro, ludi** and **gelli**) were killed over two days and given in offering at the site where the **kilinkabur** was seen. The blood was painted on the four children's bodies and strips of the peritoneum placed around their necks.

Coat-Colours That Are Usually Not To Be Sacrificed

The coat-colours of livestock that are not to be sacrificed during normal rituals are established and include **ludi, tulk'a, seroji, derdi** and **kor-cha7i**.

The reason that **ludi** are not to be killed is that they are the chief's **morare** and consequently the symbol for the entire region. There were even people who told me they had never yet seen a **ludi**-coloured animal killed. They say that if a **ludi** were killed, it would lead to war and people's deaths. If a **tula-golonyi (tulk'a + golonyi**: red with black vertical stripes) were killed, there would follow a plague of mites. This is because mites are of the same pattern as **tula-golonyi**. In the case of **tulk'a,** they are not to be killed because they are of the same colour as the disease-spreading land-turtle. **Seroji**, being a mixture of all the various patterns scattered finely overall, is said to be the same as the supernatural creature called the **orome,** so it should not be killed. The **orome** is said to look like a python and breathe fire from its mouth. They also say that the rainbow is the breath of the **orome.** Finally, **derdi** and **kor-cha7i**-coloured livestock are not killed because 'it would make lands awkward.'

Conclusion and Overview

In the preceding discussion, I have examined Bodi cognition of cattle coat-colour polymorphism, their classification and genealogical knowledge of cattle by coat-colour, their folk genetics mediated by coat-colour and how they select bulls. I have also shown how coat-colour is important in a variety of social and ritual contexts not directly connected with breeding.

It is clear that Bodi actively seek to develop and maintain diverse coat-colour polymorphisms and that this maximizes their chances of ecological and social survival. It would not be an exaggeration to say that without this diversity Bodi society itself would be hardly recognizable. In order to maintain this diversity Bodi select bulls based on coat-colour. What is interesting about the bulls listed in the livestock 'family register' in Table 12.1 is that two of them were the **morare** that their respective owners identify with. This shows that they choose bulls with specific coat-colours based on their folk genetics.

It is difficult to know whether the coat-colour polymorphisms which we observe today are the result of pastoralist groups such as the Bodi having observed the 'natural' variation that occurs in wild animals during the domestication process and made it part of their culture, or whether it is that humans, having noticed what could be called accidental genetic polymorphism, were drawn to the idea of diversification and worked to increase this diversity and to control it, at the same time as they placed it at the centre of their cognitive system and world-view. Of course, we cannot say one way or the other until we have further evidence, but whatever the case may be, there is no doubt that the phenomenon of coat-colour polymorphism is deeply entrenched in many human cultures. The phenotypes of genetic polymorphisms which arose as a result of breeding have been reinforced through cultural elaboration, which in turn has maintained selection for diversity in animal coat-colour. In short, coat-colour polymorphism in domesticated stock has co-evolved with particular cultural traditions as part of the process of domestication.

The method I have used in this chapter has been to look at coat-colour polymorphism primarily through the filter of Bodi concepts. For this reason, the terms that appear are not always those that have been developed in the field of livestock genetics. Much research has now accumulated on the subject of the genetics of cattle coat-colour polymorphism [e.g. Searle 1968], but even this cannot account for all Bodi empirical knowledge. For example, in livestock genetics the polymorphism called 'roan' is elaborated by Bodi through such terms as **cha7i**, **shimaji** nd **gidhangi**. Also, the Bodi have an extremely detailed classification of what in livestock genetic studies is referred to as a 'spot.' They elaborate this concept through such terms as **ludi**, **gelli**, **choburi**, **kordi**, **bholigaasi**, and so on. The Bodi have also identified several coat-colour polymorphisms that are manifested as parallel phenotypes, and they have their own terms for them. For instance, one is the coat-colour they call **shinde**. This refers to cattle that are orange overall, but combined with the parallel characteristics of black around the eyes and on the testicles and tail. If the animal's coat is red, but the three latter parts are black, then it is called by the different name of **wosi-idi**.

As we saw in Table 12.5, many of the coat-colour polymorphisms are used either as objects of personal identification for individuals or groups, or as objects of sacrifice in ritual. As I explained, the coat-colours of the cattle used in the rain-making ritual are chosen because they are black like the rain or the same **bholigaasi** colour as the lightening. Similarly in the seed-sowing ritual the cattle sacrificed are chosen because they are the colour **gidhangi**, which is regarded as the colour of the earth. In this way, it can be seen that there is an underlying correspondence between Bodi empirical knowledge of livestock variation and heredity, and a world view perceived through the medium of colour and pattern.

There remains the question of what kind of standard is used in the selection of coat-colours that become the objects of personal identification for individuals and groups. Looking at the examples of cattle coat-colours that actually function as **morare**, and comparing them with those that have been excluded from becoming **morare**, their characteristics become clear. For instance, if one looks at the **morare** of **ludi**-colouring, which means white with black head and rumps, one can see that coat-colouring perfectly matches those characteristics. However, among the cattle that are called **ludi**, there are those who have roan-coloured sections mixed in with the white. Although these 'indeterminate coat-colours' are still referred to as **ludi**, they are not only not regarded as **morare** but also the bulls are never used for breeding.

As we have seen, many kinds of coat-colour polymorphism exist in East Africa which cannot be seen in textbooks, a fact most prominently demonstrated by the Bodi case. Together with the Bodi's folk classification and knowledge of subtle coat-colour variations, a problem for future research in the field will no doubt be how to translate the diversity of coat-colour polymorphism into the language of modern genetics. Questions specifically arise concerning the relationship of the following with rules of heredity: coat-colour that is the object of identification for individuals and groups; the categorization of cattle coat-colours that may even serve as 'clan' names, coat-colours, unrelated to the Bodi world-view that tend to be used for rituals; and coat-colours whose use is restricted. These questions are important because their investigation will help us understand further how livestock coat-colour polymorphisms and their cultural representations have arisen in the process of domestication.

Notes

This is the revised version of a paper originally published in Japanese with an English summary [Fukui 1988b].

1. The existence of livestock coat-colour polymorphism has been pointed out for such societies as the Nuer [Evans-Pritchard 1940], Dinka [Evans-Pritchard 1934; Lienhardt 1961], Jie [Gulliver 1952], Karimojong [Dyson-Hudson 1966], Dassanetch [Almagor 1972], Nyangatom [Tornay 1973], Mursi [Turton 1980], Datoga [Umesao 1966], and recently the Turkana [Ohta 1987].
2. I lived in Bodi land for thirteen months between 1973 and 1976, and more recently visited them for five short periods. I would like to thank the Institute for Languages and Cultures at Tokyo University for Foreign Studies, the Toyota Foundation and the Ministry of Education, Science and Culture in Japan, who enabled my field research among the Bodi to be carried out. I also wish to thank Professors Emeriti Tadao Umesao and Morimichi Tomikawa for encouraging me to study in a pastoral society, and Professor Emeritus Ken Nozawa for his advice on folk genetics following my field survey.
3. See Berlin and Kay [1969: 5–7].
4. For Bodi, colour configurations based on coat-colour are an essential element of their social and world-view, and the

author has in his possession detailed data on each of these. The data provided here are illustrative rather than comprehensive.

5. Why the cattle that become objects of personal identification are oxen is a question of some interest. Beidelman [1966] has developed the theory that through the act of castration they are made neuter and are viewed as sacred, an issue I hope to return to on a subsequent occasion. There are occasionally cases of cows with particular coat-colours that are the objects of women's identification. Such cows are called **ham**.
6. All members of Bodi society belong to one of the two moieties **komo** and **koluo**. Members of the **komo** moiety are in charge of rituals relating to rain-making, those connected with the god, and they can become regional chiefs. In contrast, members of the **koluo** moiety are designated to assist them. **Komo** are regarded more highly than **koluo**, but when it comes to marriage, brides must be chosen from the opposite moiety.
7. There are cases where the Bodi have their own interpretation for why certain coat-colours are used and other cases where the explanation is unknown. In the instances where it is unknown one has simply to make an intelligent guess. What does the white of the white cattle sacrificed at rituals for recovering from sickness or installing a new chief represent? In this case it can be supposed that it comes from the metaphor of white symbolizing 'virility'. If one looks at other things regarded as white, such as the sun, milk, semen, and so on, they are all things that sustain life. There is no question that the sacrifice of white cattle is at the core of the Bodi worldview, and I plan to analyze colour metaphors in a future paper.
8. Here I am referring in particular to the Mela people, who are a subgroup of the Me7ken who lived in the area where I stayed.

Bibliography

Almagor, U., Name-oxen and ox-names among the Dassanetch of southwest Ethiopia, *Paideuma* 18, 1972, pp.79–96

Beidelman, T. O., The ox and Nuer sacrifice: some Freudian hypotheses about Nuer symbolism, *Man* (N. S.) 1, 1966, pp.453–67

Berlin, B. and P. Kay, *Basic color terms: their universality and evolution*, Berkeley and Los Angeles: University of California Press, 1969

Conklin, H. C., Hanunóo color categories, *Southwestern Journal of Anthropology* 11(4), 1955, pp.339–44

Dyson-Hudson, N., *Karimojong politics*, Oxford: Clarendon Press, 1966

Evans-Pritchard, E. E., Imagery in Ngok Dinka cattle-names, *Bulletin of the School of Oriental and African Studies* 7, 1934, pp.623–8

——, *The Nuer*, Oxford: Clarendon Press, 1940

Finch, V.A. and D., Western, Cattle colors in pastoral herds: natural selection or social preference? *Ecology* 58, 1977, pp.1384–92

Fukui, K., Color concepts and classification among the Bodi, *Bulletin of the National Museum of Ethnology* 15 (4), 1979a, pp.557–65 (in Japanese with English Summary)

——, Cattle colour symbolism and inter-tribal homicide among the Bodi, in K. Fukui and D. Turton (eds), 'Warfare among east African herders', *Senri Ethnological Studies* 3, 1979b

——, Between ecology and history, in K. Seki (ed.), *What is historical anthropology?*, Tokyo: Kaimeisha, 1986 (in Japanese)

——, The Religious and Kinship Ideology of Military Expansion among the Bodi (Mela), in T. Beyene (ed.), *Proceedings of the 8th International Conference of Ethiopian Studies*, Addis Ababa, 1988a

——, Cultural device for the diversifying selection of animal coat-colour: folk genetics among the Bodi in southwest Ethiopia, *Report of the Society for Researches on Native Livestock* 12, 1988b, pp.1–46

——, *Cognition and culture: ethnography of color and pattern*, Tokyo: Tokyo University Press, 1991 (in Japanese)

——, Conflict and ethnic interaction: the Mela and their neighbours, in K. Fukui and J. Markakis (eds), *Ethnicity and conflict in the Horn of Africa*, London: James Currey, 1994

Gulliver, P. H., Bell-oxen and ox names among the Jie, *Uganda Journal* 16 (1), 1952, pp.72–5

Lienhardt, G., *Divinity and experience: the religion of the Dinka*, Oxford: Oxford University Press, 1961

Nozawa, K., 'Genetics of animal domestication', *Ecology and genetics of domestication*, Inuyama: Institute of Primatological

Studies, Kyoto University, 1982 (in Japanese)
——, *Animal population genetics*, Nagoya: Nagoya University Press, 1994 (in Japanese)
——, Domestication and coat-colour polymorphism, in K. Fukui (ed.), *Coexistence of nature and humans: coevolution between genes and culture*, Tokyo: Yuzankaku Shuppan, 1995 (in Japanese)
Ohta, I., Classification of livestock and characteristics of husbandry among the Turkana, in S. Wada, *Ethnological Studies in Africa*, Kyoto: Douhousha, 1987 (in Japanese)
Searle, A.G., *Comparative genetics of coat colour in mammals*, London: Logos Press, 1968
Sturtevant, W.C., Studies in ethnoscience, *American Anthropologist* 66 (2–1), 1964, pp.99–131
Tornay, S., Langage et perception:la dénomination des couleurs chez les Nyangatom du sud-ouest Éthiopie, *L'Homme* 8 (4), 1973, pp.66–94
Turton, D., There's no such beast: cattle and colour naming among the Mursi, *Man* (N. S.) 15, 1980, pp.320–38
Umesao, T., Families and herds of the Dagota pastoral society: an analysis of the cattle naming system, *Kyoto University of African Studies* 1, 1966, pp.173–206
Zeuner, F. E., *A history of domesticated animals*, Stroudsburg: Hutchinson, 1963

Chapter 13

Domestic Animal as Serf: Ideologies of Nature in the Mediterranean and the Middle East

Yutaka Tani

Introduction

I once observed, in a mountain valley of Kashmir, the following event. A group of Gujar Bakkarwala (nomadic goat-herders) were about to arrive at their summer quarters after many days of migration and had to cross a river. It was raining, and the river was swollen by meltwater. To arrive at the summer campsite before the dark, they had to get their flock of goats to cross it. But the goats didn't dare jump across the river. The shepherd knows that if he succeeds by force in pushing down any individual animal into the river, all the others will begin to jump into the water following the example of the first. It was, however, very difficult to thrust it down. Despite strenuous attempts on the part of one of the shepherds to push the first goat that came to hand, it agilely slipped away. He tried another animal, only to fail again. After repeated attempts, the terrified goats got beyond the shepherd's control. He became quite desperate. Already half an hour had passed before he succeeded in thrusting down one of the animals, having called on two of his colleagues to help. The goat in the water swam across the river. Following this example, the goats jumped one after another into the river, and in a few minutes all of them had completed the crossing.

A medallion from seventeenth-century Germany depicts the same scene of sheep crossing a river [Henckel and Schöne 1967: 528]. A large ram with a bell goes ahead in the river and the other sheep follow it. One sheep is leaping down into the river from a bridge. From the accompanying commentary, we learn that the medallion is meant to illustrate the ideal relationship between the leader and his faithful followers, according to the allegory of the ewes following the bell-wether. This bell-wether is known as a kind of flock-guide, castrated and trained by the shepherd.

Sheep become timid when they face a gap or stream or anything dangerous. When the leading group stops, the followers – jostling with each other – begin to flow sideways like a stream of water that has been intercepted. On such occasions, the flock sometimes divides into two, or changes direction and does not follow the course intended by the shepherd. The flock is usually accompanied and guided by only two shepherds. For them, such splitting or deviating of the flock from the planned course must be avoided so that time and energy are not wasted in restoring order. The shepherd expects in general that a flock on the move advances smoothly. It is convenient for the shepherd to have an individual to lead the flock, especially when the flock stalls in front of some obstacle. Moreover, the shepherd occasionally requires his flock to turn back or halt or change direction during herding time or seasonal migration. On such occasions, too, it is better for the shepherd to have a flock leader that will respond to his vocal orders and that will, in turn, induce the rest of the flock to follow it.

Of course, this technique is not indispensable for herding control. The Bakkarwala, who do not know of the guide-wether, use a basic repertoire of herding techniques to keep their large flocks under control. The technique of using the bell-wether is but one of the elaborations of herding techniques.

Bell-wethers are trained and used by transhumant shepherds in the Abruzzo region of central Italy, where I carried out field research in 1970. The Abruzzo shepherd culls most male yearlings except for a number retained for reproductive purposes. In the following year, he chooses one of these and castrates it to train as a bell-wether. After it has been castrated, the shepherd puts a short woollen rope around the wether's neck and walks along with it as if he were walking a dog. When the wether becomes habituated to him, the shepherd begins to instruct the wether on how to respond to vocal commands. To teach the wether what each vocal command means, the shepherd shouts the command and pulls the rope by hand in such a way as to make the wether learn how it should react. In the next stage, after the castrated ram has mastered responding to vocal commands, the shepherd replaces the short rope around its neck with a longer one and places it in the midst of the flock in the grazing field, where he continues to train it until it behaves obediently even in the flock. Having completed this training the shepherd removes the rope. Now, without any physical control, the trained wether goes ahead or stops or returns upon receiving the shepherd's vocal commands, even from a long distance [Tani 1977: 155–7].

The shepherd generally calls the wether **manziero** or **guidarello**, and gives it a special personified name (e.g. Generale, Mussolini). The shepherd sets great value on the guide-wether as his faithful follower, and the theft of such a wether can start a vendetta. The shepherd makes use of the guide-wether not only on those occasions when the flock has to cross a river during seasonal migration or to change its direction of movement. In winter, when the sheep must move through the snow, the wether makes a path in advance of the flock. Indeed, the usefulness of the guide-wether is considerable.

Apart from the practical value of the guide-wether in herd control, this technique is based on the following three items of practical knowledge: (i) sheep and goats are predisposed to follow a precedent; (ii) human beings can control the behaviour of domesticated animals by means of vocal signs taught in the context of a bond of intimacy between human and animal; and (iii) castration causes the ram or bull to become obedient. These elements of knowledge are common to shepherds everywhere. Though their simple combination does not necessarily lead to the adoption of the guide-wether, it nevertheless suggests the following idea: in order to control a flock on the move that does not understand human vocal orders, you may rear an individual which is tamed and instructed in responding to these orders. By controlling the instructed individual, you can, in turn, control the rest of the flock, because they are disposed to follow. It is worth asking how people first arrived at such an idea.

Moreover, the guide-wether holds a unique position in terms of the relationship between human beings and domestic animals. Having been instructed in the context of a bond of intimacy established between the shepherd and the wether, the wether can understand the words of the shepherd. By virtue of this relationship, the shepherd – the dominant partner – can control the flock through the wether's obedient response. Establishing a special relationship with the shepherd which the other flock members do not have, the guide-wether acts as an agent for the shepherd. In this context, it is on the dominant shepherd's side. At the same time, of course, it remains a member of the flock that he dominates. Moreover, it is castrated. Being thus alienated from the flock's reproductive process, it plays the role of mediator between the dominator (shepherd) and the dominated (flock). Most of the flock is female except for some reproductive males. In this context, the wether can also be regarded also as a guardian of the females. In this respect, the guide-wether plays a role which is functionally analogous to that of the eunuch (see Table 13.1).

Aristotle defined the eunuch as the mediator between the emperor and the people, and the transmitter of the emperor's commands to them. Being castrated, the eunuch is also the guardian of the harem. Because of his inability to have descendants, he wins the emperor's trust and is raised as the emperor's

Table 13.1.

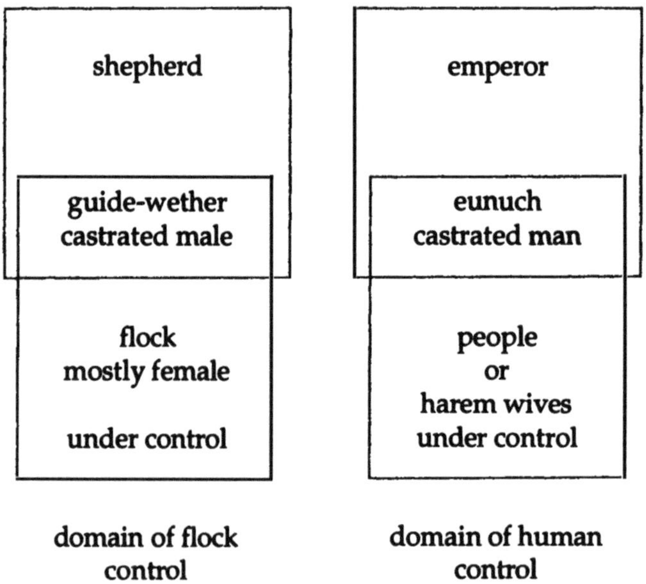

confidante. The position of the eunuch in the imperial order corresponds to that of the castrated male guide in the pastoral relationship. Very similar techniques of management are employed in the two domains, the control of domestic animals and the control of human subordinates. We know, moreover, that the eunuch appeared in the ancient empires of the Middle East before they made their appearance in China [Mitamura 1963].

In which domain, flock control or human control, was this management technique first invented? We have no historical evidence to answer this question. It is, however, most unlikely that the technique of employing the castrated male flock-guide was invented as an analogous extension of the human political technique of employing the eunuch. One would rather expect the reverse. If the geographical distribution of the guide-wether technique corresponds to the original area of the distribution of the role of the eunuch, we may suppose that the similarity between the two control techniques in the two different domains is not coincidental. And if it can be admitted that the same technique of control had been applied in the two different domains, we can suggest that they may have been ideologically equated. In this paper, finally, I would like to present an argument concerning cultural attitudes towards domestic animals and characteristic aspects of the domestication process in the Mediterranean and Middle East. However, I will first present data on the geographical distribution of the herd-control technique using the guide-wether, which I have reported elsewhere [Tani 1987a]. I shall then consider at what stage in the domestication process this technique was invented. And, taking into account the geographically restricted distribution of the technique and the parallel between guide-wether and eunuch, I would like to make some suggestions about the ideological framework which might have brought about such a parallel development.

Three Types of Caprid Flock-Guide and Their Geographical Distribution

When I first became interested in the cultural implications of the guide-wether and its geographical distribution, I could find little historical or ethnographic data on pastoral herd-control techniques involving the guide-wether outside the European region. In order to clarify the geographical distribution, I had to collect information by making extensive visits to various pastoral peoples, starting in Italy where I first met the guide-wether.

In Greece, the Sarakatsani and Vlach shepherds in Epirus and Thessaly, as well as Cretan shepherds, use guide-wethers. Sarakatsani call them **kliari ghissemia** [Campbell 1984: 14; Tani 1982: 14; 1987a: 20]. Going north through the Balkans this technique can be found up to the southern Carpathian mountains in the Romanian provinces of Sibiu and Dobrogea, where they call the guide-wether **batal**, a term derived from the Greek **batallos**, which means 'wether'. It is highly probable that the use of the guide-wether was introduced from the southern Balkan peninsula [Tani 1980, 1982; Hasedeu 1976].

Going east, among the nomadic Yöluk of Anatolia, who are of central Asian origin, it is reported that there is no evidence for the employment of the guide-wether [Matsubara 1983: 51]. On the other hand, nomadic Kurds near Hakkari in southeastern Turkey do use it. Sometimes, a castrated he-goat is also used [Tani 1989a: 189]. In Iran, the Baxtyâri nomadic people also use the guide-wether [Digard 1981: 57; Tani 1992, fieldnotes]. Baxtyâri however, use a he-goat in place of a sheep as the guide-wether. Going further east towards Afghanistan, data were collected from several different groups of pastoralists [Matsui 1980; Tani and Matsui 1980]; the Kandahari of Durrani Pashtun, the Uzbecki, the Arabi, and the Shaghni of mount Tajik, who come up every summer to the Shewa high plateau in Badakhshan. Except for the Shagni, who are rather more involved in agricultural activities, these nomadic groups use the guide-wether, not a ram, but a castrated he-goat.

Further east, in India, the Malwari in Rajasthan are ignorant of the technique of the guide-wether [Tani 1985: 78–84; 1987b: 81–7]. In Kashmir, Gujar Bakkarwala and Kashmiri shepherds also do not know of it. In Zanskar, Tibetan short-range transhumant pastoralists also do not use it. In Tibet, it is reported that the use of the guide-wether is unknown (personal communication, Matsubara). Nor is it known in Mongolia [Konagaya 1991].

Even though more evidence needs to be collected from different regions, it can be said that the use of the guide-wether (either sheep or goat) is distributed along the southern horizontal belt from Western Europe to Afghanistan, with the exception of areas where pastoralists of Turkic origin have migrated from the north to Turkey.

In the course of collecting my data, I discovered that there were two other types of flock-guide besides the castrated and trained male. I met the first in the northeastern Carpathians of Romania. It is called **fruntaşa**.

Fruntaşa literally means 'the female leader that goes in front'. To my surprise, shepherds said that there are many **fruntaşi** in one flock. This fact raised a question in my mind: supposing that the role of the **fruntaşa** is to go ahead in order to guide the flock, does this multiple leadership not weaken the guiding effect? This question was resolved by my observations of the **fruntaşa**'s behaviour and positioning during herding time. The **fruntaşa** had no disposition to go ahead.

The **fruntaşa** is tamed by rewarding it with food from the yearling stage and she comes close to the shepherd, responding to his calls. She is trained to respond to the shepherd's vocal commands to stop or return or go ahead. Only when she responds to his vocal order does she take the initiative, the rest of the flock following her [Tani 1989a].

A flock of sheep and goats tends to maintain its existing movement unless this is modified by some external force or inducement. The **fruntaşa**, like the bell-wether, is an individual who can induce a change in the movement of the flock by responding to the commands of the shepherd. The more numerous the inducers the more effective they are in causing the remainder of the flock to move. Since all the **fruntaşa** called by a vocal command make their initial move as a unanimous response, a plurality of **fruntaşa** does not bring about any problem of herd control.

The use of the female flock-guide is also found in southern Carpathia where the castrated male flock-guide is used as well. According to shepherds, the castrated and trained male guide is effective not only in the course of daily herding and seasonal migration, but also in making a path through the snow [Tani 1989a].

Simultaneous use of two different kinds of flock-guide is found in Greece. Besides the castrated male flock-guide, not the **fruntaşa** but another type of flock-guide, namely the **manari** (male) or **manara** (female), is used among Sarakatsani and

Cretan shepherds. Without regard to its sex, they select a certain male or female lamb from among the new-borns and keep it in the shepherd's hut in order to be tamed. Instead of allowing the lamb to suckle from its mother, they feed it by hand at first with milk and, after weaning, with barley. It is used to being called by name. Being fed usually with barley, it runs to the shepherd when called. The shepherds use it mainly to induce the flock to move in the morning or to return to camp in the evening. They keep at most two or three **manari/manara** in a flock. Even if it is male, the **manari** is not castrated but is later used for breeding purposes [Tani 1982].

This third type of guide-sheep could not be found among other groups of pastoralists of the Mediterranean and the Middle East whom I visited. But I cannot conclude that the third type is confined only to Greece. Further east, among the Malwari nomadic people of Jaipur and Bikaner in western India, it was found in one of four cases observed. Further northeast, the Gujar Bakkarwala, Kashimiri and Tibetans in Ladakh knew of no type of guide-sheep at all [Tani 1989b].

I have described above three types of flock-guide, along with their geographical distributions (see Figures 13.1 and 13.2). All of them have the same role: to respond to the shepherd's commands and to induce the flock to move as the shepherd wants. However, they differ functionally from each other. The male or female **manari/manara** type is tamed by provisioning to come to the shepherd when it is called. The male **manari** is not castrated. When it is grown, it must be separated – along with other sexually active males – from the main flock before the mating season. During the mating season, the male **manari**, being sexually excited, can get out of the shepherd's control and can thus cause trouble in the sexual context. The plural female type exemplified by the **fruntaşa** can be regarded as one way of eliminating the problem of the male **manari**. Another solution is to castrate the problematic male **manari** leading to the castrated wether type of flock-guide.

I shall now turn to consider the conditions which could have given rise to these alternative techniques in the long course of the domestication process.

All three types of flock-guide are based on the particular intimacy established between the shepherd and an individual sheep or goat. Human beings must have succeeded in inter-

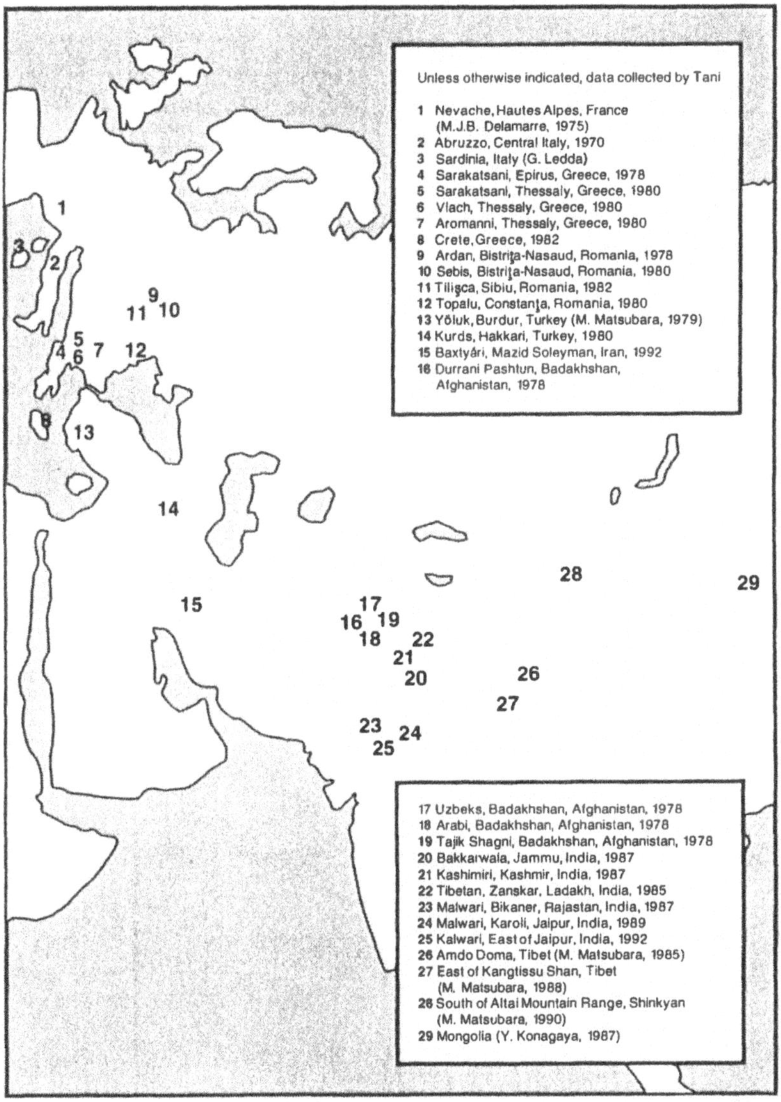

Figure 13.1. Geographical distribution of ethnic groups referred to in text

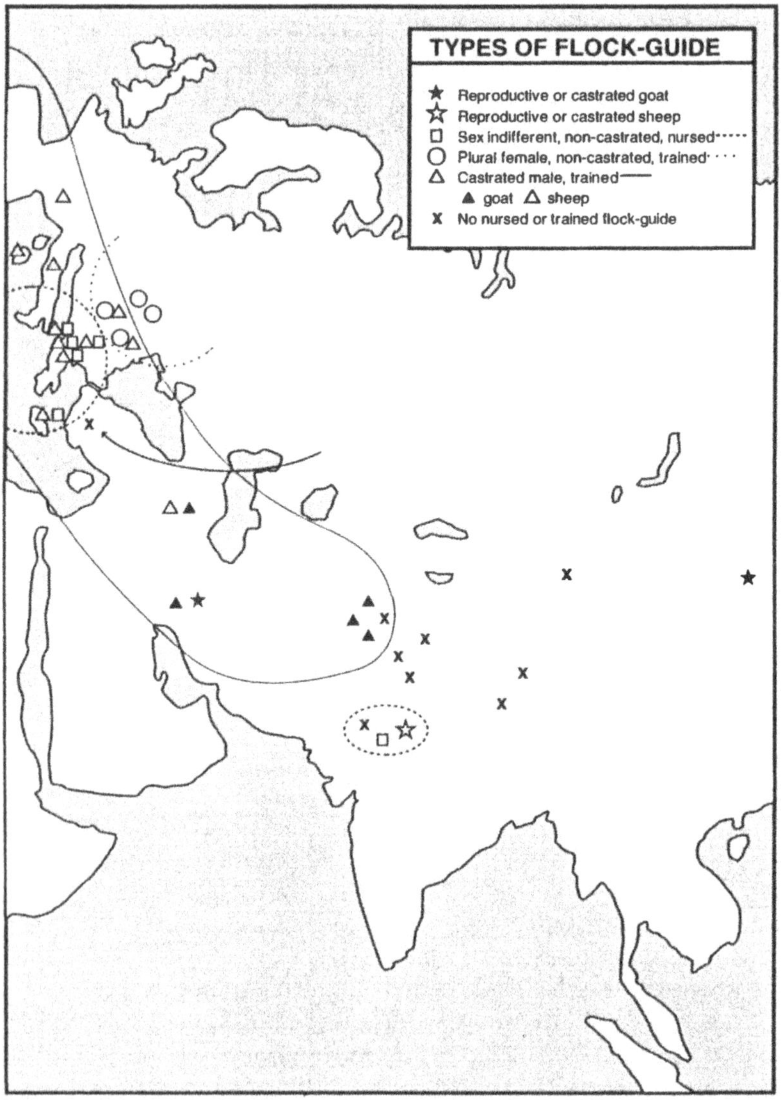

Figure 13.2. Geographical distribution of three types of flock-guide

vening on an individual level before these techniques could have arisen. The techniques generally serve to promote the docility of the flock. This means, however, that unified and homogeneous flocks must already have been in existence. As I shall go on to show, in wild and feral sheep and goats, this kind of homogeneity and agglomeration is not normally encountered. We must therefore ask at what stage and through what human intervention did the domesticated homogeneous flock arise. Bearing in mind that the Mediterranean and the Middle East, where the techniques of the flock-guide have been mainly adopted, are also archeologically the most important areas as regards the origins of caprid domestication, I shall now briefly address these questions.

Preconditions for the Flock-Guide Technique in the Domestication Process

It is through human intervention that animals have been transformed into domesticates which are physiologically and socio-behaviourally different from their wild or feral counterparts. The intimate relationship between animals and humans observed among pastoralists today is the result of a long history of such intervention. The domestication process has been cumulative, entailing the coordination of human involvement and the animal's biological constitution in each successive stage of human-animal interaction, and we find remarkable socio-behavioural differences between feral caprids and domesticated varieties.

(1) According to the reports of Rinley and Caughly and Shikano, female feral goats with a common range form home-range groups [Rinley and Caughly 1959; Shikano 1984]. Normally they disperse into small groups centred on the bond between mother and yearlings. The home-range groups come into contact with each other and sometimes an individual may leave its home-range group to join another group. Separate home-range groups do not merge into larger groups except when they are threatened by the presence of predators or humans. By contrast, the domesticated flock is unified and

homogeneous. Owing to this homogeneity, a large flock can be managed by no more than two persons. In order to establish such a flock, the 'vertical' mother-offspring bond must be weakened and a collective 'horizontal' tie somehow created by human intervention.

(2) As Ohta and Shikano have previously noted [Ohta 1982; Shikano 1984], a kind of closed group-oriented behaviour can be observed in the individuals of the domesticated flock. In other words, there is a sense in which individuals belong to the flock. In domesticated sheep and goats, individuals distinguish their own flock from other flocks and tend to keep with their own flock, avoiding mixing with different herds.

In the feral goat, individuals frequently transfer to other groups. Males wander across the boundary of female home-range group in search of mates. In the case of the circumboreal semi-domesticated reindeer, animals in one herd, under certain owners, often stray into other herds, which can cause conflict between reindeer owners. It is also reported that wild males sometimes come to mate with the semi-domesticated reindeer. In these cases, the genetic isolation that normally accompanies domestication does not occur.

(3) Individual sheep and goats in the wild state do not allow people to touch them or to approach them beyond a certain distance. By contrast, individuals in a domesticated flock make no attempt to escape from the shepherd. Moreover, they return to the human camp-site in the evening without any guidance from the shepherd. Ohta and Shikano describe this change as the transfer of the home range to the human camp-site. The fact that individual animals allow people to approach and even touch them marks the establishment of intimacy in human-animal relationships.

Comparing the socio-behavioural characteristics of the domesticated flock with those of the feral flock, there are therefore three major differences: in the size and homogeneity of flocks; in the group orientation of behaviour; and in the possibility of human intervention on the individual level. Through what kinds of human intervention did sheep and goats diverge from the wild state so as to take on these new socio-behavioural characteristics? Archaeozoologists have tried to determine the original region and period of the domestication of sheep and goats by analysing the evidence of animal bones excavated in the Middle East. They have judged the origin of domestication from the discontinuous changes generally found in these bone assemblages, from the relative proportions of different age and sex classes in the populations, from morphological characters, etc. Although the evidence is not entirely conclusive, archaeologists seem to have identified correctly the approximate date and place in which a drastic change occurred in the domestication process of sheep and goats. The archeological evidence, moreover, suggests the emergence of homogeneous flocks, closed-group-oriented behaviour and intimacy in relation with people. However, it is human intervention that has given rise to these characteristics of the domesticated flock. Bone evidence does not tell us directly what kinds of human intervention were actually taking place at the time of the origin of domestication. Unfortunately the behaviour involved in human intervention leaves no direct evidence that would enable us to reconstruct it and its effects on animal populations.

What clues can help us rediscover the nature of this intervention? If the behavioural characteristics of domesticated flocks are lost in feral populations, it must follow that it is the continuous technical activity of the shepherd that is responsible for domestication. We have some knowledge of the technical interventions of shepherds today. There are common techniques of intervention among pastoralists in different areas. Consciously or unconsciously, they repeat their daily and annual interventions in the flock and reproduce the socio-behavioural characteristics and the human-animal relationships found in present-day domesticated flocks. It is not impossible to determine which interventions contribute to the reproduction of these characteristics. Moreover, by comparing all the human interventions and techniques, we can judge to some extent whether some were preconditions for the invention of others. Through such analysis aimed at determining the sequential order of technical inventions, we may propose hypothetical stages in the development of human intervention and pinpoint those that were critical for the initial onset of domestication.

Without developing the arguments in detail [cf. Tani 1989c], I would like to describe some of the ways commonly observed among pastoralists in which shepherds intervene in the formation of the mother-offspring tie around the time of delivery, since these are crucial for domestication.

Lambing generally occurs in winter. The delivery may occur not only at night when the sheep are asleep at the camp-site, but also while they are being herded during the day. In the latter case, the mother-ewe separates from the flock on the move as she begins delivery of her lamb. The shepherds know well in advance when a pregnant ewe is about to give birth. Out on the pastures, he keeps in mind the place where the ewe will be separated for delivery.

Generally speaking, if the mother and her offspring are kept separated immediately after delivery for more than ten minutes, the mother-offspring tie cannot be established and the mother is reluctant to allow the offspring to suck [Craig 1981: 118]. After delivery, the mother ewe usually licks her new-born lamb. They cry to each other. Through these exchanges, they become able to identify one another. A short while later, the new-born lamb stands up and begins to walk. If the flock has not gone far away, they catch it up and rejoin it. If not, the mother-offspring pair remains alone and wanders around. In such a case, among the Mongolian [Konagaya 1991: 61–4], Greek Sarakatsani and Iranian Baxtyâri pastoralists [Tani 1992, fieldnotes], the shepherd picks up the new-born lamb by the legs and invites the mother to follow him by showing the lamb to the mother, eventually bringing the pair to his camp-site. In this way, he can bring both of them without severing the mother-offspring tie that is so important if the mother is to allow the lamb to suck.

In the camp-site, the shepherd separates the new-born lamb from the mother after a short while. From that time on, every morning and evening the lamb is given a chance to encounter its mother in order to suck. Through this artificially repeated daily encounter, the mother-offspring tie is maintained. This tie would, of course, be naturally maintained under wild conditions and does not depend on human intervention. Indeed, the shepherd's intervention might seem like meddling. However, it makes it possible for the shepherd to establish intimacy with the animals on an individual level. Through these interventions, the sheep come to allow people to approach and touch them directly.

The mother ewes go out daily into the pastures to graze. The lambs remain restricted to the camp-site. Separated from their mother, they are nursed collectively. I often saw a group of lambs running round the camp-site, attesting to the formation – as a result of collective nursing – of a horizontal co-generational group. This contrasts with the vertical mother-offspring relationship which predominates under wild conditions. Through the annual repetition of such intervention and collective nursing, I believe that the original disposition of sheep to disperse under wild conditions might have been lost, and the homogeneous and agglomerated flock created. At the same time, the flock would have been anchored to the camp-site, which became the focus of a new home-range for them.

I have presented above a sequence of three interventions following delivery: (1) picking up the new-born lamb and inviting its mother to the camp; (2) artificially controlling the mother-offspring encounter; and (3) collective nursing. The former two interventions promote the intimacy between human and animal on the individual level. Collective nursing weakens the mother-offspring tie and promotes the formation of the homogeneous flock, at the same time serving to anchor the flock to the human camp-site. These changes in socio-behavioural characteristics in turn promote closed-group-oriented behaviour and the consequent genetic isolation of the domesticated flock. On the basis of my observations of shepherd's interventions, and by contrast to conventional approaches to the origins of domestication, I believe that the annual repetition of this sequence of interventions is responsible for the creation of the domesticated flock.

However, the first intervention in the sequence – picking up the new-born lamb and inviting the mother to the camp-site – could not be successfully applied unless a degree of intimacy had already been established at least as great as that presently observed between the Sami reindeer herder and his semi-domesticated reindeer. It is therefore better to suppose that the first stage of domestication consists in the establishment of intimacy on the level of the herd *prior* to the origin of domestication as this is conventionally defined by archaeologists. Thus the archaeologically defined origin of caprid domestication, dated to around the tenth millennium BC, corresponds to the next stage of domestication in which the three interventions described above were already being applied. I refer to this as the second stage of domestication, which might have been initiated in the Middle East.

I have proposed a tentative hypothesis concerning the process of domestication of sheep and goats. The use of the guide animal is a technique for controlling the homogeneous flock. It is only after the advent of the homogeneous flock that the guide animal can find his or her role. Moreover, it is by virtue of the intimacy of the relationships between people and animals that the shepherd can rear the flock-guide. Both the advent of the homogeneous flock and the intimacy of the relationships with humans on the individual level which are prerequisites for the use of the guide animal can be regarded as consequences of the second stage of domestication. Hence, we can conclude that the use of the guide animal for herd-control might have been adopted after the second stage of domestication (the origin of domestication as archaeologically defined), along with a range of pastoral techniques: castration, milking and the allocation of orphan lambs to other mother ewes, all of which require the establishment of intimate and particularistic ties between the shepherd and individual animals.

I have suggested that there are three types of flock-guide. Even though they appeared after the second stage of domestication, it is impossible to determine historically the sequence of their appearance, because we lack historical data. Of course, between these types, we can find different degrees of elaboration. The **manari/manara** is separated from its mother at birth and is hand-nursed by the shepherd. At the beginning of the second stage of domestication, when humans began to intervene in the mother-offspring relation at the time of delivery and to undertake the collective nursing of lambs, it is not difficult to imagine that particular individuals (probably orphans) would have been selected for special care and nursed by hand, leading them to have relationships of exceptional intimacy with the shepherd. The prerequisites for rearing the **manari/manara** type of guide sheep seem to have existed at an earlier stage than for the other two types. On the other hand, the **manari/manara** type of technique is defective in so far as the uncastrated **manari** can create a problem in the sexual context. The plural female type could be regarded as a way of solving this problem by eliminating the trouble-making male **manari**. And the castrated male type can be regarded as another solution, achieved by castrating the problematic male **manari**. The **manari/manara** type of flock-guide is used to make the flock come close to the shepherd by virtue of the shepherd's intimate bond with the **manari/manara**. Its function is rather simple. On the other hand, the plural female type and the castrated male type satisfy many needs of the shepherd in the herding situation. And the castrated male guide-wether is not only the most stable flock-guide in the sexual context, but also makes a path in the snow. Taking these elaborations into account, the castrated male type of flock-guide seems to be the latest or the most elaborated technique of the three. And it seems to have been accepted in places where other types had previously been used, as a substitute or supplement to these other types, as is indicated by its maximum range of geographical distribution (see Figure 13.2).

Now, to return to my initial theme, we must recall the similarity between the guide-wether and the eunuch. I remarked above that the direction of analogic or metaphoric extension might have been from the domain of domesticated animals to the human domain. Regardless of which of the two domains furnished the model for analogic extension, we can now be sure that the technique of the guide-wether is found in the same areas of the Mediterranean and Middle East as where the political institution of the eunuch first appeared and from where it diffused.

Domestic Animals as Serfs

Even though metaphor has some objective background, as Lakoff and Johnson [1980] say, there are no a priori guidelines for a metaphorical extension. Metaphorical and analogical extension is often guided according to a culturally unique pattern of categorical proximity.

It may be pertinent to recall the following remark of Emile Benveniste, in *Le vocabulaire des institutions indo-européennes* [1969: 48]. He is concerned with the term **pasu** found in an ancient Vedic text, which designates movable property in the form of living domestic animals. He observes that there is one expression for quadruped **pasu** and another expression for bipedal **pasu**.

The former designates domestic animals, while the latter designates the subordinated domestic human beings as serf or slave. Under the term **pasu**, members of the two different semantic domains, domestic animals and subordinated domestic serfs or slaves, are classified in the same category. In this mode of categorization, we can discern the perspective of the dominator-owner, for whom both are identified as movable living property (see Table 13.2).

Table 13.2.

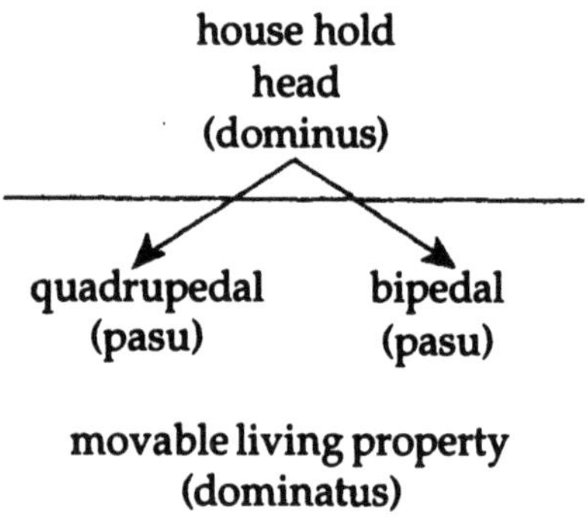

This viewpoint is **dominus**-centric. Given such a categorical view, it is easy to see how management techniques directed towards domestic animals could have been extended into the human domain.

I would like to show that such a viewpoint can be confirmed not only in the Vedic world but also in the ancient Sumerian world. Maekawa, a leading specialist on the Sumerian temple economy, has analysed certain groups of economic tablets from the Sumerian temple-state of Lagash in the third millennium BC. In his papers, he examines two different groups of economic texts, one consisting of records of the periodic provisioning of captured female slaves in the weaver's camp, and the other of records of cattle kept in the training centre to supply ploughing oxen to farmers under the temple economy [Maekawa 1979, 1980, 1982].

From the former text, Maekawa reveals the following facts. Weavers from among captured female slaves were generally not allowed to marry. However, they often kept their children.

Significantly, of these children, the daughters, when they had grown up, were recruited into the weaving groups of their mothers in order to make them, too, work as weavers. On the other hand, sons were castrated and left the groups of their mothers to work as labour-slaves pulling boats up the river. These sons were called **amarKUD**, a term whose meaning will be explained below. We can find here two different courses of life: daughters remain in the group of their mothers whereas sons, after castration, leave the group in order to be used for bodily labour.

The other set of records, of the training centre for supplying ploughing oxen, show that mainly male calves were brought to the centre annually from local cattle-breeders as tribute to the temple. In this centre, these male calves were first castrated, then trained for ploughing and distributed to the temple farmers. Maekawa observes that it was this castrated ploughing ox that was originally called **amarKUD** (=bull cut), the term only later being applied metaphorically to the castrated weaver's son.

It was preferable for cattle-breeders to offer male calves to the Temple as tribute, since it was important to retain female calves in order to secure the reproduction of the herd and future milk supplies. Otherwise, male calves, except for a few kept for breeding, were superfluous. In the Mediterranean and Middle East regions, they were generally castrated and either slaughtered for meat or used for transport or ploughing. There were thus two different courses of life according to sex, the female remaining in the maternal group, the male being castrated and leaving it. This pattern was not unique to the Sumerian world, but has been a customary feature of pastoral herd management throughout the Mediterranean and the Middle East. Note that the same pattern is attested among the children of captured female slaves in the Sumerian camps; the daughter remains in the maternal group while the son, after being castrated, leaves it. We are dealing here not only with a linguistic metaphor but also with a practical analogy (see Table 13.3).

We have no documentary evidence that would tell us when and where the technique of herd control using the guide-wether was invented. The distributional data presented above suggests an origin in the Near East around the Fertile Crescent or in the eastern Mediterranean, an area which significantly coincides with the areas where the eunuch had been adopted as a political

Table 13.3.

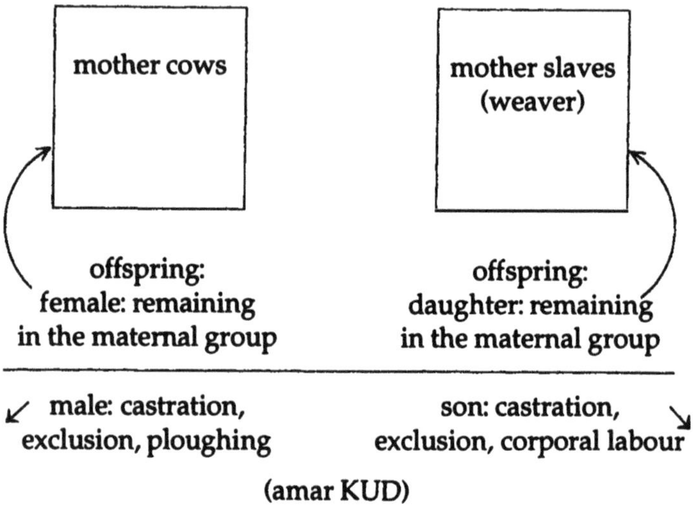

institution in the ancient imperial court.

The castrated wether is not only good to eat, but also good to control. Through this obedient mediator, the shepherd controls the flock. The idea of the guide-wether itself presupposes a managerial mentality. However, the more interesting and more important fact is that the technical practice applied in the domain of domesticated animals seems also to have been applied in the domain of human subordinates. Already from the Sumerian tablets, we find a remarkable example of the analogical extension of a practice from the domain of domesticated animals to that of human subordinates. But this categorical perspective which identifies human subordinates with domesticated animals exempts the human classifier from the class of humans to be classified as subordinate. The classifier stands over and above both subordinated humans and animals. In these examples of analogical extension, we can see an underlying attitude which places both domestic animals and human subordinates in the same category by regarding them as **dominatus** vis-à-vis the household head as **dominus**.

Now we know that this technique disappears in the western part of the Indian subcontinent, and that it is unknown to the Mongolian pastoralist. The absence of the technique from these regions may be due to the distance from the original centre of distribution. But we can also suggest that cultural-ideological barriers may have filtered out this technique on grounds of incompatibility. I would now like to argue that a management practice found among Mongolian pastoralists, documented by Konagaya [1989], suggests an attitude towards animals quite different from that found in the Mediterranean and the Middle East.

Among the management techniques of Mongolian pastoralists, we find many interventions which also have been adopted by Mediterranean and Middle Eastern pastoralists: intervention in the mother-offspring relationship at the time of delivery; daily, seasonal and annual patterns of partitioning the flock according to age, sex and other individual features; adoption of orphan lambs; visual and acoustic signalling when flocks are on the move, and so on. Many of these interventions may be independent of ethnic variation. However, apart from these basic techniques, Mongolian pastoralists have a special practice which appears to be meaningless from both the economic and the management points of view: they choose an individual from among the flock or herd, and neither shear it nor slaughter it, but leave it to die a natural death [Konagaya, 1989]. The individual is selected on the basis of some memorable or remarkable character: a reproductive male, the first ewe to give birth after a large loss caused by a winter with heavy snow fall, the first lamb to be born after the division of animal property, etc. The people believe that this auspicious individual assures the future prosperity of the flock or herd.

According to Lot-Falk, a similar practice is observed among Siberian hunters [Konagaya 1989; also Lot-Falk 1953]. They do not hunt indiscriminately. With regard to sable, for example, they believe that there is a chief among them that is especially large or beautiful. They take care that this individual is not hunted and killed, because it is thought to assure the future reproduction of the species. For each species, they believe that there is such a chief individual who is responsible for prosperous reproduction and should not be killed. Siberian hunters also believe that certain species (bear or tiger or deer) are responsible for the prosperous reproduction of the entire animal world.

Konagaya suggests that there are two levels of hunting avoidance: the first on the level of the whole animal world, i.e. the sum of all species; and the second on the level of each particular species [1989]. According to such beliefs, hunters who depend for their food on hunted animals should not kill the chief who is responsible for prosperity in order not to exhaust their food resources. But they may hunt all the other animals of each class at their disposal since their reproduction can be guaranteed so long as they do not kill the chief. Konagaya argues that the ritual prohibition on slaughtering a selected animal from the herd as practised among Mongolian pastoralists might be traced back to the tradition of the Siberian hunters. In the case of the Mongolian pastoralists, of course, the reproductive unit for whose prosperity the special individual is responsible is the flock owned by each pastoralist rather than the whole species to which it belongs.

Now leaving aside Konagaya's further argument, I would like, on the basis of these ethnographic data, to consider a further aspect of the underlying belief behind the ritualistic avoidance of slaughtering special species or special individuals among Siberian hunters and Mongolian pastoralists.

First, however, I would like to propose that another, third, level of avoidance should be added to the two levels proposed by Konagaya. From the evidence of enormous quantities of unbroken bones, Lot-Falk suggests that Siberian hunters did not break the bones of animals in consuming them. And she argues that the purpose of this avoidance of breaking bones was to ensure the revival of the animals they consumed. Such a ritual avoidance of destroying certain crucial parts of the individual body can also be found in a Mongolian practice on the occasion of their collective slaughtering in the autumn. They hang the intestine uncut, and keep the carcass in its original form. It is believed that this ritual procedure will ensure the rebirth of the domestic animal. It seems that at the level of the individual, as well as at those of the species and the animal kingdom as a whole, the same basic proposition is applied, namely that a specific part (or parts) is responsible for the reproduction of the whole to which that part belongs.

If we accept this argument, then three levels of existence may be proposed, corresponding to the three kinds of ritual avoidance of destruction in order to assure reproduction: (1) the level of animality in which animals of every species participate; (2) the level of species specificity shared by all animals of a particular kind (or flock or herd); and (3) the level at which all the components of each animal participate. And corresponding to each level, there is a special member responsible for the prosperous reproduction of the whole: (1) a special species of animal; (2) a special individual of the species (or herd); and (3) a special bodily part of the individual. Hunters should ensure the reproduction of their prey. Of necessity, therefore, they must avoid slaughtering the special members responsible for the reproduction on each respective level.

I am not suggesting that the ideas of Siberian hunters about the natural world can be reduced to such a schematic form as that set out above. It does seem, however, that such a belief concerning the reproductive process is at least one of the core features of their ideology of nature.

Conclusion

Every animal species has its own self-reproductive system. Human beings, who depend on natural resources, intervene in the life process of other species and consume them as foods. This means killing them. If so many are killed that the reproduction of the species is threatened, then human life itself is endangered. People must follow certain practices, based on a certain belief system, in order to ensure the perpetuation of the natural resources despite unavoidable destruction of animal and plant lives in the course of their exploitation of food. Not only Siberian hunters, but also various pastoralists and farmers, have devised systems of belief and ritual practice with precisely this end in mind.

The shared belief of Siberian hunters and Mongolian pastoralists can briefly be summed up in the following proposition: in each self-reproductive unit there is at least one representative member which is capable of, and responsible for, reproducing the whole unit of which it is a part. Each representative special member, whether special species (bear, tiger or deer), or a special individual which is especially large or beautiful, a special body part (bone or intestine) or special individual in the flock or herd (breeding male, first-bearing ewe or first-born kid), confers either animality or speciality or individuality, according to each level, on the other members that it serves to reproduce. The special member takes on a significant status and role in the group concerned. It is one of the constitutive members of the group, just as much as are other ordinary members. At the same time, however, it has a unique reproductive competence to perpetuate the group in the following generation. The idea is reminiscent of the contemporary notion of cloning. At any rate, according to this belief, in return for observing the prohibition on destroying the representative species, individual or body part, people are authorized to kill and consume what remains as a gift of nature.

This kind of ideology of nature presupposes that the reproduction of the animal world is independent of human control and protection, even if this independence might be vulnerable to the human desire to violate it. It might be argued that this interpretation of the animal world, and the accompanying rituals, naturally ensue from the predicament of the hunter who pursues prey which do not readily reveal themselves and their world to people and hardly surrender themselves willingly to human control. Yet – and this is important for our argument – Mongolian pastoralists, who do keep their domestic animals under their control, nevertheless maintain a set of beliefs and practices which are essentially similar to those of Siberian hunters.

It has been proved archaeologically that Mediterranean and Middle-Eastern people achieved the domestication of sheep and goat. Before the domestication of animals, of course, these people had been hunters. We do not know what kind of ideology of nature was held by the hunting ancestors of the people of these regions. However, among the descendants of the pastoralists who had domesticated sheep and goats, we find no trace of any ritual prohibition on slaughtering a special individual, as found among Siberian hunters and Mongolian pastoralists. Middle-Eastern and Mediterranean peoples seem to have developed their ideology of nature along quite different lines from their Mongolian counterparts. My discussion in previous sections suggests several lines of evidence for reconstructing the basic attitudes towards domestic animals of the peoples of the Mediterranean and Middle East. Given such a view, no possibility is recognized for the animal's life-process to be in any way independent of human control. Interestingly, the discourses concerning animals in the Old Testament seem to accord in their underlying tone with this view.

According to the Bible, in the first period of Creation, when man was entrusted by God with control over the animals, he was offered all the seed-bearing parts of plants and trees, whereas animals were offered the foliage. According to the first dietary prescription, God imposed vegetarianism on man.

In the next stage, however, after the landing of Noah from the ark, God lifts the vegetarian prescription of Eden and allows man to eat all animals as food, but with the reservation, accompanied by the words that man's heart contrives evil from infancy, that man is not to eat any blood along with the flesh. God declares that any man who eats 'life', that is to say blood, will be required to pay with his own life. This demand is reiterated in Leviticus. Here it is said that the blood of life is given to man to be redeemed before an altar and not to be eaten with flesh. To eat blood implies a sacrilegious human appropriation of the property of God. These declarations may be summed up in the following propositions: (1) the blood of life belongs to God; (2) man should not eat blood with flesh: (3) blood is given to man to be redeemed before an altar.

In the third dietary prescription found in Leviticus, where the rules of the edibility and inedibility of animals are presented, we find the principle that to eat any flesh along with blood is impure and sacrilegious. Now, setting aside my previous attempt [Tani 1991] to analyse these dietary regulations, together with my critical comments addressed to the analyses of Mary Douglas [1966, 1975], we may observe that all carnivorous beasts and birds were judged to be impure, just as is the man who eats blood with flesh, because these animals do not preserve the life-blood when they eat the flesh of the animals they consume. Underwriting the prescriptions of Leviticus is the general proposition that man is allowed to eat the flesh of animals with the reservation that the blood must be returned to God because, as the essence of life, it belongs to him.

Siberian hunters and Mongolian pastoralists justify their killing and consumption of animals by prohibiting the slaughter of a special individual held to be responsible for the herd's prosperity. By contrast, people in the Old Testament world were permitted to consume the flesh of animals so long as they gave back the life-blood of the animals to their original owner-creator, God. The key role of assuring the reproduction and prosperity of life, which is carried by the special individual of each species in the Siberian hunter's belief system, is taken by the transcendent God (**dominus**) in the pastoral world depicted in the Old Testament. In the Creation, God created man and animals and entrusted control of animals to man because man was created in the image of God. This is the Biblical interpretation. But an anthropological interpretation inverts this logic, suggesting that God was created in the image of the human household head (**dominus**), who keeps domestic animals and human domestic serfs (**dominatus**) under his control. As the unique God of the Hebrew, he may take a transcendent position over all animals and man. He will be viewed as having entrusted the job of controlling animals to his subordinate, man, because he is created in the image of the household head, who often entrusts his flock management to his subordinate shepherd. And likewise, he will be viewed as having an original proprietary right to his subordinated domestic animals and serfs.

In the discourses of the Old Testament concerning animals, we find a similar attitude to domestic animals as to human domestic serfs. Given such an attitude, it is not surprising that the same technique of control has been applied equally to both of these domains.

I have tried, in this paper, to demonstrate a characteristic feature of the process of animal domestication in the Mediterranean and Middle East and to bring out its ideological implications. These implications, I suggest, are of some significance for understanding the origin of the dichotomy between nature and culture that lies at the heart of the Western tradition of thought.

Bibliography

Benveniste, E., *Le vocabulaire de institutions indo-européenes*, Tome 1. Paris: Editions de Minuit, 1969

Campbell, J.K., *Honour, family and patronage*, Oxford: Clarendon Press, 1984

Craig, J. V., *Domestic behavior-causes and implications for animal care and management*, New Jersey: Prentice-Hall, 1981

Delamarre, M.J.B., *Techniques de production: l'élevage*, Paris: Edition de Musee Nationaux, 1975

Digard, J-P., *Techniques des nomades baxtyâri d'Iran*, Cambridge: Cambridge University Press, 1981

Douglas, M., *Purity and danger*, London: Routledge and Kegan Paul, 1966

——, *Implicit meanings*, London: Routledge and Kegan Paul, 1975

Hasedeu, B.P., *Etymologicum Magnum Romaniae*, Bucresti: Editura Minerva, 1976

Henckel, A. and A. Schöne, *Emblemata, Handbuch zur Sinnbild Kunst des XVI und XVII. Jahrhunderts*, Stuttgart: J.B. Mertzlersche Verlag Buchhandlung, 1967

Konagaya, Y., *Concepts of animal resource among the Mongolian people* (oral report in Japanese), in *Symposium on the domestication process* (organized by Ken Matui, 16th December 1989 in Kyoto), 1989

——, *Spring of Mongol-Anthropological sketch book*, Tokyo: Kawadeshobo shinsha, 1991 (in Japanese)

Lakoff, G. and M. Johnson, *Metaphors we live by*, Chicago: University of Chicago Press, 1980

Ledda, G., *Padre padrone: L'educazione di un pastore*, Milan: Feltrinelli, 1975

Lot-Falk, E., *Les rites de chasse*, Paris: Gallimard, 1953

Maekawa, K., 'Animal and human castration in Sumer, I', *Zinbun* 15, Research Institute for Humanistic Studies, Kyoto: Kyoto University, 1979, pp.95–140

——, 'Animal and human castration in Sumer, II', *Zinbun* 17, Research Institute for Humanistic Studies, Kyoto: Kyoto University, 1980, pp.1–56

——, 'Animal and human castration in Sumer, III', *Zinbun* 18, Research Institute for Humanistic Studies, Kyoto: Kyoto University, 1982, pp.95–122

Matsubara, M., *Ethnography of the nomadic Yöluk in Anatolia*, Tokyo: Chuoukoronsha, 1983 (in Japanese)

Matsui, K., *The pastoral life of the Durrani Pashtun nomads in Northeastern Afghanistan*, Institute for Research in Humanities, Kyoto: Kyoto University, 1980

Mitamura, T., *The eunuch*, Tokyo: Chuokoron, 1963 (in Japanese)

Ohta, I., 'The man-animal interaction complex in the goat herding of the pastoral Turkana', *African Study Monographs, Supplementary Issue* 1, Kyoto: Kyoto University, 1982, pp.13–42

Rinley, T. and G. Caughly, 'A study of home range in a feral goat herd', *New Zealand Journal of Science* 2, 1959, pp.150–70

Shikano, K., 'On the stability of the goat herd in the pastoral Samburu', *African Study Monograph, Supplementary Issue* 3, Kyoto: Kyoto University, 1984, pp.59–69

Tani,Y., 'On sheep flock management techniques among the transhumant shepherds of central Italy', *Society and Culture of Europe*, in Y. Aida and T. Umesao (eds), Research Institute for the Humanistic Studies, Kyoto: Kyoto University, 1977 (in Japanese)

——, 'Man-sheep relationship in the flock management techniques among north Carpathian shepherds', in Y. Tani (ed.), *Preliminary report of comparative studies on the agro-pastoral peoples in southwestern Eurasia*, Research Institute for Humanistic Studies, Kyoto: Kyoto University, 1980

——, 'Implications of the shepherd's social and communicational interventions in the flock, from the field observation among the shepherds in Roumania', in Y. Tani (ed.), *Preliminary report of comparative studies on the agro-pastoral peoples in southwestern Eurasia*, Research Institute for Humanistic Studies, Kyoto: Kyoto University, 1982

——, 'Two types of human intervention in the sheep flock: intervention into mother-offspring relationship and raising of the flock leader', in Y. Tani (ed.), *Domesticated plants and animals of the southwestern Eurasian agro-pastoral culture complex*, Research Institute for Humanistic Studies, Kyoto: Kyoto University, 1987a

——, 'Preliminary notes on the flock management techniques of the Bakkarwala and the Ladakhi shepherds in northwestern India', in S. Sakamoto (ed.), *A preliminary report of the studies on millet cultivation and its agro-pastoral culture complex in the Indian subcontinent-1985*, Kyoto: Kyoto University, 1987b

——, 'The geographical distribution and function of the sheep flock leader', in J. Clutton-Brock (ed.), *The walking larder*, London: Unwin Hyman, 1989a

——, 'Group organization and herding techniques of the Bakkarwala in Kashmir', in S. Sakamoto (ed.), *Preliminary report on the studies on millet cultivation and its agro-pastoral culture complex in the Indian subcontinent*, Kyoto: Kyoto University, 1989b

——, 'Two stages in the domestication process' (oral report in Japanese), *Symposium on the Domestication process* (organized by Ken Matui, 16th December 1989 in Kyoto), 1989c

——, 'Mode analysis of the dietary narrative in the Pentateuch', *Zinbun*, Institute for Research in Humanities, Kyoto: Kyoto University, 1991, pp.315–53

——, and T. Matsui, 'The pastoral life of the Durrani Pashtun nomads in Northeastern Afghanistan', in Y. Tani (ed.), *Preliminary report of comparative studies on the agro-pastoral peoples in southwestern Eurasia*, Research Institute for Humanistic Studies, Kyoto: Kyoto University, 1980

Chapter 14

Crops, Techniques, and Affordances

François Sigaut

For about twenty years now it has become fashionable in many parts of France, as in many other so-called developed countries, to organize rural festivals where bygone works and crafts are demonstrated before a public of mostly city-dwellers. The number of such festivals in France must now be in the order of several hundred each year. Most of them are of course held in summer, the tourist season, although some must of necessity take place at another time of the year, as for example cider-making festivals in October or November. In most festivals, ploughing is performed with old wooden ploughs harnessed to horses or oxen, grain is harvested with scythes and harvester-binders, threshed with flails or in steam-driven threshing machines, and so on. There are sheep festivals, where the animals are shorn with hand-shears, the wool washed and combed in the old way, spun by hand, and woven or knitted on hand looms or old-fashioned machines. There are flax festivals, donkey festivals, blood-sausage festivals. Last year, a horse-carriage race was organized between the seaport of Boulogne and Paris (250 km) to revive the **chasse-marée** of old, when fish had to be carried from the Channel ports to Paris as quickly as possible before the coming of the railways. There is even a **Fête de la Bouse** (Cow-dung Festival), where cow droppings are collected and kneaded into cakes for use as fuel, as was still regularly done in the wood-deficient parts of the Poitou marshes in the 1950s.

One of the most interesting festivals I have attended is the **Fête du Millet** that has taken place in the village of Aizenay since 1989.[1] I knew that millets, mainly *Panicum miliaceum* L, had been a crop of some importance in western France in the nineteenth century and before. I also knew that many museums in the region kept a few wooden mortars and pestles said to have been used for the husking of millet; but I had never met anyone who had used such a mortar, and it appeared that no one alive had ever done so. So it came as a surprise to learn that at Aizenay and in the surrounding area, there were indeed a number of such persons alive and well, not even very old but in their sixties and seventies. The tasks they demonstrated were also much more than just the use of the pestle and mortar. The ears of millet were first harvested by hand, one at a time, with the help of a pocket knife. The grains were then separated from the ears, not by threshing but by rubbing the ears between the naked feet. Afterwards came a first winnowing in a hand-driven winnowing machine, the husking in the pestle and mortar, a second winnowing with the help of a flat basket named **guenotte**, and finally cooking in a mixture of water and milk. The resulting **bouillie**, locally **pilaïe** (a kind of porridge), was sold to onlookers.

Of course, all this was not exactly as it was in older times. The standing crop of millet had received a dose of herbicide to make it the proper yellowish colour one month before maturity; for in September, the normal time when millet ripens, the tourists would be gone. And the **pilaïe** was in fact prepared not with the grains just harvested, but with millet imported from overseas (Colorado and Argentina were mentioned) and husked and cleaned beforehand in a nearby industrial mill, the only one processing millet in France. It was not an attempt to cheat, however. These adaptations to changed times were necessary to make the show possible. And if the show was indeed a show, there was an incontrovertible authenticity to it, because the tasks were performed in the only way people knew how to perform them, i.e. as they were performed in the 1950s and early 1960s, just before millet cultivation was abandoned.

The main reason why I was so excited by the **Fête du Millet** was that it featured, right in the middle of a typically French village, techniques of grain processing that, if observed out of context, would undoubtedly have been described as typically African or Asian. Of course, any such exotic influence was out of the question. And so was any argument of backwardness, for as far as grain-mills are concerned, the region was neither especially advanced nor backward. Water- and windmills had developed there in the early Middle Ages like everywhere else in western Europe, and people had long been in the habit of carrying their breadgrains to the mill to have it ground into flour every other week or so. Still, those people perfectly acquainted with mills for centuries stuck to the pestle and mortar for the husking of millet.

There are a number of possible explanations. Millet was grown in comparatively small quantities and was preferred by the poor because its cultivation required more manpower and less capital than bread cereals. In addition, husked millet does not keep well, so that the husking cannot be done in advance in large quantities as with the grinding of breadgrain into flour (good and well-cleaned flour can be stored from a few weeks to a few months, according to the quality of the grain, the weather, etc.). All these explanations are valid, and others may be valid too, but it is not my intention to discuss them here. What I found fascinating in the Aizenay show was that it looked like a kind of thought-experiment, as they say in physics. Just imagine that for some reason, millet and not bread cereals had become the staple food of Europe for the two last millennia. What would the development of mills have been like? And what of the development of industry, so dependent on the mill's waterwheel for power until well into the nineteenth century? The whole history of Europe would obviously have been different. Just replace wheat and rye with millet, and Europe is not Europe any more.

As far as Europe is concerned, this is a historical fiction of course. Things did not happen this way. But something of the sort did happen in China, since millets and rice have been the staple cereals there for millennia. Indeed, the development of mills and of machinery and of many other things has been very different in China from the West. Joseph Needham was perhaps not the first to ask why this was so, but he was certainly the first to address that question seriously and to devote his entire life to trying to answer it. I have a boundless admiration for the works of J. Needham and his followers [e.g. Needham and Wang Ling 1965]. I must confess, however, that I have never found their answers quite satisfying. Perhaps the Needham Question is simply too global to be really answerable. Since it is by no means futile or meaningless, however, we ought to be able to transform it or split it up into smaller, more manageable, questions. The **Fête du Millet** was for me a live demonstration of what one of these manageable questions could be. If we follow up through time and space the techniques of grain processing for each cereal, including mechanics and biochemistry, what do we get?

Hulled Grains and the Winnowing Machine

It is well known that a first generation of machinery made its appearance toward the beginning of our era both in the West and in China. One of these machines was the rotating mill, an extremely important innovation, since it made possible the harnessing, first of animals, then of water-power, in grinding grain. The first evidence for watermills in the West dates back from the last half-century before the Christian era, and the first evidence for the adaptation of the waterwheel to industries other than milling dates from the tenth and eleventh centuries. For nearly one millennium afterward, the water-wheel was to prove an increasingly important source of power for all European industry. This development did not occur on the same scale in China, one possible reason being that the husked grains of rice and millet do not keep well. The consequences are that the husking has to be done day in, day out, using small quantities at a time, and so the processing of millets and rice tends to be more firmly retained inside the household, which clearly is an obstacle to the development of larger machinery.

But if the use of husked grains tends to limit the development of mills, it tends to enhance the development of another machine, namely the winnowing machine. This is at least one possible explanation of the fact that the winnowing machine, although used in China from the Han dynasty (2nd century BC to 2nd century AD), was not known in the West before the first years of the seventeenth century. For contrary to what occurs with wheat and other free-threshing cereals, for which winnowings are done only between threshing and milling to clean the grain in bulk, the processing of paddy and other hulled grains requires a number of winnowings, not only after threshing, but also after each of the several poundings or millings deemed necessary to get a thoroughly husked and cleaned grain.

As a problem in the development of machinery, the history of the winnowing machine is both important and ancient. It is important because in the West, the winnowing machine was the first agricultural machine proper to be invented after the water-mill itself, although much later. It thus opened the way for the development of modern agricultural machinery: the Scotsman Andrew Meikle, who designed the first successful threshing machine in 1785, was the son of a millwright who had been the first maker of winnowing machines in Britain around 1710. And the problem is ancient because as early as the 1780s, the origins of the winnowing machine has already become a mystery. One of the few things known with any certainty was that it had come to Britain from the Netherlands. But beyond this, people could only speculate, and one of their first preferred speculations was that the machine was borrowed from the East Indies.

This mystery has now been largely solved. A German linguist, Uwe Meiners [1983], has shown that there were at least two different types of early winnowing machines in Europe, appearing in two different places: the Netherlands, where a first patent for it was obtained in 1604, and Switzerland, where the machine was in common use among peasants in the canton of Zurich by 1664 and may have been listed in a German-Latin dictionary of 1592. There is no hard evidence for whether the Swiss and Dutch machines were invented independently or not. But the chronology, together with the fact that the early Swiss and Dutch models did not resemble either each other or Chinese machines, are strong arguments against the Far Eastern theory. On the other hand, no evidence has been found that Europeans became acquainted with the Chinese winnowing machine before the 1730s, a time when European models had begun to be actively extended throughout the continent. What made a Far Eastern theory so attractive in the 1780s was probably that engineers and scientists only became aware of the winnowing machine over a century after it had first been put to use among craftsmen and peasants. It was not referred to in the literature before 1709 and 1717, which suggested an introduction not much earlier that 1700 and made a direct borrowing from China quite plausible. It remained plausible until the publication of Meiners' book in 1983.[2]

Thus, I believe that China and the West developed a first set of winnowing machines at approximately the same time, toward the beginning of the Christian era, because their economic and cultural conditions were then similar. But these machines were different because of different crops, food habits, and so on. With an economy based on hulled grains (millets, rice and barley, requiring more and more frequent winnowings), the Chinese had every incentive to contrive a machine for winnowing, which was not very difficult to achieve, either by Chinese or by Roman engineering standards. On the other hand, with an economy where wheat and bread became more and more dominant, the Greeks and Romans were interested in other machines, especially the flour-mill; they also made significant innovations in oven-building, for instance. By the sixteenth century, however, the situation was changing fast. Wheat bread was then firmly established as the food of the rich in most of Europe and as the food of most townspeople, including workers, in countries like France and England. In many countries, however, the peasants and the urban poor had increasingly to content themselves with less expensive staples: rye, barley, oats, millets, chestnuts, buckwheat, to which were later added maize, and later still potatoes. Some areas remained comparatively untouched by this change, as for example German-speaking Switzerland and the neighbouring region of Germany, where spelt, an ancient cultivar of hulled wheat, remained the main food cereal. There is now some pretty good evidence that in Switzerland and in southwest Germany, winnowing machines were used mainly for cleaning spelt after husking in the mill, whereas in the Netherlands, they were used mainly for cleaning buckwheat and pearl-barley. Thus in early modern Europe as in ancient China, there is little doubt that the winnowing machine was a specific invention answering the specific needs of processing hulled grains.

In technology as in other fields of anthropology, I do not believe in one-sided explanations, and I am not trying to propose one here. I am not trying to say that hulled grains 'explain' the invention of the winnowing machine: that would obviously be absurd, if only because hulled grains were the main crop in numerous countries where nothing of the sort ever happened. As is the case with every invention, a host of factors were involved. Hulled grains were only one among many, but they *were* one factor, and it is by no means absurd to say that without hulled grains the winnowing machine would not have been invented, at least not when and where it was. Neither would the threshing or harvesting machines, since the way for each invention was cleared by the preceding one. In this sense, the fact that grains are hulled or naked is a factor without which the history of machinery and of industry cannot be fully understood. What I want to emphasise here is (1) that it is an environmental factor, even if not usually recognized as such, and (2) that technological studies were the only means of identifying it.

Grinding Stones and Wooden Mortars

There is of course much more to say on grains than whether they are hulled or not, and hulledness itself is not that simple: it is not physically the same in barley and spelt, in emmer and rice, in oats and buckwheat, and in the score of different cereal species put together under the name of 'millets'. But to say anything more would be beyond the scope of this chapter. I now want to address another question: the relations, if any, between the morphology of the 'primitive' tools used for husking or crushing grains, and the techniques of grain-processing.

Here is a paradox. In most villages of tropical Africa, the sound and sight of women pounding grain is ubiquitous; and it is or was much the same in most tropical countries. Still we have very few studies on what happens in and around the mortars. It is as if grain-processing was so self-evident a part of the daily routine that nobody ever thought it worthwhile to have a closer look. The same neglect seems to have long prevailed among archaeologists. In the 1930s Cecil Curwen had already denounced our 'blissful ignorance' of the subject; according to Kraybill [1978: 511] things had not changed much by the 1970s. They have begun to improve since as far as archaeologists are concerned; the work of Gordon Hillman [1984a, 1984b] in Turkey is one of the best examples. But among students of recent societies, mainly anthropologists, historians and geographers, the change remains timid. Comparative studies are rare and have not been really followed up.[3]

Emil Meynen was probably the first, in 1927, to notice that the wooden mortar was not universal. According to him, it was absent from the whole of Australia, from the southern tip of South America and from the highlands extending from the Andes to New Mexico. California should probably be added to the list, since, according to Carter [1978], 'The metate is known to have preceded the mortar in much of California [. . . .]. The data suggest an introduction of the mortar somewhere near central California and its slow gain of dominance over the metate in adjacent areas'. In Africa the wooden mortar may not have been as ubiquitous once as it seems today. In Nubia and northern Sudan for example, the traveller F. Caillaud [1826] noticed only grindstones in the 1820s, although the main cereal processed was durrah (sorghum). And following hints obtained from students or colleagues, it is a question whether the wooden mortar did not arrive in some areas of the Western Sahel only after the colonial period.

By itself, the geography of the wooden mortar may seem an obsolete way of doing research; but as a way of finding out meaningful differences and changes it is not. Anyway, the field is so vast and our ignorance so deep that I cannot see why any means of posing useful questions should be scorned. The absence of the wooden mortar may be a matter of materials, for example. It seems to have been the case in Australia and California, where mortars and 'anvils' of stone were in common use, as well as 'pounding pits' hollowed out of the solid rock, or any conveniently hard surface on which it suffices to put some bottomless container for pounding something therein. The absence of wood may thus refer to a lack of the necessary wood-working techniques. It may also refer to the utilization of space, especially in Australia. Australian aborigines did not ordinarily carry with them heavy utensils like grindstones or mortars but used to leave them near the main food-gathering places where they returned season after season and where they expected to find them again, ready for use. It can easily be imagined that this would not have been possible with wooden utensils because they would not have long resisted the effects of weathering and insects.[4]

In other regions, the absence of wooden mortars suggests quite different explanations. In the Central American highlands, for example, there was a quite specific method of maize processing called 'nixtamalization'. Details differ somewhat, but one of the main features of the process was to soak the grains for one hour or so in a mixture of nearly boiling water mixed with lime or ashes. The whole was then left to cool for some hours (typically overnight), after which a simple washing in cold water was enough to separate the grains from their envelopes. The cleaned wet grains were then ground on the **metate** (saddle-quern) with the **mano** (upper grindstone) into a kind of paste, to be cooked afterwards as tortillas on a hot plate. Nixtamalization has remarkable nutritional properties [Katz et al. 1974; Muchnik 1981]. Tortillas have for a long time been the preferred food of the Mexican peasants, who have always insisted that they must be eaten freshly made. This has made the whole process very difficult to change and especially to mechanize [Bauer 1990]. The two points to be made here are (1) that while **metates** may be similar to Old World saddle-querns morphologically, they are very different functionally; and (2) that wooden mortars are completely out of context with such a method of grain-processing.

Or are they? I have been informed of another method used in Venezuela. Maize grains are soaked in hot water, ground wet into paste and cooked into tortillas as in Mexico. The difference is that no lime or ashes are added to the hot water, so that the grains have to be cleaned of their husks and germs by pounding in a wooden mortar before soaking.

Other factors may also be relevant. In California, the saddle-quern is said to have been used to grind grass-seeds only, whereas acorns, the other staple food of the area, were pounded [Testart 1982: 95–6]. But in France, the recently excavated chalcolithic village of Boussargues (Hérault) has yielded a number of saddle-querns for which no evidence of use could be found other than for the grinding of acorns [Colomer et al. 1990]. Although acorns have been more important for human food (and until quite recently) than is generally realized, including in Europe, acorn-processing techniques have rarely been described and our information about them is poor. However, what we do know is enough to remind us that grindstones and mortars have been used for many other purposes than merely processing grain. In West Africa especially, the prevalence of the wooden mortar may be due to its plurifunctionality. It is everywhere used for husking and breaking grains into grits or flour. But it is also often used with small quantities to separate the grains from the ears instead of threshing. And chiefly, the wooden mortar is also currently used in the processing of yams, bananas, etc. In the southern Ivory Coast, for example, they are prepared into a kind of hard mash called **foutou** in the following way. The yams or bananas are peeled, cleaned, cut into pieces, and boiled in water. If merely crushed afterwards, one obtains a slightly fluid mash called **foufou**, which is consumed on certain occasions but is not a main dish. To obtain **foutou**, the mash has to be pounded for a fairly long time in the wooden mortar, and it hardens considerably as a result. I have no idea why pounding makes yam or banana mash harder, nor why Africans should prefer a hard mash; but this preference is a very definite fact, with consequences possibly as far-reaching as the preference of Mexican peasants for freshly made tortillas. My point here, however, is only that the wooden mortar has probably more uses than the grindstone in West Africa, which may go some way toward explaining why it seems to have in part superseded it there.[5]

It goes without saying, but it goes better still by saying it, that our knowledge of the basic processing techniques of yams, bananas and many other starchy tubers and fruits currently used by millions of people today is hardly better than our knowledge of the processing of acorns. Perhaps because it is so open to view, that part of everyday life seems all but invisible, like *The Stolen Letter* of Edgar Allan Poe.

Threshing, Harvesting, Sowing, etc.

It is a long time since Eduard Hahn [1896] proposed a distinction between hoe agriculture (*Hackbau*) and plough agriculture (*Ackerbau*). His proposal was questionable, and it was indeed soon criticized by another German geographer, Karl Sapper [1910], who rightly pointed out that many so-called 'hoe agricultures' did not use hoes at all, but digging sticks, spades, etc. Nobody now, if asked, would explicitly endorse Hahn's theory anymore. And yet a large majority of anthropologists endorse it unwittingly when they speak of 'horticulture', not in the current sense of the term (gardening, specialized production of fruit, vegetables, etc.) but to designate agriculture without animal-drawn implements as it was or is practised in pre-contact America and in many tropical regions. Worse still, the societies concerned are often labelled 'horticultural' as others are 'hunting-gathering', 'pastoral nomadic', 'agricultural', etc., which refers to an implicit classification of societies after arbitrarily selected cultural traits.

There are a lot of reasons to contest the validity of such labelling. Mine are technological. I believe that all technical elements of a culture are relevant, so that to select one and to declare it crucially important to the exclusion of others is flawed logic. Hahn was right to decide that the plough is important; but so are the wooden mortar, the flail, the sickle. And he was wrong to ignore the fact that the hoe cannot be the basic tool of ploughless agricultures because most hoes have iron blades, or at least iron-shod blades, so that typical 'hoe-agricultures' could not really have evolved before the coming of iron.

It is always a little unfair to criticize an author working a century ago, and I would not have done so if Hahn had not been criticized by Sapper in his own time. Moreover, I certainly do not want to hint that Hahn's work is worthless. It still makes interesting reading today. Indeed Hahn's error, to use technology before being acquainted well enough with it, can be seen as a warning for us. For technological studies look so little rewarding by themselves that many people are tempted to 'use' them in support of supposedly more interesting aims as soon as they believe it possible. I am not immune to this bias myself. But there are rules to overcome it. The rules I am trying to follow may look contradictory; they are (1) make sure you have gathered all that it is possible to know about a technique before using it for any 'theoretical' purpose, and (2) do not hesitate to multiply hypotheses, because it is the only means we have to guard against attaching too much significance to any one in particular.

Following this line, we can say today, I believe, that Hahn's geographical approach was not intrinsically bad, it was only premature and lacking in accuracy. Hahn failed to distinguish between tillage hand tools, especially those having iron parts and those not. He failed to distinguish between ploughs and ards and did not attempt to identify their functions accurately (something rarely done even today, however). And he probably did not ascribe enough importance to the use of animal power in other agricultural tasks such as threshing, transport, water-hauling, etc. In a sense, Hahn's programme was sound, but led astray by naive intuitions and insufficient scholarship. The same can be said, with qualifications, of a later American geographer, Carl O. Sauer [1952]. Here, I only want to add a few remarks to substantiate this opinion.

Maize is a good case in point. We have just seen how the food-processing techniques associated with maize could be conservative, i.e. impervious to change (nixtamalization). The same can be said of nearly all the other operations of maize-growing and processing. Thus, there is no threshing proper, but each cob is rubbed against the edge of some hard and fixed object to separate the grains. The cobs themselves are harvested one by one by hand and maize may be the sole food plant in the world for which no harvesting tool other than the bare hand was ever developed before the corn-picker in the twentieth century. Maize-planting and weeding are also mostly done by hand using the simplest implements. The size of the plant and of the seeds make elaborated techniques of sowing and tillage both impracticable and unprofitable. With seed-yield ratios easily reaching 200:1 and more, there is little incentive to go beyond dibbling, which does not require more than a pointed stick. And except where irrigation is practised, dibbling only requires a minimum of tillage. It has often been remarked that maize is one of those plants whose morphology has been most transformed by artificial selection. It can also be said, I believe, that until very recent times maize has been the most adverse to the development of new tools and mechanical devices. The domestication of maize in the New World is much less ancient than the domestication of Old World cereals (except rye and oats probably). But I do not think the difference is enough to account for the fact that the tool-kit of the maize growers was always so limited and changed so little. This cannot have been without consequences for the evolution of Amerindian societies.

A corroboration of this can very probably be seen in the absence of the ard and plough in Sub-Saharan Africa. This absence has long puzzled scholars, who proposed a number of not very convincing explanations, including a cultural 'refusal' to adopt ploughs, more mysterious than the fact itself. I think that the real explanation may be, quite simply, that the main cereals of Sub-Saharan Africa are bulrush millet and sorghum, two plants which, like maize, are large, give high seed-yield ratios and are most usually dibbled. Ards are indeed irrelevant in agricultures based on dibbling. Ethiopia is a nice counter-example, because small-sized cereals are grown there where the ard is in use (barley, emmer, teff, etc.). And there is also the much more recent but paradoxically little-known example of colonial French Guinea. Contrary to what happened in most other parts of tropical Africa, the plough, when first introduced there in the early 1920s, met with immediate success. Later events (the 1929 recession, World War Two, and administrative nonsense thereafter) reduced this success to little. But for a time, at least, success was real, although in one region only, the Central Guinean plateau. The reasons were (1) a pretty regular and moist climate, without too severe a dry season, (2) a large deforested area with grass cover, (3) a relative plenty of cattle, and (4) main crop rice, not sorghum, bulrush millet, yams or bananas.[6]

I am pretty sure that owing to the diversity of the practices still to be observed there, Africa can give us a large number of similar examples as soon as we are prepared to conduct some serious investigations. One such example concerns the sickle. I have no Hahn-like theory based on the sickle, although it would not be more improbable than the original. The sickle proper is rare in tropical Africa, and even rarer is its use in the harvesting of cereals. Its main use is in the harvesting of straw, so extensively used in the Sahel in housing and furniture. But this straw is harvested in the bush from wild Gramineae such as *Andropogon gayanus*. Now, if looked at in comparative perspective, it is by no means obvious that the sickle should be the most ancient and the most general of harvesting tools, as archaeologists usually assume. There are, in fact, a number of simple and efficient techniques requiring no sickle, indeed no cutting implement at all, for harvesting grains – but grains only, not the straw, and this is probably the nub. When people do not have much use for straw they do not need sickles. When they do need straw (or grass) but obtain it from wild plants, they use sickles, but not in the harvesting of grain, and so the sickle does not become especially important. Only when straw is needed *and* obtained from the same plants as grain does the sickle become of primary importance. This is probably what happened very early in the Near East with cereals like barley, emmer and wheat, and much later in the Far East with rice. These may rightly be called **céréales à paille** ('straw cereals') according to the current French usage. Large-size cereals like maize, sorghum, bulrush millet, etc., give no 'straw' in the technical sense of the term, and so sickle-like implements never developed with them.

My last example will take us back to Europe. It concerns harrows. Harrows, like sickles or wooden mortars, are usually looked upon as too common or simple to be of interest. But harrows in non-Mediterranean Europe have something quite specific about them: they are used to bury the seeds after broadcast sowing. This is not the case in the Mediterranean and in western Asia, where the seeds are commonly buried with an ard; elsewhere, it is infrequent too, if only because broadcast sowing is not the dominant mode of sowing.

To my knowledge, harrowing in the seed is a comparatively late innovation. It is only mentioned by Plinius the Elder in the first century AD, not by earlier Roman agricultural writers, and the first archaeological evidence for modern harrows in that sense dates from the third or fourth century AD. The reasons why harrowing in the seed developed and eventually supplanted their ploughing under are not quite clear to me, and anyway it would be impossible to discuss them here. But there can be little doubt either about the fact or about its importance. The point I want to make is that the practice of harrowing in the seed led to the use of horses, and especially of large horses, in agriculture.

There are in the world a number of cases where horses, donkeys, mules or camels were used to draw ards and carts instead of oxen. These cases are rather scattered and did not lead, to my knowledge, to any significant technical change. One gets the impression that in the absence of oxen, any other animals could do as a second choice, but it did not make much difference. In northern Europe too by, say the early Middle Ages, the oxen were the preferred plough-drawing animals; horses were only used for the saddle and the pack, and also for drawing carts where there were passable roads. So there developed a situation where farmers, at least well-to-do farmers, had two sets of animals: oxen for the plough, and horses for transport.

There was no reason why this situation should not have lasted indefinitely. Indeed, in many areas it lasted until the nineteenth century, and for good reasons. Only oxen were bred to make powerful draught animals. Horses were too small to be of real use for the plough; their use in ploughing was infrequent and usually a consequence of impoverishment. There was one agricultural operation, however, where horses were better than oxen because they were more rapid: harrowing. Medieval miniatures quite often show ploughs drawn by a team of oxen, whereas harrows are drawn by one horse, often mounted by its driver, in the same field where the seed is being scattered broadcast by another worker. As I see it, the use of horses in harrowing tilted the balance. It created conditions under which the two sets of animals, horses for transport and oxen for fieldwork, could begin to be mixed up. With horses increasingly being used in the field, a demand for heavier animals arose, until it was realized that oxen could be entirely dispensed with. In the more advanced regions of northwest Europe, i.e. northwest Germany, the Netherlands, northern France and southeast England, the replacement of oxen by horses was completed by the early seventeenth century, if not earlier; in backward regions like Scotland, it did not begin before the second half of the eighteenth century. My contention is that without the practice of harrowing in broadcast seed, this process would not have taken place, or at least not where and when it did.[7]

Conclusion: Technology, Environment and the Concept of Affordance

Is grain to be husked before being pounded into grits or milled into flour? How many winnowings does it take to clean it thoroughly? How well do grits and flour keep in storage? What difference does it make to add lime or ashes to the hot water for soaking maize as far as the adherence of the envelope to endosperm is considered? What happens to banana or yam mash when it is pounded vigorously for half an hour? How exactly can acorns be made edible? Which species of oaks yield tannin-free acorns? What is the size of the seeds of this or that cereal crop, to what depth must they be planted, how much do they yield when they are dibbled, drilled or sown broadcast? Of what speed are horses and oxen capable, and for how long?

This list of odd questions could be made much longer. It is in fact interminable. For what it is, however, this one gives a pretty good idea of what 'environment' is from a technological point of view. Environment is an interminable list of unanswered (and sometimes unanswerable) questions.

This is nothing new. Everybody knows that the concept of environment has no content itself, it can only be given one by reference to the thing or things environed. However, if everybody pays lip service to this truism, most environmental studies tend to ignore it, either because environment is confused with nature, or because the mere fact of focusing one's attention on the environment is already a first step toward reifying it. I feel quite at ease with the natural sciences, in so far of course as I am able to understand what is going on. I often feel ill at ease with environmental studies because I usually cannot find answers to my odd questions among the numberless data they accumulate. It is only when we know something precise about how a society works that we can ask relevant questions about its environment.

So any environmental study that does not begin by looking for relevant questions within society itself runs the risk of being futile.

The idea of relevance is therefore crucial, and it certainly has to be included in the concept of environment if the latter is to make any sense. This is why I welcomed with enthusiasm the concept of 'affordance', of which I was made aware by Ad Smitsman quite recently.[8]

As a matter of fact, the concept of affordance was developed in psychology about fifteen years ago. Affordances are defined as 'environmental resources for behaviour': a flat and smooth surface of ice 'affords' physical opportunities that are put to use by a skater; air 'affords' properties that will be exploited by a young bird as soon as its wings reach a sufficient size; the earth affords us all a surface on which we can walk or run, etc. I cannot develop the implications of the concept further here, if only because it is still too fresh knowledge to me. Of course, this concept belongs to psychology and refers to the behaviour of individuals, whereas my interest is with techniques as social facts. But I think it is not a real obstacle. There is nothing in the concept of affordance that prevents it being adapted to the use of anthropologists. And if it is only a word, it is a very useful one. The fact that wheat and rice produce both edible grains and usable straw, whereas reeds and rushes produce only usable straw, and maize or sorghum only edible grains, points to different affordances, with which different cultural traditions have evolved. To be able to give them a name is something.

Notes

1. Aizenay is a large village or small town of the département of Vendée (western France), situated 90 km due south of Nantes. On the growing and uses of millet in the Vendée see Hongrois [1991]. See also Hörandner [1995].
2. I have presented elsewhere evidence for the early history of winnowing machines and its relevance for the history of mechanization [Sigaut 1989a, 1989b].
3. I do not want to suggest that the immense literature on the history, prehistory and ethnology of grain-processing tech-

niques is worthless: quite the contrary. But this literature is so heterogeneous and scattered that it can hardly be taken as a corpus of usable data. The only partial but true attempts at a synthesis I know of are the antiquated papers of Meringer [1909] and Meynen [1927] in an ethnogeographical perspective, and those of Carter [1978] and Kraybill [1978] in an archaeological perspective, to which must be added an interesting if not quite successful attempt at classifying grain-milling devices by Anderson [1938]. With two colleagues at the EHESS, Rolande Bonnain and Françoise Sabban, we have set up a seminar on the food uses of cereals and other starchy plants in order to explore the possibility of a meaningful synthesis.

4. For data on Australia and California, I have relied mainly on Carter [1978], Hamilton [1980], Heizer and Elsasser [1980: 91–101, 114–16] and Testart [1982: 95–7].
5. On acorns in the western Mediterranean, the main recent paper is by Lewthwaite [1982]. My information on West Africa comes from a number of sources, of which the most informative have been a mimeographed study by Chateau [1973] and an African student at the EHESS, O. Gnabro.
6. This is a very short summary of an argument that I have presented with some more substance elsewhere [Sigaut 1985, 1989c]. The 'refusal' theory is more often implied than stated, but it has been stated at least once [Paulme 1961: 122]. Ethiopia is one of the few regions of Africa where the geography of agricultural implements has been extensively studied; see e.g. Alkämper [1971] or Westphal [1975] for the literature. For the history of the plough in French Guinea in colonial times, see Bigot [1989].
7. On the relation between sowing techniques on the one hand and the uses of ards, ploughs and harrows on the other, see Sigaut [1988]. On the relation between harrowing in the seed and the replacement of oxen by horses in agriculture, see Sigaut [1982].
8. The concept of affordance was proposed by Gibson [1979]. See also Smitsman et al. [1987].

Bibliography

Alkämper, J., 'Die Pflüge Ethiopiens', *Zeitschrift für Agrargeschichte und Agrarsoziologie* 19(2), 1971, pp.137–59

Anderson, R.H., 'The technical ancestry of grain-milling devices', *Agricultural History* 12, 1938, pp.256–70

Bauer, A.J., 'Millers and grinders: technology and household economy in Meso-America', *Agricultural History* 64(1), 1990, pp.1–17

Bigot, Y., 'Un siècle d'histoire d'une technologie agricole: la traction animale en Guinée', in G. Raymond et al. (eds), *Economie de la mécanisation en région chaude*, Montpellier: CIRAD, 1989

Caillaud, F., *Voyage à Méroé, au Fleuve Blanc, au-delà de Fâzoql dans le midi du royaume de Sennâr, à Syouah et dans cinq autres oasis, fait dans les années 1819, 1820, 1821 et 1822*, Paris, 1826, 4 vols, atlas

Carter, G.F., 'The metate: an early grain-grinding implement in the New World', in C.A. Reed (ed.), *Origins of agriculture*, Paris and The Hague: Mouton, 1978

Chateau, J.P., 'Les produits vivriers de base dans l'alimentation en Côte d'Ivoire, multigr', République de Côte d'Ivoire, Ministère du Plan, Direction des Etudes de Développement, 1973

Colomer, A. et al., *Boussargues (Argelliers, Hérault): un habitat ceinturé chalcolithique; les fouilles du secteur Ouest*, Paris: Editions de la M.S.H., 1990

Gibson, J.J., *The ecological approach to visual perception*, Boston: Houghton Mifflin, 1979

Hahn, E., *Demeter und Baubo, Versuch einer Theorie der Entstehung unsres Ackerbaues*, Lübeck, 1896 [Published by the author, 'In Commission bei Max Schmidt']

Hamilton, A., 'Dual social systems: technology, labour and women's secret rites in the Western Desert of Australia', *Oceania* 51(1), 1980, pp.4–19

Heizer, R.E. and A.B. Elsasser, *The natural world of the California Indians*, Berkeley: University of California Press, 1980

Hillman, G., 'Interpretation of archaeological plant remains: The application of ethnographical models from Turkey', in W. Van Zeist and W.A. Casparie (eds), *Plants and ancient man, studies*

in palaeoethnobotany, Rotterdam: A.A. Balkema, 1984a
——, 'Traditional husbandry and processing of archaic cereals in recent times: the operations, products and equipment which might feature in Sumerian texts', *Bulletin on Sumerian Agriculture* 1, 1984b, pp.114–52
Hongrois, C., *Si t'aimes pas le meuille . . .*, Aizenay: OMAC [Office municipal d'Action Culturelle], 1991 ['meuille' is the local vernacular for millet]
Hörandner, E. (ed.) *Millet-Hirse-Millet* Frankfurt au Main: Peter Lang, 1995
Katz, S.H. et al., 'Traditional maize processing in the New World', *Science* 184, 1974, pp.765–73
Kraybill, N., 'Pre-agricultural tools for the preparation of food in the Old World', in C.A. Reed (ed.), *Origins of agriculture*, Paris and The Hague: Mouton, 1978
Lewthwaite, J.G., 'Acorns for the ancestors: the prehistoric exploitation of woodland in the West Mediterranean', in S. Limbrey and M. Bell (eds), *Archaeological aspects of woodland ecology*, Oxford, 1982 [British Archaeological Reports International Series 146]
Meiners, U., *Die Kornfege in Mitteleuropa*, Münster: F. Coppenrath, 1983
Meringer, R., 'Die Werkzeuge der pinsere-Reihe und ihre Namen', *Wörter und Sachen* 1, 1909, pp.3–28
Meynen, E., 'Die Verbreitung des Holzmörsers, eine vergleichende Studie', *Ethnologica* 3, 1927, pp.45–122
Muchnik, J., *Technologies autochtones et alimentation en Amérique latine*, Massy and Paris: ALTERSIAL, 1981
Needham, J. and Wang Ling, *Science and civilisation in China*, vol.4, *Physics and physical technology*, Part II, *Mechanical engineering*, Cambridge: Cambridge University Press, 1965
Paulme, D., *Les civilisations africaines*, Paris: P.U.F., 1961
Reed, C. (ed.), *Origins of agriculture*, Paris and The Hague: Mouton, 1978
Sapper, K., 'Der Feldbau der mittelamerikanischen Indianer', *Globus* 97, 1910, pp.8–10 [with a reply by Hahn to Sapper's criticism and a rejoinder by Sapper in the same volume, pp.202–4 and 345–7]
Sauer, C.O., *Agricultural origins and dispersals*, New York: The American Geographical Society, 1952
Sigaut, F., 'Les débuts du cheval de labour en Europe', *Ethno-*

zootechnie 30, 1982, pp.33–46
——, 'Une discipline scientifique à développer: la Technologie de l'agriculture', in C. Blanc-Pamard and A. Léricollais (eds), *A travers champs, agronomes et géographes,* Paris: ORSTOM Editions, 1985
——, 'L'évolution technique des agricultures européennes avant l'époque industrielle', *Revue archéologique du Centre de la France* 27, 1, 1988, pp.7–41
——, 'La naissance du machinisme agricole moderne', *Anthropologie et Sociétés* 13, 2, 1989a, pp.79–102
——, 'Les spécificités de l'épeautre et l'évolution des techniques', in J.P. Devroey and J.J. Van Mol (eds), *L'épeautre* (Triticum spelta), *histoire et ethnologie,* Treignes: Editions Dire, 1989b
——, 'Coup d'oeil sur l'histoire à long terme de la mécanisation en agriculture', in G. Raymond et al. (eds), *Economie de la mécanisation en région chaude,* Montpellier: CIRAD, 1989c
Smitsman, A. et al., 'The primacy of affordances in categorization by children', *British Journal of Development Psychology* 5, 1987, pp.265–73
Testart, A., *Les chasseurs-cueilleurs ou l'origine des inégalités,* Paris: Société d'Ethnographie, 1982
Westphal, E., *Agricultural systems in Ethiopia,* Wageningen: Centre for Agricultural Publishing and Documentation, 1975

Chapter 15

Domesticatory Relationships of People, Plants and Animals

David R. Harris

In a book which brings together scholars from East and West, and which seeks to rethink nature and culture, it is apposite to start by questioning the conceptual distinction between 'wild' and 'domestic' which has permeated Western thought on 'man's place in nature'.

Wild Versus Domestic

The dichotomy between wild and domestic has a long pedigree in European philosophy which reaches back to those Greek and Roman writers who speculated on the earliest periods of human history. They supposed that mankind had progressed through successive stages of development from an initial Golden Age to the ultimate achievement of city life based on agriculture. Only in the Golden Age did man live in a state of nature, using such products as the earth voluntarily afforded. People were happy and contented then because their wants were few and easily satisfied; they had no cares and sorrows because they did not labour physically to try to 'improve' their lot.

This idealized picture of man's earliest condition can be traced back to the views of a philosopher and student of Aristotle, Dicaearchus, whose work influenced many writers including, in particular, the Roman author Varro, as evidenced in his well-known treatise on farming. What is perhaps the earliest description of the Golden Age, when humans lived naturally, like other creatures, occurs in the work of Hesiod, who probably lived in the eighth century BC. He divided human history into five ages: golden, silver and bronze, followed by the time of the demigods or epic heroes, and lastly the iron age (his own time), when 'men never rest from labour and sorrow'; in the Golden Age, by contrast, people lived 'remote and free from toil and grief' and never suffered old age, while 'the fruitful earth unforced bore them fruit abundantly and without stint' [Glacken 1967: 132–3].

The concept of a Golden Age is inextricably linked with another ancient strand in Western thought: the loss of innocence when man ceased to live in a state of nature, or, in Biblical terms, when God expelled Adam and Eve from the Garden of Eden. The belief that humanity had developed through successive stages – in its simplest expression a threefold sequence from hunting through herding to agriculture – implied a progressive separation of man from nature, a process epitomised in Varro's aphorism: 'divine nature made the country, but man's skill the towns' [Glacken 1967: 32]. It likewise implied an increasingly sharp distinction between the 'wild' and the 'domestic', a distinction which was assumed to have applied earlier to animals than to plants. Thus, in Dicaearchus's conception of the history of humanity, the Golden Age was succeeded by a pastoral stage when people continued to eat wild plant foods while capturing, confining and taming animals [Glacken 1967: 140–1].

However, not all Classical authors envisaged animal domestication as a one-way process effected exclusively by people. Lucretius, who was a contemporary of Varro, ascribes to the animals an element of choice: they 'volunteer' to be domesticated because it offers them a safer and easier life than existence in the wild [Glacken 1967: 138–9]. Regardless of differences of opinion among Classical authors on the process of animal domestication, their recognition of it firmly establishes the distinction between domestic and wild in man's relations to nature. And at the next stage – that of agriculture – the distinction becomes even more definite when crops are domesticated and come to share with the domestic animals the role of food providers for humanity.

The perceived separation of man from nature, and of domestic from wild animals and plants, was reinforced in the West in post-Classical times by Christian theologians who argued that man had been created in the image of God and was inherently different from and superior to all other living creatures. It was further strengthened during the Renaissance by the development of theoretical and practical science, which was associated with a growing conviction that man could control nature and modify natural processes for human benefit. By the end of the eighteenth century, a diverse but coherent body of thought had crystallized around the conceptual dichotomy of man and nature. It focused on three distinct but interrelated ideas: that the earth was purposively designed for human habitation; that the moral and social characteristics of its inhabitants were influenced by the environments in which people lived; and that humans had changed the earth from its pristine state, notably by domesticating animals and plants [Glacken 1967: vii–viii].

Most early-nineteenth century thinkers concerned with man's relationship to nature continued to accept the dichotomy, but there were exceptions. Foremost among them was Alexander von Humboldt, who was profoundly influenced by his travels in the American tropics and who, in 1805, delivered his highly original *Essai sur la Géographie des Plantes*. In this remarkable document (which he dedicated to Goethe), Humboldt not only discusses the relations of plants to such environmental factors as altitude, temperature, rainfall, animals and other plants, but also speculates on the effects of plants on people's lives and on the ways in which, through the centuries, man has altered plant communities. He thus replaces the conception of man *and* nature with that of man *in* nature and perhaps deserves to be recognized as the first exponent of what we now think of as the science of human ecology [Humboldt c.1807].

The Concept of Domestication

There is no need, for the purposes of this chapter, to follow, through the nineteenth and into the twentieth century, the complex philosophical trail of changing ideas about man's relation to nature, and, in particular, the 'stage theories' of cultural history which continued to exert a strong influence on Western scholars [Harris 1981: 6–10; Kramer 1967]. Having emphasised the antiquity in Western thought of the dichotomy between culture and nature, between the domestic and the wild, we can now narrow the framework of discussion and focus specifically on the concept of domestication and on such closely associated terms as cultivation, agriculture and husbandry.

Although the conceptual distinction between wild and domestic has been firmly established in Western thought since Classical times, it has, until recently, been used by most archaeologists, anthropologists and historians in a static sense to denote states of being. Biologists, on the other hand – most notably, in the nineteenth century, Charles Darwin [1868] – have discussed domestication in dynamic terms, regarding it more as a process than a state of existence. Even such a perceptive archaeologist as V. Gordon Childe, when proposing his seminal hypothesis of the Agricultural or Neolithic Revolution, did not suggest, or even speculate on, the interactive processes between people, plants and animals that led to the domestication of particular species in prehistoric Southwest Asia. For him, it was sufficient to suggest that a climatic shift toward greater aridity about 10,000 years ago had the effect of crowding plants, animals and people together in river valleys and oases where domestication then occurred, although precisely how it did so he never discussed [Childe 1934, 1936, 1942].

One exception to the lack of interest in the processes of domestication shown by most archaeologists during the first half of the twentieth century is seen in the work of Frederick Zeuner [1963] on animal domestication. His initial training as a zoologist and palaeontologist probably accounts for his more dynamic view of domestication. He regarded it as a particular type of biologically defined symbiotic relationship, comparable with other examples of symbiosis that exist between non-human species. This led him to distinguish between various categories of relationship, such as scavenging, parasitism and taming, and to propose five 'stages of intensity' of animal domestication (Table 15.1a) as well as to suggest a probable chronological order in which animals were domesticated. Seventeen years after Zeuner published his *History of Domesticated Animals*, the view that domestication should be defined in terms of types of symbiotic relationship between men and animals was critically examined by the anthropologist Tim Ingold. He argued that the term symbiosis should be restricted to inter-specific associations in which one or both species benefit and neither is disadvantaged, i.e. relationships of, respectively, commensalism and mutualism. According to this line of reasoning, parasitism and predation, which are part of the complex of man-animal relationships,

Table 15.1. Classifications of man-animal and man-plant relationships according to Zeuner [1963], Jarman et al. [1982] and Ford [1985]

(a) Five 'stages of intensity' of animal domestication according to Zeuner [1963: 63]

1 Loose contacts, with free breeding
2 Confinement to human evironment, with breeding in captivity
3 Selective breeding organized by man to obtain certain characteristics, and occasional crossing with wild forms
4 Economic considerations of man leading to the planned 'development' of breeds with certain desirable properties
5 Wild ancestors persecuted or exterminated

(b) Evolutionary classification of 'pre-domestication' animal husbandry according to Jarman et al. [1982: 51–2]

1 Random predation
2 Controlled predation
3 Herd following
4 Loose herding
5 Close herding

(c) Evolutionary classification of man-plant relationships according to Jarman et al. [1982: 53–4]

1 Casual gathering
2 Systematic gathering
3 Limited cultivation
4 Developed cultivation
5 Intensive cultivation

(d) The 'stages and methods of plant food production' according to Ford [1985: 2–7]

Stages	*Methods*	*Activities*
1 Foraging		Tending
2 Food production		
(i) Cultivation	Incipient agriculture	Tilling Sowing
	Gardening	Transplanting
(ii) Domestication	Field agriculture	Plant breeding

would not be regarded as types of symbiosis [Ingold 1980: 27–8].

It was not until the 1960s that a significant number of archaeologists and scholars in related disciplines began to focus their attention on processes of domestication. It was during that decade that ecological concepts, and also systems theory, came strongly to influence the ideas of archaeologists and others investigating plant and animal domestication and the origins of agriculture. This trend was particularly evident in the USA and Britain, where its exponents included Lewis Binford [1968], Kent Flannery [1968], Eric Higgs [Higgs and Jarman 1969] and myself [1969]. The overall effect was a paradigmatic shift in how hunter-gatherers, early agriculturalists and the phenomenon of domestication were perceived. The previously accepted distinction between hunter-gatherers who depended on wild foods and agriculturalists who depended on cultivated crops and domestic animals gave way to a view that emphasised the behavioural and evolutionary continuities that linked, rather than the contrasts that separated, hunter-gatherer 'food-procurement' from agricultural 'food-production'. Higgs and his associates even argued that the term 'domestication' should be replaced by 'husbandry' to encompass all the ways in which humans had intervened in the biology and behaviour of animals and plants, a process which, they suggested, did not start in the Neolithic but reached back far into Palaeolithic times [Higgs 1972, 1975].

During the last two decades students of prehistoric subsistence have largely abandoned the formerly rigid distinction between hunter-gatherers and agriculturalists and have increasingly focused their attention on the diverse and complex ways in which humans have interacted with plants and animals in the past. In the 1980s, several new theoretical contributions were made within the now established ecological-evolutionary paradigm, which further refined the conceptualization of 'domestication' and 'husbandry'.

In 1982 the third and last volume of the series of *Papers in Economic Prehistory*, which Higgs had inaugurated in 1972, appeared [Jarman, Bailey and Jarman 1982]. In it, his successors reaffirmed his distinction between domestication, in the strict sense of morphological change in plants and animals resulting from their selective breeding by humans, and husbandry. They also proposed evolutionary classifications both of 'pre-domestication' animal husbandry (Table 15.1b) and of man-plant relationships (Table 15.1c). The latter endorsed the distinction between plant cultivation and plant domestication which had originally been made by Hans Helbaek and which has been used more recently by Gordon Hillman and others investigating the phenomenon of 'pre-domestication cultivation' in prehistoric Southwest Asia [Helbaek 1960; Hillman 1975].

In the early 1980s too, the concept of domestication was examined and elaborated, in the framework of Darwinian evolutionary theory, by David Rindos. Like Zeuner two decades earlier, Rindos viewed domestication as a type of symbiosis and he described it as a process of co-evolution, defined as 'an evolutionary process in which the establishment of a symbiotic relationship between organisms, increasing the fitness of all involved, brings about changes in the traits of the organisms' [Rindos 1984: 99]. His interest was in plant domestication and the evolution of agricultural systems, and he proposed three categories of domestication which comprise an evolutionary sequence but are not mutually exclusive, i.e. incidental, specialized, and agricultural domestication, the last category being equivalent to the orthodox definition of domestication as a process involving the selective breeding of an organism by humans which results in morphological change in the organism [ibid.: 152–66].

The concept, and classification, of domestication advocated by Rindos is so all embracing that it could be said to encompass three other recent theoretical statements, those by Hynes and Chase, by Yen, and by Ford. Hynes and Chase proposed the term 'domiculture' to describe the interactions of hunter-gatherers (in this case Aboriginal people in northeastern Australia) with plants and animals in local hearth-centred environments or 'domuses'. Ecologically, their concept of domiculture closely resembles Rindos's category of incidental domestication (and in part also his specialized domestication), but they argued that domiculture results from intentional human action focused on culturally recognized plants and animals and is not part of the more general biological phenomenon of symbiosis between non-human organisms [Hynes and Chase 1982].

Working with similar hunter-gatherer data from Australia, but considering also agricultural subsistence in New Guinea and elsewhere, Yen proposed a still more comprehensive concept than domiculture, that of 'the domestication of environment'. He deliberately extended the term 'agronomy' to include methods of plant-food procurement used by hunter-gatherers and argued that the only feature that clearly differentiates the 'subsistence structures' of Australian hunter-gatherers from their agricultural neighbours in New Guinea is the existence among the latter of genetically domesticated species (of crops) [Yen 1989].

The meaning of domestication and of such closely associated terms as food production and agriculture was refined by Ford [1985] with particular reference to data from prehistoric North America. He proposed a classification of 'the stages and methods of plant food production' which distinguished first between foraging and food production and then subdivided the latter into successive stages of cultivation and domestication. He further suggested that three main methods of food production – incipient agriculture, gardening and field agriculture – had succeeded one another, and he distinguished several types of human interaction with plants which represented a continuum from tending and weeding, through tilling, transplanting and sowing, to plant breeding (Table 15.1d).

Ford's classification helps to clarify many of the concepts and terms commonly used by archaeologists and anthropologists in discussions of plant domestication and the emergence of agriculture. It is descriptive rather than explanatory because the 'stages' it delineates are not explicitly related to any independent variables other than time. It resembles quite closely the essentially descriptive model of an evolutionary continuum of people-plant interaction which I proposed – independently of Ford's model – and which was published first in 1989 and in a revised form in 1990 [Harris 1989, 1990].

Domesticatory Relationships: People-Plant Interactions

In Figure 15.1 a further revision, and simplification, of the above model is presented. It designates the main ways in which humans have intervened in the ecology of plants to obtain food (and other products), to which many plant species and populations have responded opportunistically, increasing their biological fitness as a result. Although the model is presented as a continuum, it is descriptive only and does not imply any unidirectional or inevitable progression through time from one

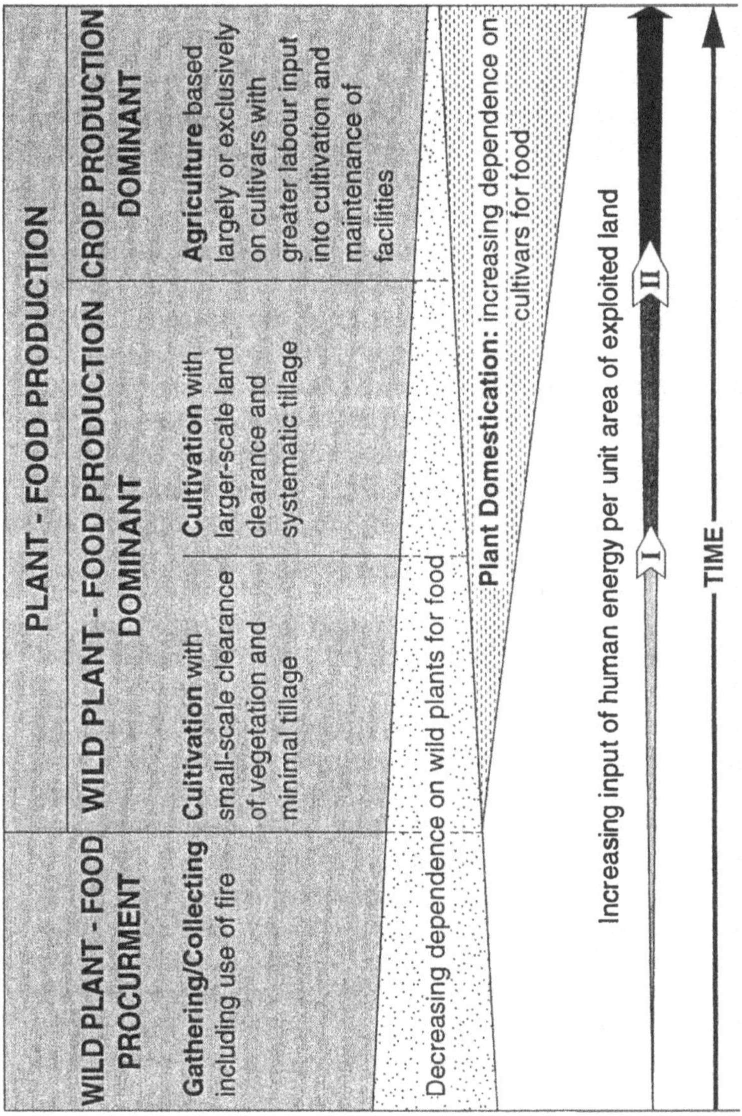

Figure 15.1. A classification and evolutionary model of systems of plant exploitation. The Roman numerals indicate postulated thresholds in the input of human energy.

plant-exploitative activity to another. Knowledge of these human activities is derived largely from ethnographic observations and historical accounts rather than from archaeobotanical evidence, although they may, to varying degrees, be archaeologically detectable.

The interactions with plants that result from these human activities form a continuum which may, for analytical convenience, be divided into three major categories. *Wild plant-food procurement* incorporates the (usually) spatially extensive effects of burning vegetation and the more localized activities of gathering, collecting and protectively tending wild plants. *Wild plant-food production* adds to those activities planting, sowing, weeding, harvesting, storing, and even the drainage and/or irrigation of undomesticated 'crops'. It can be subdivided into *cultivation with small-scale land clearance and minimal soil tillage* and *cultivation with larger-scale land clearance and systematic soil tillage. Agriculture* is denoted when domesticated plants (cultivars) are the main or exclusive components of systems of crop production (domestication meaning that genetic and/or phenotypic selection has led to morphological change and a degree of dependence on human actions for the plant's survival), and when more human labour is invested in cultivation and the maintenance of agricultural facilities (field systems, tools, storage, etc.) (Figure 15.1).

The general analytical distinction between domesticated and undomesticated crops is of course not absolute, but it focuses attention on the distinction between the cultivation of predominantly wild plants and agriculture based mainly on crop production. It thus highlights the phenomenon of 'pre-domestication cultivation'; and it encourages us to try to detect archaeologically the presence or absence of plant remains that show unequivocal evidence of domestication [see e.g. Zohary and Hopf 1993]. Land clearance and tillage are likely to be more difficult to detect archaeologically, although the techniques of pollen, phytolith and wood-charcoal analysis do sometimes yield evidence of the former, and experimental investigations of microwear on flint sickle-blades suggest a possible method of detecting the latter [Unger-Hamilton 1985, 1989].

Because this chapter attempts an overview of people-plant domesticatory relationships, it is not my intention to examine evidence for all the activities by which humans have intervened in the ecology of the plants concerned. But it is worth alluding to some examples drawn from ethnographic and historical sources which include descriptions of wild plant-food procurement and wild plant-food production. (The much more abundant and better known literature on agricultural methods of plant-food production need not be discussed here.)

There are many scattered references to the effects on vegetation that result from the systematic use of fire by hunter-gatherers, which tends to increase the abundance of herbaceous plants, especially grasses, relative to woody taxa, and also enhances the productivity of many fruit- and seed-bearing trees and shrubs [see e.g. Beaton 1982; Jones 1969; Lewis 1973]. Gathering and collecting has frequently resulted in the casual dispersal of seeds and other propagules and has extended the distribution of the plants concerned [Hynes and Chase 1982]. The replacement planting of yams and other tuberous plants is well documented, especially in Australia [Hallam 1989; Harris 1977a: 433–7; Yen 1989: 59–60], as is the planting of acorns and other tree seeds in North America [Shipek 1989]. The sowing, harvesting, storage and processing of wild grass seeds is widely reported [see e.g. Allen 1974; Cane 1989; Harlan 1989; Shipek 1989], and there are some accounts of drainage and irrigation being used in non-agricultural contexts to regularize and increase harvests of wild-grass seeds and roots and tubers, for example by the Paiute people of western North America [Harris 1984]. These ethnographic and historical examples suggest that intervention by humans in the ecology and reproductive biology of many different plant communities was commonplace and widespread before the advent of agriculture, but we largely lack archaeological proof of this contention. The investigation of pre-agricultural people-plant interactions constitutes a major challenge for archaeobotanists, to which some are beginning to respond [e.g. Hansen 1991; Hillman 1989; Hillman, Colledge and Harris 1989; Mason, Hather and Hillman 1994].

Domesticatory Relationships: People-Animal Interactions

As for plants, so for animals, it is possible to define a series of activities by which humans have intervened in the ecology and reproductive biology of animals, and to view these activities as an evolutionary continuum (Figure 15.2). However, like the model of people-plant interaction (Figure 15.1), this second model is intended to be descriptive rather than explanatory. It is not deterministic and does not imply that transitions from one type of activity to another, such as from hunting to herding, are either inevitable (given sufficient time) or irreversible. Again, as for the plants, knowledge of the types of activity described is drawn mainly from ethnographic and historical sources rather than from archaeological evidence. Figure 15.2 groups the main systems by which humans have exploited animals for food (and other products) into three categories of people-animal ecological relationship: predation, protection and domestication.

Hunting (including scavenging) and fishing are forms of *predation* in which humans compete with other predators and scavengers for the prey. The distinction suggested in Figure 15.2 between generalized and specialized hunting rests mainly on two criteria: the diversity of animals hunted, and the importance of the resulting animal products in the hunters' domestic economy. Generalized hunting is – or was – particularly characteristic of tropical and temperate ecosystems of high plant diversity, mainly forest and woodland environments, where 'hunter-gatherer' groups derive most of their food supply from gathering and fishing and where hunting is a varied but relatively minor activity. Specialized hunters, on the other hand, have occupied temperate, sub-arctic and arctic ecosystems of low plant diversity, mainly grassland and tundra environments, where hunting focused on a single, abundant species of prey which provided a high proportion of the food supply and many other domestic needs, as did the bison on the North American grasslands, the guanaco on the South American pampas, and the caribou/reindeer on the circumpolar tundra.

The ecological relationship of *protection* is less self-evident than predation. The intended implication of the term is that humans intervene in various ways to modify the predator-prey relationship in favour of a particular species by affording its population some degree of protection from other predators as well as increased, or more assured, access to its food supply. The overall effect of such intervention is to enhance the reproductive potential of the species concerned through manipulation of the environment it occupies.

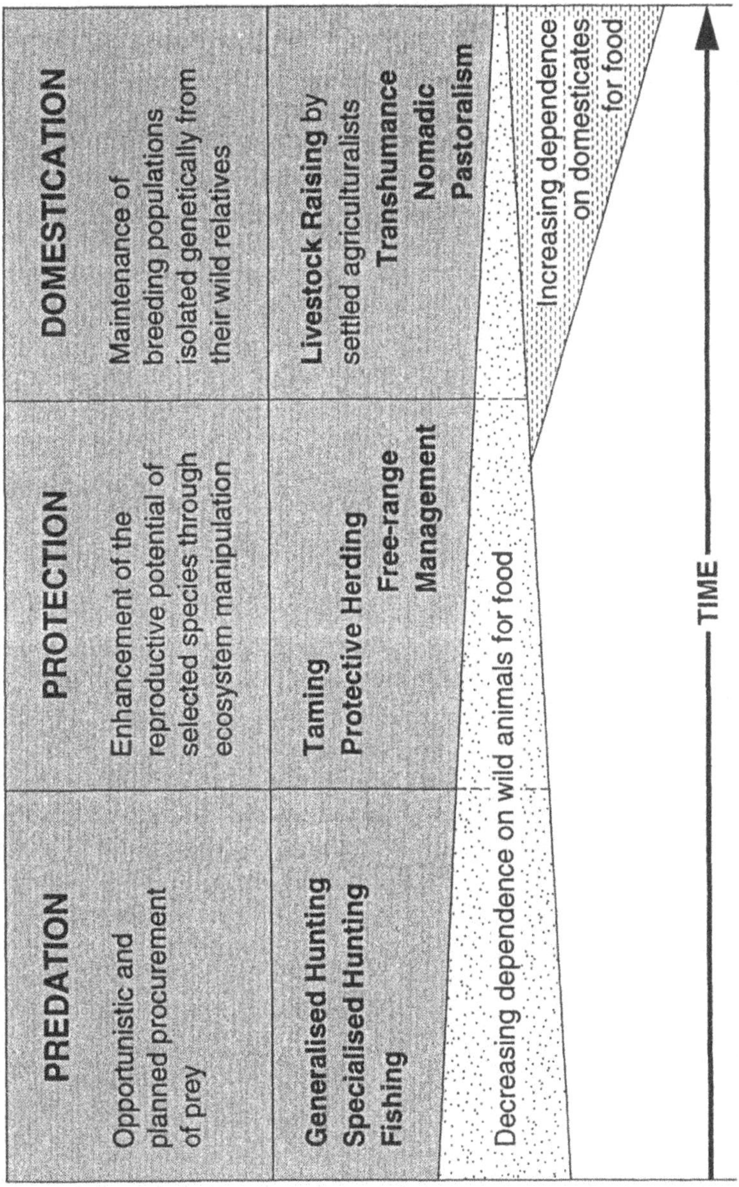

Figure 15.2. A classification and evolutionary model of systems of animal exploitation.

The most ubiquitous, and probably most ancient, means by which hunter-gatherers have manipulated ecosystems to confer competitive advantage on animals they exploit for food and other purposes is fire [Simmons 1989: 38–42]. Deliberate seasonal burning of vegetation by hunter-gatherers has been part of the ecology of most terrestrial ecosystems for many millennia, the only major exceptions being the equatorial rain forests, which remain too moist throughout the year for burning to be generally effective, and the perennial deserts and tundra, where low plant biomass and wide spacing between patches of burnable vegetation inhibit the spread of fire. Elsewhere, especially in tropical savanna and temperate grassland and forest environments, hunter-gatherers have systematically fired vegetation during the driest season of the year to promote fresh plant growth which, in turn, attracts the herbivores that are the hunters' prey. Grazing and browsing animals, such as deer, cattle, sheep and goats, are attracted by the salty ash that is left after a burn, as well as by the new growth of herbaceous plants and the buds and leafy shoots that many shrubs and trees then produce. Post-burn vegetation tends to be richer in proteins and minerals than old growth, and experimental data show that both net primary production and the diversity of plant species often increase after a burn, especially in understorey vegetation. Some understorey species, such as hazel in European temperate forests, respond to fire with particularly vigorous growth.

Overall, seasonal burning increases the capacity of the habitat to support herbivores by improving both the quantity and the quality of forage, while at the same time increasing the spatial and temporal predictability of the animals' movements and thus enhancing the hunters' chances of success. This manipulation of habitats by burning does not literally afford 'protection' to the individual animals concerned, but it does favour their populations, which gain a competitive advantage over other species in the ecosystem and tend to increase disproportionately (unless, of course, predation by the hunters becomes too intensive and results in overkill).

Hunter-gatherers have sometimes provided their prey animals with protection in the much more literal sense by reducing the populations of other, non-human predators. Perhaps the best known examples of this are the efforts of the Inuit and other native peoples of northern North America to kill wolves that prey on the herds of wild caribou on which the people depend for a large part of their food supply. Similarly, hunting peoples of northern Eurasia have sought to reduce the populations of wolves that prey on wild reindeer herds. These examples constitute a form of protective herding (Figure 15.2), as does the pre-Hispanic exploitation of llamas and alpacas in the central Andean highlands [McGreevy 1989; Murra 1965]. In Eurasia this process was later intensified when reindeer were domesticated, probably as a result of contact with pastoralist peoples of the central Asian steppes, and tundra dwellers gradually increased the proportion of tame animals in the wild population, which continued to be exploited by means of protective herding [Ingold 1980: 95–133].

There is also evidence that some fishing peoples have manipulated aquatic ecosystems in ways that enhance the populations of the species they exploit for food. For example, Aboriginal people in southeastern Australia extended the river systems up which eels came seasonally to breed by linking them to swamps by means of artificial channels [Lourandos 1980], and some native peoples of northwestern North America are recorded as having occasionally restocked streams with spawn to try to regularize the supply of salmon, which was one of their staple foods [Forde 1949: 78]. A still more 'interventionist' method of manipulating aquatic ecosystems is represented by the creation of artificial fish ponds in which selected species were raised, thus affording the fish a high degree of protection from non-human predators and increasing their populations. This mode of fish 'production' is historically, and perhaps prehistorically, associated particularly with southern and eastern Asia, where fish ponds are often an integral part of a subsistence system based on wet-rice cultivation; and it appears to have provided part of a unique if minor evolutionary pathway that led to some species of fish becoming domesticated, as for example both food-producing and ornamental varieties of carp.

The ecological relationship of protection is most explicit, and approaches domestication most closely, when it manifests itself in the taming of animals. The capturing – usually in infancy – and taming of animals is an activity common to most ethnographically and historically known hunter-gatherers and agricultural societies. It is less widely practised among seasonally mobile hunter-gatherer groups than among settled foragers, fisherfolk or farmers; but that some mobile hunter-gatherers do routinely capture, rear and tame wild and/or feral animals is demonstrated by the well-known example of Australian Aboriginal people raising dingoes and also, occasionally, wallabies, opossums and even young cassowaries [Meggitt 1965; Zeuner 1963: 39–40]. Tamed animals frequently attain the status of pets, in the sense that they come to be regarded as members of the human groups in which they live, but many also perform useful tasks as, for example, hunting aids, messengers or pest controllers. Indeed, the distinction between tamed useful animals, such as Indian elephants used for riding and haulage, and pets is a very hazy one, because even when the animal is kept principally for utilitarian purposes, it is still likely to be named and regarded affectionately as a member of the household rather than just a 'beast of burden' [Serpell 1989]. Although such domestic status would seem to imply that the animal was, quite literally, domesticated, it is preferable to restrict this term to animals which routinely breed under human control. Tamed animals, captured in infancy and raised to sexual maturity, usually rejoin the wild population to breed, or do not breed at all and die in captivity. When the animal returns to the wild to breed, it is likely to lose its tameness and remain in the wild for the rest of its life, its place being taken in the human group by another captured infant of the same or a different species.

The relationship between wild and domestic sometimes takes an even closer form when a nucleus of tamed animals is maintained which reproduces under partial human control by the tame females leaving the group temporarily to mate with wild males and then returning to the domestic domain with, or to give birth to, their infant young. Such has been the pattern of reproduction among, for example, tame pigs in the highlands of New Guinea [Rappaport 1967] (as well, probably, as in prehistoric Eurasia), and it can be said to represent a form of free-range management which still comes within the ecological relationship of protection but which approaches closely that of domestication (Figure 15.2).

The ethnographic and historical evidence suggests that free- (or open-) range management has been practised particularly with such herbivorous social ungulates as sheep and goats, reindeer and cattle, as well as with pigs. There are numerous references to sheep and goats being managed under systems of loose control, grazing, browsing and breeding in semi-wild habitats and being rounded up as required, but most of these accounts refer to feral rather than truly wild animals. Although it cannot be demonstrated from archaeological evidence, it is in my view probable that sheep and goats – which appear to have been domesticated earlier than other social ungulates – first entered into a protective relationship with humans through a loose association akin to free-range management before breeding populations were established under direct human control. The process may well have been initiated by the capture and rearing of lambs and kids which, through regular feeding and familiarization with their keepers, would quickly have become tame and would have continued to associate with their human providers when mature, even if they were unconfined.

Descriptions of reindeer management among the peoples of northern Eurasia provide other examples of the ease with which social ungulates can enter into, and be maintained in, a 'loose herding' or free-range association with humans. For example, Ingold cites earlier accounts of reindeer management among the northern Tungus of Siberia, where the reindeer are allowed to forage freely near the settlement, usually returning of their own accord even after absences of several days and seldom defecting to join the wild population, a particular reason for this behaviour being the reindeer's craving for salt, which is provided at settlements for them to lick [Ingold 1980: 97–8]. They share this craving with sheep and cattle, and it has been suggested that the provision of salt may have been one of the principal means by which people first tamed and established some control over the movements and breeding of these social ungulates [Harris 1977b: 226–7].

There are ethnographic and historical descriptions too of cattle being managed under free-range systems, the most instructive of which concerns the mithan (*Bos frontalis*), a bovid which inhabits the forests of northeastern India and northern Burma. Frederick Simoons [1968] has presented detailed evidence relating to the use and management of the mithan by swidden cultivators in the region, from which it is clear that, although the animal is phenotypically distinct from its wild relative and probable ancestor, the gaur (*Bos gaurus*), it is not confined but browses and breeds in the secondary forests that regenerate on abandoned swidden land. Contact is maintained with the animals by providing them with salt, and the frequency with which they voluntarily return to a village varies from regular visits every evening to occasional encounters after weeks or months in the forest.

These examples of the free-range management of social ungulates lead us to consideration of the third category of ecological relationship between people and animals: *domestication*. We have anticipated the defining distinction between the relationships of protection and of domestication in previous paragraphs, so it need only be briefly re-stated here. The essential criterion for domestication is the maintenance by humans of a self-perpetuating breeding population of animals isolated genetically from their wild relatives, with resulting behavioural, and usually also phenotypic, changes in the domestic stock.

As the preceding discussion of free-range management and protective herding implies, the transition from a situation in which there is gene flow between wild and tamed animals to one in which the tamed population is breeding in genetic isolation will not occur unless the tamed animals are physically unable to mingle with the wild population. This can come about in various ways, such as by confinement in or near a settlement, by removal of the tame animals out of range of the wild ones, or by elimination of the wild population. Probably all three of these processes have, to varying degrees, contributed in the past to the isolation of animal populations undergoing domestication. But none is likely to have occurred, and been sustained long enough, to establish and maintain the necessary isolation unless the human population controlling the domesticates was already living in permanently occupied settlements. Seasonally mobile hunter-gatherers may, as we have seen, capture, raise and tame animals, but without continuous occupation of a base camp, they could not control and confine the animals sufficiently to establish an isolated breeding population and bring about the transition to domestication. Sedentary life is thus a prerequisite for the first steps in controlled breeding that transformed tamed into domesticated animals.

Theoretically, one could postulate that the process began in settlements occupied year round by hunter-gatherers who practised no agriculture whatever, and such a possibility cannot be entirely excluded in some situations, but it is far more likely that domestication first took place among settled agricultural communities, albeit ones that were also still engaging in both hunting and gathering. This assertion is justified not only by the archaeological evidence, which – at least in western Southwest Asia – appears to indicate that village settlements supported by agriculture were established before herd animals were first domesticated [see chapters by Garrard et al., Hole, Legge and Uerpmann in Harris 1996], but also by the greater ease with which cultivators could have fed their domestic animals on stored agricultural waste products, especially after harvests during the winter. Indeed, it has been suggested by Flannery [1969:87] that the earliest farmers in Southwest Asia may have used herds of domestic sheep and goats as a means of 'banking' agricultural surpluses in order to even out the effects of fluctuations between years of good and bad harvests.

The general conclusion to be drawn from this line of reasoning is that it was sedentary agriculturalists who first bred small herds of livestock and in so doing effected the transition from taming to domestication. They may have done so to add assured supplies of protein and fat to a carbohydrate-rich diet derived from cereals and to provide useful raw materials such as hides and hair, especially if populations of wild animals had by then been reduced by hunting and by land clearance for agriculture. As already implied, it is highly improbable that seasonally mobile hunter-gatherers, even when practising protective herding, segregated breeding populations of tamed animals from their wild counterparts for sufficient lengths of time to establish domestic breeds. It has been argued that the domestication of reindeer represents an exception, the hypothesis being that hunters first domesticated them as decoy animals to improve hunting efficiency, but in his comprehensive discussion of the question Ingold [1980: 103–12] has argued convincingly that reindeer were first domesticated in the boreal coniferous forest zone or taiga as a result of contact with equestrian pastoralists of the steppe grasslands, and that only later did pastoralism based on domesticated reindeer develop across the northern tundra zone.

If this interpretation is correct, and reindeer domestication did not follow a unique pathway from hunting direct to pastoralism, then all types of nomadic pastoralism can be seen as having derived, directly or indirectly, from livestock that had initially been domesticated by settled farmers and incorporated into systems of agricultural production (Figure 15.2). The various types of pastoralism which then evolved and which focused on different principal ungulate species – sheep and goats, cattle, camels, horses and reindeer – appear as a sequential development in time and space around the dry periphery of the Southwest Asian core area of settled village farming.

The earliest expression of this process is likely to have been a pattern of transhumance whereby herdsmen migrated with their flocks of sheep and goats from their homes in farming villages in the lowlands to summer grazing grounds in the uplands. Nomadic pastoralism in the strict sense, in which the whole pastoral population travels with and depends (for milk and to a lesser extent meat) on their animals, while not becoming wholly independent of settled farmers, evidently developed later. It did so as small groups on the arid margins of cultivable land abandoned agriculture in favour of a purely nomadic way of life as herders: on the central Asian steppes, where horses and Bactrian camels were ridden; in the deserts of Arabia and northern Africa, where dromedaries and to a lesser extent horses were ridden; and in the dry savanna environments of eastern and southern Africa, where cattle were central to the pastoral-nomadic life. Finally, reindeer pastoralism developed as a northern offshoot of central Asian pastoralism in which the reindeer were used nutritionally primarily for their meat (carnivorous pastoralism) in contrast to the nutritional exploitation of horses, camels, sheep and goats primarily for their milk (milch pastoralism) (Figure 15.2).

Mixed Farming in Southwest Asia and the Place of Domestic Animals in Other Regional Agricultural Systems

Having analysed the spectrum of domesticatory relationships which people have developed with plants and animals and delineated them in terms of two evolutionary continua (Figures 15.1 and 15.2), I conclude by examining the varying extent to which animals were integrated into systems of agricultural production following domestication. It is instructive to consider the evolution of agricultural systems in this way because it highlights the fact that the full integration of domesticated animals into systems of crop production as providers of food, fertilizer and other products, and as beasts of burden and traction, is the exception rather than the rule; indeed, I contend that mixed crop-livestock farming originated in prehistoric times in one region only – Southwest Asia – from which it later spread to other parts of Eurasia, becoming modified in the process as it adapted to different ecosystems, and ultimately, in modern times, being introduced by Europeans to other continents.

It is convenient to begin this summary survey in the Americas which, in contrast to Asia and Africa, conspicuously lacked indigenous domesticated animals. Despite the abundance of many taxa of wild ungulates, including deer, sheep and bison, the only herd animals to be domesticated were the llama and alpaca in the central Andean highlands. However, although the llama was used to transport goods, and both it and the alpaca provided meat, wool and hides, they were not incorporated into agricultural production as draft animals, consumers of crop wastes and providers of fertilizer. In other words, mixed farming did not develop in South America as it did in Southwest Asia, despite the presence of two species of domesticated ungulate.

Nor were any other animal domesticates in the Americas incorporated fully into agricultural systems, which remained almost exclusively focused on crop production until the Europeans introduced domestic livestock in the sixteenth century AD. Indeed, only four other domesticated animals were present in the Americas in pre-European times: the dog, which had been introduced originally from Asia via the Bering Strait; the guinea pig, which fulfilled a useful function as a household pet, scavenger and source of food, especially in the Andes; the Muscovy duck, valued for its eggs, flesh and feathers from southern South America to Mexico [Donkin 1989]; and the turkey, valued likewise, but of more northerly distribution from northern South America to the southwest of what is now the United States. With that last minor exception, no indigenous animals were domesticated in the whole of North America.

Tropical Africa, too, appears to have totally lacked indigenous animal domesticates, although domestic sheep, goats, cattle and camels were introduced into northern Africa in prehistoric times and spread across the Sahara far into western, eastern and southern Africa [Sealy and Yates 1994; Shaw 1977: 106–10]. Their role in African subsistence was, however, as pastoral rather than agricultural animals and no fully integrated systems of mixed farming developed following their introduction.

As has already been suggested, it was in prehistoric Southwest Asia that domestic animals were first incorporated into agricultural production as beasts of burden and traction, as consumers of crop surpluses and waste products, and as providers of meat, milk and manure, horn, hair and hides. Their meat and milk provided animal protein and fat which complemented the mainly carbohydrate-yielding cereals and pulses of the Southwest Asian crop complex; their power as draft animals harnessed to ploughs, carts and other agricultural implements allowed cultivation to be extended and intensified; and their manure enhanced soil fertility and promoted the development of annual regimes of cropping.

From Southwest Asia, the animals associated with the mixed-farming system appear to have spread to other regions of Eurasia and (as already noted) of Africa. The westward spread into and across Europe evidently involved all the staple crops and herd animals of Neolithic Southwest Asian agriculture, but the eastward spread into central, southern and eastern Asia appears to have been more selective. Domesticated goats, sheep, cattle and horses, as well as wheat and barley, reached northern China across central Asia in Neolithic times, but the spread of Southwest Asian domesticates into tropical southern and southeastern Asia was more limited. In those regions, indigenous species of cattle, such as the water buffalo and the mithan, were domesticated, as were the chicken, duck and goose. Pigs too may have been domesticated in southern and/or eastern Asia independently of their domestication in western Eurasia. Although pork and chicken flesh are widely eaten in southern and eastern Asia, the cattle in these regions have traditionally been valued more for their ritual and exchange value than as providers of meat and milk. The water buffalo has an important role also as a draft animal and as a provider of manure with which wet-rice fields are often fertilized, but neither in South and Southeast Asia nor in China and Japan were domestic animals integrated as fully into systems of agricultural production as they were in Southwest Asia.

Conclusion

By examining, and exemplifying, the varied patterns of interaction that have linked people, plants and animals in the past, I have sought to clarify our understanding of the phenomenon of 'domestication'. Comparative study of the ethnographic, historical and (as yet less informative) archaeological records on a world scale demonstrates how diverse have been the domesticatory relationships established between humans and other organisms; and yet we can discern functional regularities in these relationships and classify them for analytical purposes in terms of continua or gradients of interaction. Such an analytical framework has, I suggest, theoretical validity, but my hope is that it will also serve a more pragmatic purpose by helping to inform, and direct, future research on the complex web of interrelationships that have progressively bound people to particular plant and animal species since the emergence, over 100,000 years ago, of anatomically modern humans.

Bibliography

Allen, H., 'The Bagundji of the Darling Basin: cereal gatherers in an uncertain environment', *World Archaeology* 5, 1974, pp.309–22

Beaton, J.M., 'Fire and water: aspects of Aboriginal management of cycads', *Archaeology in Oceania* 17, 1982, pp.51–8

Binford, L.R., 'Post-Pleistocene adaptations', in S.R. Binford and L.R. Binford (eds), *New perspectives in archeology*, Chicago: Aldine, 1968

Cane, S., 'Australian Aboriginal seed grinding and its archaeological record: a case study from the Western Desert', in D.R. Harris and G.C. Hillman (eds), *Foraging and farming: the evolution of plant exploitation*, London: Unwin Hyman, 1989

Childe, V.G., *New light on the most ancient East*, London: Kegan Paul, Trench, Trubner, 1934

——, *Man makes himself*, London: Watts, 1936

——, *What happened in history*, Harmondsworth: Penguin, 1942

Darwin, C., *The variation of animals and plants under domestication*, 2 vols, London: John Murray, 1868

Donkin, R.A., *The Muscovy duck, Cairina moschata domestica*, Rotterdam: Balkema, 1989

Flannery, K.V., 'Archeological systems theory and early Mesoamerica', in B.J. Meggers (ed.), *Anthropological Archeology in the Americas*, Washington D.C.: Anthropological Society of Washington, 1968

——, 'Origins and ecological effects of early domestication in Iran and the Near East', in P.J. Ucko and G.W. Dimbleby (eds), *The domestication and exploitation of plants and animals*, London: Duckworth, 1969

Ford, R.I., 'The processes of plant food production in prehistoric North America', in R.I. Ford (ed.), *Prehistoric food production in North America*, Ann Arbor: University of Michigan Museum of Anthropology, Anthropological Paper 75, 1985

Forde, C.D., *Habitat, economy and society*, 7th edition, London: Methuen, 1949

Glacken, C.J., *Traces on the Rhodian shore: nature and culture in western thought from ancient times to the end of the eighteenth century*, Berkeley: University of California Press, 1967

Hallam, S.J., 'Plant usage and management in Southwest Australian Aboriginal Societies', in D.R. Harris and G.C. Hillman (eds), *Foraging and farming: the evolution of plant exploitation*, London: Unwin Hyman, 1989

Hansen, J.M., *The palaeoethnobotany of Franchthi cave, excavations at Franchthi cave, Greece*, vol. 7, Bloomington: Indiana University Press, 1991

Harlan, J.R., 'Wild grass-seed harvesting in the Sahara and Sub-Sahara of Africa', in D.R. Harris and G.C. Hillman (eds), *Foraging and farming: the evolution of plant exploitation*, London: Unwin Hyman, 1989

Harris, D.R., Agricultural systems, ecosystems and the origins of agriculture, in P.J. Ucko and G.W. Dimbleby (eds), *The domestication and exploitation of plants and animals*, London: Duckworth, 1969

——, 'Subsistence strategies across Torres Strait', in J. Allen, J. Golson and R. Jones. (eds), *Sunda and Sahul: prehistoric studies in Southeast Asia, Melanesia and Australia*, London: Academic Press, 1977a

——, 'Alternative pathways toward agriculture', in C.A. Reed (ed.), *Origins of agriculture*, The Hague: Mouton, 1977b

——, 'The prehistory of human subsistence: a speculative outline, in D.N. Walcher and N.Kretchmer (eds), *Food, nutrition and evolution: food as a factor in the genesis of human variability*, New York: Masson, 1981

——, 'Ethnohistorical evidence for the exploitation of wild grasses and forbs: its scope and archaeological implications', in W. van Zeist and W.A. Casparie (eds), *Plants and ancient man: studies in palaeoethnobotany*, Rotterdam: Balkema, 1984

——, 'An evolutionary continuum of people-plant interaction', in D.R. Harris and G.C. Hillman (eds), *Foraging and farming: the evolution of plant exploitation*, London: Unwin Hyman, 1989

——, 'Settling down and breaking ground: rethinking the Neolithic Revolution', *Twaalfde Kroon-Voordracht*, Amsterdam: Stichting Nederlands Museum voor Anthropologie en Praehistorie, 1990

——, (ed.), *The origins and spread of agriculture and pastoralism in Eurasia*, London: UCL Press, 1996

Helbaek, H., 'The palaeoethnobotany of the Near East and Europe', in R.J. Braidwood and B. Howe (eds), *Prehistoric investigations in Iraqi Kurdistan*, Chicago: Chicago University Press Studies in Oriental Civilization 31, 1960

Higgs, E.S. (ed.), *Papers in economic prehistory*, Cambridge: Cambridge University Press, 1972

——, (ed.), *Palaeoeconomy*, Cambridge: Cambridge University Press, 1975

Higgs, E.S. and M.R. Jarman, 'The origins of agriculture: a reconsideration', *Antiquity* 43, 1969, pp.31–41

Hillman, G.C., 'Appendix A. The plant remains from Tell Abu Hureyra: a preliminary report', in A.M.T. Moore, 'The excavation of Tell Abu Hureyra, Syria, a preliminary report', *Proceedings of the Prehistoric Society* 41, 1975, pp.70–7

——, 'Late Palaeolithic plant foods from Wadi Kubbaniya, Upper Egypt: dietary diversity, infant weaning, and seasonality in a riverine environment', in D.R. Harris and G.C. Hillman (eds), *Foraging and farming: the evolution of plant exploitation*, London: Unwin Hyman, 1989

Hillman, G.C., S.M. Colledge and D.R. Harris, 'Plant-food economy during the Epipalaeolithic period at Tell Abu Hureyra, Syria; dietary diversity, seasonality, and modes of exploitation', in D.R. Harris and G.C. Hillman (eds), *Foraging and farming: the evolution of plant exploitation*, 1989

Humboldt, A. von, *Essai sur la Géographie des Plantes*, Paris:

Institut de France, c.1807 (facsimile published by the Society for the Bibliography of Natural History, London, 1959)
Hynes, R.A. and A.K. Chase, 'Plants, sites and domiculture: Aboriginal influence upon plant communities in Cape York Peninsula', *Archaeology in Oceania* 17, 1982, pp.38–50
Ingold, T. *Hunters, pastoralists and ranchers: reindeer economies and their transformation*, Cambridge: Cambridge University Press, 1980
Jarman, M.R., G.N. Bailey and H.N. Jarman (eds), *Early European agriculture: its foundations and development*, Cambridge: Cambridge University Press, 1982
Jones, R., 'Fire-stick farming', *Australian Natural History* 16, 1969, pp.224–8
Kramer, F.L., 'Eduard Hahn and the end of the "Three stages of man"', *Geographical Review* 57, 1967, pp.73–89
Lewis, H.T., *Patterns of Indian burning in California: ecology and ethnohistory*, Ballena Press Anthropological Papers 1, Menlo Park, California: Ballena Press,1973
Lourandos, H., 'Change or stability? Hydraulics, hunter-gatherers and population in temperate Australia', *World Archaeology* 11, 1980, pp.245–64
Mason, S.L.R., J.G. Hather and G.C. Hillman, 'Preliminary investigation of the plant macro-remains from Dolní Vestonice II and its implications for the role of plant foods in Palaeolithic and Mesolithic Europe', *Antiquity* 68(258), 1994, pp.48–57
McGreevy, T., 'Prehispanic pastoralism in northern Peru', in J. Clutton-Brock (ed.), *The walking larder: patterns of domestication, pastoralism, and predation*, London: Unwin Hyman, 1989
Meggit, M.J., 'The association between Australian Aborigines and dingoes', in A. Leeds and A.P. Vayda (eds), *Man, culture and animals*, Washington, D.C.: American Association for the Advancement of Science, 1965
Murra, J.V., 'Herds and herders in the Inca state', in A. Leeds and A.P. Vayda (eds), *Man, culture and animals*, Washington, D.C.: American Association for the Advancement of Science, 1965
Rappaport, R.A., *Pigs for the ancestors: ritual in the ecology of a New Guinea people*, New Haven: Yale University Press, 1967
Rindos, D., *The origins of agriculture: an evolutionary perspective*, New York: Academic Press, 1984
Sealy, J. and R. Yates, 'The chronology of the introducation of pastoralism to the Cape, South Africa', *Antiquity* 68(258), 1994,

pp.58–67

Serpell, J., 'Pet-keeping and animal domestication: a reappraisal', in J. Clutton-Brock (ed.), *The walking larder: patterns of domestication, pastoralism, and predation,* London: Unwin Hyman, 1989

Shaw, T., 'Hunters, gatherers and first farmers in West Africa', J.V.S. Megaw (ed.), *Hunters, gatherers and first farmers beyond Europe,* Leicester: Leicester University Press,1977

Shipek, F.C., 'An example of intensive plant husbandry: the Kumeyaay of southern California', in D.R. Harris and G.C. Hillman (eds), *Foraging and farming: the evolution of plant exploitation,* 1989

Simmons, I.G., *Changing the face of the earth: culture, environment, history,* Oxford: Basil Blackwell, 1989

Simoons, F.J., *A ceremonial ox of India; the Mithan in nature, culture, and history,* Madison: University of Wisconsin Press, 1968

Unger-Hamilton, R., 'Microscopic striations on flint sickle-blades as an indicator of plant cultivation: preliminary results', *World Archaeology* 17, 1985, pp.121–6

——, 'The Epi-Palaeolithic southern Levant and the origins of cultivation', *Current Anthropology* 30, 1989, pp.88–103

Yen, D.E., 'The domestication of environment', in D.R. Harris and G.C. Hillman (eds), *Foraging and farming: the evolution of plant exploitation,* London: Unwin Hyman, 1989

Zeuner, F.E., *A history of domesticated animals,* London: Hutchinson, 1963

Zohary, D. and M. Hopf, *Domestication of plants in the Old World: the origin and spread of cultivated plants in West Asia, Europe, and the Nile Valley* 2nd edn, Oxford: Clarendon Press, 1993

Part III

Nature, Co-evolution and the Problem of Cultural Adaptation

Chapter 16

The Co-existence of Man and Nature in the African Rain Forest

Mitsuo Ichikawa

The Zaire Basin Forest is the largest in Africa, comprising almost one fifth of the total tropical rain forest area in the world. Various products have been extracted and exported from this forest since first contact with Western societies in the fifteenth century [Vansina 1977]. Particularly in the early colonial period, ivories, wild rubber and other forest products comprised the major export items, accounting for almost 95 per cent of the total export [Jewsiewicki 1983]. However, because of the difficulty of transportation on the lower Zaire River, timber has not been intensively harvested from the interior of the central basin until recently. We can still find in this region a vast stretch of dense forest which accommodates many rare or endangered animal species, such as gorillas, chimpanzees and okapis.

Despite their appearance, most of the present-day tropical rain forests in Africa have been formed comparatively recently. There was in Africa an extremely dry period during the last ice age of the Quaternary Period, when most of the present forested areas of central Africa consisted of a drier, savanna type of vegetation. Increasing evidence suggests, however, that even in this period there remained forest refuges along the Gulf of Guinea, from Cameroon to Gabon and in eastern Zaire [Hamilton 1983; Kadomura 1990; Mayr and O'Hara 1986]. If this is true, these forests are the oldest in Africa, and with the richest biological diversity.

These old forests of Africa have been inhabited by hunting and gathering peoples for many centuries. They are called Mbuti or Efe in northeastern Zaire, Batwa in central Zaire, Aka or Bayaka in northern Congo and Central Africa, Baka in Cameroon and western Congo, and BaGieli in the coastal area of Cameroon (Figure 16.1). We do not know exactly how long they have lived in these forests, since no archaeological evidence has been found in these regions. However, their physical characters are adapted to the forest environment [Hiernaux 1972]. There are historical records from ancient Egypt which mention them [Gusinde 1956; Turnbull 1961]. Their present distribution is almost entirely confined to the forest areas. All these facts suggest that they are genuine forest people.

These hunter-gatherer societies, widely distributed in central Africa, share a number of features. They form small residential groups of several tens of people and lead, at least partly, a nomadic life in the forest, changing camp-sites every one to two months. They have no chiefs, nor any other central authority. In their social life, egalitarianism and generalized reciprocity are the norm. Their major food-procuring method is a collective hunt with nets or bows aimed at small to medium-sized mammals, especially forest duikers [Bahuchet 1985; Harako 1976; Tanno 1976; Terashima 1983]. Spear-hunting for larger mammals (elephants in particular) became common after contact with Arab or Western traders seeking ivory. In addition to hunting, forest plants are used in a variety of ways, which will be described below in more detail.

Hunter-gatherers are not the only inhabitants of the forest. There are also various agricultural peoples of Bantu, Ubanguian and Sudanic-speaking origin. In some areas, the ancestors of the hunter-gatherers came into contact with these agriculturalists as early as 1,000 to 2,000 years ago [Bahuchet 1983]. Since then, they have co-existed with agricultural villagers, forming a so-called symbiotic relationship, which has influenced various aspects of their life. In the Ituri forest of Zaire, the Mbuti hunter-gatherers provide the villagers with labour, meat and other forest products in exchange for iron implements, salt, clothes and agricultural products. Such an exchange relationship is efficient ecologically. The major forest crops, cassava and plantain, are adapted to the humid forest environment and are excellent sources of carbohydrate, though they contain little protein compared with the seed crops of the savanna areas. On the other hand, animal meat obtained from hunting is a good source of protein, though inefficient as a source of energy. Moreover, hunting and cultivation require different, often incompatible modes of life;

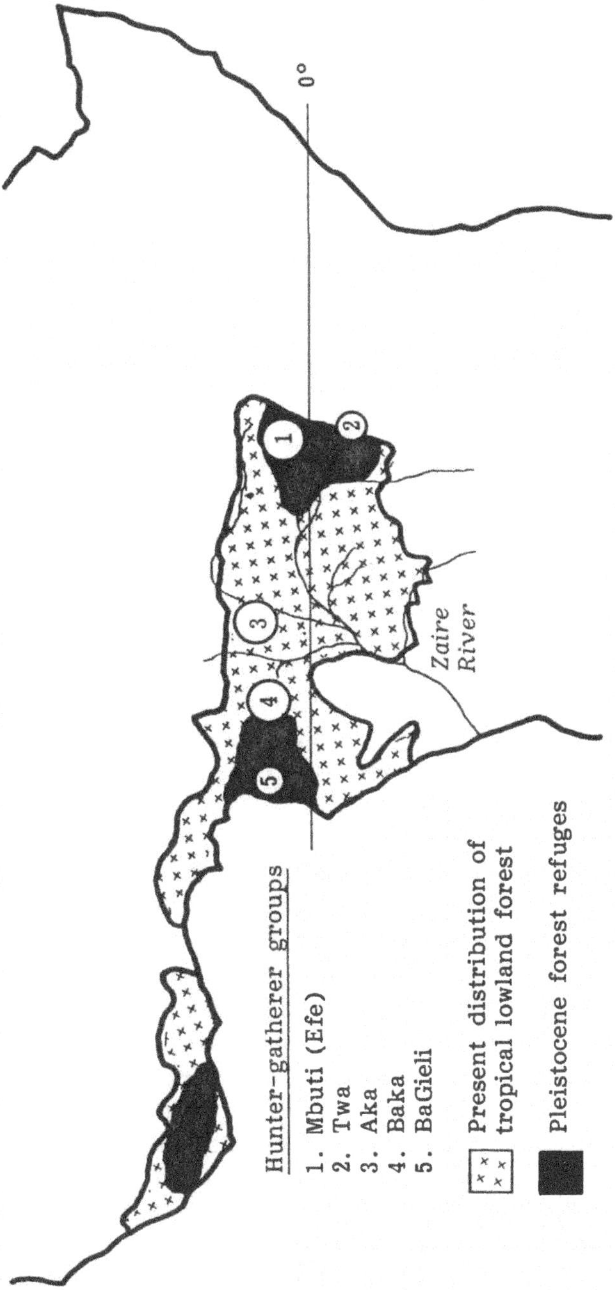

Figure 16.1. Ancient African rain forest and hunter-gatherer distribution

hunting requires a mobile life over an extensive area of the forest, while cultivation must be concentrated on fields and secondary forest near the villages. A form of coexistence has thus developed, an interdependent system of different peoples who would otherwise be competing for the same resources [Ichikawa 1986]. Within such a framework, the Mbuti have specialized as hunter-gatherers, and as such integrated, though loosely, into a regional society. It is largely because of such long-standing interdependent relationships that they have remained as hunter-gatherers until the present day.

Diversity and Multiplicity in Forest Use: The Case of the Mbuti in Zaire

The Ituri forest is situated in the northeastern part of the Zaire Basin, covering approximately 100,000 km^2. A large part of the forest remains primary, characterized by high trees of the family Caesalpiniaceae. Unlike the lower Zaire Basin, large-scale swamp forest is rarely found in this region, whereas secondary forest and cultivated land are expanding along the major roads and in areas with high population densities. Today, the primary forest has almost disappeared along the major roads.

There are a total of fifty-seven large and medium-sized wild mammals recorded from the Ituri Forest, including fourteen species of primates, nine carnivores and fourteen artiodactyles. All of these are considered as food by the Mbuti,[1] although the bulk of their meat consumption comprises forest duikers [Ichikawa 1983]. Duikers are caught by collective net-hunting among the Bantu-speaking Mbuti in southern and western parts of the forest, and by bow-and-arrow hunting among the Sudanic-speaking Mbuti (called 'Efe') in the northern and northeastern parts. Guns were not introduced to the area until recently, and traps are mainly used by villagers and are not common among the Mbuti. Net-hunting has intensified in recent years, as it provides a reliable yield, which can be traded and brought to the local towns.

On average, the Mbuti today depend for as much as 60 per cent of their diet on agricultural food, cassava and plantain [Ichikawa 1986]. They obtain these foods from the villagers in exchange for the labour on miscellaneous (mainly agricultural) works and for meat, mushrooms, honey, palm lianas (*Eremospatha* sp.), thatch and other forest resources.

The Mbuti use the plants themselves in a variety of ways. I and my colleagues from Kyoto University have so far collected from the Ituri Forest more than 1,100 specimens comprising nearly 450 species along with their vernacular names, local usage and other ethnographic information. The Mbuti use more than 100 species as food, 124 for medicine, 260 as materials for construction, binding, and making various instruments and ornaments, 74 for arrow and fish poisons, and 10 for narcotics. Another 114 are used for ritual and other non-material purposes [Ichikawa 1992]. Among the wild food plants, there are *Dioscorea* roots, nuts of *Gilbertiodendron* and *Irvingia*, and fruits of *Canarium* and *Landolphia*, which are frequently eaten, even as the major food in the peak fruiting season [ibid.]. The use of plants for material culture deserves special mention. Of the 83 items recorded in their material culture, 69 (83 per cent) are made, either totally or in part, of plant material, whereas those with animal or metal (which is of external origin) components are as few as 14 and 12 respectively [Tanno 1981]. Their material culture is thus characterized by plant products which are abundant in the forest.

In addition to those plants directly used, hundreds are useful in indirect ways, as nectar sources and as the food of animals which are hunted, fished and collected by the Mbuti. Mbuti have precise knowledge of the food plants of wild animals, and ambush them from the trees which animals are approaching to feed from. Many high canopy trees seem to have no direct use, but they are important sources of honey. Mbuti depend on honey for as much as 80 per cent of their calorific intake in the peak honey season [Ichikawa 1981]. Another favourite food of the Mbuti, various insects and their larvae, also feed on forest plants. Many plants are used in multiple ways, such as *Canarium schweinfurthii*, the fruit of which is eaten both by people and animals and the resin of which is used for making torches. The large leaves of Marantaceae are particularly useful. Beside the seeds eaten roasted, the stems are used for binding, the leaves for thatching, wrapping, making sleeping mats and many other purposes. If we take into consideration the diversity and multiplicity of Mbuti use of plants, it is evident that the value of plants cannot be reduced to their specific material uses as food or medicine.

The Mbuti believe that the forest is imbued with a supernatural being called Apakumandura,[2] who controls all the life in the forest. The Mbuti attribute the continued failure in hunting or gathering to Apakumandura, saying that he has made the forest 'cool' or 'closed'. In order to make the forest 'hot' or 'open' again, it is necessary to please him by a **sulia** ritual followed by intensive singing and dancing. Sometimes, Mbuti make a small shrine, **endekele**, at the exit of their camp, in which some offerings of cola nuts or tobacco are placed for Apakumandura. There is also a small cauliflorous tree species called **akobisi** (*Uvariopsis congolana*, Annonaceae), which grows only in the dense forest [Le Thomas 1969]. This tree is one of the visible agents[3] of Apakumandura and is distributed extensively in the dense forest, though not commonly seen. It is strictly forbidden to cut or break this tree. If someone (mostly villagers who do not know the tree) carelessly cuts it, the Mbuti must sing and dance on the spot, beating a buttress root in place of a drum, in order to appease the anger of Apakumandura.

Turnbull [1965] states that the Mbuti call the forest 'father' or 'mother' and consider it to be their guardian.[4] While I have not witnessed them addressing the forest by such kinship terms, they might do so particularly when they need something from the forest. However, the forest for them is not the source of goodness alone. It has negative influences on their life as well. For example, although forest animals are important sources of food, they may become the source of dreadful diseases and other misfortunes if eaten carelessly. There are wild animals, generally called **kuweri**, which, according to the Mbuti, cause diseases (or deformation of the newborn) in the person who eats them, or in their dependent children. Mbuti parents are, therefore, usually careful when giving small children food which they have not previously tasted, taking into consideration whether or not their children have acquired the strength to resist the dangerous power of **kuweri** [Ichikawa 1987].

Human Impact on the Forest Environment

As described above, Mbuti culture is forest-oriented, characterized by and dependent on the use of a diversity of forest resources. But heavy dependence on the forest also means that Mbuti activities have an impact on the forest environment.

Some Mbuti cultivate fields on a small scale. They clear an area around their base camp near the village and grow cassava and plantains. The size of the fields is generally one tenth that of the villagers, from 300 to 400 m^2 on average, which is by no means enough even for their own consumption. As they seldom cultivate around the hunting camp, their present cultivation[5] has little impact on the ecology of the interior forest.

Honey-collecting may also cause some disturbance to the forest vegetation. However, when Mbuti collect honey they usually climb a tree which has a natural hive and cut a hole in the trunk to take out the honey. Only when they are prevented from reaching the hive by spiny lianas around the trunk or extremely aggressive bees do they fell the tree with an axe. But even in these cases only fast-growing trees with soft wood are cut down. The major tree species of the primary forest, the high trees of Caesalpiniaceae or Meliaceae, have hard wood and are difficult to cut down with the small axes available.

The Mbuti band is composed of fifty–sixty people on average and has a territory of 150 to 250 km^2 extending along the forest paths which connect several hunting camps situated at intervals of three to six km (Figure 16.2). They move from one camp to another every two weeks to two months, depending on the availability of game and vegetable food. Sometimes a new camp is established in the forest or an old one abandoned. When they establish a new camp, they clear the underbrush and small to medium-sized trees, but leave large trees which provide shelter from intense sunlight. When a camp is abandoned, herbaceous plants of Zingiberaceae and Marantaceae grow rapidly, followed by light-demanding trees, which soon transform the camp-site into secondary forest.

Thus, small-scale disturbances by humans are observed to some extent, and their traces are found in scattered patches of secondary vegetation in the forest. We should note, however, that such moderate disturbances have improved the resource base, rather than leading to a deterioration. Some Mbuti food plants do not regenerate well in the shade of the primary forest [Hart and Hart 1986; Ichikawa 1992]. *Canarium, Antrocaryon, Ricinodendron, Landolphia, Tetracarpidium* and *Dioscorea,* for example, require light for germination or for certain stages of growth (see Table

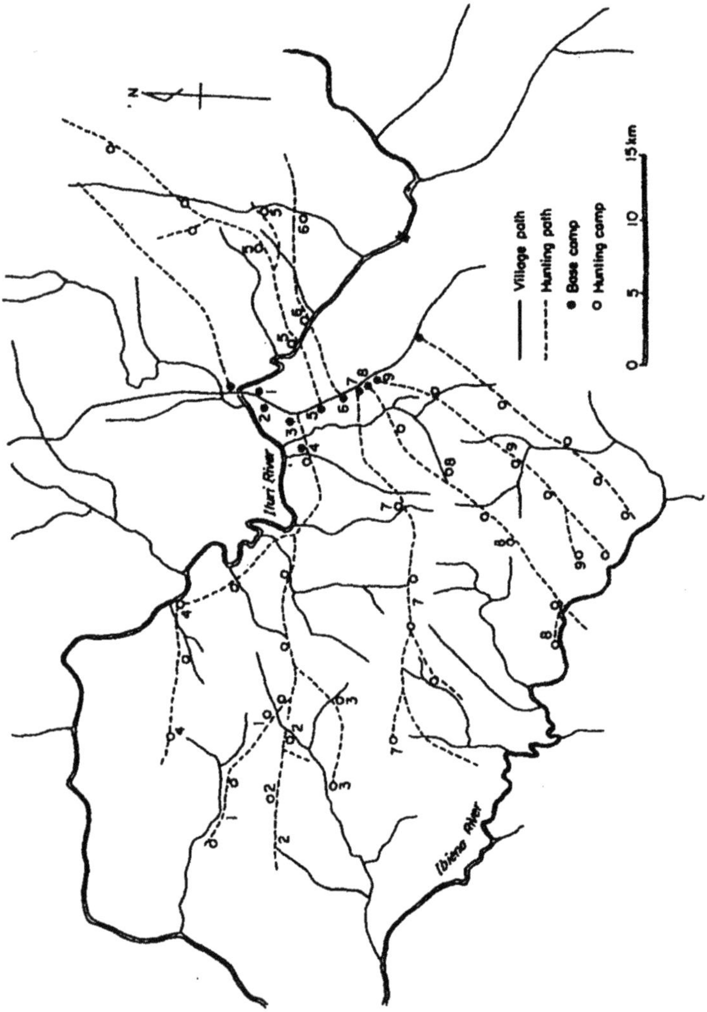

Figure 16.2. Distribution of hunting camps in the Teturi area, Central Ituri

16.1). These plants are now found in scattered patches in the forest, but they must have regenerated in gaps where there was sufficient light. While storms, rain and lightning undoubtedly played a part in making such gaps, human influences were probably more important in this area than these natural phenomena. Particularly in the abandoned camp-sites, *Canarium*, *Landolphia*, *Dioscorea* and other light-demanding food plants must have grown from discarded food waste.[6]

More importantly, there had been scattered villages of shifting cultivators and their fields in the forest before these were

Table 16.1. Habitats of important food plants

Species	Habitat (source)
Nuts and seeds	
# *Antrocaryon nannani*	old secondary forest (1, 2)
Gilbertiodendron dewevreii	evergreen forest (3)
Irvingia robur	evergreen forest (4)
I. gabonensis	evergreen forest and mixed deciduous forest (4, 5)
# *Ricinodendron heudelotti*	deciduous and secondary forests (3, 5)
Treculia africana	evergreen and deciduous forest often by streams (5)
# *Tetracarpidium conophorum*	low bush (5)
Fruits and berries	
# *Elaies guineensis*	secondary forest and abandoned fields (1, 3)
# *Canarium schweinfurthii*	secondary forest, regenerates only in gaps and roadside (2, 6)
Anonidium mannii	closed forest (5)
# *Landolphia owariensis*	deciduous and secondary forest (5)
Tubers and bulbils	
# *Dioscorea praehensilis*	secondary and hill forest (1, 4) regenerates in shade, but grows best in forest margins and gaps (6)
# *Dioscorea bulbifera*	secondary forest and abandoned fields (1, 4)

#: Indicates that the species is mainly found in secondary forest.
Source: 1) Observation by Ichikawa; 2) UNESCO 1983; 3) T. Hart 1985; 4) Hart and Hart 1986; 5) Irvine 1961; 6) Hall and Swaine 1981.

concentrated along the major roads by the Belgian Colonial Government in the 1930s and after (Figure 16.3; see also Ichikawa 1991b). Along the old trade route there were trading posts and large villages even in the early 1900s, such as Mawambi on the bank of the Ituri River, which are now gone and covered with thick vegetation [see Powell-Cotton 1907]. During these times, Mbuti camps and their activities had been even more widely distributed in the forest. The traces of these remain as old secondary forest, distinguished from the primary forest by the low frequency of mature Caesalpiniaceae trees and the high proportion of deciduous trees of Meliaceae, Euphorbiaceae, Ulmaceae and Anacardiaceae [T. Hart 1985]. According to Hart and Hart [1986], the density of food trees is significantly higher in the secondary forest than in the primary forest (13 trees/ha on the average in the former as compared with 6/ha in the latter).[7] Therefore, the resource base for wild plant food has been improved by the activities of villagers as well. Moreover, the plants on which edible caterpillars and grubs rely are more commonly found in secondary forest, as shown in Table 16.2.

Mbuti dependence on these light-demanding plants shows the importance of secondary vegetation to their subsistence. The regeneration of this vegetation, however, is facilitated by human activities, both those of the villagers and those of the Mbuti. We should also consider the potential of abandoned fields and the resulting secondary vegetation for attracting animals [see Wilkie 1989] which are hunted by the Mbuti and exchanged with the villagers. The Ituri Forest as a human habitat has been gradually modified through the interaction of vegetation, animals, Mbuti and villagers, and cannot, therefore, be called a 'pristine forest'.

Apart from these modest influences, Mbuti have not made an impact which might lead to the serious deterioration of forest vegetation. Except for the area around large villages and towns, Mbuti are distributed fairly evenly, at a density of 0.4 to 0.5 person/km^2, throughout the forest. There is no area in Ituri which has never been visited by Mbuti. Indeed, their hunting and gathering territory has shifted over a period of several decades [Ichikawa 1978, 1986]. For example, bands in the Teturi area moved, over fifty to sixty years, as much as 200 km from Koki on the main road to their present location in Teturi.[8] Despite an even distribution of the Mbuti population and their migration, more than 80 per cent of Ituri still remains dense tropical forest.

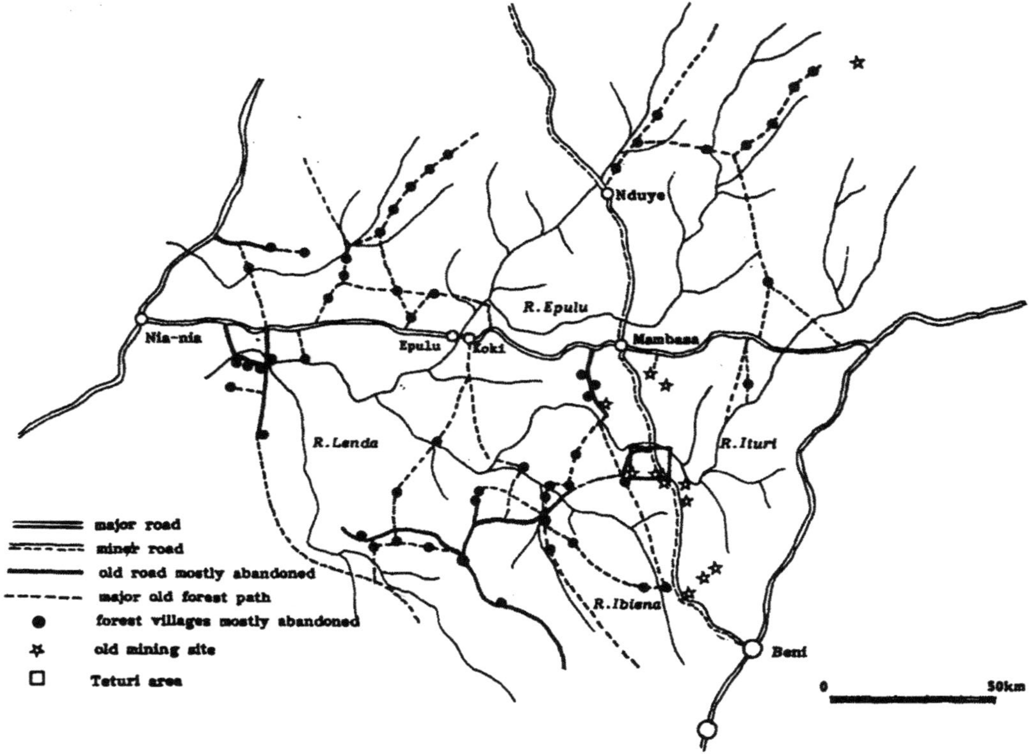

Figure 16.3. Distribution of old village sites in the Central Ituri Forest (based on the map of 'Zone de Mambasa')

Table 16.2. Food plants of major edible insects and their habitat

	Scientific name of the plant (Family)	Vernacular*	Habitat (Source)
#	*Bridelia micrantha* (Euphorbiaceae)	enjeku[a]	Fringing and secondary forests (3)
#	*Ricinodendron heudelotii* (Euphobiaceae)	songo[b]	Fringing, deciduous and secondary forests (3); regenerates only in gaps (4)
#	*Tetracarpidium conophorum* (Euphorbiacea)	tobye[b]	Low bush (3)
	Entandrophragma cylindricum (Meliaceae)	poyo[b]	Deciduous and evergreen forest (3); regenerates in shade (4)
#	*Triplochiton scleroxylon* (Sterculiaceae)	toko[c]	Common in deciduous forest; grows rapidly in secondary forest (3)
	Petersianthus macrocarpus (Lacythidaceae)	hoyo[b]	Drier parts of closed forest (3); regenerates in shade and small gaps (4)
#	*Celtis adolfi-friderici* (Ulmaceae)	kene[d]	Deciduous forest (3); regenerates in shade (4)
#	*Elaies guineensis* (Palmae)	ngasi[e]	Secondary forest and abandoned fields (1, 2)
#	*Raphia* sp. (Palmae)	libondo[e]	Secondary forest (1, 2); often protected by man (1)

#: Indicates that the species is mainly found in secondary forest. Source: 1) Observation by the author; 2) Hart and Hart 1986; 3) Irvine 1961; 4) Hall and Swaine 1981

*: Edible caterpillars are generally called **ba-soko**, whereas each species is called by the same name as its food plants. a) A species of silk-moth (*Anaphe* sp.); b) Unidentified species of butterflies or moths; c) Another species of *Anaphe*; d) The grubs of beetles feeding on this wood are called **pela**; e) The grubs of the giant elephant beetle feeding on rotten palm wood are called **sholewa**.

Turnbull [1972; 1983] stated that there is a 'no man's land' which is not utilized by the Mbuti. In fact, the farthest part of their territory is seldom used for hunting and gathering, except for hunting big game with a spear. According to the Mbuti, this is simply because such remote areas are too far to carry farm food, the major source of calories, and no ritual protection seems to be involved in maintaining this no man's land. It consequently serves as a 'natural reserve' for animal and plant populations. Such a centre for resource regeneration has shifted with the band migration which has taken place over a period of several decades. It is largely due to the high mobility of Mbuti bands, along with their low population density, that the forest resource base has been conserved, despite their long history of forest occupation. The frequent movement of camp-sites and the shifting of territories have both resulted in dispersed land use, through which a sort of equilibrium has been maintained between Mbuti and forest resources.

However, there is increasing hunting pressure by the Mbuti on animal populations. The hill region to the east of Ituri forms one of the major agricultural centres of Zaire, supporting a high population density. After Belgian colonization, plantations and small mines were opened up in this region and several commercial centres established. Since the agriculturalists in this region do not keep livestock, except chickens and some goats which are seldom slaughtered except on ritual occasions, meat from the forest is an important source of animal protein. Moreover, as these people are rapidly losing contact with the forest wilderness, the meat is also valued as a source of 'wild power' which cannot be obtained from fish or domestic animals. The meat traders, mostly Nande people from the eastern hill country, take clothing, rice and cassava flour to the Mbuti hunting camps, where they exchange them for duiker meat at fixed rates. Such meat trading has brought the Mbuti into direct contact with a market economy for the first time.[9] The meat obtained from the Mbuti is carried back to the eastern hill region and sold there at a price five to six times that which was paid for it in the forest. Meat-trading was first introduced by the Nande in the 1950s [J.A. Hart 1978] and has also involved some of the Bira, the Bantu-speaking forest villagers. It has accelerated the intensification of Mbuti hunting activities to some extent [Ichikawa 1991a].

In the middle 1970s when my first field research was

conducted, the annual catch from net-hunting by a band of sixty-seven people was estimated at 7,000 kg [Ichikawa 1986], that is about 100 kg per person. Their hunting territory covered an area of about 150 km^2, in which the biomass of duikers (the major target of net-hunting) was estimated at 45,000 to 75,000 kg. This means that the annual catch comprised only 10 to 15 per cent of the standing stock. If the hunting pressure remained at this level, the resource base will not deteriorate.

Despite involvement in commercial hunting, Mbuti hunting did not result in excessive pressure on animal populations until recently. This is largely due to the character of economic exchange in which they participate. Meat-trading is a means of obtaining food and other consumer goods which directly satisfy their daily requirements. Once their desire is satisfied, they are no longer interested in hunting for additional meat. Consequently, hunting pressure is kept at a low level. Moreover, stable exchange rates have been maintained. Figure 16.4 shows the recent changes in the wage and commodity price in Kinshasa, the capital of Zaire. While the commodity price rose at a tremendous rate, wages did not actually increase. In the Teturi area, prices of commodities rose as high as 500 to 1000 times during the twelve years from 1975 to 1987 (Figure 16.5); see also Ichikawa 1991a. The relative prices also changed during the same period. Despite such changes in cash prices, the exchange rates did not change substantially; one unit of meat (weighing from 2 to 2.5 kg) exchanged for 1.5 to 2 kg of cassava flour, etc.[10] Therefore, the Mbuti did not need to hunt additional animals even when cash prices changed considerably. The stable exchange thus contributed to maintaining the balance between their needs and the resource base. I consider some further implications of this stable exchange system below.

The Influences of Development and Conservation Projects

The people in the Ituri Forest are now facing serious problems which may lead to a fundamental change in their way of life. One of the problems is the destruction of forest associated with the economic development of the region. While the Ituri Forest itself has a very low population density, it is surrounded by densely populated areas (Figure 16.6), which threaten the prospects for effective forest conservation. Particularly to the east of the forest lies the hill region which, blessed with fertile soil of volcanic origin, forms one of the most productive agricultural areas of the country. Land shortage has recently become a serious problem here owing to rapid population increase and the expansion of cash-crop cultivation. Increasing numbers of people are now migrating westward into the thinly populated lowland forest, which causes further destruction of forest and degradation of an already poor soil. It has now become difficult for the Mbuti to find good hunting ground, particularly around the large villages. At the same time demand for labour in the area is increasing as agricultural production expands. As a result, some Mbuti have abandoned their traditional hunting life and become wage labourers working for cash-crop cultivators [Bailey 1982; Ichikawa 1991b].

Another factor adversely affecting the ecological balance of the forest is gold-mining. In 1981 the Zairean government liberalized the law relating to the mining of gold, which had been prohibited since independence. Since then, the population of the Ituri has been rapidly increasing. People have been migrating, particularly from the eastern hill region, to prospect for gold and seek new land for cultivation. In some areas, the village population more than doubled during the early 1980s. Villages with more than a hundred houses are now to be found, something which was unknown before 1981. The gold miners have even begun to enter the 'no man's land' which had been seldom used by Mbuti. Mbuti have also begun to hunt duikers in this area to sell to the miners.

With the destruction of the forest ecosystem has come a growing concern for its conservation. In the centre of the Ituri Forest, a nature reserve of 1.4 million ha is planned (about half of it to be declared a national park) for the protection of wildlife, including the famous okapis. In this plan, the Mbuti are to be allowed to hunt and gather for their own consumption. Even so, their life-style would be considerably restricted by this project, as they have long been exchanging forest products for farm food, clothes and other commodities of external origin. The Mbuti, like other contemporary hunter-gatherers, are now placed in a dilemma, caught between 'development' on the one hand and

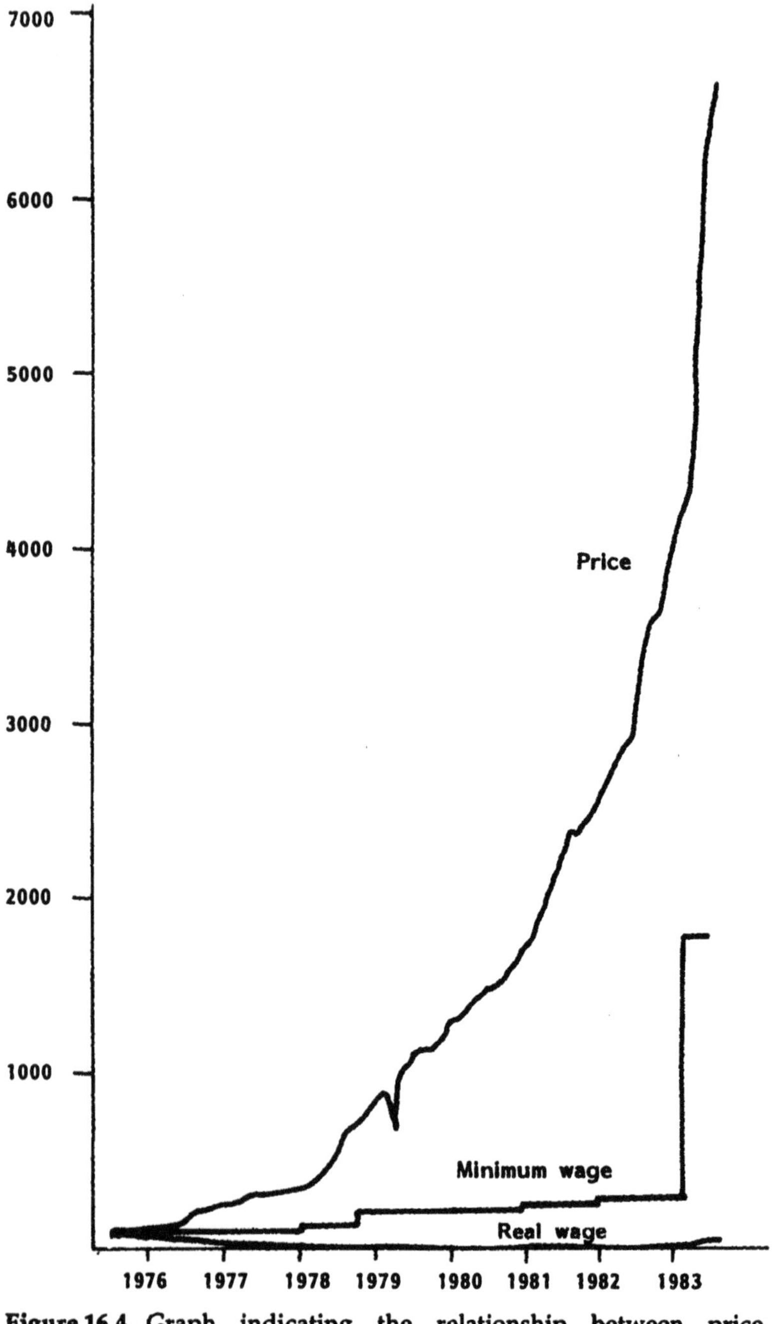

Figure 16.4. Graph indicating the relationship between price, minimum wage and real wage for the period 1975–84 (1975=100, adapted from [Ohbayashi 1986])

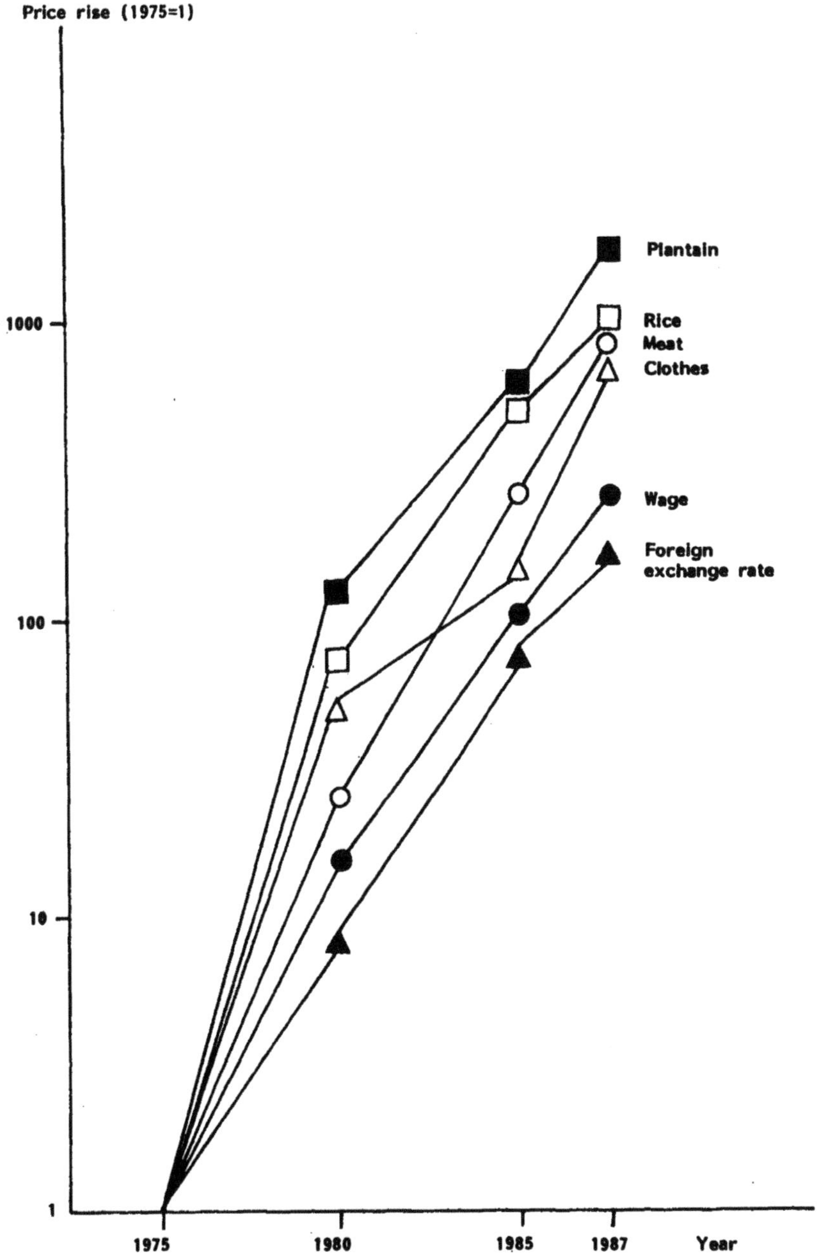

Figure 16.5. Price increases in the Teturi area during the twelve years from 1975 to 1987. The rise in the foreign exchange rate for US$ is also shown for comparison

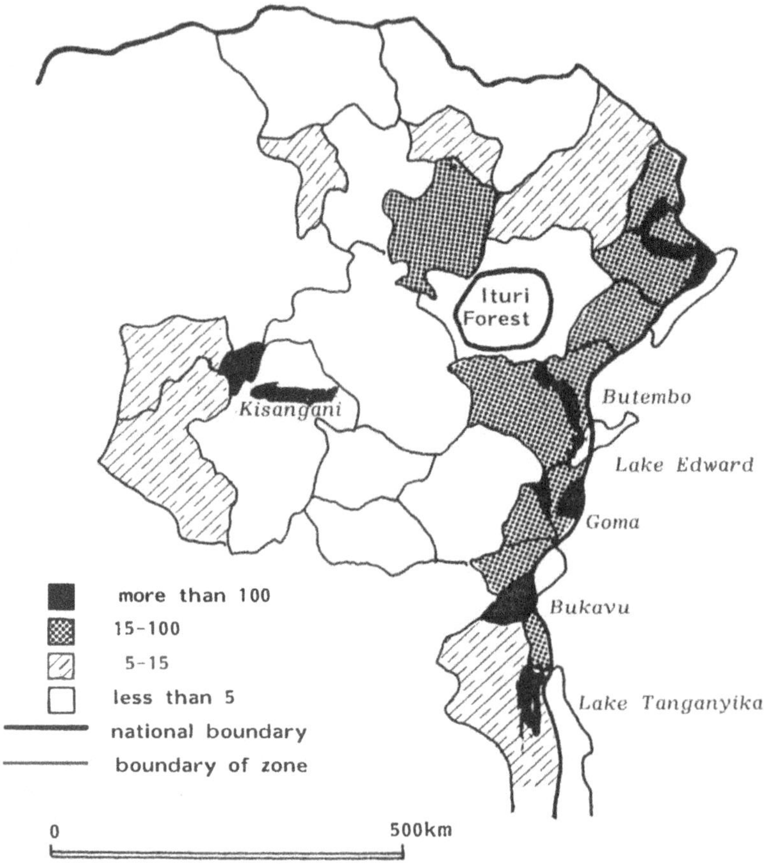

Figure 16.6. Population density (person/km^2) in Northeastern Zaire. Average population density for Zaire is 11 person/km^2. Modified from Ohbayashi, 1986

'conservation' projects on the other.

As a result of development projects, huge trees grown over hundreds of years are cut, and an intricate and diverse forest ecosystem is destroyed. Also disappearing is the basis of a unique human forest culture, which has evolved with the forest over many centuries. The problem cannot be solved by simply establishing a natural sanctuary from which all human activities are excluded. Along with the conservation of the forest, some measures must be taken to preserve the culture of the forest people. It is necessary to design conservation schemes 'with a human face' [Bell 1987], which include human activities within protected areas.

Conventionally, development is represented as a human attempt to conquer nature, while conservation is the converse, an attempt to counteract the excesses of development. Both seek to control nature. As anthropologists, we must examine attitudes towards nature which treat it as an external object to be controlled, either through exploitation or protection. We need a model which does not reinforce the simple dichotomy of man and nature, and in this regard, hunter-gatherer societies may provide us with some important clues. In the final section of this chapter, I examine how we might use the Mbuti case to establish such a model, focusing on the sustainable use of forest.

The Mbuti Case as a Model for the Sustainable Use of Rain Forest

Despite a long history of forest occupation, the Mbuti have not significantly depleted plant and animal resources. There are important lessons to be learned from the Mbuti case for our understanding of the relationship between humans and other species in the tropical rain forest. The Mbuti case also provides us with a model, as well as the fundamental data, for planning the sustainable use of the forest in both ecological and socio-cultural terms.

First, the value of the forest should not be measured only by a small number of resources with commercial values, such as timber and rubber. The forest provides the Mbuti with many useful resources which are not commercially exploited, but which are none the less important to their spiritual as well as material life, both directly and indirectly. It is necessary to appreciate the potential of the forest as a source of such use values, though not, for the moment, of exchange values. Thus, Mbuti need the forest as a whole, rather than as a source of specific items actually used. We should see the Mbuti case as a critique of our use of the forest, which is utilitarian and commercially oriented.

There is an increasing interest in tropical forest as a repository of new medicines, foods, condiments, dyes and other useful products. This is understandable, if we take into consideration the great diversity of forest plants which have accumulated various secondary chemical compounds. The exploitation of such potential is expected to contribute to the conservation of the forest, since it utilizes non-timber forest resources which can only be produced by living trees. However, commercialization usually involves intensive exploitation of resources, which might lead to the degradation of the resource base. More importantly, it would also lead to the distortion of the unique forest culture characterized by the rich assets of use values.

Secondly, Mbuti society has a mechanism for avoiding the over-extraction of forest resources. Their nomadic life-style should be re-examined in this regard. Nomadism is generally viewed as an ancient way of life, typical of the period prior to the introduction of agriculture. In fact, the Zairean government once tried to sedentarize the Mbuti along the major roads. This was not successful, probably because the government underrated their need to range widely over an extensive area. Nomadism is perhaps the best way to utilize resources which are thinly distributed over an extensive area. Moreover, there is a 'no man's land' which is seldom utilized by the Mbuti. This area consequently serves as a natural reserve for the regeneration of plant and animal populations. This 'no man's land' shifts according to the migration of the band, which also contributes to the dispersion of land use. Through a dispersed extraction of resources the hunting and gathering pressures were kept, at least until recently, at a lower level. Even twenty years after the introduction of the meat trade into this area, the annual catch of the Mbuti was estimated at 10 to 15 per cent of the standing stock, which is well within the range of sustainable use.

Thirdly, the Mbuti maintain a sort of buffer against the national market economy. Meat hunted by the Mbuti is usually exchanged directly for agricultural food or clothes. The rates of exchange have not changed much over the last twelve years. This is surprising if we take into consideration the tremendous price rise in commodities and the change in their relative prices during this period. Such stability in the exchange rate demonstrates the relative independence of the local economy from the national cash economy. Because of this dual economic structure, the Mbuti did not need to produce additional meat even when the relative prices in cash terms changed.

As stated before, the Mbuti participate in the meat trade in order to acquire their daily necessities. They sell meat for cash only when they need cash for some specific purpose, such as paying bridewealth, taxes or fines. In this sense, cash to them is no different from other commodities obtained through exchange. When they acquire cash without any specific purpose in mind, it is immediately spent. The purpose of exchange is not to acquire profit itself; Mbuti are interested neither in storing nor in accumulating money for its general purchasing power. By maintaining their exchange orientation towards 'the present' [Meillassoux 1973; Woodburn 1980], they keep themselves free from the inverted desire for profit which is common among people involved in the market economy.

Those deeply involved in the market economy pursue maximization of the production or productive efficiency of a few resources with commercial importance. This is essentially incompatible with the forest ecological system, which is sustained by a delicate balance of its numerous elements. It is also incompatible with the life of a people who utilize a variety of forest resources and maintain a stable relationship with other components of the forest ecosystem. Mbuti, however, restrict themselves to acquiring concrete use values while maintaining a stable exchange in the local economy. Such a buffer may be necessary to avoid the hazardous effects of the market economy on both their society and forest resources. We should therefore re-examine the potential of the existing dual economy, in particular the role of a local economy based on the exchange of use values, in the increasing prevalence of the national market economy.

Finally, it is emphasised that the tropical rain forest of Ituri has long been influenced by human activities. Scattered in the primary forest of Caesalpiniaceae trees, there are patches of light-demanding trees, or trees growing in disturbed areas. Some of these comprise important sources of Mbuti food and also attract animals to feed on them. The regeneration of these trees has been facilitated, at least partly, by human intervention, both by the Mbuti and by villagers. We may, however, recall here the ritual protection of **akobisi** trees which are extensively distributed in the dense forest. As this tree is small (usually only several metres in height) and easily damaged while clearing the surrounding vegetation, it is no exaggeration to say that ritual protection has contributed to the preservation of the forest vegetation.

Most of the Ituri Forest has long ceased to be a 'vast stretch of pristine forest' in the strict sense. On the contrary, in this context 'nature' must be understood as the outcome of a complex ecological history involving the interaction of humans and other species. Therefore, if a wildlife sanctuary is established and if human activity is excluded from this area, the forest might become different from what it is at present. What is required is not conservation simply through the cessation of human extraction, but 'conservation with a human face' [Bell 1987]. That is, conservation of an ecosystem which includes humans (both the Mbuti and the villagers) as indispensable components.

Acknowledgements

The field research on which this study is based was carried out in the Ituri Forest of Zaire from 1974–5, 1980–1, and in 1985, 1987, 1989 and 1990. I express my sincere thanks to Dr Zana Ndontoni, the director of the Center de Recherche en Sciences Naturelles (CRSN), who kindly accepted me as research collaborator. The study is financially supported in part by the Nippon Life Insurance Foundation.

Notes

1. The only mammal species that Mbuti do not eat are goats, house rats and small insect-eating bats.
2. Apakumandura literally means 'master of the forest', and is also called Kalisia in some areas. **Apa** originally meant 'father' or 'the owner', and **ndura** means 'forest'.
3. Another visible agent of Apakumandura is a small mushroom-shaped termite tower called **pakira** which is also found extensively in dense forest. When the Mbuti catch a forest francolin (a quail-like bird) in their nets, they throw its head into the forest as an offering to Apakumandura.
4. Turnbull also insists that the Mbuti conceptualize the forest under the canopy as a 'womb'. However, this is too simplistic a generalization. When an Mbuti dies he is buried in the forest

behind the camp. Mbuti say that the corpse remains under the ground, whereas the person 'himself' becomes **tore** or **keti** (spiritual beings including ancestors) and roams around the forest. According to Sawada [1990], some Mbuti songs are said to have been inspired by **tore**, during dream encounters with them in the forest. The forest to the Mbuti is therefore the place from which people come (the womb) and to which they go after death.

5. According to some Mbuti, cultivation was practised more widely in the past than it is now.
6. A *Canarium* tree is often said to be an indicator of an old village or camp-site.
7. This generalization excludes the case of forest dominated by *Gilbertiodendron dewevreii.*
8. Waehle [1986] reports a similar example of band migration among the Efe in northern Ituri.
9. Although some Mbuti hunting products, ivory in particular, have been traded for a long time, the exchange has been mediated by villagers [Ichikawa 1991b].
10. The exchange rate, however, differs according to the location of the hunting camp, i.e. the distance from the towns or large villages.

Bibliography

Bahuchet, S., 'Aka-farmer relationship in Central African rain forest', in R.B. Lee and E. Leacock (eds), *Politics and history in band societies*, Cambridge: Cambridge University Press, 1983

——, *Les Pygmees et la Foret Centralafricaine*, Paris: SELAF, 1985

Bailey, R., 'Development in the Ituri forest of Zaire', *Cultural Survival Quarterly* 6(2), 1982, pp.23–5

Bell, R., 'Conservation with a human face: conflict and reconciliation in African land use planning', in *Conservation in Africa*, D. Anderson and R. Grove (eds), Cambridge: Cambridge University Press, 1987

Gusinde, M., *Die Twiden: Pygmaen und Pygmoide in Tropischen Afrika*, Wien: Wilhelm Braumüller, 1956

Hall, J. B. and M. D. Swaine, *Vascular plants in a tropical rain forest*, The Hague: Dr. W. Junk 1981

Hamilton, A., 'African Forest', in F.B. Golley (ed.), *Tropical rain forest ecosystem (ecosystems of the world)*, Amsterdam: Elsevier Scientific Publishing Company, 1983

Harako, R., 'The Mbuti as hunters', *Kyoto University African Studies* 10, 1976, pp.37–99

Hart, J. A., 'From subsistence to market: a case study of the Mbuti net hunters', *Human Ecology* 6(3), 1978, pp.325–53

Hart, J. A. and T. Hart, 'The ecological basis of hunter-gatherer subsistence in the Ituri forest of Zaire', *Human Ecology* 14, 1986, pp.29–55

Hart, T., 'The ecology of a single-dominant forest and a mixed forest in Zaire, Equatorial Africa', unpublished Ph.D. thesis, Michigan State University, 1985

Hiernaux, J., *The people of Africa*, London: Weidenfeld and Nicolson, 1972

Ichikawa, M., 'The residential groups of the Mbuti pygmies', *Senri Ethnological Studies* 1, 1978, pp.131–88

——, 'Ecological and sociological importance of honey to the Mbuti hunter-gatherers', *African Study Monographs* 1, 1981, pp.55–68

——, 'An examination of a hunting-dependent life of the Mbuti pygmies', *African Study Monographs* 4, 1983, pp.55–76

——, 'Ecological bases of symbiosis, territoriality and intraband cooperation of the Mbuti pygmies', *Sprache und Geschichte in Afrika* 7(1), 1986, pp.161–88

——, 'Food restrictions of the Mbuti pygmies', *African Study Monographs*, supplementary issue 6, 1987, pp.97–121

——, 'The impact of commoditisation on the Mbuti of Zaire', *Senri Ethnological Studies* 30, 1991a, pp.135–62

——, 'Barter and cash transactions in the Ituri forest of Zaire: with special reference to the inconsistency of exchange rates', in *Interpretations of other cultures*, Y. Tani (ed.), Kyoto: Jinbun-shoin, 1991b (in Japanese)

——, 'Traditional use of tropical rain forest by the Mbuti hunter-gatherers in Africa', in *Topics in primatology: behavior, ecology and conservation*, N. Itoigawa et al. (eds), Tokyo: University of Tokyo Press, 1992

Irvine, F. R., *Woody plants of Ghana*, Oxford: Oxford University Press, 1961

Jewsiewicki, B., 'Rural society and the Belgian colonial economy', in D. Birmingham and P.M. Martin (eds), *History of Central*

Africa, London: Longman, 1983, pp.95–125
Kadomura, H., 'Environmental change in tropical Africa', *Global Environmental Problems* 2, 1990, pp.6–93 (in Japanese)
Le Thomas, L., *Flore du Gabon: Annonacees*, Paris: Museum National d'Histoire Naturelle, 1969
Mayr, E. and R. J. O'Hara, 'The biogeographic evidence supporting the pleistocene refuge hypothesis, *Evolution* 40(1), 1986, pp.55–67
Meillasoux, C., 'On the mode of production of the hunting band', in P. Alexandre (ed.), *French perspectives in African studies*, Oxford: Oxford University Press, 1973
Ohbayashi, M., *Agriculture in Zaire: present situation and problems for development*, Tokyo: Association for International Cooperation of Agriculture and Forestry, 1986 (in Japanese)
Powell-Cotton, P.H.G., 'Notes on a journey through the great Ituri forest', *Journal of the African Society* 7(25), 1907, pp.1–12
Sawada, M., 'Two patterns of chorus among the Efe, hunter-gatherers in northeastern Zaire: why do they love to sing?', *African Study Monographs* 10(4), 1990, pp.159–95
Tanno, T., 'The Mbuti net-hunters in the Ituri forest, eastern Zaire: their hunting activities and band composition', *Kyoto University African Studies* 10, 1976, pp.101–35
——, 'Plant utilization of the Mbuti pygmies: with special reference to their material culture and use of wild vegetable food', *African Study Monographs* 1, 1981, pp.1–53
Terashima, H., 'Mota and other hunting of the Mbuti archer's, *African Study Monographs* 3, 1983, pp.71–85
Turnbull, C., *The forest people*, New York: Simon and Schuster, 1961
——, *Wayward servants: the two worlds of the African pygmies*, New York: Natural History Press, 1965
——, 'Demography of small-scale societies', in G. Harrison and A. Boyce (eds), *The structure of human population*, Oxford: Clarendon Press, 1972
——, *The Mbuti pygmies: change and adaptation*, New York: Holt, Rinehart and Winston, 1983
UNESCO, *Vegetation of Africa*, Paris: UNESCO, 1983
Vansina, J., 'Long-distance trade-routes in Central Africa', in Z.A. Konczacki and J. M. Konczacki (eds), *An economic history of tropical Africa*, London: Frank Cass, 1977
Waehle, E., 'Efe (Mbuti pygmy) relation to Lese Dese villagers in

the Ituri forest, Zaire: historical changes during the last 150 years', *Sprache und Geschichte in Afrika* 7(2), 1986, pp.413–32

Wilkie, D.S., 'Impact of roadside agriculture on subsistence hunting in the Ituri forest of northeastern Zaire', *American Journal of Physical Anthropology* 78, 1989, pp.485–94

Woodburn, J., 'Hunters and gatherers today and reconstruction of the past', in E. Gellner (ed.), *Soviet and western anthropology*, London: Duckworth, 1980

Chapter 17

Image and Reality at Sea: Fish and Cognitive Mapping in Carolinean Navigational Knowledge

Tomoya Akimichi

Introduction

The Carolineans of Micronesia are known as expert navigators who have created and manipulate a system of unique knowledge and technical skills in order to conduct ocean voyages. For the indigenous populations of these areas this has not only become an empirical tool for moving from one island to another, but also carries socio-economic significance in sustaining life in atoll environments [Alkire 1965, 1978].

Although the art of navigation itself has come to be employed only on certain islands of the Central Carolines, enquiries into the system of Carolinean navigational knowledge seem to have gained more and more attention from those who are interested in the cognitive psychology and development of space perception [Hutchins 1983, 1988; Neisser 1976] as well as Pacific culture history [Bellwood 1979; Gladwin 1970; Lewis 1975; Riesenberg 1976]. Recent discussion on spatial cognition also suggests the importance of this problem to wider theoretical issues [Frake 1985; Gell 1985].

Such debates are clearly associated with and focused on how navigators recognize location at sea without using modern navigational apparatus and maps. Hitherto, the **etak** or reference island has been considered to represent practical knowledge enabling location-finding [Alkire 1970; Gell 1985; Gladwin 1970; Lewis 1975]. Given the primary importance of **etak** in practical navigation, how are we to understand the role of other kinds of navigational knowledge? As Riesenberg notes, on the basis of the Puluwatese case, there are other types of knowledge in which imaginary objects and esoteric existences are frequently cited. Unfortunately, these have been treated as non-technical, mnemonic devices, or as ways of maintaining secrecy among members of the same matrilineal clan or those belonging to a particular navigational school [Riesenberg 1976: 125–8]. Such a dichotomy between utilitarian and non-utilitarian, however, leads to a disregard of cultural achievements in navigational knowledge and a failure to understand spatial cognition and navigational skills which, I believe, are more abstract, anomalous and complex than might be thought.

In this chapter, using my own field data from Satawal island, I will first examine navigational knowledge which has been considered non-effective for practical navigation. I will then illustrate the importance of that knowledge in which fish and other sea creatures are employed as metaphorical tools to facilitate orientation and place-finding at sea. Finally, I introduce an image-mapping theory as a new way of understanding Carolinean navigational skills.

Satawalese Navigation

Satawal is located in the middle of the Caroline Islands of Micronesia (7 degrees N, 143 degrees E). It is a low raised coral island of about 0.06 km^2. Like other atoll inhabitants in Micronesia, the (as of 1980) 500 or so Satawalese depend on the cultivation of root and tree crops and on marine extraction for subsistence [Akimichi 1986; Alkire 1978]. The socio-economic significance of navigation in this area has already been reported [Akimichi 1986; Alkire 1965, 1978; Lessa 1966; Sudo 1979], and Satawalese navigators are famous for having retained extraordinary navigational skills. The data presented here are derived either from Satawalese or Puluwatese informants, the latter being neighbouring atoll dwellers on whom we already have considerable information [Gladwin 1970; Lewis 1975; Riesenberg 1976]. The Satawalese data were obtained during fieldwork on Satawal island in 1980 as part of a research project into the traditional navigational knowledge of the Central Caroline Islands headed by S. Ishimori of the National Museum of Ethnology and funded by the Monbusho Overseas Grant-in-Aid Research Programme of the Japanese Government.

Sidereal Compass and Basic Knowledge

At sea, where no landmarks are visible, it is indispensable for the navigators to know where they are, and how to get to their destination. Even in modern sailing conducted with the aid of satellites, a magnetic compass, marine charts and even a computer, knowing the proper orientation of a vessel is of the utmost concern. Without such modern apparatus, however, Carolinean navigators have invented a unique space-reckoning technique. As we shall see below, in the navigational knowledge retained by the Carolineans, the most fundamental technique is determining orientation using a locally developed sidereal compass.

The Sidereal Compass

A sidereal compass or **nááng** (lit., 'heaven' or 'sky') is the most fundamental tool of Satawalese navigation, used widely in learning navigational knowledge. It is a round compass, along the circumference of which thirty-two different positions as viewed from the centre are marked (Figure 17.1). Fifteen selected stars and constellations are used for allocating these thirty-two positions [Goodenough 1953]. As each position coincides with the rising or setting of particular stars on the horizon, it is basically symmetrical along the north-south axis. For instance, the rising and setting positions of alpha Altair (**máyinap**) face each other along the east-west line. Hence, each star position is conceived in terms of **táán máyinap** versus **tupwun máyinap**. **Táán** and **tupwun** here denote 'to rise' and 'to set' respectively. Exceptionally, Polaris occupies only one position, i.e. north, whereas Crux provides five positions. As each position is ideally equidistantly spaced from neighbouring positions, some directional discrepancies of as much as fifteen degrees result, if compared with the actual positions of rising and setting observed with a magnetic compass. In this chapter, the numbers one to thirty-two are used to represent star courses, as shown in Figure 17.1.

Four Basic Lessons

Four basic lessons learned at the elementary stage of knowledge acquisition are **paafúú, yárhowumw, yamas** and **woofanúw** [Akimichi 1981a].

Paafúú is 'to count and memorize stars'. It is used for memorizing star courses based on thirty-two named positions of the sidereal compass. Usually, citation of each star course commences with **táán máyinap** (P7), moves counter-clockwise, and then back to the same position (see Figure 17.1).

Yárhowumw means 'partnership of stars' and is used for memorizing the position of a particular star with its reciprocal partner located on the opposite side of a sidereal compass; for instance Polaris (P1) is coupled with Crux upright (P17), and the rising of Vega (P5) is coupled with the setting of Antares (P21). There are sixteen such pairs in total.

Yamas is used for locating four different star courses as a set. It links four star courses in four different directions with reference to four canoe parts; bow, stern, outrigger-side and platform-side. This set corresponds to two pairs of stars which cross-cut each other (Figure 17.2).

Woofanúw means 'to see the island if it is a due course', and comprises a set of data on island orientation as observed from one island, using the thirty-two direction terminology. **Woofanúw** is specific to each island. It refers not to imaginary things, but, as a rule, to real islands and reefs. The **woofanúw** of Satawal, for instance, includes eleven islands, two reefs, one channel and one area of sea. Everything referred to in this set is likely to be real. However, even esoteric beings are occasionally mentioned with respect to **woofanúw**, as found in the case of that for Gaferut island, where the mythical island of Fanúwánkuwen is included [Akimichi 1984: 666–7].

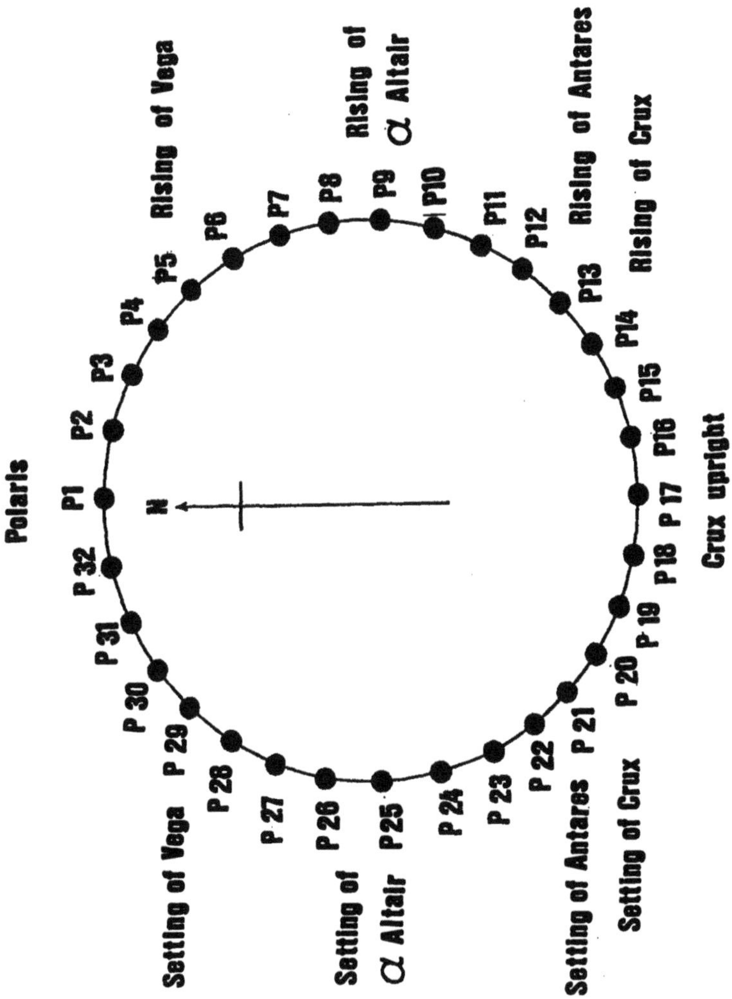

Figure 17.1. Sidereal compass of the Caroline Islands

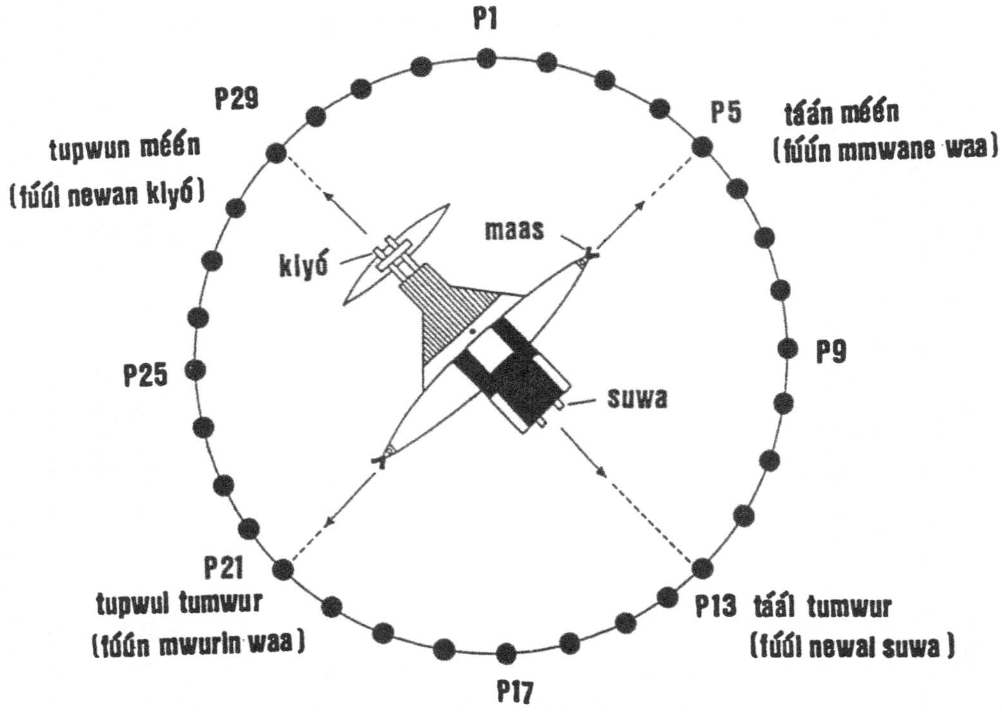

Figure 17.2. Diagram to show the learning of *yamas*

Advanced and Specific Lessons

Besides the four types of navigational knowledge mentioned above, more advanced and special knowledge is transmitted. This includes various kinds of information and skills related to navigation, such as specific methods of island orientation, landfall techniques, knowledge of sea birds and fish, names for areas of sea [Akimichi 1984], waves and currents, and ways of forecasting storms [Akimichi 1980; Burrows and Spiro 1957]. Taboos and restrictions, related to ideas concerning the supernatural existence of the canoe, navigation, ocean and sky, and reflected in much magic and ritual, also form an important part of navigational knowledge [Ishimori 1985].

Not all of these kinds of knowledge are known to everyone. Some is often the secret knowledge of particular individuals and certain groups. Because of this, it is almost impossible to get a precise view about who knows most and who knows what . In this chapter, I try to establish a generalized model of spatial representation rather than describe exhaustively the navigational knowledge shared by contemporary navigators.

Fish in Navigational Knowledge

Several kinds of navigational knowledge are presented in this section to give some idea of how fish are labelled and identified, and how these are associated with practical navigational skills. The knowledge is discussed here under three different headings, according to type of fish, and how these are cognitively and culturally manipulated: fish encountered, fishing techniques and fish-configuring geographical spaces.

Fish Encountered

Pwukof

Pwukof is systematic knowledge concerning the appearance of things, animate and inanimate, at sea. Such things are encountered if one sails in a particular direction from one island.

Pwukof is derived from the word **pwukopwuk** 'to bind', possibly referring to the association between an island and its surrounding environment.

Information on **pwukof** is island-specific, like **woofanúw**, and each set of knowledge is composed of various named things occasionally linked to particular behavioural features, under a particular star course. For instance, under star course P11 (see Figure 17.1), two male whales, designated as **napwichenuk**, appear at a certain point. Under star course P18, a tern whose name is **mesórumeyéér** appears, but not singing; while under star course P19, another tern, **mesórumeyefáng**, appears, but singing. Similarly, under thirty-two different star courses around Satawal island, sharks, frigate birds, whales, tuna, reefs and islands are designated in this way. The difference between **pwukof** and **woofanúw** should be noted.

Despite being acknowledged as a basic item of Carolinean navigational knowledge, **pwukof** has not yet been examined thoroughly, although it is mentioned briefly in the literature [Alkire 1970; Gladwin 1970; Riesenberg 1976]. As there are individual differences in content, more than one instance are recorded here from data for several islands. I collected thirty-six instances in total, from eighteen islands. The kinds of things mentioned in **pwukof** are diverse. Biological things include fish, birds, marine mammals, insects, driftwood, sea algae, crustacea, shellfish, sea snakes and sea turtles. Oceanographic phenomena include whirlpools, foam and waves. It includes forty-seven identifiable Micronesian islands and atolls, the other islands and reefs being barely identifiable on a marine chart [Akimichi 1988: 127–73].

Biological phenomena occur most frequently in **pwukof**. This is one reason why **pwukof** is translated as 'sea life' by Riesenberg [1976: 128]. If we focus on biological things, it becomes evident that fish and birds appear in about the same numbers. Of the eighteen cases selected from eighteen islands, 258 instances of biological phenomena are recorded. For birds, the frequencies are as follows; thirty-three frigate birds (12.9 per cent), fifteen tropic birds (5.9 per cent) and ten boobies (3.9 per cent) in order of appearance. For fish and marine mammals, the frequencies are: twenty-one sharks (8.2 per cent), eighteen whales (7.0 per cent), fourteen sailfish (5.5 per cent) and ten porpoises and rays (3.9 per cent).

A variety of behavioural and morphological features of animals are given. For birds, singing, fighting and stopping-over are frequently mentioned, while for fish, jumping is frequently mentioned. Morphological features and colours seem to be more metaphoric than real, as with the sailfish with a tail tied by a coconut leaf frond, a whale with its back rotted, a fish without a dorsal fin, or yellow fish such as shark, trevally and unicornfish.

The most fascinating feature of **pwukof** is that each item, animate or inanimate, has it own name. For instance, under the direction **tupwui sárepwén** in the **pwukof** of Satawal, one barracuda (**seraw**) is called **serawánimwár**, where **ni-** denotes 'of' and **mwár** a suffix attached to the thing. Even the reef, which is called **worhánipwóópw** (lit. 'reef of the shark'), is said to be the habitat of sharks called **pwóópw** in the **pwukof** of West Fayu island.

In the **pwukof** of Pulusuk atoll, barracudas designated as **yúnúmwárátá** and **yúnúmwárútiw** are said to appear under the directions **táánupw** and **tupwunupw** respectively. The former points its head upwards, the latter downwards. The suffixes **-tá** and **-tiw** denote 'towards this way' or 'up' and 'towards that way' or 'down'. The two items are binarily opposed: the rising of the Crux, upward, and **-tá**, versus the setting of the Crux, downward, and **-tiw**.

Senong

There is a similar system of knowledge in which sea creatures are said to be encountered under a particular star course direction radiating out from an island or a reef. **Senong** concerns twelve whales which are said to line up like coconut skids used for launching a canoe. **Senong** denotes 'to build coconut skids', but in this context **nong** also represents a whale with its head pointing north to where a particular island is located.

For instance, to the south (**wenewenenupw**) of Pulusuk atoll, there is a huge whale with its head pointing north. In the **tupwun máyinap** direction of this whale there also lies another whale with its head pointing north (**fúúsumwakút**), to where Pikelot island is located. A navigator then takes the north route until he reaches Pikelot. From Pikelot he returns to the south to meet the whale again, and then moves to the west. Following a similar course, the voyage lasts until he reaches Palau islands in the Western Caroline Islands. In brief, twelve whales are sighted due south of particular islands, and all face north.

The same system of knowledge is also reported from Puluwat atoll [Riesenberg 1976: 105–6], although the islands referred to in Riesenberg's text are not the same as in the Satawalese case, and only six whales are mentioned.

Fishing Techniques

Yarhuwow

Yarhuwow literally denotes 'to poke a hole with a stick'. This knowledge evokes an image in which a fisherman of Puluwat atoll dives for a kind of parrotfish, **wurha** (*Scarus jonesi*) [Akimichi and Sauchomal 1982: 19]. When he pokes the fish in the reef hole with a stick, it flees to another hole in the reef of Pikelot island. He tries to poke the fish again in the hole: then it escapes to Tamtam island. In this way he continues to pursue the fish from island to island until he eventually returns to Puluwat atoll and catches it.

The islands mentioned in connection with **yarhuwow** knowledge cover most of the inhabited islands and atolls of the Central and Eastern Carolines. The specific name of a reef hole, as well as information concerning the next destination of the fish, are given in terms of sidereal compass bearings. A case collected by Riesenberg is slightly different from the one I collected, referring to an unidentified parrotfish species, **meruput** [Riesenberg 1976: 94].

Rheániy

Rheániy is similar to **yarhuwow**, making use of fishing activity to evoke images of an imaginary voyage. **Yániy** is a generic name for grouper or rock cod (Epinephelidae), and in this context refers to fishing for **yániy** by hand when the fish is hidden in a hole. As one man tries to catch the fish by hand, it flees to a hole in another island. The chase after this fish starts from Pulusuk atoll until the navigator reaches the channel of the mythical island of Fanúwánkuen (lit. 'lizard island'). Not only reefs but also birds, fish and whales are mentioned as being encountered on the way. In this sense, it is similar to the information contained in **woofanúw** and **pwukof**.

It is clear from these two examples (Yarhuwow and Rheániy) that fishing and associated fish behaviour create images useful in continuous navigation from island to island.

Fish that Configure a Geographical Space

Pwuupwunapanap

Pwuupwunapanap (lit. 'large triggerfish') is a system of locating islands using an image of a large triggerfish as if it were lying in the water, with five different islands, reefs or other objects corresponding to the head, dorsal fin, ventral fin, tail and spine [Akimichi 1987: 279–98; Alkire 1970; Riesenberg 1976]. The word **pwuupw** originally means either triggerfish or the Crux. Triggerfish is one of the important food sources for islanders, not only in ordinary reef-fishing, but also when these fish accompany driftwood [Akimichi 1981b]. Driftwood is often associated with hugh quantities of certain species, such as trevally, leather jackets, jacks, and large predatory fish such as tuna, wahoo, rainbow runners and sharks.

Pwuupw is also applied to the Crux as used in the sidereal compass, and it is the perceived similarity in shape which results in the polysemic use of the term **pwuupw** for both triggerfish and the Crux. Furthermore, the triggerfish is regarded as cognitively identical with the Crux since triggerfish in the sea is transformed into the Crux in the sky when it emerges on the eastern horizon at sunrise. When it sinks down under the water, the star is transformed once again into a fish. Thus, **pwuupw** circulates counter-clockwise between the celestial and the aquatic worlds [Akimichi 1987].

In **pwuupwunapanap** knowledge, several sets of **pwuupw** are arranged which overlap longitudinally by half, as shown in Figure 17.3. When in motion a navigator visualizes his canoe as being within a large triggerfish. Certain islands and reefs correspond to the head, dorsal fin, ventral fin and tail of the triggerfish. There are individual variations in **pwuupwu-napanap**, and not only real islands and reefs but also imaginary ones are mentioned as forming a part or parts of the **pwuupw**.

Féfén perhan Ayufan

Féfén perhan Ayufan is an imaginary navigation from island to island. The navigator starts from Puluwat atoll and sails to different islands and reefs. When he approaches Truk from Mwirino atoll, he enters through a channel, **Mwokucchis**. The spot is regarded as the tail (**perhan**) of **Ayufan**, a mythical fish the shape of which is created by successive inter-island sailings. **Féfén perhan Ayufan** thus means 'to bind the tail of fish'.

The way in which sailing routes trace out the form of this fish are strikingly shown in Figure 17.4. The fish has a mouth, dorsal fin, caudal fin, body and ventral fin. Thus, the mental map of the fish enables navigators to locate themselves, for example by stating that we are now at the tip of the dorsal fin (Pulusuk atoll), or the tip of the caudal fin (Magur island).

Serhakini masaccha

Serhakini masaccha means 'Navigation of Spiny Squarrelfish' (**serhak**: navigation; **masaccha**: spiny squarrelfish, *Sargocentron spiniferum*). This refers to a body of knowledge also referred to as **semááy**, literally meaning a stone weir. This stone weir is made by piling coral in a V-shape so as to lure fish into the central trap. The course which the fish take is focused on the islands of Ulul and Pikelot, and they follow a series of reefs which extend from Puluwat atoll. The course taken also resembles the V-shaped stone weir or **mááy**, as shown in Figure 17.5, and is therefore also named **semááy**.

From the above examples, we can extract three distinct types of relationship with sea creatures, particularly fish and whales, used in navigational orientation. The first concerns those creatures 'encountered' during sea voyages, even though some may strictly be imaginary, or have features that are imaginary. The second clearly concerns the application of the notion of an

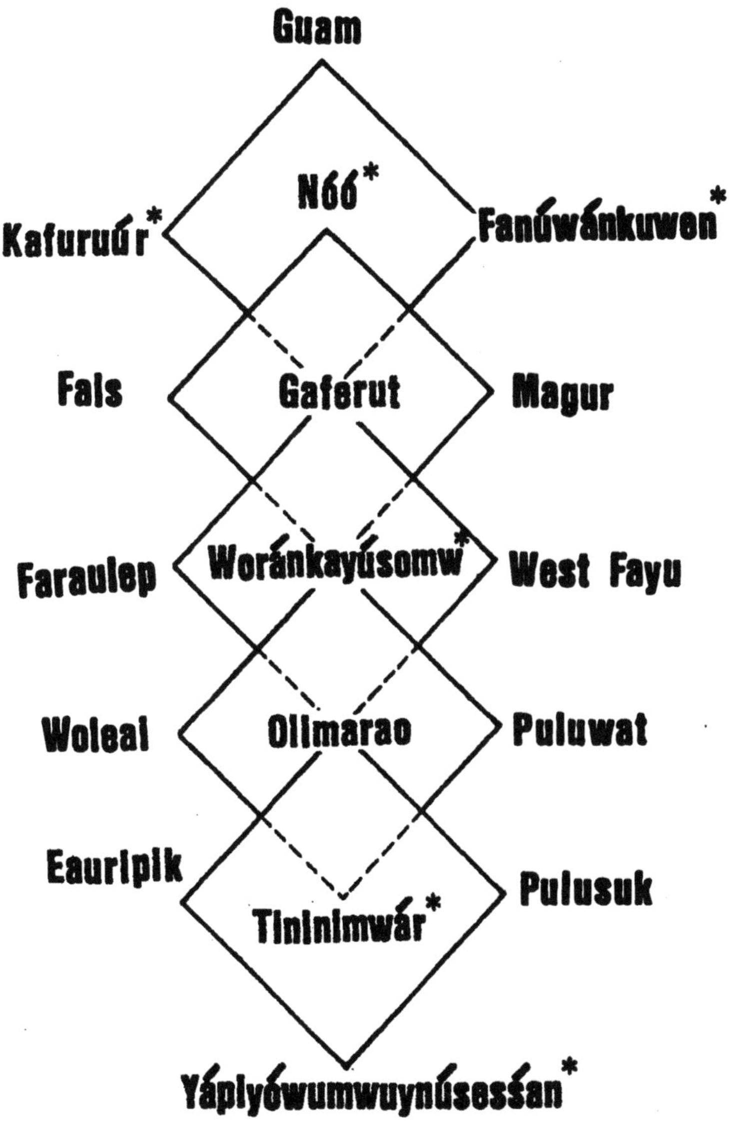

Figure 17.3. Model of **pwuupwunapanap.**
* Shows imaginary or unidentifiable existence

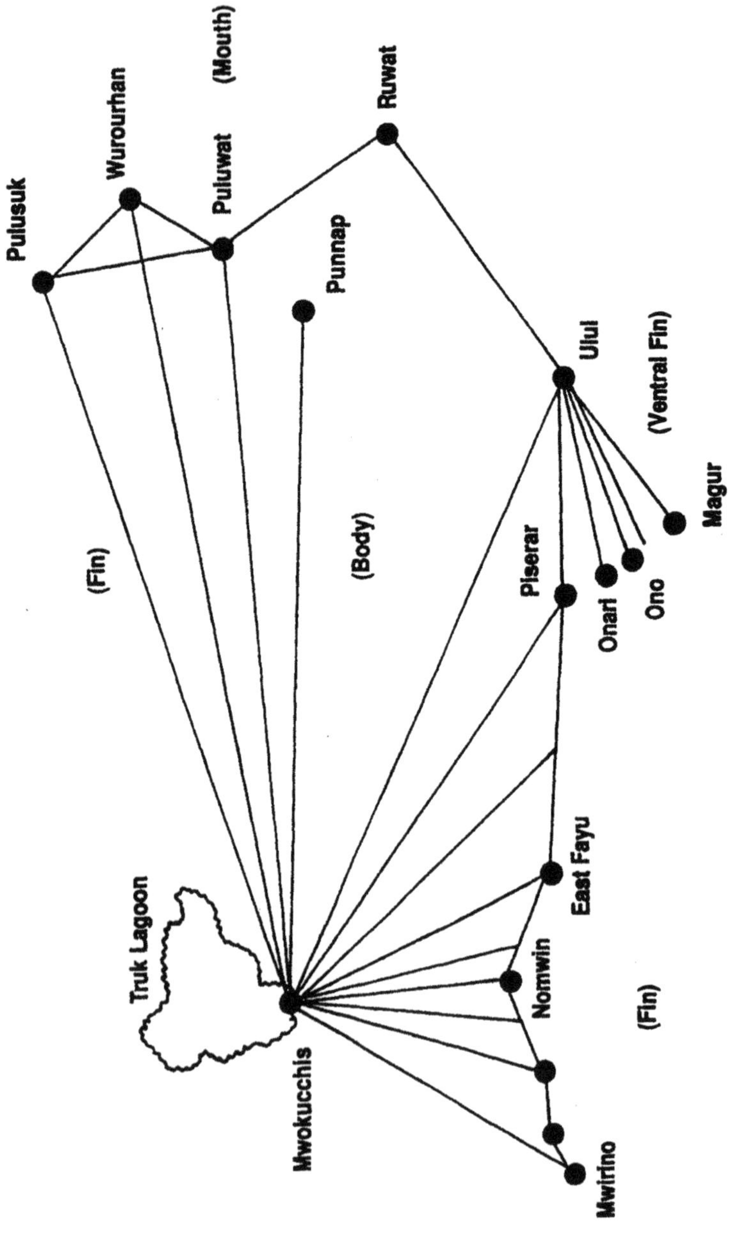

Figure 17.4. Diagram to show the féféu perhan Ayufan knowledge

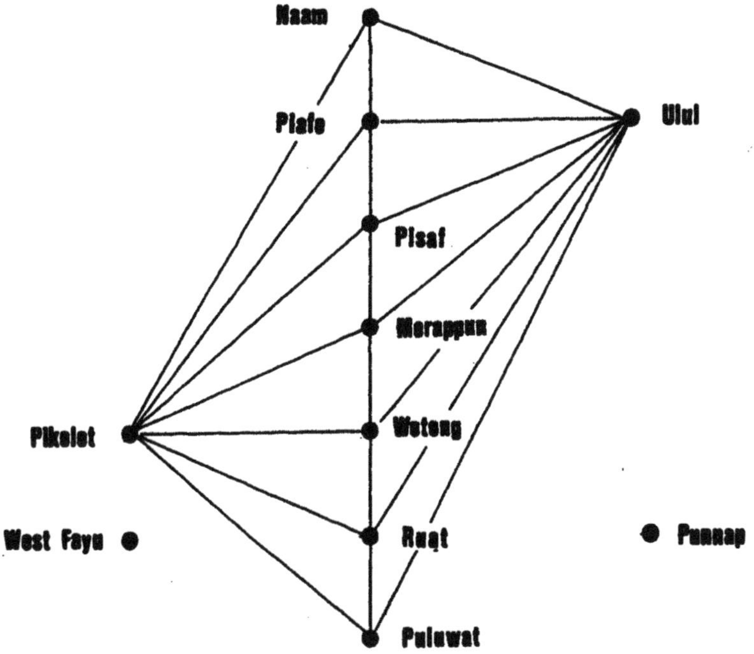

Figure 17.5. Diagram to show the navigation of **masaccha**

imaginary fish that swims from one reef hole to another. Unlike those in the first category, this sea creature is chased and not passively encountered. The third is quite different again. Here sea creatures are used to evoke a shape: a huge fish lying in the sea. On the face of it, all three fail to provide an adequate basis for practical navigation.

Folk-Interpretation of Navigational Knowledge

From the several bodies of navigational knowledge described above we can further infer certain important aspects of their folk-interpretation.

Replacement Theory in Pwukof

As mentioned in the account of **pwukof**, a navigator expects to meet particular kinds of phenomena at sea, such as islands, birds, fish, butterflies and whirlpools. To encounter sea creatures in particular is termed **serfi**. When lost at sea due to bad weather or technical problems, navigators try to find their way to the nearest island or reef for safety. The importance of **pwukof** is reinforced in daily conversation, where it is used metaphorically and allusively.

Even if one grants that the animals which appear in **pwukof** are real, it is, of course, most unlikely that they would remain at the same spot, being subject to capture by humans or predation by large fish. When questioned on this point one of the navigators' gave a unique answer. **Sorom** is a kind of shark which is said to occur in the **pwukof** of Ulul island. It was once killed by a Puluwatese navigator but it appeared again. It is likely, therefore, that animals that are said to have been captured or killed are replaced by the same kind of animal. **Ye suwen sefan** is the precise phrase used to signify this idea, meaning, 'it is replaced again'. How particular fish or birds are constantly present in a certain direction is explained here in terms of the replacement of natural kinds.

There still remains the question as to the distance of such encounters from an island. According to the navigators, one area where sea creatures mentioned in the **pwukof** are encountered is the open sea, **neemetaw**, which is located in the middle zone between any two islands. This suggests the uncertainty of such encounters.

A Concentric Model of Fish and Bird Classification

On Satawal, animals, both wild and domesticated, are classified in the following ways. **Maan**$_1$ is the most inclusive category and includes animals and human beings in contrast to **miin**, which includes such inanimate objects as water, stones, fire and immovable trees and plants. **Maan**$_1$ is further divided into two: human beings (**yaramas**) and the other animals (**maan**$_2$). **Maan**$_2$ includes a variety of animals: birds, fish, bugs, and even special kinds of human beings, such as pregnant women, new-born babies, a person who has fallen from a tree, a thief, a person who can dive very deep [Akimichi 1981b], in other words abnormal and anomalous kinds of human. Fish and other marine creatures, such as octopus, whale, porpoise and turtle, are termed **yiik**, while birds and flying insects are designated **maan**$_3$. Thus, **maan** has a polysemic status in the folk-classification of Satawal.

In contrast to this general folk classification, animals referred to in **pwukof** relate to each other in another way. First, animals appearing in **pwukof** are usually divided into **yeparh** and **manure**. **Yeparh** refers to sea creatures while **manure** refers to those creatures inhabiting the sky. The former includes fish, porpoises and whales, while the latter includes birds, butterflies and dragonflies.

However, according to one navigator, the difference between **manure** and **yeparh** is that the first applies to animals that occur in inshore waters, while the latter applies to animals found twenty to forty kilometres offshore. Furthermore, areas of sea between **manure** and **yeparh** are called **nepetan manure me yeparh**, an area equivalent to ten kilometres offshore from the spot where **manure** are found (**nepetan**: between). **Manure** is composed of two morphemes: **maan** (animals) and **rerhi** (to reach). In any case, these two ideas indicate a sort of concentric model of the distribution of sea life as viewed from an island. Whichever of the two distinct interpretations is appropriate in any one situation, the distinction between **manure** and **yeparh** is perceived either as the habitat segregation of animals at sea or the difference of medium in which creatures dwell, e.g. sea and sky.

Practical Knowledge of Fish and Birds

Encounters with birds and fish during navigation are not incidental, but a regular occurence. Indeed, navigators are aware of the importance of sea creatures as indicators of landfall and orientation.

In the case of fish, those found between inshore and offshore waters are distinguished in terms of habitat and species. Wahoo, rainbow runner and barracuda are regarded as good indicators of proximity to an island, while tuna, skipjack and dolphinfish are seen as being unreliable for estimating the distance of the nearest island, as they are found everywhere. Flying fish are also distinguished in terms of those kinds occurring in coastal and offshore habitats.

Birds are distinguished in terms of coastal and offshore species. The white tern or **kiyakiy** is regarded as the most reliable for determining landfall. When it flies low, the island can be expected to be seen soon, while when it flies high the island is judged to be further away. The noddy or **kurukaak** is also regarded as important, said to be **nayur yanú**, the child of the supernatural, which implies a good indicator of landfall. By contrast, frigate birds, tropic birds, boobies and sooty terns are often unreliable as a means of direction finding and are termed **maanfaisopw**, 'tricky birds'.

Practical knowledge of fish and birds applicable for determining landfall is thus species-specific, and evaluation depends on individual kinds of fish and birds. Interestingly, sea creatures that appear in **pwukof** are considered unreliable for such purposes.

Discussion

I have attempted to demonstrate in this analysis of navigational knowledge the effectiveness of information on island orientation, although to some extent this is variable between informants. Memorizing such knowledge is fundamental to practical navigation.

As a navigator embarks on a sailing career, he starts to learn **pwukof**. **Pwukof** embodies an immense stock of information about the environment. The frequent appearance of supernatural phenomena, however, suggests that much of it is not for direct practical application but for memorizing knowledge. If learning about a new environment is initially dependent on recognizing landmarks, as psychologists suggest [Siegel and White 1975], then many of the creatures and reefs which are mentioned in **pwukof** are likely to be kinds of 'seamark', regardless of whether their existence can be empirically verified.

Learning **woofanúw** and **pwukof** occurs in a similar way: it is subject-oriented and the data are organized sequentially as straight lines radiating from a particular island. The difference between them depends on whether the things under a particular star course are real or imaginary. Hence, learning a new environment is achieved smoothly by memorizing information which includes both the real and the imaginary. The sea creatures mentioned in the **pwuupwunapanap** must be treated as imaginary, as it is hardly possible to demonstrate their existence empirically. It is, however, unlikely that navigators are disturbed by the juxtaposition of real and imaginary. On the contrary, from the **pwuupwunapanap**, navigators can use the image of a diamond-shaped space effectively for the navigation in which they are sailing. It does not matter to them whether one or two parts of a large fish are imaginary or not.

Cognitive mapping is thus employed as a useful means to figure out the environment. A space formed by the image of **pwuupw** is cognitively symmetrical, but it bares little resemblance to the cartographic map (Figure 17.6) [Blades and Spencer 1986: 343–4; Frake 1985; Gell 1985: 271–86]. Although the two navigational models discussed by Gell are critical to the issue discussed here, Satawalese navigational knowledge also suggests a more complicated way of creating particularly cognitive maps. In **yarhuwow** and **rheániy**, by using a metaphor of fishing activity, navigation becomes something more like continuous pursuit.

Knowledge of the voyages of **serhakini masaccha** and **féfén perhan Ayufan** may seem hard to memorize. However, the mastery of such knowledge overlays the images of the cartographic map with that of a huge fish or the dorsal fin of a fish. In this way, the image of a fish or a part of a fish reinforces recognition of the geographical location of islands at sea and helps to memorize the knowledge itself.

What needs to be stressed here is that by using both empirically real and imaginary features, navigators are able to create cognitive maps which make sense to them. Despite disagreement concerning the empirical validity and dependability of techniques for landfall determination and orientation during navigation, encounters with sea creatures still provide a good general indicator for navigators. The views of the master navigators (**panúw**) confirm this. According to one master navigator, the individual birds and fish that accompany him on a particular navigation are regarded as especially significant. Such animals are said to be **manni panúw** (the master navigator's own

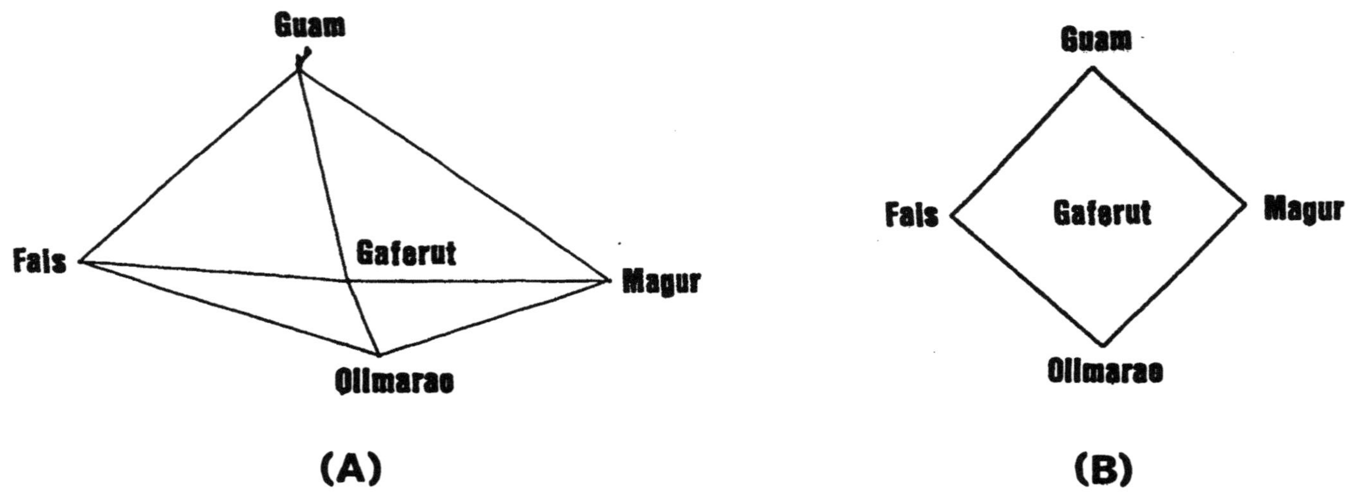

Figure 17.6. Comparison of cartographic and cognitive models of **pwuupwunapanap** (A): cartographic model, (B): cognitive model

birds) and yikiwe nayurh panúw (the master navigator's child or fish). In other words, only master navigators are privileged to accompany fish and birds that can guide them safely to a destination. The notion of possession with regard to these animals suggests a powerful familiarity and mastery of knowledge of sea creatures and of the environment.

Bibliography

Akimichi, T., 'Storm Stars and the Ethnometeorology on Satawal', *Kikanzinruigaku* 11(4), 1980, pp.3–51

——, 'Fundamentals of Satawalese navigational knowledge', *Bulletin of the National Museum of Ethnology* 5(3), 1981a, pp.617–41

——, 'Bad fish or good fish - the ethnoichthyology of the Satawalese (Central Carolines, Micronesia)', *Bulletin of the National Museum of Ethnology* 6(1), 1981b, pp.66–133

——, 'Island orientation and the perception of sea areas in Satawal (Central Caroline Islands)', *Bulletin of the National Museum of Ethnology* 9(4), 1984, pp.651–709

——, 'Conservation of the sea: Satawal, Micronesia', in A.J. Anderson (ed.), *Traditional fishing in the Pacific: ethnographical and archaeological papers from the 15th Pacific Science Congress*, 1986

——, 'Triggerfish and the Southern Cross: cultural associations of fish with stars in Micronesian navigational knowledge', *Man and Culture in Oceania* 3 (Special Issue), 1987, pp.279–98

——, 'Navigational knowledge of sea life (Pwukof) in Satawal, Central Caroline Islands, Micronesia', *Bulletin of the National Museum of Ethnology* 13(1), 1988, pp.127–73

——, and S. Sauchomal, 'Satawalese fish names', *Micronesica* 18, 1982, pp.1–34

Alkire, W.H., *Lamotrek Atoll and inter-island socioeconomic ties*, Urbana: Illinois University Press, 1965

——, 'Systems of measurement on Woleai Atoll', *Anthropos* (65), 1970, pp.1–73

——, *Coral islanders*, Arlington Heights: AHM Publishing Corporation, 1978

Bellwood, P., *Man's conquest of the Pacific: the prehistory of Southeast*

Asia and Oceania, New York: Oxford University Press, 1979

Blades, M. and C. Spencer, 'Maps and wayfindings', *Man* (n.s.) 21(2), 1986, pp.343–4

Burrows, G. E. and M. E. Spiro, *An Atoll culture – ethnography of Ifaluk in the Central Carolines*, New Haven: Greenwood Press, 1957

Frake, C. O., Cognitive maps of time and tide among medieval seafarers, *Man* (n.s.) 20(2), 1985, pp.254–70

Gell, A., How to read a map: remarks on the practical logic of navigation, *Man* (n.s.) 20(2), 1985, pp.271–86

Gladwin, T., *East is a big bird: navigation and logic on Puluwat Atoll*, Cambridge, Massachusetts: Harvard University Press, 1970

Goodenough, H.W. *Native astronomy in the Central Carolines*, Philadelphia: University of Pennsylvania, 1953

Hutchins, E., 'Understanding Micronesian navigation', in D. Gentner and A. L. Stevens (eds), *Mental models*, Hillsdale, N.J. and London: Lawrence Erlbaum Associates Inc., 1983

——, 'The technology of team navigation', *ICS Report* 8804, La Jolla: University of California, San Diego, 1988

Ishimori, S., *Crisis and cosmology in a Micronesian society*, Tokyo: Fukutake-Shoten, 1985

Lessa, W. A., *Ulithi: a Micronesian design for living*, New York: Holt, Rinehart and Winston, 1966

Lewis, D., *We, the navigators: the ancient art of landfinding in the Pacific*, Honolulu: The University Press of Hawaii, 1975

Neisser, U., *Cognition and reality: principles and implications of cognitive psychology*, San Francisco: W.H. Freeman, 1976

Riesenberg, S. H., 'The organization of navigational knowledge on Puluwat', in B.Finney (ed.), *Pacific navigation and voyaging*, (Polynesian Society Memoir 39), Wellington, 1976

Siegel, A.W. and S.H. White, 'The development of spatial representations of large-scale environments', in H.W. Reese (ed.), *Advances in child development and behaviour* (10), New York: Academic Press, 1975

Sudo, K. 'Canoes and social relations on Satawal', *Bulletin of the National Museum of Ethnology* 4(2), 1979, pp.251–84

Chapter 18

Long-term Adaptation of the Gidra-Speaking Population of Papua New Guinea

Ryutaro Ohtsuka

Introduction

This chapter summarizes findings from earlier studies among Gidra-speaking Papuans [e.g. Ohtsuka 1983; Ohtsuka and Suzuki eds 1990] and considers these in the context of temporal and spatial adaptation. The central concern is with the long-term survival of Gidra-speakers from the pre-colonial era, about a hundred years ago, to the present time. Human-environment relations have both changed and diversified among Gidra throughout this period and the importance of cultural factors in these processes is discussed.

An understanding of adaptive (or survival) processes in time and space and in environmentally heterogeneous territory is central to the study of human population ecology [Ohtsuka and Suzuki eds 1990]. Temporal adaptation of traditional populations, which are under modernizing influences, can be readily assessed in the field as part of ongoing historical processes. Where modernizing influences are few, or where the focus of study is on past changes, the task is more difficult. In this paper I use genealogy-based demographic analysis to estimate population increase rates and take these latter to reflect the extent of adaptive response. The spatial adaptation of human populations may be examined by recording environmental variation within the inhabited territory and assessing the extent to which intra-population variation (i.e. diversification) correlates with this.

The Gidra and Their Environment

There are 1,850 Gidra-speaking Papuans (1980 census) who live in thirteen villages and occupy 4,000 square kilometres in the lowlands of the Western Province, Papua New Guinea (Figure 18.1). In addition to language, these people share a creation myth, totemic clan organization and a superimposed moiety system. Ownership of land is vested in clans. An analysis of contemporary Gidra villagers' birthplaces revealed that 97 per cent (N = 1,850), or 95 per cent of married villagers (N = 798), were born within Gidra territory [Ohtsuka et al. 1985a]. The Gidra, therefore, may be identified as a distinct 'population' that has persisted, both biologically and socioculturally, through many generations [cf. Bayliss-Smith 1977].

Gidraland is at the eastern end of the Oriomo Plateau which stretches from the Digul River in the west to the Fly River in the east. At its highest point the plateau rises about sixty metres above a low-lying, swampy plain; the Morehead Ridge separates the northern and southern parts of the plateau.

The thirteen Gidra villages may be divided into two groups: six inland villages and seven riverine villages (Figure 18.1). The former occupy less than 1,000 square kilometres whereas the latter are scattered over an area of more than 3,000 square kilometres. Of the riverine villages two are classed as northern, four as southern and one as coastal. This last village (Dorogori) moved from a riverine to a coastal location half a century ago but is included within the riverine group for purposes of analysing long-term adaptive trends.

Three environmental variables condition patterns of local adaptation by people who reside at different Gidra villages. First, while monsoon forest predominates throughout Gidraland, savanna habitats are better developed in the inland region. Shifting cultivation and the burning of grass for hunting and other unspecific purposes have contributed to the development of savanna which, in turn, provides inland villagers with access to many game animals (e.g. agile wallaby, *Wallabia agilis*). Secondly, the starch of *Metroxylon* sago palms is a staple plant food among Gidra, but these palms are more numerous in inland and northern riverine areas than in southern riverine and coastal areas. The palms are not tolerant of saline environments. Thirdly,

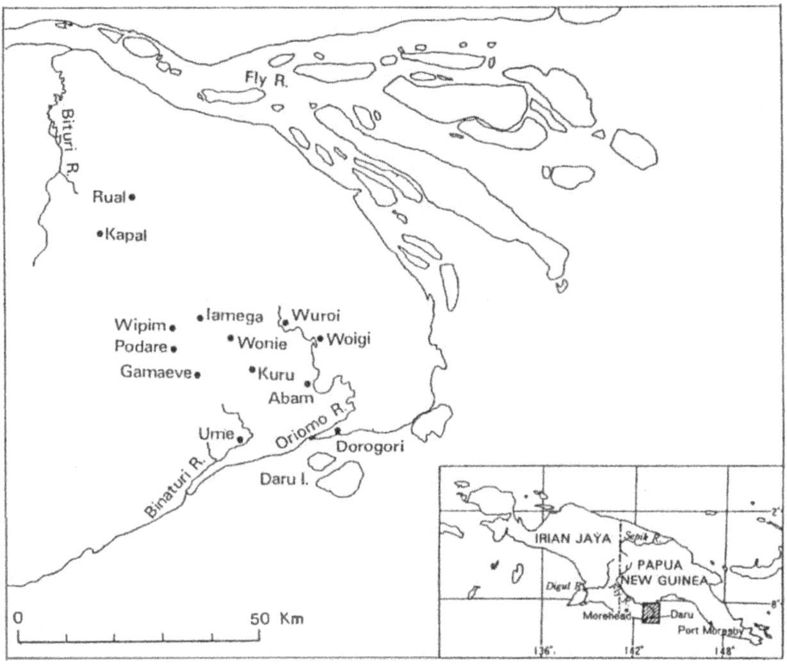

Figure 18.1. Location of the thirteen Gidra-speaking villages in Papua New Guinea

the density of mosquitoes varies throughout Gidraland, and the prevalence of malaria, which is a major mortality factor among Gidra, is higher in riverine and coastal areas.

Daru, the capital of the Western Province, is an island immediately south of Gidraland. In 1980 Daru was inhabited by 7,000 people most of whom were Kiwai-speakers. By 1967, when I first visited Gidraland, people from southern riverine and coastal villages had more frequent contact with Daru than did those from inland and northern riverine villages. Modernization, as evidenced by consumption of purchased food, acceptance of medical services and migration to urban areas, has been significant since the 1950s or 1960s among southern Gidra and since the 1970s among the inland and northern Gidra.

Gidra Ethnohistory

There is little reference to Gidra people in early ethnographic literature. Beaver [1920], who studied coastal people living to the southwest of the present Gidraland, reported that in the local language (a dialect of Kiwai) the people of the interior were called 'bushmen'. This designation may be connected with the local understanding that Gidra-speakers learned how to make river-going canoes from Kiwai-speakers. Even today, Gidra-speakers living on the coast do not make sea-going canoes. The absence, or recent appearance, of a canoe-making tradition among Gidra suggests that their original homeland was inland. Their own oral history supports this view.

The Gidra creation myth tells that the founder of Gidra emerged from a small hole in the ground in the westernmost part of present-day Gidraland. The ancestors of all Gidra clans, of which there are more than forty, apparently came from a small area near the founder's birthplace. After their birth the clan ancestors dispersed, travelling in different directions and eventually settling on land that became their own. These early ancestral journeys took place only within the inland zone of the present Gidraland. That is, expansion into the riverine environment was a secondary event. The oral traditions of some clans, whose lands are located within the riverine area, report the subsequent travels of ancestors, along creeks and rivers, from inland to their present locations.

Some reports, based on early European patrols into Gidraland in the 1920s and 1930s, are deposited in the National Archives of Papua New Guinea in Port Moresby. Although the available information is limited these reports reveal the presence of Gidra-speaking villages (hamlets) along the Oriomo River, though perhaps only along its upper reaches. It is possible that movement by Gidra from inland to riverine locations has occurred gradually and over the past century.

Rate of Population Increase in the Pre-European Period

Base-line demographic information is absent for the Gidra and it is difficult to know the exact ages of contemporary villagers. For these reasons a new analytical tool was needed to estimate population increase rates. In this study the measure of intergenerational replacement of females was the average number of daughters who survived to marry that were born to cohorts of married women during the lifetime of those women. (Cohorts of married women were identified on the basis of their generations.) This index is called the 'daughter-mother ratio' (DMR). It may be substituted for 'net reproduction rate (R)' in the calculation of 'rate of natural increase (r)' (i.e. $r = R^{1/y} - 1$, where y is the mean age of childbearing or the length of a generation. In this study the mean age of childbearing was estimated at 22.5 years; see Ohtsuka 1986, for details).

Genealogical records were collected by interview in all Gidra villages in 1980 and 1981. The ideal Gidra marriage pattern entails simultaneous exchange of two women from clans of different moieties. Marriages that did not conform to this ideal occurred, for example, where one woman died soon after the exchange took place. To maintain balanced relations between clans it was important that people memorized the marital relations of ascendant generations (both living and deceased) and, for these reasons, genealogical records obtained in this study are, in general, reliable. These records permit estimation of DMR values.

Table 18.1. Daughter-mother ratio (DMR) and annual increase rate by mother's group

Group (N)	No. of married female children	DMR	Annual increase rate (%)*
C (176)	184	1.0455	0.198
B (211)	217	1.0284	0.125
A (266)	282	1.0602	0.260

* The mean childbearing age was estimated at 22.5 in this calculation.

Details from 653 females are used in the present analysis (Table 18.1). These females had all married and were either dead or, if living, had passed reproductive age. They are separated into three cohorts: Group A, in which all female offspring have married; Group B, in which all female offspring have completed reproduction; and Group C, in which all female children born to daughters of the subject women have married. It is assumed that the years of birth of women classed in these ways were 1900–20, 1880–1900 and 1860–80, respectively, for Groups A, B and C. Most of the children of these women would have been born during the years 1880 to 1945.

DMR values and corresponding annual increase rates are shown in Table 18.1 for the three cohorts of females. There is little variation between cohorts and there is no consistent temporal trend. Thus, for the period under consideration, a time when Gidra were essentially uninfluenced by modernization, the rate of population increase was low and relatively constant at about 0.2 per cent per year.

Table 18.2. Daughter-mother ratio (DMR) by mother's birth and death or dwelling place and the corresponding increase rate per year

Group (N)	DMR	Increase rate (%)
Birthplace*		
Inland (332)	1.160	+0.658
Riverine (287)	0.885	–0.543
Death or living place**		
Inland (319)	1.110	+0.463
Riverine (318)	0.987	–0.056

* Excluded are 34 mothers who were born outside Gidraland and whose married daughters numbered 44.
** Excluded are 16 mothers who died or live outside Gidraland and whose married daughters numbered 15.

In Table 18.2 details concerning the same set of females are examined in another way. Here, the records are classed according to the birthplace (inland or riverine) and place of death (or primary dwelling place) of those females. Because sample sizes are small, information from the three cohorts has been combined.

Estimates of DMR and of population increase rate were much higher for females born in inland villages than for those born in riverine villages. In fact, estimates based on females born in riverine villages indicate population decline. In most cases, females are likely to have born children at their place of death (or dwelling place) rather than at their own birthplace. Differences between inland and riverine villages are reduced when comparison of females is based on values derived from the place of death (or dwelling place) of females. Presumably this is a result of inter-village migration by women. The rate of endogamous (within village) marriage among Gidra is about 0.7 and, where marriage is exogamous, it is brides that tend to move to the villages of their spouses [Ohtsuka et al. 1985a].

These data suggest a change in demographic parameters with expansion of the Gidra population from inland to riverine areas. In particular, groups moving into the riverine area were apparently subject to population decline. Since it seems that in the past all Gidra lived in inland areas it is likely that the overall rate of population increase declined as the population expanded the area and range of environments occupied.

Some Evidence for Adaptive Diversification

Long-term demographic trends among Gidra are connected with differences in rates of reproduction and population increase between people living in inland and riverine villages. Studies of nutrient intake and of the prevalence of malaria among contemporary populations suggest reasons for these demographic differences.

Nutrient Intake

Food consumption and nutrient intake by Gidra people were studied in 1971–72, 1981 and 1989 [Ohtsuka et al. 1985b]. In 1971–72 one inland village was surveyed once during the dry season and once during the wet season. In 1981 and 1989 four villages were surveyed during the dry season by different workers; these were the inland village studied in 1971–72 and one village from each of the northern riverine, southern riverine and coastal areas. At each village, survey data were usually obtained during fourteen successive days and from the members of six to eight households. Data for 1989 are available for only the inland village.

Energy-rich plant foods such as sago starch and garden crops were readily available to Gidra and their energy intake exceeded levels recommended by FAO/WHO [1973]. Protein intake also exceeded safe levels identified by FAO/WHO [1973] but was judged to be marginal for the inland village in 1971–72 and the northern riverine village in 1981. On the basis of available survey data I have estimated energy and nutrient intake patterns during the period preceding modernization [Ohtsuka 1993]; this is the period to which the estimates of rate of population increase in Tables 18.1 and 18.2 refer. Estimates of protein intake by adult males were 47 g/day for the inland village and 43–44 g/day for the riverine villages. These levels are judged to be marginal and the people, especially those living in riverine villages, may well have been vulnerable to malaria or other endemic diseases.

Malaria Infection

In 1989 blood samples were collected from about 80 per cent of adults and adolescents in all thirteen Gidra villages. The 183 samples were analysed for malarial infection. Serum antibodies against *Plasmodium falciparum* and *P. vivax* were measured using the indirect fluorescent antibody test (IFAT), which detects the frequency and strength of malarial infection for several months after the fever has abated [Nakazawa et al. 1994].

Table 18.3. Per cent distribution of villagers according to the stronger antibody titre of two species of *Plasmodium* (males and females combined)

Group (N)	<1:4	1:16	1:64	1:256	1:1024	1:4096
Inland (36)	36.1	30.6	16.7	13.9	2.8	0
Riverine (147)	6.8	8.8	22.4	29.3	21.1	11.6

Results from these tests demonstrate a marked difference in antibody titre levels of *Plasmodium* species between inland and riverine villagers (Table 18.3). Individuals whose titre level is 1:64 or higher are regarded as malaria-positive. Thus, 33 per cent of inland villagers and 84 per cent of riverine villagers were malaria-positive. Malaria not only kills (particularly infants and young children) but, by reducing age-specific fertility rates or shortening the life span of females, it depresses fertility.

Therefore, the lower DMR values of riverine villagers in the past may be strongly connected to their more frequent exposure to malaria.

The density of mosquitoes – the vectors of *Plasmodium* species – is higher in riverine than in inland villages. Gidra people were well aware of this and inland villagers sometimes avoided visiting the riverine area because they hated the increased level of biting from the insects. The people did not, however, consciously recognize a causal relationship between mosquito bites and malarial infection.

Spatio-temporal Implications of Long-term Survival

Adaptedness in the Homeland

People who inhabit inland villages say that their 'proper' foods are sago starch and the meat of game animals, particularly of agile wallabies. Our records show that more than half their food energy comes from sago starch and almost all their animal protein comes from land animals. Among riverine villagers these foods are less important in quantitative terms: garden crops contribute more to food energy and fishing contributes more to the intake of animal protein.

Exploitation of sago palms has several advantages over horticulture. Stability of supply is essential in hot, humid environments where food storage is undeveloped. Sago starch is available throughout the year but many garden crops are available only in particular seasons. Again, annual vegetables, such as tubers, are more vulnerable to damage by blight or insect pests or more at risk from environmental hazard than long-lived, robust plants like sago palms. My own estimates from the inland village are that sago palms are sufficiently abundant and that they could satisfy the bulk of energy needs of the people.

Again, with regard to the production of animal protein, inland villagers satisfy their requirements by hunting with bows and arrows. Although most arrow heads are now made by modifying steel bush-knives, Gidra hunting practices have not otherwise been strongly influenced by modernization. This contrasts with fishing procedures, which are important to riverine villagers, where small-mesh gill nets, diving goggles and electric torches are now essential technical aids. Although, at present, fishing is almost as productive as hunting in terms of weight (or animal protein) produced per unit time, it is likely that in the past, when modern apparatus was unavailable, returns from fishing were much lower.

These apparent advantages of sago exploitation and hunting, the primary food-getting activities among inland villagers, are consistent with the earlier interpretation that in the past protein intake was somewhat higher among inland than among riverine villagers. Even without reference to differing rates of malarial infection the inland environment was favourable for survival. It is likely that the people's original survival strategies were established in this environment, which they themselves identify as their homeland.

Spatial Expansion and Adaptedness

The rate of population increase among Gidra of the inland area was estimated at about 0.5 per cent per year (Table 18.2). If this rate of increase applied when all Gidra occupied the inland then the population doubled roughly every 140 years. An increase rate of 0.5 per cent per year is lower than estimates from many contemporary anthropological populations but higher than those for prehistoric populations (e.g. 0.1 per cent for Neolithic populations [Hassan 1981]). The Gidra population may well have increased slowly over a period of several centuries until, at a threshold density, it expanded from the homeland area into the surrounding riverine environment.

It could be expected that people who moved into the riverine environment changed their subsistence strategies. First, fishing should have partially replaced hunting. Secondly, reliance on sago starch should have been reduced and the importance of horticulture should have increased. This latter change would be more important in the southern riverine environment than in the northern riverine environment (see Table 18.4).

An important requirement for people adapting to riverine environments is the need to raise more varieties of crops and/or enlarge the area under cultivation. Garden crops, in particular tubers and bananas, contain larger amounts of various nutrients

Table 18.4. Dependence on major food-obtaining activities by Gidra village group

Village group	Plant food		Animal food	
	Sago extraction	Horticulture	Hunting	Fishing
Inland	++	+	++	-
Riverine				
Southern	+	++	+	+
Northern	++	+	+	+

(including protein) than does sago starch; in fact, sago starch, though energy-rich, contains negligible amounts of nutrients. The protein content of garden crops grown by Gidra is, per unit of food energy, five to ten times that of sago starch [Ohtsuka et al. 1984]. Thus, a switch from sago starch to garden crops by riverine dwellers would have partially compensated for reduction in access to animal foods. Without this switch the reduction in available protein to riverine dwellers would have been more marked than we have estimated.

The higher rate of malarial infection among riverine people has been already noted. This may have been the principal factor causing depopulation in the riverine area and there was little that people, through their own efforts, could do to reduce its impact.

Coexistence in a Heterogeneous Environment

Ellen [1982] has suggested that 'ecological versatility' (the number of environments within which a population can survive) provides one measure of the success of human adaptation. In this context the territorial expansion of Gidra people may be regarded as adaptive. But, at the same time, the rate of increase of riverine villages has been negative, and depopulation is indicative of maladaptation. It is necessary, therefore, to consider the process of territorial expansion during the pre-modernization phase and to do so with reference to the rate of population increase. The process may be thought of in two ways.

The relationship between inland and riverine populations, treated separately, parallels the 'population sink' model proposed by Stanhope [1970] in a discussion of highland and lowland populations of Papua New Guinea. Under this model slow population increase in the malaria-free highlands results in expansion into lowland areas where the impact of malaria and some other diseases causes high mortality and population decline. The lowland area is regarded as an ever-dying periphery of human occupation, and the movement of people into this region is non-adaptive. The focus of this interpretation, however, is short-term; it ignores adaptive responses of the population as a whole.

In the long-term humans may make adaptive adjustments to new and less advantageous environments. A tentative goal is that the migrating group attains a level of at least zero population increase. Indeed, in the Maprik area of the Sepik Province of Papua New Guinea, where malaria is a major mortality factor, population density is relatively high. In the case of the Gidra, improvement in their genetic resistance to malaria was unlikely, but improvement in nutritional status, as a consequence of technological changes, was at least theoretically possible. In the long-term, therefore, Gidra people may have adapted to the riverine environment.

Modernization and Adaptedness

For several decades Gidra people, particularly of coastal and southern riverine villages, have been subject to modernizing influences. Systematic estimates of population increase rates in recent years are not available. But information concerning changing fertility patterns among women, mortality patterns of offspring, and changes in *de facto* population from 1980 to 1989, suggest that in recent years the annual rate of population increase has been about 2.0 per cent [Ohtsuka 1990]. This is attributable to both rising birth rates and declining death rates and the improvements have been apparently greater in southern riverine and coastal villages than in inland and northern riverine villages.

Recent improvements in the rate of population increase have resulted from the impact of modernization. Gidra people have shared in a variety of benefits flowing from the development of both Governmental and non-Governmental sectors in the

Western Province as a whole. Three developments are directly implicated in changes to survival strategies. First, people have had access to cash through selling garden crops at the Daru market and, in turn, have been able to purchase imported foods (e.g. rice, flour and tinned fish) which contain more nutrients than do local foods. Secondly, new subsistence techniques have become available both as a result of direct efforts by Governmental agricultural officers and as a result of the elevated purchasing power of the villagers themselves. The impact of these changes has been primarily to horticulture and fishing; the early introduction of shotguns greatly increased hunting efficiency but, since the 1980s, the use of these weapons has been restricted by law. Thirdly, medical services, centred on a Government hospital in Daru, have been active in vaccinating small children and providing antimalarial tablets to those in need of them. These beneficial impacts are more easily and frequently received by coastal and southern riverine Gidra, who visit Daru more often using canoes.

Some disadvantages following modernization have been observed, for example, the appearance of obese and/or hypertensive individuals [Ohtsuka 1993]. But the overall effect has been that the rate of population increase, which is used here as a key criterion of adaptedness, has been raised in response to modernization. The characteristics of particular village localities, which have an important influence on survival patterns, have changed drastically since the pre-modernization period.

Cultural Manipulation of the Human-Environment Relationship

In this final section I discuss the effects of environmental and cultural factors on Gidra adaptive strategies. With reference to food-producing activities I examine levels of production as set by the environment (i.e. stable production), alterations to productive potential arising through the people's efforts using local (traditional) technologies, and alterations to productive potential arising through the implementation of modern technologies (see Table 18.5 for summary).

A comparison of sago extraction and horticulture reveals the

Table 18.5. Characteristics of major food-obtaining activities

	Stability of production	Manipulation with local technologies	Manipulation with modern technologies
Sago	+	–	–
Horticulture	±	±	+
Hunting	±	–	–*
Fishing	±	–	+

* Except when shotguns are used.

former to be superior in stable production but inferior with respect to both traditional and modern technologies. This is because sago palms thrive almost exclusively in permanently swampy environments and are not amenable to intensive manipulation by people. Hunting and fishing rank equally on all dimensions provided the potential impact of shotguns on hunting efficiency is included in the analysis. But where shotgun hunting is unavailable to the people (as, with Gidra, since the early 1980s) the availability of modern technological aids results in some advantage in fishing over hunting.

The primary subsistence strategies of inland villagers, sago extraction and hunting result in a stable supply of food but are currently not amenable to enhancement by cultural (technological) means. These strategies, therefore, which are judged to represent the original Gidra subsistence pattern, have proven less amenable to adaptive response in the circumstance of modernization than horticulture and fishing, even though the latter, in the recent past, may have provided a less secure subsistence base. The original Gidra subsistence strategies apparently proved less flexible than their historical successors under modernization.

In the past, the disease pattern of the Gidra, especially as reflected in malaria-related mortality, has been conditioned exclusively by environmental circumstances: it has not been amenable to modification through the efforts of the people. In recent decades, however, the development of medical services has greatly reduced death rates from malaria, some other infectious diseases and malnutrition. Improvement in the subsistence base, combined with a reduction in death rates, has shifted survival patterns in favour of riverine (and coastal) villagers.

Conclusion

The expansion of population into the riverine environment was an epoch-making event in the early history of Gidra people. In a theoretical sense, this change may be interpreted as a successful adaptive response at the population level. However, within the riverine areas themselves the process of modernization intruded before local populations had adjusted, such that rates of population increase were zero. Until modernization the survival strategies of riverine villagers lagged behind that of inland villagers.

Modernization has altered intrapopulation differences in survival patterns of Gidra. Present-day adaptation is characterized by rapid change in response to town-centred stimuli and developmental programmes that, in turn, take their source from a broader economic and political system. It remains to be seen whether newly emergent Gidra survival strategies can maintain the favourable relationship between humans and their environment that characterized the past.

Bibliography

Bayliss-Smith, T.P., 'Human ecology and island populations: the problem of change', in T.B. Bayliss-Smith and R. Feachem (eds), *Subsistence and survival: rural ecology in the Pacific*, London: Academic Press, 1977

Beaver, W.N., *Unexplored New Guinea*, Edinburgh: Seeley, Service and Company, 1920

Ellen, R., *Environment, subsistence and system: the ecology of small-scale formations*, Cambridge: Cambridge University Press, 1982

FAO/WHO, *Energy and protein requirements*, Rome: Food and Agricultural Organization of the United Nations, 1973

Hassan, F.A., *Demographic archaeology*, New York: Academic Press, 1981

Nakazawa, M., R. Ohtsuka, T. Kawabe, T. Hongo, T. Suzuki, T. Inaoka, T. Akimichi, S. Kano and M. Suzuki, 'Differential malaria prevalence among villages of the Gidra in lowland Papua New Guinea', *Tropical and geographical medicine* 46, 1994, pp.350–4

Ohtsuka, R., *Oriomo Papuans: ecology of sago-eaters in lowland Papua*, Tokyo: University of Tokyo Press, 1983

——, 'Fertility and mortality in transition', in R. Ohtsuka and T. Suzuki (eds), *Population ecology of human survival: bioecological studies of the Gidra in Papua New Guinea*, Tokyo: University of Tokyo Press, 1990

——, 'Changing food and nutrition of the Gidra in lowland Papua New Guinea', in C.M. Hladik, A. Hladik, O.F. Linares, H. Pagezy, A. Semple and M. Hadley (eds), *Tropical forests, people and food: biocultural interactions and applications to development*, London: Parthenon, 1993, pp.257–69

——, T. Kawabe, T. Inaoka, T. Suzuki, T. Hongo, T. Akimichi and T. Sugahara, 'Composition of local and purchased foods consumed by the Gidra in lowland Papua', *Ecology of Food and Nutrition* 15, 1984, pp.159–69

——, T. Kawabe, T. Inaoka, T. Suzuki, T. Akimichi and T. Suzuki, 'Inter- and intra-population migration of the Gidra in lowland Papua: a population-ecological anlaysis', *Human Biology* 57, 1985a, pp.33–45

——, T. Inaoka, T. Kawabe, T. Suzuki, T. Hongo and T. Akimichi, 'Diversity and change of food consumption and nutrient intake among the Gidra in lowland Papua', *Ecology of Food and Nutrition* 16, 1985b, pp.339–50

—— and T. Suzuki (eds), *Population ecology of human survival: bioecological studies of the Gidra in Papua New Guinea*, Tokyo: University of Tokyo Press, 1990

Stanhope, J.M., 'Patterns of fertility and mortality in rural New Guinea', *New Guinea Research Bulletin* 34, 1970, pp.24–41

Chapter 19

Nurturing the Forest: Strategies of Native Amazonians

Emilio F. Moran

Introduction

Our scientific understanding of the relation between people and their habitat has gone through a number of phases. In this century, many of the debates over the causal priority of nature or culture which have been dominant in anthropological analyses generally have used Amazonia as a testing ground. In this chapter, I will examine these shifting views, focusing on recent and ongoing studies that seek to discover folk and native Amazonians' cognitive and behavioural treatment, and transformation, of the environment within which they live. For a very long time the nature/culture debate was dominated by a view that privileged a single directionality of causation. Most important among these was the view which privileged the environment as a determinant force in human-habitat relations [see reviews in Ellen 1982; Moran 1979, 1993; Thomas 1925]. Cycles of rejection of this view privileged human culture and dismissed environment as a significant influence on the course of human affairs [Boas 1911; Goldenweiser 1937].

A more interactive approach followed, most impressive among which was the cultural ecological approach proposed by Julian Steward [1938]. Steward broke with both determinism and possibilism, focusing instead on the interaction of social organization with those features of the environment cognized by a cultural group. The crucial element to Steward was neither nature nor culture but the *process* of resource utilization, and how this process reflected features of given environments (e.g.

dispersed resources) and historical processes such as local market demand for animal furs or rubber.

Steward has been faulted by some scholars because he seemed to neglect a number of variables dear to anthropologists, ritual, political economy and demography among them [Geertz 1963; Rappaport 1968; Vayda and Rappaport 1976]. On the other hand, Steward privileged neither environment nor culture, and he gave particular analytical significance to the culturally defined, or 'cognized', environment. Unfortunately, it was to be several decades before a number of scholars pursued this balanced, processually oriented agenda [Lees and Bates 1990; Orlove 1980] that is the central interest of this volume.

The international concern with the environment in the 1960s led to a period of interest in the role of environment, not only in how it shaped culture but, by focusing on larger scales than before, using units such as the ecosystem and the biosphere. Vayda and Rappaport [1976] advocated the use of biological units (such as population, community and ecosystem) as units of study that were more analytically inclusive that culturally-defined ones. Vayda [1974, 1976] related changes in man/resource ratios, population fluctuations, and competition for gardens and pigs to cycles of warfare. Rappaport, working in the same region, showed how the ritual cycles regulated the size of pig herds, the frequency of warfare, the availability of horticultural land, the length of the fallow cycle, and the dynamic equilibrium of the system through time [1968].

These views have led to a vigorous critique since the mid 1970s that has sought to underprivilege the environment and to place causal priority on individuals' cultural constructions. In its more extreme forms, post-modernists negate the possibility of studying how people relate to their physical and social surroundings using scientific methods capable of leading to generalizations – given that each cultural construction is unique and non-comparable. In its less extreme forms, it privileges individual constructions above the social, cultural and population levels of analysis. The post-modern approach has merit in so far as it brings attention to the decision-making processes of individuals – but it reduces the understanding of how individuals, as members of social units, produce and reproduce themselves, how they make decisions about the use of what they call nature or natural resources, and how they are transformed by such use into given kinds of social beings who seek to maintain access to familiar resources.

Research in recent years has brought to the fore the importance of history and politics to an understanding of human-habitat relations. At the nexus between society and habitat stands the individual. It is the individual who makes decisions that affect the habitat and the society of which he or she is a part. However, individual decisions are not taken in isolation. To understand human decisions one must understand the historical traditions of individuals within a society. What forces have brought them to this historical moment? What past forms of resource use are customary to different segments of this society? What is their historical demography? Are they just coming out of a period of decimating mortality due to epidemics, or are they at a point of high demographic density and unparalleled health and nutritional well-being? Is it sustainable? The questions could go on, but the point is that individuals do not make decisions about the environment in a decontextualized historical setting. They come to the present with customary ways of behaving toward the physical environment that may, or may not, reflect current conditions. This human dilemma requires a capability to study both cultural constructions and physically observable realities such as soil chemistry, if we are to understand human situations in which cultural knowledge is out of phase with current physical environmental conditions, population growth rates and other noncognitive phenomena. Individuals decide whether the cost of changing customary behaviour exceeds the return – a calculus which is daunting in its complexity and evolutionary in its implications (Boster, this volume). The information coming to the individual from the environment may or may not be about the need to change strategies. Historical experience informs the individuals and affects their decisions in a highly variable manner, depending on the individual's gender, age, social class and current access to resources.

Not only are individuals informed by history, but the freedom of individuals to choose is influenced by the political system within which they exist and by external political and economic forces with which they may be more or less articulated. No individual acts alone. Each one is a member of a household, whose internal structure of social relations and distribution of power affects what different individuals may or may not customarily do, without experiencing sanctions. Moreover, individuals belong to families, lineages, clans, moieties and other forms of kinship group, whose inherent allocation of rights and duties affects individuals differently. Individuals are also part of local communities whose relative structure of rank and stratification provides differential access to the resources of the physical environment. Perhaps most often forgotten in the past has been the role of external political forces coming from world economic transformations. The individual who may be inclined to plant manioc, for example, may find herself influenced by the role of the state who, for its own reasons, decides to give a certain region or certain types of farmer access to capital if, and only if, they plant rice. Depending on the authority of the state, and the attractiveness of the credit terms, the individual may very well choose to plant rice rather than manioc against their own best ecological judgment if the external incentive, or threat, is large enough.

Thus, the most recent views on human ecology no longer talk about adaptation to physical environment as the most important dimension of ecological analysis, thereby assuming that this is always the case. Nor do they focus on nature or culture to determine which is causally prior. Rather, the ecological behaviour of individuals is taken to be a product of multiple sources of information and influence: history, demographic experience, the cognized physical environment, social membership and political forces. The individual takes all these into account in making a decision which, if consistent and effective, may become 'an adaptive strategy'. Adaptive strategy does not in itself imply 'success'. It refers to the development of a plan of action which attempts to balance the conflicting forces pushing and pulling on the individual-as-member-of-a-social group. It represents a 'best', but still contingent plan for dealing with these competing influences as a member of a given kin group, of an ethnically identifiable population, within often fractious nation-states composed of competing ethnicities and social classes.

Amazonian Landscapes

Research in Amazonia has reflected many of these trends in the discipline. Disagreement over the influence of environment on cultural development has been frequent and has focused on simple environmental limitations. Lathrap [1968] brought attention to the influence of finding enough animal protein. Over the years, this insight grew into what came to be known as the 'protein debate' which pitted a large number of scholars against each other as each sought to prove or disprove the role of low animal biomass on the size of settlements, their warfare practices, and the development of political units above that of acephalous villages [see reviews in Sponsel 1986]. Meggers [1954, 1970] pointed out that it was soils rather than protein that were most likely to serve as a limiting factor to the development of complex polities. She argued that the poor soils of Amazonia could not support cultivation by means other than slash-and-burn techniques and that this doomed the populations to politically acephalous societies and to materially simple conditions. This view was faulted quickly by Carneiro [1957] and Ferdon [1959]. Carneiro pointed out that weed invasion, was more likely to lead to field abandonment than declining soil fertility. Ferdon, on his part, showed that most soils can be managed if costs are justifiable. Thus, tropical soils, while poor, were shown to be cultivated intensively and successfully when populations have seen the necessity to invest the labour required for their sustainable cultivation.

These views giving causal priority to one or another environmental feature have been followed in recent years by more comprehensive views that see not single but multiple patterns of causality. Moreover, they have been informed by the research of those who have pointed out the significance of cognition in the use of the environment [Boster 1983]. In giving emphasis to the knowledge of native Amazonians one should beware of 'noble savage' or 'conservationist' assumptions that one might impose on these populations [Johnson 1989]. Like us, they differ widely in their attitude towards the physical world. Different ethnic groups have different ideologies, social structures, and concerns with their environment. If they are conservationists, they practice an utilitarian version of it, rather than a romantic or 'liberal' version such as may be common in our own society.[1] Most groups conserve because they still interact closely enough with their physical environment to understand that they have to be concerned with the impact of their actions upon the long-term productivity of their habitat. Unlike us, they have not become entirely dependent on imports from distant regions to support their current levels of consumption, a procedure that allows us to overlook the impact that consumerism may have on those distant habitats from which we get our materials.[2] Most indigenous groups are still able to see the cutting of the forests beyond their current needs for what it is – a loss of future farm land, a loss of future wild game, a loss of wood and vine products for house construction, a loss of medicinally valuable plants, loss of a comfortable place through which to move in search of other resources, and an aesthetic loss of wild landscape. Thus, if they clear less land, it may be because they do not yet share our consumerist assumptions, our access to capital, and our distance from ecosystem production and regeneration.

Despite the virtual explosion of research in Amazonia over the past fifteen years [cf. Barbira-Scazzocchio ed. 1977; Hames and Vickers eds 1983; Hemming ed. 1985; Moran 1981, 1983; Prance and Lovejoy eds 1985; Schmink and Wood eds 1984; Sioli 1984; Wagley ed. 1973, to name some recent collections of research reports], few advances have been made in our ability to compare human ecologies. Findings from one site are viewed as generalizable to the entire region, or conversely, findings are presented as having unique site-specific characteristics that defy comparison. Most anthropological colleagues accept the *terra firme/varzea* dichotomy as expressive of the important differences to focus on, and place data from areas as ecologically different as the Xingu Basin, the Rio Negro Basin and Central Brazil in the same category – i.e. that of *terra firme*, or the even more aggregating 'lowland South America'. Thus, evidence from ecosystems with widely different soils, above-ground biomass and water regimes are used to support radically opposing views explaining cultural development, village size and population mobility. The distinction between *terra firme* and *varzea* glosses over important differences, especially within the vast *terra firme*.

The first thing to note about Amazonian landscapes is that they are far more complex than has generally been acknowledged in the anthropological literature. The simpler dichotomies of nature versus culture and the division of Amazonia into two habitat types (*terra firme* and *varzea*) have persisted long after their usefulness has expired. The Amazon Basin, a region the size of the continental USA, is extremely varied (see Figure 19.1). It includes areas of montane rain forest, lowland rain forest, moist

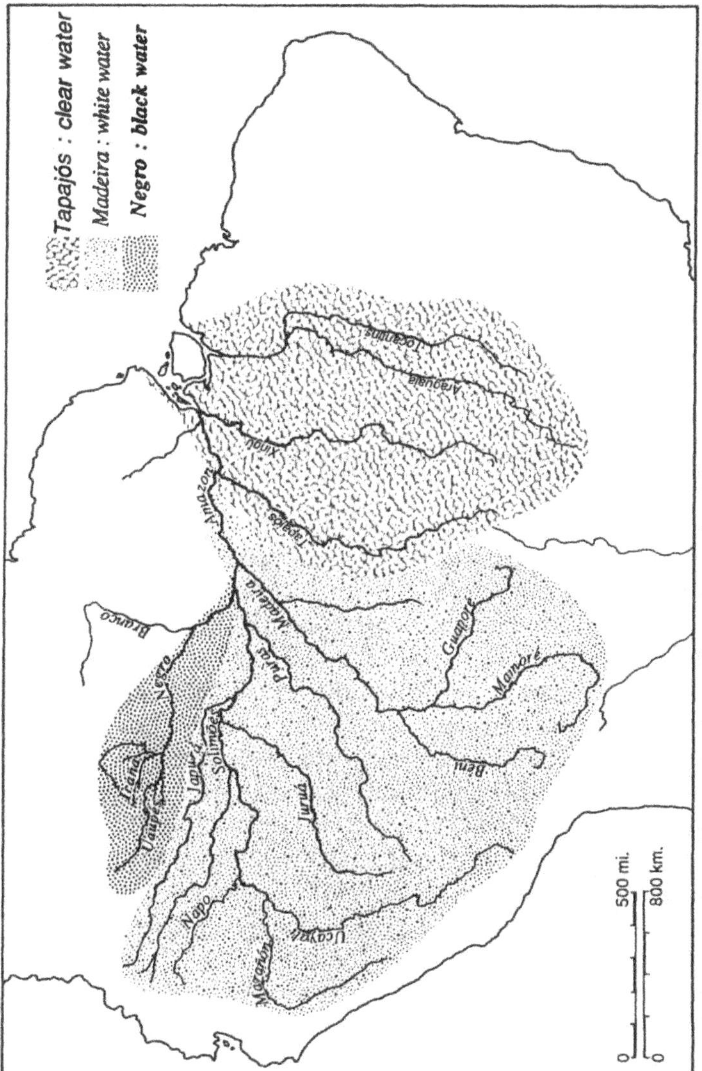

Figure 19.1. Amazon-Basin habitats

forest, vine forest, palm forest, bamboo forest and seasonal forest. It also includes both poorly drained and well-drained savanna, xeromorphic vegetation (i.e. *Caatingas amazonicas*) and species-dominant regions. The image most people still hold – of heterogenous forests with hundreds of species per hectare – is not wrong, but it fails to describe adequately most of the Amazon Basin. The Amazon is as diverse as one should expect an area of continental size to be.[3] This diversity appears to be both 'natural' and humanly induced. Some of these extant vegetations are increasingly suspected of being the product of long-term human activities – e.g. the vine forests of eastern Amazonia, some of the savannas on better soils, much of the transitional area between forest and savanna in Central Brazil [Balée 1989; Moran 1993].

Indigenous Alteration of Amazonian Landscapes

While it is attractive to think of the Amazon Basin as pristine or virgin forest, such a view says more about our need to have pristine landscapes to establish a link with a long-lost ecological past than it describes the Amazon. Balée [1989] contends that at least twelve per cent of the *terra firme*'s interfluvial forests are anthropogenic. This is not to say that such humanly-induced forest came about through conscious design. The issue of indigenous management is whether alteration of plant communities led to the establishment of plant communities markedly different from 'naturally occurring' communities – and whether such changes in plant frequencies resulted in greater net returns to human communities in terms of usable plant products such as vines, fruits, medicines, thatch and construction materials, to name but some economic uses.[4] A great deal of research is needed focusing on this issue. A recent attack by Parker [1992] on the existence of managed forests among the Kayapó, the now well-known *apetê* [Posey 1985], serves to remind us that until such time as one can demonstrate a statistical difference in the plants growing in 'managed areas' vis-à-vis the naturally occurring adjacent secondary successional forests, acceptance of the role of people in nurturing the forest can be brought into question.

To date, interpretation of vegetation types in Amazonia using radar and satellite images aggregate most of the ground-level diversity which is found in the Basin – most of the region being lumped together as 'high forest' [RADAM 1973]. In part, this reflects a problem of scale [Moran 1990b] and also an 'epistemological problem in the interpretation of what is pristine and what is not' [Balée in press: 6]. Clearly, what is needed are detailed studies of plant communities related in time and space to remotely sensed images and to cognitive studies of how local peoples conceive of these diverse plant communities.[5]

Although distinguishing between anthropogenic forest and 'natural' forest is difficult from satellite and aerial platforms at this time,[6] it is possible to use ground-level criteria to differentiate between them [cf. Anderson, May and Balick 1991]. Of particular value are differences in basal area, floristic composition, species richness, and the presence of pottery in the soil profile. Balée finds that old fallows are typically within the range of 18 to 24 square metres, whereas high forests are in the 25 to 40 square metre range [Balée, in press: 13; Balée and Campbell 1990; Boom 1986; Pires and Prance 1985; Saldarriaga and West 1986: 364].

Notable differences between old fallows and 'natural' forest may be seen in the coefficient of similarity between the two. Balée found that when comparing the thirty ecologically most important species (i.e. as measured by frequency), that the two forest types shared but a single species – *Eschweilera coriacea* – yielding a coefficient of similarity of only 1.7 per cent as compared with a coefficient of 16.4 per cent when comparing high forest plots and 11 per cent when comparing fallow plots [Balée, in press: 17]. Moreover, it is important to note that of the thirty most frequent plants in a plot of fallowed forest, fourteen were significant food species for indigenous populations of the area, as compared with only six in the high forest plots. This difference in the frequency of economically useful plants suggests human-induced alteration of forest composition, a feat achieved without significantly affecting the heterogeneity of the forest (139 spp. vs 199 spp. in high forest).[7]

The range of indigenous alteration of forest species composition is probably quite large. Some populations may be engaged in conscious planting of species that concentrate resources into 'forest islands' [Posey 1985], managing their fallows so that they concentrate fruits of economic value (see Figure 19.2), while

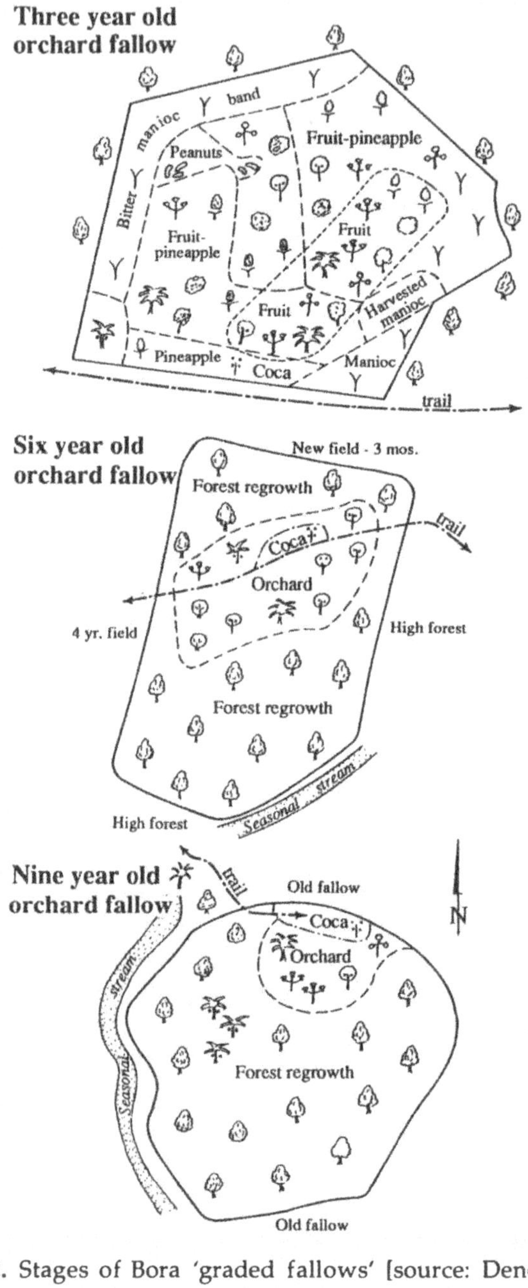

Figure 19.2. Stages of Bora 'graded fallows' [source: Denevan et al. 1984]

others may plant forest species minimally or not at all, aware that a consistent result of making gardens is that seed dispersers will bring certain favoured species into the garden and thereby concentrate species of value. The planting of certain plants in the gardens attracts given animals to them which have food preferences among forest species, as well as in garden crops. Thus, by manipulating the planting density in gardens it is possible indirectly to influence the re-seeding of a garden with forest species of economic value.

Human populations of Amazonia have not only influenced the species composition of contemporary forests, they have also altered the soils upon which plants grow. Anthropogenic black soils or anthrosols [Kern 1988; McEwan 1983; Smith 1980] drain well, are dark in colour and relatively fertile. They have an abundance of potsherds, are reported to be coincident with areas of relatively dense pre-Columbian settlements, and have been reported as occurring over large areas of the Amazon [Falesi 1974]. Phosphorus concentration, sometimes used as an indicator of former human habitational sites [Eidt 1977], is high in these soils. While Amazonian soils tend to average about 5 milligrams per 100 grams (mg/100 g) of phosphorus, the black earths average 40.1 mg/100 g and in one case reached 315 mg/100 g [Smith 1980; Sombroek 1966]. Ash from fires, bones from fish and game, turtle shells, and even human bones might account for the high levels of phosphorus and calcium in anthrosols. The superior quality of black earths is recognized by Amazonian populations [Frikel 1959, 1968; Smith 1980]. Sites tend to occur within easy reach of water sources and average 21.2 hectares along the major tributaries.

Indian black earths have been found overlying a variety of soil types and on a range of geomorphological surfaces. This suggests that they are formed as a result of human activity rather than as a product of geologic processes [Smith 1980: 555]. As some layers are as deep as two metres, it is unlikely that they represent burning of garden areas or the accumulation of organic matter in gardens. Thus, while these soils are excellent for agriculture, once formed, there is no evidence that they are the product of agricultural activity [Eden, Bray, Herrera and McEwan 1984: 137]. There is some controversy over what minimum degree of blackness and what physical or chemical properties constitutes an unambiguous black earth. This judgment must await more competent descriptions of soils and complete sets of analytical data. Eden et al. [1984: 127] note that few archaeologists, ethnologists or geographers offer such data in their reports of black earths. Black earths are not only evidence of prolonged settled life in both *terra firme* and the floodplain in prehistory, but constitute today an important high-quality soil estimated to cover some 50,000 hectares in the Brazilian Amazon alone. When located, Amazon populations commonly give them priority for their most nutrient-demanding crops.

Despite widespread appreciation of the contribution made by Conklin [1957] in describing indigenous systems of soil-plant interrelations and their cognitive dimensions, it is remarkable how very little has been published on the ethnopedology of indigenous populations worldwide, including that of native and folk Amazonians. Only four articles to date have both examined the cultural categories and verified by soil analyses the accuracy of Amazonian systems of naming soils [Behrens 1989; Hecht 1989; Hecht and Posey 1989; Moran 1977].

Behrens [1989] reports that Shipibo distinguished between soils used for making ceramics and those used for gardens. Six different types of named soils were used for gardens (see Table 19.1). An important vegetative indicator of good soils for them seemed to be the presence of *Heliconia* and its association with soils good for growing plantains, the major staple. The preferred soil was 'sandy', probably an Entisol, of silt-loam texture that had a pH of 7.5, whereas the other soils had a pH of 5.4 to 5.8. The better pH and texture of **mashimai** made it suitable for all crops, particularly for plantains and bananas which only did well on this soil.

Moran [1977] found that local Amazonian peasants along the Transamazon highway used vegetation indicator species to identify high base status soils with pH above 6.0, virtually no aluminium saturation, and relatively high amounts of phosphorus, potassium, calcium and magnesium. These soils contrasted with surrounding soils with pH around 4.5, 70 to 85 per cent aluminium saturation, and low levels of macronutrients (see Figure 19.3 and Table 19.2).

Recent studies of the Gorotire Kayapó [Hecht 1989; Hecht and Posey 1989] suggest the active manipulation of soils by some groups. They apply termite and ant nests, bones and leaf mulches to crops to enhance their growth. Fallows are managed to favour

Table 19.1. Cell Means on nine soil characteristics for forty-five topsoil samples taken near Nuevo Edén, Peru, during 1981*

Dependent variable	Cell									
	Nii/ máshimai (n = 5)	***Nii/ maikon*** (n = 5)	***Nii/ mapumai*** (n = 5)	***Huai/ máshimai*** (n = 5)	***Huai/ maikon*** (n = 5)	***Huai/ mapumai*** (n = 5)	***Nahuë máshimai*** (n = 5)	***Nahuë maikon*** (n = 5)	***Nahuë mapumai*** (n = 5)	Grand mean (N = 45)
SP	80.40	72.20	84.40	46.80	60.60	50.20	42.40	60.80	63.60	63.39
pH_s	7.52	5.34	5.50	7.66	5.86	5.14	7.38	5.54	5.58	6.17
EC_c	0.89	0.73	0.46	0.70	0.72	0.37	1.09	0.73	0.42	0.68
p**	10.34	6.78	18.13	10.56	13.40	8.40	6.74	5.26	2.32	8.79
K	119.0	180.8	209.8	91.6	354.80	94.6	80.8	143.6	134.4	156.9
Ca + Mg	9.82	5.66	3.56	6.42	5.88	2.22	9.78	6.20	2.72	5.81
Na	1.18	1.00	0.83	1.26	1.34	1.14	2.24	0.90	0.93	1.21
Cl	1.22	1.10	0.82	1.24	1.34	1.87	2.74	1.05	0.96	1.20
NO_3N	19.20	35.60	15.32	6.20	23.82	0.30	9.00	35.00	12.30	18.33

* ***Nii*** = primary forest, ***huai*** = garden, and ***nahuë*** = fallow garden.

** Two outliers were eliminated from the phosphorus data, a ***huai/maikon*** sample with a value of 35.0 and a ***nii/ mapumai*** sample with a value of 74.0.

Source: Behrens 1989

Table 19.2. Forest vegetation indicative of agricultural soils

a. Forest Vegetation Indicative of Good Agricultural Soils

Local Term	*Scientific Name*
Pau d'arco or ipé (yellow variety)	*Tabebuia serratifolia*
Pau d'arco or ipé (purple variety)	*Tabebuia vilaceae*
Faveira	*Piptadenia* spp.
Mororó	*Bauhinia* spp.
Maxarimbé	*Emmotum* spp.
Pinheiro preto	(unidentified)
Babaçú	*Orbignya martiana*
Açaí	*Euterpe oleracceea*

b. Forest Vegetation Indicative of Poor Agricultural Soils

Local Term	*Scientific Name*
Acapú	*Vouacapoua americana*
Jarana	*Holopyxidium jarana*
Sumaúma	*Ceiba pentandra*
Melancieira	*Alexa grandiflora*
Sapucaia	*Lecythis paraensis*
Piquí	*Caryocar microcarpum*
Cajú-Açú	*Anacardium giganteum*
Massaranduba	*Manilkara huberi* (or *Mimusops huberi*)

Source: Moran 1977

nitrogen-fixing tree species, like *Inga*, as well as planted-in fruit trees that attract seed dispersers from the forest. The Bora Indians of Peru use ashes to fertilize peanut mounds and a basil-leaf infusion as an insecticide [Denevan and Padoch eds 1988]. The productivity of some native systems is evident in the thirty tons per hectare of manioc reported by Frechione [1981] and the 23.16 tons of plantains recorded by Smole [1976].

Not only is the forest landscape affected by human activities, so is the riverine landscape. The indigenous peoples along the floodplain of the Amazon experienced the brunt of the early contact, with its disastrous epidemiological consequences and mortality from warfare and enslavement [Moran 1993]. Over the centuries, the floodplain was resettled by mestizos, often

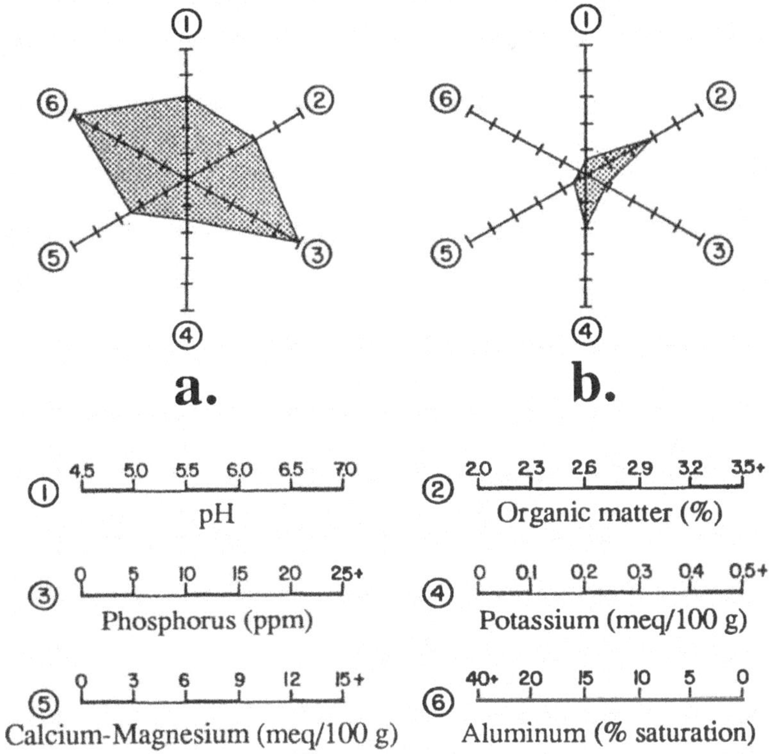

Figure 19.3. Comparison of *terra firme* soils

practicing forms of resource use which approximate that of earlier populations. Along the floodplain, the activities of these populations are synchronized to the complex fluctuations in river level. As soon as waters begin to recede, the population begins to plant on the soggy beaches. In some years, when the floods return due to upperwatershed rains, the crop may be lost. However, the returns can be high if the drop in water level is consistent and ensures a better distribution of labour over the growing season. Rice is the most common crop planted on these beaches, while crops less able to deal with waterlogging are planted on levees. Ecologically, most students of Amazonia agree that the floodplain environment is more resilient than the interfluves [cf. Hiraoka 1985: 245]. The very volume of the floodplain has provided protection from contemporary forms of management that could have altered the habitat. Some signs of coming modification are the growing presence of bauxite plants and other mining concerns, paper-pulp manufacturing plants, and other industrial firms whose potential to pollute could put an end to the rich fisheries of the region. Studies by Goulding [1983] noted that the catch at Manaus had already peaked and that total catch was declining yearly as a result of overfishing.

Fisheries appear to have been protected from excessive human exploitation by the belief that the flooded forests were inhabited by spirits. A rich mythology surrounds the aquatic environment of Amazonia which has had as an indirect effect the conservation of those regions now known to be important natural hatching areas for Amazonian species. Most important in regional folklore is the 'mother of the fish' – a supernatural protector of fish that adopts many guises, among them that of the giant watersnake. She is said to punish overzealous fishermen [Smith 1983] and to be responsible for making them become disoriented so as to lose their way, and even to drown. This is particularly undesirable since according to some local lore those who lose their lives in the water will never get to heaven. The existence of 'supernatural reserves' provides a safe haven for many river creatures, but these beliefs have been eroded in some places as newcomers unfamiliar with such beliefs proceed to use them without adverse effects. However, in other places these beliefs have been reproduced as newcomers experience the unexplainable and local populations share with them explanations which not only give meaning to these experiences but bring back to order the complexities of the Amazonian riverine habitat. Smith [1983] has argued for the creation of fish reserves through legislation coterminous with those areas sanctified by local lore, and biologically verified to represent effective hatching areas.

Use of ethnoecological methods provides a window into the complexity of this riverine world. At Marajó island, at the mouth of the Amazon, preliminary research on the cultural knowledge of the fishermen of the island uncovered a complex taxonomy of the domain fish that we are just beginning to understand (see Figure 19.4). Cultural categorization can vary along many dimensions. It may be concerned with mapping the diversity of the physical world for a non-literate population, or it may be one of many strands that link the physical world to the cosmological one, the social one, and other aspects within our daily lives. The use of physical attributes to put order into an infinitely varying

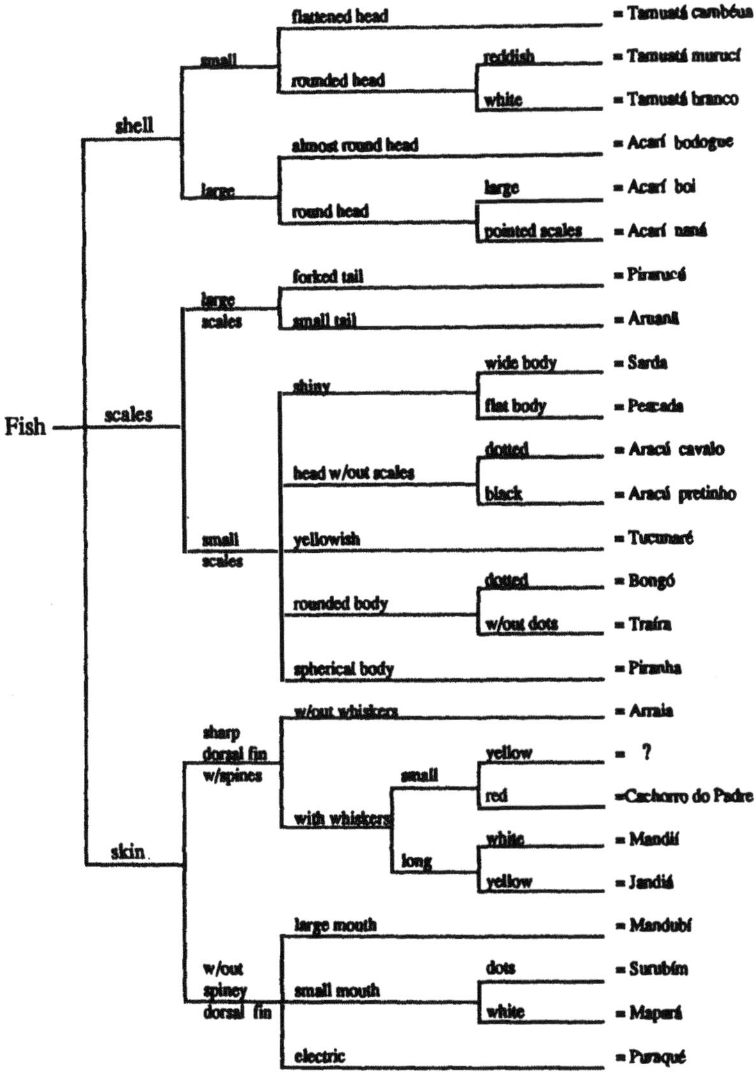

Source: Fábio de Castro & Emilio Moran, field notes, 1989

Figure 19.4. Ethnoecological fish classification from Marajo

world is one of the important processes in human action which may be tied to action, cognition, or both.

Conclusions

Contemporary deforestation in the Amazon Basin has become a matter not merely of international concern but of alarm: in 1987 alone it is reported that eight million hectares of forest were burned. The Amazon is host to about half of the world's biota and its continental size and high evapotranspiration rates make it a notable influence on world climate. Despite this, relatively little attention has been devoted to how native Amazonians use and conserve their physical environment. Are they conservationists? Are they utilitarians? Are they 'at one with nature', or against it if need be when utilitarian concerns come to the fore?

This paper has tried to show how research in the Amazon has reflected the theoretical debates in the discipline and other trends in Euro-American societies (such as concern with the state of our physical environment). It shows a clear cycling between views which privilege single causal factors to views which are more processual and multicausal. Particular emphasis has been placed on the non-destructive transformation of Amazonian landscapes by native and folk populations. Some 'natural' forests have been changed so that economic species are more dominant, while still maintaining high levels of diversity. Soils which are generally poor have been changed to make them more productive for particular crops, and the rare soils of high fertility have been selected using indicator species as a guide. Even the productivity of the fisheries has been influenced by associating some habitats, like flooded forests, with the presence of supernatural spirits capable of bringing harm to overzealous fishermen.

There are some signs of new versions of old dichotomous debates rearing their simple-minded heads. One of the most insidious seems to be a new form of the old nature versus culture debate. It has taken the form of 'are native peoples of Amazonia conservationists or are they simply decapitalized forest destroyers like ourselves?' Like past nature versus culture debates it imposes a uniformitarianist view on the diverse behaviour of hundreds of cultural groups with widely differing attitudes towards the physical environment [cf. Redford 1990; Redford and Robinson 1985]. These new versions of old and simplistic generalizations must be rejected. It is all too easy to find one or a few groups in a region that have a certain attitude towards forest mammals. This is not the same as saying that 'indigenous peoples' or pre-capitalist social systems are 'conscious of the sacredness of nature' or 'utilitarian towards their environment'. Both views err in trying to generalize from an insufficient number of cases and in not recognizing the stochasticity of complex systems. As we have seen in this chapter, to date the number of studies that probe into the environmental understanding of native Amazonians have been few. There is evidence for some populations having certain cognitive preferences at the time they were studied. Whether these preferences are inherent in that social system, whether they represent recent adjustments to changes in their political economy and demography, or whether they are undergoing the pains of transitions in modes of production are empirical considerations that must be settled by more inclusive human ecological research than those commonly undertaken.

This paper presents a view which is in stark contrast to that most of us were raised on – dependent as those earlier views were on a static notion of how pre-industrial peoples adapted to environment by accepting the limitations of the areas where they found themselves. Instead, we find that like populations living in more complex political and economic systems, native Amazonians have brought about notable transformations in the regional landscape. Unlike our destructive forms of transformation, however, native systems seem to have been able to transform the environment while preserving some of the features that we all value in tropical rain forests, in particular their rich species diversity. Anthropologists need to focus further on recovering this wealth of knowledge, and how it can be a basis upon which future forms of resource use can be constructed that balance use and conservation, provide dividends to local peoples for their past and current knowledge, rewards their future preservation of germplasm that make infertile areas more fertile, could ensure the sustainability of net yields, and could give local populations economic returns for their labour.

Acknowledgements

I would like to thank the J.S. Guggenheim Foundation for a Fellowship during 1989–90 that provided time to reflect and research this topic. Breakthroughs in the identification of managed forests from adjacent floodplain forests at Marajo island have been recently achieved thanks to a grant from the National Science Foundation to E. Moran, SES 9100526, whose assistance is gratefully acknowledged. These foundations should not be held responsible for the views expressed in this chapter. I also want to thank the MOA Foundation and the organizers of this Symposium for the invitation to address these issues and the opportunity to share and discuss them with colleagues from throughout the world.

Notes

1. Although it is not inconceivable that we may find some groups with not wholly 'utilitarian' views. Indeed, we may find some with views that approximate 'romantic' and religious ones [cf. Reichel-Dolmatoff 1970].
2. Some indigenous groups have, in fact, moved quickly to adopt consumerist ways and have been noted to be more concerned with their own wealth than with the integrity of their physical environment or with internal social equity. Whether this represents a temporary compromise in their world-view, as in the case of those needing cash quickly to travel to capital cities to lobby for their ethnic group's rights to land, or a wholesale abandonment of their world-view and the adoption of national ideologies will be settled only by long-term study [cf. Clay 1988].
3. The drainage basin is estimated to be about six million square kilometres in extent.
4. Even this requirement may not be necessary, according to Boster in this volume. It may be evolutionarily advantageous to select plants that are physical variants even if they are not better from a productivity point of view. Such an attitude tends to preserve a richer gene pool, while still permitting a narrow portion of the gene pool to provide the bulk of the

consumption needs of the population.

5. This is one of the tasks that my current research focuses on, and is also the main mission of the recently created Anthropological Center for Training and Research on Global Environmental Change at Indiana University.
6. Recent breakthroughs in our own research in the Amazon point to the possibility of spectral differentiation between native forests and secondary growth areas up to about fifteen years using Landsat TM digital image analysis when combined with field studies. See paper presented at the Ecological Society of America meeting in Hawaii [Moran, Brondizio, Mausel and Wu 1994].
7. This finding is in contrast to what Parker [1992] claims he found in the areas denominated by apetê, among the Gorotire Kayapó, where no significant differences in frequencies were found.

Bibliography

Anderson, A., P. May and M. Balick, *The subsidy from nature: palm forests, peasantry, and development on an Amazon frontier*, New York: Columbia University Press, 1991

Balée, W., 'The culture of Amazonian forests', *Advances in Economic Botany* 7, 1989, pp.1–21

——, 'Indigenous transformation of Amazonian forests: an example from Maranhão, Brazil', *L'Homme*, in press

Balée, W. and D.G. Campbell, 'Evidence for the successional status of Liana forest, Xingu river basin', Amazonian Brazil, *Biotropica* 22, 1990, pp.36–47

Barbira-Scazzocchio, F. (ed.), *Land, people and planning in contemporary Amazonia*, Cambridge: Centre for Latin American Studies, Cambridge University Occasional Publication, 1977

Behrens, C., 'The scientific basis for Shipibo soil classification and land use, *American Anthropologist* 91, 1989, pp.83–100

Boas, F., *The mind of primitive man*, New York: Macmillan, 1911

Boom, B.M., 'A forest inventory in Amazonian Bolivia', *Biotropica* 18, 1986, pp.287–94

Boster, J., 'A comparison of the diversity of Jivaroan gardens with that of the tropical forest', *Human Ecology* 12, 1983, pp.343–58

Carneiro, R.L., 'Subsistence and social structure: an ecological study of the Kuikuru', Ph.D. Dissertation, University of Michigan, Department of Anthropology, 1957

Clay, J., *Indigenous people and tropical forests*, Cambridge: Cultural Survival, 1988

Conklin, H.C., *Hanunóo agriculture*, Rome: FAO, 1957

Denevan, W. and C. Padoch (eds), *Swidden-fallow agroforestry in the Peruvian Amazon*, New York: New York Botanical Garden Monograph Series 5, 1988

Denevan, W., J.T. Treacy, J. Alcorn, C. Padoch, J. Denslow and S.F. Paitan, 'Indigenous agroforestry in the Amazon: Bora Indian management of fallows'. *Interciencia* 9(6), 1984, pp.346–57

Eden, M.J., W. Bray, L. Herrera, C. McEwan, 'Terra Preta soils and their archeological context in the Caquetá basin of SE Colombia', *American Antiquity* 49(1), 1984, pp.125–40

Eidt, R.C., 'Detection and examination of anthrosols by phosphate analysis', *Science* 197, 1977, pp.1327–33

Ellen, R., *Environment, subsistence and system*, New York: Cambridge University Press, 1982

Falesi, I.C., Soils of the Brazilian Amazon, in C. Wagley (ed.), *Man in the Amazon*, Gainesville: University of Florida Press, 1974

Ferdon, E., 'Agricultural potential and the development of cultures', *Southwestern Journal of Anthropology* 15, 1959, pp.1–19

Frechione, J., 'Economic self-development by Yekuana Amerinds in S. Venezuela', Ph.D. Dissertation, University of Pittsburgh, Department. of Anthropology, 1981

Frikel, P., 'Agricultura dos Indios Munduruku', *Boletim do Museu Paraense Emilio Goeldi*, n.s. Antropologia (1), 1959, pp.1–35

——, *Os Xikrín*, Belém: Publicações Avulsas do Museu Goeldi, 1968

Geertz, C., *Agricultural involution*, Berkeley: University of California Press, 1963

Goldenweiser, A., *Anthropology*, New York: Crofts and Company, 1937

Goulding, M., 'Amazonian fisheries', in E. Moran (ed.), *The dilemma of Amazonian development*, Boulder: Westview Press, 1983

Hames, R. and W. Vickers (eds), *Adaptive responses of native Amazonians*, New York: Academic Press, 1983

Hecht, S., 'Indigenous soil management in the Amazon basin', in

J. Browder (ed.), *Fragile lands of Latin America*, Boulder: Westview Press, 1989

Hecht, S. and D. Posey, 'Preliminary results on soil management techniques of the Kayapó Indians', *Advances in Economic Botany* 7, 1989, pp.174–88

Hemming, J. (ed.), *Change in the Amazon Basin*, 2 vols, Manchester: Manchester University Press, 1985

Hiraoka, M., 'Mestizo subsistence in Riparian Amazonia', *National Geographic Research* 1, 1985, pp.236–46

Johnson, A., 'How the Machigenga manage resources: conservation or exploitation of nature?', *Advances in Economic Botany* 7, 1989, pp.213–22

Kern, D.C., 'Caracteristicas Pedológicas de Solos com Terra Preta Arqueológica na Região de Oriximina', Pará, MA Thesis, Universidade Federal do Rio Grande do Sul, Department of Agronomy, 1988

Lathrap, D., 'The "hunting economies" of the tropical forest zone of South America', in R. Lee and I. DeVore (eds), *Man the hunter*, Chicago: Aldine, 1968

Lees, S. and D. Bates, 'The ecology of cumulative change', in E.F. Moran (ed.), *The ecosystem approach in anthropology: from concept to practice*, Ann Arbor: University of Michigan Press, 1990, pp.247–77

McEwan, C., 'Amazonian Terra Preta soils: clues to prehistoric population dynamics in the Amazon basin', paper presented at the XI Midwestern Conference on Andean and Amazonian Archeology and Ethnohistory, Bloomington, Indiana, 1983

Meggers, B., 'Environmental limitations on the development of culture', *American Anthropologist* 56, 1954, pp.801–24

——, *Amazonia: man and culture in a counterfeit paradise*, Chicago: Aldine, 1970

Moran, E.F., 'Estratégias de sobrevivencia: o uso de recursos ao longo da Rodovia Transamazônica', *Acta Amazônica* 7, 1977, pp.363–79

——, *Agricultural development along the Transamazon highway*, Bloomington: Center for Latin American Studies, Indiana University, Monograph Series, 1979

——, *Developing the Amazon*, Bloomington: Indiana University Press, 1981

—— (ed.), *The dilemma of Amazonian development*, Boulder: Westview Press, 1983

——(ed.), 'Levels of analysis and analytical level shifting: examples from Amazonian ecosystem research', in *The ecosystem approach in anthropology: from concept to practice*, Ann Arbor: University of Michigan Press, 1990

——, *Through Amazonian eyes: the human ecology of Amazonian populations*, Iowa City: University of Iowa Press, 1993

——, E. Brondizio, P. Mausel and Y. Wu, 'Integrating Amazonian vegetation, land-use, and satellite data', *BioScience* 44(5), 1994, pp.329–38

Orlove, B., 'Ecological anthropology', *Annual Review of Anthropology* 9, 1980, pp.235–73

Parker, E., 'Forest Islands and Kayapó resource management in Amazonia: a reappraisal of the Apetê', *American Anthropologist* 94(2), 1992, pp.406–28

Pires, J.M. and G. Prance, 'The vegetation types of the Brazilian Amazon', in G. Prance and T. Lovejoy (eds), *Key environments: Amazonia*, Oxford: Pergamon Press, 1985

Posey, D., 'Indigenous management of tropical forest ecosystems: the case of the Kayapó Indians of the Brazilian Amazon', *Agroforestry Systems* 3, 1985, pp.139–58

Prance, G. and T. Lovejoy (eds), *Key environments: Amazonia*, Oxford: Pergamon, 1985

RADAM, *Levantamento da Região Amazônica*, 12 vols, Rio de Janeiro: Ministerio de Minas e Energia, 1973

Rappaport, R., *Pigs for the ancestors*, New Haven: Yale University Press,1968

Redford, K., 'The ecologically noble savage', *Orion Nature Quarterly* 9(3), 1990, pp.24–9

Redford, K. and J. Robinson, 'Hunting by indigenous peoples and conservation of game species', *Cultural Survival Quarterly* 9(1), 1985, pp.41–4

Reichel-Dolmatoff, G., *Amazonian cosmos: the sexual and religious symbolism of the Tukano Indians*, Chicago: University of Chicago Press, 1970

Saldarriaga, J.G. and D.C. West, 'Holocene fires in the northern Amazon basin', *Quaternary Research* 26, 1986, pp.358–66

Schmink, M. and C. Wood (eds), *Frontier expansion in Amazonia*, Gainesville: University of Florida Press, 1984

Sioli, H., *The Amazon: limnology and landscape ecology of a mighty tropical river and its basin*, Dordrecht: Junk Publication, 1984

Smith, N.J.H., 'Anthrosols and human carrying capacity in

Amazonia', *Annals of the Association of American Geographers* 70, 1980, pp.553–66
——, 'Enchanted forest', *Natural History* 92(8), 1983, pp.14–20
Smole, W., *The Yanomama: a cultural geography*, Austin: University of Texas Press, 1976
Sombroek, W., *Amazon soils*, Wageningen: Centre for Agricultural Publication and Documentation (the Netherlands), 1966
Sponsel, L., 'Amazon ecology and adaptation', *Annual Review of Anthropology* 15, 1986, pp. 67–97
Steward, J., *Basin plateau Aboriginal sociopolitical groups*, Washington DC: Smithsonian Institution, Bulletin 120. Bureau of American Ethnology, 1938
Thomas, F., *The environmental basis of society*, New York: the Century Company, 1925
Vayda, A.P., 'Warfare in ecological perspective', *Annual Review of Ecology and Systematics* 3, 1974, pp.107–32
——, *Warfare in ecological perspective*, New York: Plenum, 1976
Vayda, A.P. and R. Rappaport, 'Ecology, cultural and con-cultural', in P. Richerson and J. McEvoy (eds), *Human ecology*, N. Scituate: Duxbury Press, 1976
Wagley, C. (ed.), *Man in the Amazon*, Gainesville: University of Florida Press, 1973

Chapter 20

Process Versus Product in Bornean Augury: A Traditional Knowledge System's Solution to the Problem of Knowing

Michael R. Dove

Introduction

The practice of augury in Borneo has received at least passing mention from nearly every natural scientist or ethnographer to visit the island since the nineteenth century (for a representative sampling see [Geddes 1956; Hose and McDougall 1912; and Low 1848]). The idea that human behaviour might be directed by such a seemingly fanciful system caught the imagination of foreign observers. The apparent lack of connection between the system and empirical reality added to the curiosity and often indignation with which it was viewed. Detailed descriptions of augury were eventually provided by Harrisson [1960], Jensen [1974], Richards [1972] and Sandin [1980]. King [1977] and Metcalf [1976] debated whether augural interpretation is fixed or subject to individual interpretation. Freeman, whose work among the Iban in the 1950s set the standards for contemporary ethnography in Borneo, carried out the most important analytical study of augury to date. He concluded that augury has an essentially psychological explanation. He writes [Freeman 1960: 97]:

> This leads me to an hypothesis – basically sociological – concerning augury. . .. Human behaviour is essentially purposive. In augury, as in other comparable systems of divination, we have, I would

> suggest, an extension of this notion of purpose to animal species, so that their behaviour in the presence of man and in respect of man is treated as though it were purposive in human terms. This attitude, I would contend, is one which would arise naturally from primitive man's experience as a member of society: the kind of thinking found in augury is fundamentally an extension of the way in which man thinks about his fellows.

Freeman therefore 'reads' Bornean augury as saying something not about birds and men but about something else, about purposiveness in human society. I suggest that while augury does indeed have much to do with purposiveness, we would do better to begin by reading this ritual text more literally. In an application of 'Ockham's razor', which operates on the principle that a simple explanation is superior (cet. par.) to a complex one, I suggest that in the absence of compelling evidence to the contrary, we should accept that augury is about what it purports to be about.[1] Thus, augury appears to be 'senseless', and I suggest that it is; it appears to randomize decisions and remove human judgement from decision-making, and it does; it appears to be about relations with the physical environment, and it is.

Augury is also about the presence of purpose in human society and the lack of purpose outside it, but in a more literal sense than that intended by Freeman. He suggests that through augury people see purpose where there is none, because there is purpose in society. I suggest that through augury people see a *lack* of purpose where there is indeed a lack, in *spite* of the purpose in society. I suggest that augury, far from extending human purposiveness to the natural system, makes comprehensible in human terms the natural system's lack of purposiveness. The dramatization in augury of what is (in empirical terms) a lack of purpose is, in fact, the key to understanding the role that augury plays in a purposeful human society.

There is an important difference between Freeman's approach and mine, in the way that the relationship between society and environment, between nature and culture, is treated. Freeman reifies the natural setting of augury: he treats nature as 'available' for use in expressing ideas about society. I suggest, in contrast, that nature and nature–culture relations are compelling in their own right. I suggest that augury is less a projection on to the environment of what society thinks about itself than a reflection (and operationalization) of what society has learned about its environment, and about the relationship between itself and its environment [cf. Dove 1992]. I suggest that the problem of nature and culture – the dichotomy between the two – is the fundamental problematic of augury. I will argue that this ultimately reduces to the problem of ego and 'other', the problem of a consciousness that tries to encompass itself. This returns us, in a sense, to Freeman's original thesis. While I carry my analysis in different directions from his, I can only agree that augury basically has to do with the problem of purposeful people in a less purposeful world.

I will present my interpretation of the augural system, drawing on data from a number of the peoples of Borneo, including a group amongst whom I have done extended fieldwork (1974–6), the Kantu' of West Kalimantan (Figure 20.1). The Kantu' cultivate dry rice and assorted other crops in swiddens to meet subsistence needs, and they cultivate rubber and pepper to meet market needs. During a two-year stay in one Kantu' longhouse (comprising fifteen 'doors' or households), I compiled extensive data on all the swiddens cultivated, consisting of thirty-four separate swiddens the first year and thirty-five the second, giving a total of sixty-nine.[2] My data set includes data on all omens observed and honoured throughout the swidden cycle – but especially during the all-important first stage of site selection – in all sixty-nine swiddens. These data include first-hand observations, as well as data reported by field assistants and data obtained through interviews. Other data on augury were gathered during focused interviews with Kantu' farmers, elders and shamans. Additional comparative data on augury were gathered during four subsequent, briefer field trips (between 1979 and 1985) to other parts of Kalimantan.

Kantu' Augury

Augury

Kantu' divination is based on the belief that the major deities of the spirit world have foreknowledge of events in the human world and that, out of benevolence, they endeavour to

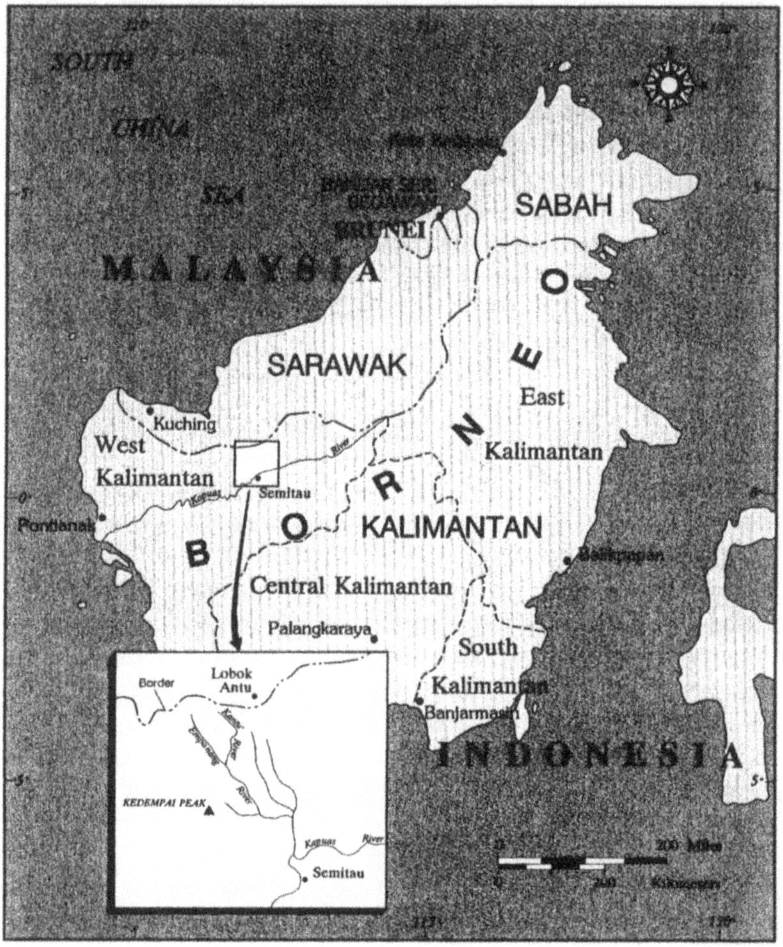

Figure 20.1. The Kantu' territory in Kalimantan, Indonesia

communicate this knowledge to the Kantu'. If the Kantu' can read the intended meaning of these communications correctly, they too can possess this foreknowledge. The deities can express themselves through a wide variety of phenomena, including dreams, cock-fights and ordeals. But the most common and also most ritually important intermediaries are seven species of forest birds, which are believed to be the sons-in-law of the major deity Singalang Burong. The birds vary in age and hence in authority. In ascending order they are: the **nenak** 'White-rumped Shama'

(*Copsychus malabaricus*), **ketupong** 'Rufous Piculet' (*Sasia abnormis*), **beragai** 'Scarlet-rumped Trogon' (*Harpactes duvauceli*), **papau**, 'Diard's Trogon' (*Harpactes diardii*), **memuas** 'Banded Kingfisher' (*Lacedo pulchella*), **kutok** 'Maroon Woodpecker' (*Platylophus galericulatus*), and **bejampong** 'Crested Jay' (*Blythipicus rubliginosus*). In practice, the Kantu' take most of their omens from the first three birds, or from three variant ritual practices called **beburong besi** 'taking omens from iron', **betenong kempang** 'taking omens from the kempang stick' (Figure 20.2) and **beburong pegela'** 'taking omens from an offering' (Table 20.1).

Figure 20.2. The augural practice, **betenong kempang** 'taking omens from the **kempang** stick'

The Kantu' deem omens from these birds to be relevant to all facets of life, including travel, litigation and rites of passage, but they are especially relevant to swidden cultivation. Omens are observed through most of the stages of the swidden cycle and are typically honoured by proscription of swidden work on the day received. The most important omens, however, are those received during the first stage of the cycle, selection of the proposed swidden site. This stage of the swidden cycle, called **beburong** 'to take birds or omens', consists of traversing a section of forest

Table 20.1. Incidence of different types of omen taken in each stage of omen taking, expressed in percentage terms

Incidence in omen-taking	Type of Omen								
	None	Shama	Piculet	*Iron	*Omen stick	S-Rumped Trogon	*Offer-ing	Other birds	
First stage	19	27	18	16	13	3	4	0	100
Second stage	34	21	31	–	–	9	–	5	100
Average	26	24	24	9	7	6	2	2	100

* These types of omens are taken in one stage only, not two.

proposed for a swidden and seeking favourable bird omens. The character of the omens received at this time – **burong badas** 'good birds', versus **burong jai'** 'bad birds' – is believed to be a major determinant of the character of the eventual swidden harvest. Accordingly, if a sufficiently bad omen is received, the site should be rejected for farming that year.

Ecological Interpretation

The key to interpreting site-rejection in particular, and the system of augury in general, is (as I have argued in detail elsewhere [Dove 1993a]) the indeterminacy of the physical environment of Borneo, the impossibility of correctly predicting critical agro-ecological conditions, and the consequent need to devise pluralistic rather than deterministic agricultural strategies.[3] For example, during my fieldwork the members of one household observed that rice-destroying floods had not occurred for several years. They reasoned on this basis that the likelihood of such a flood occurring during the coming year was relatively high, so they decided to locate all of their coming year's swiddens on high ground. This reasoning is in fact flawed: there is no evidence of regular variation in rainfall or flooding in Kalimantan, and thus anticipation of a flood (or a drought) will more often be wrong than right.[4] The best strategy for the afore mentioned household was to prepare neither for a year with a flood nor a year without one, but for an average year – that is, a year with some percentage likelihood of a flood.

The system of augury promotes such 'averaging' strategies by making agricultural decision-making less deterministic. Augury does this by systematically severing empirical linkages between the environment and human decision-making.[5] The rules of augural interpretation disrupt four key linkages in the agroecology of the Kantu' [see Dove 1993a]: between the environment and the swidden system, between one household and another, between one swidden and another belonging to the same household, and between the swidden strategy of a household in one year and its strategy the succeeding year. The result is randomization of swidden-site selection.[6]

One way of characterizing augury is to say that it stands (as a disjunction) between the environment and human decision-making; another way is to say that it stands between the future and the past. The relationship between past, present and future is problematic for farmers. Ortiz [1979: 64] writes: 'Farmers can speculate about the likelihood of future events only on the basis of information about the past.' The problem, as Ortiz [ibid.: 75] goes on to explain, is that farmers tend to concentrate on the recent past,[7] and the recent past shows more determinism than is apparent from a longer time perspective. Augury mitigates the tendency to follow this false determinism. According to Ortiz [ibid.: 78n.]: 'If . . . we realize that a statement about prospect is not a statement about the future but a summary of past experiences, the difficulty is not to generate a theory of expectations but a theory of memory.' The 'theory of memory' that we can deduce from the augural system is this: the past is characterized by indeterminacy, a 'recency' bias in human memory portrays this as more determinate than it was, and the system of augury corrects for this bias.

The corrective role of augury is based upon differentiation among different pasts: while augury breaks ties to the immediate past, it strengthens ties to an ill-remembered, more distant past. To return to the example of the flood. I noted earlier that preparation for the presence or absence of flooding is bad, while preparation for all eventualities is good. From one perspective, this represents delinkage from the past, but from a longer perspective, it does not: it represents linkage to society's experience of both floods and no floods over a long period of time. Timmerman [1986: 452 n.49] draws an analogy between the slow, medium and fast variables in an ecosystem on the one hand, and on the other the three 'times' identified by Fernand Braudel (1980): *la longue durée* (geographical or large ecosystemic time); *l'histoire sociale* (economic and social history); and *l'histoire événementielle* (the history of events). Following this analogy, the system of augury neutralizes the impact on decision-making of the fast variable of 'last year's flood', in an adaptation to the slow variable of both the presence and absence of flooding. Medium variables of socio-economic history, as I will argue in a later section, are responsible for changes over time in the augural system itself.

Adherence to the Rules of Omen-Taking

The way that the augural system articulates with fast, medium and slow variables is reflected in the differing ways that the Kantu' honour, and violate, both the rules of the system and the system as a whole.[8]

Adherence to the Augural Rules

In order to understand the significance of rule-breaking, it is necessary to know the extent to which rules are followed. To what extent do the Kantu' ignore omens or simply recollect them in post-hoc explanation of swidden outcomes (as Freeman [1960: 79n.] suggests)? The evidence suggests that, on the contrary, the Kantu' do honour most important omens – such as those observed during swidden-site selection – and that these omens thus have an independent and coercive impact on Kantu' behaviour. The Kantu' say that if they hear an omen and do not honour it, ***nadai isi' asai agi'*** ('they have no feeling left'). This reaction of anomie to rule-breaking suggests that it represents a violation of deeply felt beliefs.[9] There are many examples of the strength of these beliefs.

For example, when Gayan, the elder of one Kantu' household, was in his first day of omen-taking at the planned site of a swidden during one of the study years, he heard a Diard's Trogon. After discussing the meaning of the omen with elders in the longhouse, he decided to abandon the site for that year and select another. The Kantu' may also find less onerous ways to honour unfavourable omens, however. When Guyak, the elder of another Kantu' household, was seeking omens at the site of one of his planned swiddens during the same year, he too heard a Diard's Trogon. Instead of rejecting the entire **belah** 'section' of forest and selecting another, however, he rejected just that part of the section where he had heard the Trogon, moved to another part, and continued his omen-taking and site preparation.

While Guyak's response to the ill omen appears to violate the spirit of the augural system, the fact that it was deemed necessary reflects the coercive nature of omens just as much as Gayan's more orthodox response. This is similarly true of other methods for coping with the consequences of ill omens. Particularly noteworthy in this regard is the practice, on important occasions after favourable omens have already been secured, of beating drums or gongs to forestall the subsequent hearing of unfavourable omens [cf. Hose and McDougall 1902: 176]. Freeman [1970: 89] writes:

> Deliberate avoidance of possibly bad auguries is also common practice. Thus, whenever a group of Iban are engaged in an undertaking of consequence . . . they take care to see that certain of their number – usually two or three girls – are equipped with bronze gongs which are beaten without cease until the business in hand is concluded. . . .. Its purpose is to forestall the hearing of inauspicious omens. Such behaviour may seem flagrantly inconsistent with the basic tenets of augury, but in such cases it is usual for auguries to have been taken in advance and for a series of propitiatory offerings to have been presented to the gods. The attitude is rather that of a man who, having taken good counsel and every safeguard, decides for better or for worse to shut his ears to all further admonition.

This institution would have little meaning if omens were not perceived to be, and treated as, coercive.

Inter-Household Variation in Adherence to the Augural Rules

There is considerable variation among households in the practice of augury (Table 20.2). Even within the same household, there is some variation from one year to the next (Table 20.3); and there is great variation from one swidden to another (in households with multiple swiddens) in the same year (Table 20.4). The variation among households reflects, in part, cultural beliefs about the individual character of the augural process: the Kantu' say,

Table 20.2. Inter-household variation in omens

		Percentage of households taking the most popular omen birds in one or more swiddens in the same year
Stage in omen-taking:	first ('stem')	51 took the White-R. Shama
	second ('root')	44 took the Rufous Piculet

Table 20.3. Inter-year variation in type of omen

		Percentage of households taking same omen bird in the **umai pun** 'stem swidden' two years in a row:
Stage in omen-taking:	first ('stem')	82
	second ('root')	44
	Average	65

Table 20.4. Inter-swidden variation in type of omen

		Percentage of households taking same omen bird in two (or more) dryland swiddens made in same year
Stage in omen-taking:	first ('stem')	42
	second ('root')	36
	Average	38

burong nadai tau' kuntsi 'omens cannot be shared'. But the variation, among as well as within households, also reflects a high incidence of violation of the augural rules.

For example, the augural rules prescribe that two separate, sequential omens should be sought in every swidden: first the **pun burong** 'stem omen', and then the **pun nyuror** 'stem that gives off roots'. The rules further prescribe that, because the first omen is believed to 'select' the second, the second must be 'older' than the first. For example, the first omen can come from a White-rumped Shama and the second from a Rufous Piculet, but not the other way around. In practice, however, the proper age-order is achieved less than fifty per cent of the time (Table 20.5).

The augural rules also prescribe that the 'temperature' of the omen bird must balance that of the swidden environment in which it is taken. The Kantu' accord 'hot' or 'cold' properties to the omen birds, and they believe that the auspiciousness of their omens varies according to whether these properties mitigate or exacerbate the 'heat' or 'coolness' of particular swidden sites. Thus, the White-rumped Shama, which is considered to be 'cool',

Table 20.5. Correct age-order between first and second omens

	Age of second omen bird relative to first omen bird: [Less correct - - - - - - - More correct]			Total
	Younger	Same	Older	
Percentage of swiddens	14	45	41	100

is deemed auspicious in dry, highland swiddens (whose heat it will mitigate) and inauspicious in wet flood-zone swiddens (whose coldness it would simply exacerbate); and precisely the reverse is the case with the Scarlet-rumped Trogon, which is deemed 'hot'. In practice, again, this rule is violated more often than not: there is no observable association between the swidden environment selected (referring specifically to the selection of land within the flood zone versus selection above it) and the 'heat' of the omen bird taken (Table 20.6).

Table 20.6. Correct fit between 'heat' of environment and omen bird, expressed in percentage terms*

		**Location vis-a-vis floodzone: [Cool] - - - - - - - - - - - - - - [Hot]	
		Within	Above
Type of omen-bird taken: [Younger/Cooler]	Shama	44.5 [Incorrect]	39 [Correct]
	Piculet***	44.5 [Neutral]	45 [Neutral]
[Older/Hotter]	Others	[Correct] 11	[Incorrect] 16
	Total	100	100

* This association is *not* statistically significant. For n = 69 omens, $X^2 = .28$ and $P < .75$.

** Excludes swiddens made in swampland.

*** Includes the Piculet's inauspicious and auspicious calls.

These violations of the rules of augury do not appear to have any effect on the success or failure of the swidden, cultural beliefs notwithstanding. Thus, there is no observable association between correct or incorrect age-order between the first ('stem') and second ('root') omens, and the success of the swidden harvest (as measured against a longhouse mean) (Table 20.7). Nor is there any such association with respect to the matching of omen bird 'heat' and swidden environment (Table 20.8).

Table 20.7. Age-order of first and second omens, and swidden success expressed in percentage terms*

		Age of second omen-bird relative to first omen-bird: [Less correct - - - - - - - - - - - - More correct]		
		Younger	Same	Older
Swidden	< Mean	71	52	62
Harvest	> Mean	29	48	38
	Total	100	100	100

*This association is *not* statistically significant. For n = 31 swiddens, $X^2_c = .06$ and $P < .975$.

Table 20.8. Fit of bird 'heat' and environment, and swidden success, expressed in percentage terms*

		Relation of bird 'heat' to flood-zone location: [Less correct - - - - - - - - - - - - More correct]		
		Exacerbates	Neutral**	Offsetting
Swidden	< Mean	60	65	58
Harvest	> Mean	40	35	42
	Total	100	100	100

* This association is *not* statistically significant. For n = 54 swiddens, $X^2 = .22$ and $P < .75$.

** The 'cool' Shama offsets the heat above the flood zone but exacerbates the coolness within it; the neutral Piculet is acceptable in either zone; all other birds (because 'hot') are exacerbating above the zone and offsetting within it.

Omen-Taking Versus Not Omen-Taking

There are circumstances in which it is not the augural rules but the system itself that is violated. For example, taking omens from 'iron' – referring to the iron in the brush-sword with which a person may start slashing the forest even in the absence of a good omen – is considered to be an abbreviated, less conscientious adherence to the rules of augury. Abbreviated omen-taking, or no omen-taking at all, is done in about one-third of all swiddens selected, most of which are made in swampland (Table 20.9, cf. Table 20.1). And whereas omens are taken from birds in over 90 per cent of swiddens made on dryland, the comparable figure for swiddens made in swampland is just 13 per cent (Table 20.10). This contrast is due to a fundamental difference between the two types of swidden in the need for the randomizing impact of augury.

Table 20.9. Incidence of omen-taking, expressed in percentage terms

Omens taken from birds:	Omens taken from iron:	*Omens not taken:
64	16	20

* All of these cases involve swampland swiddens.

Table 20.10. Incidence of omen-taking in dryland versus swampland swiddens, expressed in percentage terms*

Commitment to omen-taking:			Drainage: Swampland	Dryland
[Lesser]	Omens taken from:	Not taken	69	4.5
|		Iron	18	4.5
[Greater]		Birds (etc.)	13	91
		Total	100	100

* This association *is* statistically significant. For n = 135 omens, $X^2 = 150.9$ and $P < .001$.

The agro-ecological dynamics of swampland swiddens are quite different from those of normal dryland swiddens, as is the level and nature of their risk. Two of the principal hazards in dryland swiddens, the failure to obtain a thorough burn and prolonged inundation by flood waters, pose no threat at all in swampland swiddens. The vegetative cover on most of the swampland in the Kantu' territory, by virtue of relatively frequent cropping, is grassland. In the event that the grasses cannot be burned, they are simply cut and mulched into the swamp soil, where water plays the role that fire plays elsewhere (as in the dryland swiddens) in breaking them down into their elemental components. As for flood waters, the special rice varieties that are planted in the swampland swiddens are resistant to water stress by virtue of their unusually deep roots, so they can survive most normal inundations. Thus, the selection of a site for a swampland swidden is less problematic than selection of a site for a dryland swidden: there are fewer critical variables to consider and fewer chances of failure due to incorrect selection. In consequence, there are fewer risks from deterministic farming strategies and thus less need for the randomizing impact of augury.

Swidden Success and Adherence to the Augural Rules

While adherence to the rules of omen-taking does not affect swidden success, the reverse is not true: swidden success or failure *does* affect adherence to the rules. A household whose previous year's swidden harvest fell short of its needs is more likely to break augural rules during the coming year than a household whose previous harvest met or exceeded its needs. This relationship is, to some extent, culturally recognized: thus, while the Kantu' generally regard the **bacar** call of the Rufous Piculet (see Appendix I) as inappropriate for the 'stem omen' for a new swidden, they say that it can be taken by a household that is poor, one that has never reaped a good harvest before. In general, households that have just experienced harvest failure are more likely to 'try their luck with' the less favoured omen birds than other households (Table 20.11). Similarly, households whose previous harvests were unsuccessful appear to be more likely than other households to violate the age-order prescription

Table 20.11. Prior swidden success and type of omen-bird, expressed in percentage terms*

			Adequacy of household's prior harvest for one year's needs: < Sufficiency	> Sufficiency
Type of omen-bird taken following year:	[Younger/Cooler]	Shama	33	44
	|	Piculet	39	50
	[Older/Hotter]	Others	28	6
		Total	100	100

* This inverse association is not quite statistically significant. For n = 36 omens, X^2 = .3.18 and $P < .10$.

for the stem and root omens during the coming year (Table 20.12), as well as the prescription for matching the 'heat' of the omen bird and the physical environment (Table 20.13).

While the success or failure of previous harvests affects adherence to the rules of augury, however, it does *not* affect adherence to the augural system itself. Thus, success or failure of the prior year's swidden does not appear to be associated with

Table 20.12. Prior swidden success, and age-order of first and second omens, expressed in percentage terms*

			Adequacy of household's prior harvest for one year's needs: < Sufficiency	> Sufficiency
Age of first relative to second omen following year:	[More correct]	Older	17	43
	|	Same age	33	57
	[Less correct]	Younger	50	0
		Total	100	100

* There appears to be an inverse association here, but the small sample (13 swiddens) makes testing for significance problematic.

Table 20.13. Prior swidden success, and match of omen bird 'heat' and environment, expressed in percentage terms*

			Adequacy of household's prior harvest for one year's needs:	
			< Sufficiency	> Sufficiency
Match of omen 'heat' and environment following year:	[More correct]	Offset	17	50
		Neutral	33	44
	[Less correct]	Exacerbates	50	6
		Total	100	100

* This direct association *is* statistically significant. For n = 34 omens, $X^2 = 4.58$ and $P < .05$.

the extent to which the following year's omen-taking is done thoroughly, less thoroughly, or not at all (Table 20.14). Nor does it appear to affect the decision whether to practice omen-taking in the swampland swiddens or not (Table 20.15). Commitment to the process of omen-taking appears, in short, to be unaffected by swidden success or failure.

Table 20.14. Prior swidden success and completeness of omen-taking, expressed in percentage terms*

Commitment to omen-taking following year:			Adequacy of household's prior harvest for one year's needs:	
			< Sufficiency	> Sufficiency
[Lesser]	Omens taken from:	Not taken	23	21
		Iron	15	11
[Greater]		Birds etc.	62	68
		Total	100	100

* There appears to be no association here, with $X^2 = 0.16$ and $P < .95$, but the small sample (31 swiddens) makes testing problematic.

Table 20.15. Prior swidden success and omen-taking in swampland, expressed in percentage terms*

		Adequacy of household's prior harvest for one year's needs: < Sufficiency	> Sufficiency
Omens taken in swampland swiddens following year:	No	58	50
	Yes	42	50
	Total	100	100

* There appears to be no association here, with $X^2 = 0.08$ and $P < .90$, although the small sample (11 swiddens) makes testing problematic.

Choice, Error and Learning

This commitment to the augural system, but not to the augural rules, has special pedagogical and epistemological implications.

Abrogation of Choice, and Bateson's Dolphins

Gregory Bateson, in an essay first published in 1966, drew on his observations of dolphin behaviour to make some stimulating observations regarding learning, choice, and the necessary limitation thereof. He writes [Bateson 1972: 368]:

> A simple discrimination experiment, such as has been run in the Lilly laboratories, and no doubt elsewhere, involves a series of steps: (1) The dolphin may or may not perceive a difference between the stimulus objects, X and Y. (2) The dolphin may or may not perceive that this difference is a cue to behaviour. (3) The dolphin may or may not perceive that the behaviour in question has a good or bad effect upon reinforcement, that is, that doing 'right' is conditionally followed by fish. (4) The dolphin may or may not choose to do 'right', even after he knows which is right. Success in the first three steps merely provides the dolphin with a further choice point. This extra degree of freedom must be the first focus of our investigations.

The choice to do right in the fourth step is the focus of all efforts to train the dolphins. Bateson [ibid.: 369] continues as follows:

> The first requirement of a trainer is that he must be able to prevent the animal from exerting choice at the level of step 4. It must continually be made clear to the animal that, when he knows what is the right thing to do in a given context, that is the only thing he *can* do, and no nonsense about it. In other words, it is a primary condition of circus success that the animal shall abrogate the use of certain higher levels of his intelligence.

The counter-intuitive nature of developing an animal's intelligence to a point when it must be abrogated intrigued Bateson. He writes [ibid.: 370]: 'What is amazing about circus tricks is that the animal can abrogate the use of so much of his intelligence and still have enough left to perform the trick.'

Bateson presents these data to say something not just about dolphins, but about human beings. In the very next sentence he writes [Bateson ibid.: 270]: 'I regard the conscious intelligence as the greatest ornament of the human mind. But many authorities, from the Zen masters to Sigmund Freud, have stressed the ingenuity of the less conscious and perhaps more archaic level.' Moving from dolphins to people, that is, Bateson says that while he admires the human exercise of choice that is analogous to the dolphins' step 4, he acknowledges that the prior behavioural levels have their own virtues (and champions).

I believe, like Bateson, that the extra degree of freedom provided by step 4 learning is in some ways problematic for human society, and I suggest that preventing the exercise of choice is critical to the success of some human strategies, especially for exploiting the natural environment. I further suggest that the cultural wisdom that can abrogate this choice in pursuit of more abstract ends, and in the face of a 'human nature' that is ill-suited to such abrogation, merits our attention.

The abrogation of human choice in the pursuit of indeterministic outcomes is not as rare as might be supposed. It is present in such varied endeavours as submarine warfare (based on random number tables), investment strategies (based on dollar-cost-averaging [Malkiel 1990]), and in activities more akin to the ones discussed in this paper, such as the chicken oracles of the Azande (based on the fine calculation of non-lethal doses of poison [Evans-Pritchard 1937]), and scapulimancy among the Naskapi (based on the random action of fire on bone [Moore 1957]). The intention in all these cases is the same as in Kantu' augury, namely to counter the tendency to base future decisions on past, unrelated actions. This is achieved, in all of these cases, through insertion of a randomizing mechanism between human thought and action. The principal difference between them is that the secular cases attribute this mechanism to mathematics, while the ritual ones attribute it to the supernatural.

The Lack of Choice and Learning in Kantu' Augury

In the Kantu' system of augury, as in the dolphin training analysed by Bateson, there is a critical level where choice and learning are systematically suppressed. Choice is suppressed whenever augural dictates overrule individual judgements; and learning is suppressed whenever the response to agricultural outcomes is the revision, or even violation, of specific augural rules. The latter appears to involve a process of learning but it is specious, because there is no systematic empirical linkage between augury and agro-ecology. There is no feedback and in the absence of feedback nothing can be learned, and nothing can be changed for the better.

Specific agricultural outcomes are, in any case, of no more significance in this system than the specific augural decisions that purport to give rise to them. It is not the apparent product of augury that is important, but the augural process itself. According to the thesis of this study, augury confers an advantage upon those who follow the augural system, regardless of what augural rules they follow, modify or violate (so long as the randomizing impact of augury is not compromised). Thus, it does not matter if a Kantu' household decides that it does not have to reject the 'cool' White-rumped Shama in the flood zone; but it *does* matter if a household decides that it does not have to practice augury at all. There are echoes of this stress on process versus product in the emphasis on 'movement' in the cosmologies of other Bornean peoples. Thus, Tsing [1984: 261] writes of the tribal inhabitants of the Meratus Mountains in Southeastern Borneo: 'Meratus stress the "walking" aspects of rice farming as the essential route to a good harvest: site selection and omen reading, which involve walking around in the forest; ritual, which involves the "travels" of the ritual experts as well as the rice, metaphorically.'[10]

The character of the augural rules is a key to this processual focus. Their arbitrariness ensures that there is no meaningful product, while their existence supports continued commitment to the process. The rules permit people to see, and to act upon, instrumental meaning where meaning (in this sense) is absent.[11] They permit people to observe feedback and make corrections where, in fact, feedback is absent and corrections are impossible. They permit people, in the wake of swidden failures, to break the rules but remain within the system. This distinction is crucial to achieving the indeterminism that I have argued is at the heart of the augural system. Following any set of augural rules, even a non-orthodox set, will produce indeterministic outcomes; but opting out of the system altogether will produce just the opposite. Denying the choice of rejecting the system is the key: augury permits people to exercise an irrelevant choice regarding augural rules which keeps them from exercising the only meaningful choice, which is to opt in or out of the system. Over time, however, some system-level choice also occurs.

Choice, Learning, and the Evolution of Kantu' Augury

The Kantu' system of augury allows the Kantu' to take steps 1, 2 and 3 (to return to the dolphin analogy) but limits their freedom to apply the lessons learned in step 4 by limiting their freedom to make changes at the level of the system itself. A step 4 change is likely to be an error – the only one possible in the augural system. In addition to being the only error possible, however, system-level change is also the only source of learning possible.[12] True feedback and learning, regarding the role of the augural system in the Kantu' adaptation to their environment, cannot occur at the level of the augural rules but only at the level of the system itself. While the system largely promotes specious feedback and learning at the level of the rules as opposed to true feedback and learning at the level of the system, there are exceptions.

Two examples of true learning, and two examples of attendant changes in the augural system, can be observed in the recent past.[13] The first pertains to the scope of the augural system. This has been diminished by largely excepting from the practice of augury the swampland swiddens. The lesser indeterminacy of these swiddens – as described earlier – makes deterministic agricultural decision-making less risky and the randomizing intervention of augury correspondingly less necessary. Swamp-land cultivation is a relatively recent development in Kantu' agriculture, and the corollary modification of the scope of Kantu' augury is thus an equally recent example of learning. This represents an adaptation to changes in the micro-environment; the next case represents an adaptation to changes in the macro-environment.

The second example of learning that can be observed in the augural system pertains to the intensity of its application. The Kantu' say that today the only swidden stages in which they really pay attention to omens are (in addition to the initial stage of site-selection) **nebas** 'slashing' and **nebang** 'felling'. They say that formerly they observed omens even during such relatively inconsequential sub-stages as the re-burning of poorly burned swiddens (**mandok**) and the transplanting of swamp rice seedlings (**nambak**). The recent evolution of institutions for abrogating the impact of inauspicious omens – such as the **begela'** 'offering ceremony', in which chickens and sometimes pigs are sacrificed to the spirits – suggests that the system's authority has also weakened. This is further indicated by the disappearance of the institution and person of **tuai burong** 'omen experts', who are said to have had the power to abrogate bad omens.[14] Their disappearance may reflect the fact that the ability to abrogate omens has become so commonplace as no longer to require a ritual expert.

These contractions in the scope and authority of the augural system may be associated with the expansion of the Kantu' economy beyond its traditional confines. The development this century of pepper and especially rubber cultivation has significantly increased the ability of the Kantu' to survive swidden harvest failures by providing them with a secure source of tradeable commodities with which to purchase rice from the market [see Dove 1993b]. The level of uncertainty in rubber cultivation, the volatility of market prices aside, is very low. These developments have increased, therefore, the overall stability and sustainability of the Kantu' agricultural economy. This, in turn, has arguably permitted the Kantu' to raise the level of risk in the swidden system in exchange for larger harvests in good years (albeit alongside greater shortfalls in poor years), through less diversification and, accordingly, less omen-based randomization.

These developments support one of the key theses of this study, namely that the augural system of the Kantu' helps them to adapt to environmental uncertainty through randomization of agricultural strategies. If this thesis is true, we would expect to see changes in the augural system if the basic agro-ecology changes, and this is just what we have seen. The way that the changes in the augural system have come about supports a second key thesis of this study, namely that commitment to the augural rules is less important than commitment to the augural system. If this is true, we would expect any changes to be embraced by all participants in the system, and again this is what we find. Although there is great variation among individual households in interpretation of and commitment to the augural rules, there is almost no variation in acceptance of the previously described contractions in the scope and authority of the augural system.

Development Implications

Single Versus Plural Vision

The principles that underpin the Kantu' system of augury differ in some fundamental respects from those of the modern, international, scientific tradition. Analysis of these differences can help to defamiliarize and thus make more accessible to critical review some of our own conceptual structures, in a 'repatriation' of anthropology [Marcus and Fischer 1986: 135, 137]. The structure that stands out in greatest relief by comparison to Kantu' augury, and which might benefit most from such a review, is the linear, deterministic and monistic character of modern science. Holling, Taylor and Thompson write [1991: 21]:

> Whenever we encounter rationality in the singular (economic rationality, for instance, or scientific rationality) or single metrics (Cost:Benefit Analysis, for instance, or Probability Risk Assessment) or policy analysts and decision makers who demand to know which definition of the problem is the right one before they can make a start, we are in the presence of the visually impaired. Blake, we

> believe, was right all along: the world can never be narrowed down to just one perception, just one direction of change, one starting-point, one end-point. And, if we insist on doing this to it, it will surely escape our grasp and do something that we have not, as they say, bargained for.

They end their essay with one of Blake's poems, the last, memorable line of which reads 'May God us keep, From Single vision and Newton's sleep!' While unusually astute scholars like Holling, Taylor and Thompson are distancing themselves from this paradigm, it continues to be embraced by scientists, international donors/lenders and national planners working in the field of development.[15]

The difference between the deterministic development paradigm, and the paradigm that lies behind Kantu' augury, can be seen in the ways they distinguish (or fail to distinguish) between systems and rules. Almost all development strategies for exploiting the tropical forest (for example) are written at a level analogous to that of augural rules as opposed to the augural system. It is as if the development strategies were intended to inform people about the meaning of particular omen birds, but not the meaning of the overall system. Development strategies tend to promote whatever was associated with the most recent success as *the* 'bird' for the future. Since no single solution is ever successful for long, development strategies become serially committed to one 'bird' after another. This very process is folly, however.[16] My analysis of Kantu' augury shows that the Kantu' are protected from bad decision-making by looking for the right bird; they become vulnerable when they start looking for the right system. The wisdom of the augural system, therefore, lies in the way that it focuses such efforts on the birds as opposed to the system itself. It is precisely this added dimension that is missing from the development paradigm: it typically lacks an over-arching framework within which the role, and limitations, of individual strategies can be seen. The lesson for development planners, therefore, is to write at the level not just of rules, but also of systems; to look not just for new omen birds, but for a system of omen-taking within which the individual birds make sense (if you will, to look for a systemic understanding of the tropical forest environment that can encompass individual strategies).

Denying Versus Embracing the Unknown

Another important distinction between the augural and development paradigms is in their ways of coping with uncertainty (a salient characteristic of the tropical forest ecosystem of the Kantu').[17] Both paradigms represent responses to uncertainty, but while the development paradigm tries to eliminate it, the augural paradigm embraces it. The latter approach is receiving increasing support in scientific quarters (as the previous citation from Holling et al. suggests). Converging studies in a variety of fields (perhaps the most dramatic of which is chaos theory[18]) suggest that we need to come to better terms with the limits of our ability to know, in a deterministic way, the unknown.[19] Common to all such work is the belief that embracing our ignorance is, paradoxically, the best way to overcome it. Thus, Ludwig, Hilborn and Walters [1993: 36] write: 'Confront uncertainty. Once we free ourselves from the illusion that science or technology (if lavishly funded) can provide a solution to resource or conservation problems, appropriate action becomes possible.' Norgaard [1984: 170] similarly suggests that conceptual approaches that emphasise the uncertainty of the future actually reduce this uncertainty. I suggest that Kantu' augury, by not just emphasising but indeed celebrating uncertainty, does in fact reduce the uncertainty of agro-ecological futures in the tropical forest. Augury, by trying to make sense of the limits to knowledge, transcends them; while development, by assuming that there are no limits, circumscribes itself.

Systems theory provides an explanation for this apparent paradox. Following Maruyama [1963], the augural paradigm is seen to be a 'deviation-counteracting' process, while the development paradigm is seen to be a 'deviation-amplifying' one. The essential difference between the two types of process is the presence or absence of feedback. In the development paradigm, any success tends to be rewarded with ever-greater commitments of resources to the particular strategy that produced this success: there is positive feedback to success. The augural rules of the Kantu', in contrast, minimize any alteration in resource allocations based upon the past success (or failure) of a particular omen or swidden strategy: there is no feedback, or there is neutral feedback, to success. The purpose of the positive feedback in the develop-ment paradigm is to find the *right* solution; the purpose of the neutral feedback in the augural paradigm is to avoid finding *wrong* solutions, that is solutions based on apparent but invalid ecosystemic patterns.[20] The positive feedback paradigm exacerbates the volatility of relations between society and environment, while the neutral feedback paradigm dampens it. The former generates successive, ever-more deterministic systems for managing environmental relations, which are ever more productive if successful but ever more disastrous if unsuccessful; while the latter contributes to a pluralistic strategy, which can cope with a wide range of conditions and minimizes extremes of either success or failure.

This represents, on the part of the augural paradigm, more than just 'respect' for the unknown. Rappaport [1979: 100–1] writes: 'Because knowledge can never replace respect as a guiding principle in our ecosystemic relations, it is adaptive for cognized models to engender respect for that which is unknown, unpredictable, and uncontrollable, as well as for them to codify empirical knowledge.' While I agree with the spirit of this passage, I suggest that it is no more valid to say that the augural paradigm is limited to respect than to suggest that chaos theory is about respect (although it certainly is about that as well). I suggest that the augural paradigm contains, in addition to (and perhaps because of) its respect for the unknown, a systemic *understanding* of the unknown. I suggest that in this sense the augural paradigm, no less than other cognized models, codifies empirical knowledge *about* the unknown.[21]

Perspective and System Limits

A critical difference between the augural and development paradigms, therefore, is the ability of the former to differentiate between the rules and the system, between decision-making based on short-term perspectives and longer-term and more systemic models of reality. How is augury able to do this, when most development paradigms – and, indeed, much of modern science – are not?

The Problem of Consciousness and Self-Contained Systems

Bateson was preoccupied with the same question. He believed that human consciousness is so permeated by purpose and product that the processual character of reality is invisible to it. He writes [Bateson 1972: 444]: 'The cybernetic nature of self and the world tends to be imperceptible to consciousness, insofar as the contents of the "screen" of consciousness are determined by considerations of purpose.'[22] He concluded, as a result, that the capacity to comprehend and deal effectively with the systemic logic of the environment largely exceeds human reach. He writes [ibid.: 434]:

> On the one hand, we have the systemic nature of the individual human being, the systemic nature of the culture in which he lives, and the systemic nature of the biological, ecological systems around him; and, on the other hand, the curious twist in the systemic nature of the individual man whereby consciousness is, almost of necessity, blinded to the systemic nature of the man himself. Purposive consciousness pulls out, from the total mind, sequences which do not have the loop structure which is characteristic of the whole systemic structure. If you follow the 'common-sense' dictates of consciousness you become, effectively, greedy and unwise.

The challenge is thus to avoid (at least some of the time) the dictates of common sense. This is precisely what is accomplished by augury: augury requires the Kantu' to *not* employ all of their common-sense, instrumental knowledge. Augury places limits on this knowledge and, through acceptance of these limits, also offers a way to transcend them. This transcendence depends on the Kantu' being able to distance themselves from the subjective context of their agricultural decision-making, on being able to remove themselves from that context.

Any such 'distancing' or 'removal' is dependent, in turn, on the concept of interconnected as opposed to 'self-contained' systems. The latter concept, which is another manifestation of deterministic thinking, dominates development paradigms and has been responsible for many development imbroglios. Most of the systems relevant to human resource management are, in fact, *not* self-contained.[23] Belief in the concept of self-contained systems is widespread, however, in part because it is difficult to think of non-self-contained systems (in the same way that it is difficult for consciousness to fully comprehend the nature of its being). The difficulty is one of trying to think beyond a system from within it. I suggest that the Kantu' manage to achieve this through, in part, ritual analogy.

The Solution of the Ritual Metaphor

A senior augurer of the Kantu' told me that the source of most human misfortune is that the spirits 'see' human beings as wild pigs and, accordingly, try to hunt them. (Thus, humans are to spirits as wild pigs are to humans.) The identification of humans as pigs is thought to be inadvertent on the spirits' part, and it is in order to avert its ill-consequences that the spirits send omens to the Kantu'. The cosmological basis for augury therefore involves a belief that Kantu' behaviour has one meaning in the world of everyday reality and quite a different meaning in another plane of existence, which is beyond the Kantu's immediate comprehension. This recognition of a wider context to behaviour is enhanced by the further belief that the spirits themselves practice augury. Freeman [1960: 77n.] writes:

> It is significant that the gods themselves are said, by the Iban, to practice augury. In the invocation (**timang**) that accompanies head-hunting rituals, a description is given of Singalang Burong, his wife, daughters and sons-in-law leaving their celestial longhouse to attend the *gawai* to which they have been invited. At this point Singalang Burong says: 'Here let us halt and listen for auguries'(**Ditu kitai ngetu mending ka burong**). Only after the augury heard by Ketupung has been interpreted as favourable by Singalang Burong do the gods continue their journey.

While the Kantu' appear as pigs to the spirits, therefore, the spirits themselves appear as pigs to some other class of entities. The Kantu' universe thus contains widening circles of causation and meaning. Of most importance, these circles are not independent, but rather interlinked: what happens in one affects what happens in others.

I suggest that this cosmological vision is a key to the capacity of Kantu' augury to cope with one of the central conundrums of human existence: how to transcend the limits of consciousness, how to escape the conceptual tyranny of the self-contained system. On the one hand, this cosmological vision demonstrates to the Kantu' that the principles that apply to them also apply to other spheres of existence. On the other hand, it demonstrates that the immanent (if not transcendent) meaning of these principles may not be the same in each case. The Kantu' cosmology thus provides for a distinction between common sense and systemic meaning; and it thereby provides a way to escape the lack of wisdom of the former. This cosmological vision gives the Kantu' some of the humility (owing to their lack of complete knowledge of the broader system) – but also systemic understanding (owing to their recognition that there *is* a broader system) – that Bateson found so scarce, but also so necessary, in human society.

Conclusions

The expression for the search for omens when attempting to select a swidden site for the coming year is **ninga tanah** or 'listen to the earth'. It should now be clear that the augural system does indeed allow the Kantu' to listen – at a number of different levels – to the earth. A concern for this listening is not limited to Bornean swiddeners. With increasing appreciation of the complexity of our environment and the need to be receptive to its patterns (or lack thereof), scholars are coming to recognize that a variety of different phenomena, in many different cultures, represent attempts to 'listen to the earth'.[24] For example, Diamond [1990: 27] writes:

> When, late in life, Bach wrote his Lord's Prayer, I suspect that he was trying to express the view he had reached of nothing less than life itself, and of his own struggles to hear God's voice despite the obstacles that life poses . . .. It's as if Bach were praying: yes, yes, by all means forgive us our trespasses, and all those things – but above all, God, give us the will and ability to hear Thy voice through this world's confusion. With this metaphor, Bach also unwittingly captured better than any other metaphor I know, the sense of what it's like to come to know the rain forest. This conclusion is neither blasphemous nor trivializing, because to biologists the rain forest is

> life's most complex and wonderful creation. It overwhelms us by its detail. Underneath that detail lie nature's laws, but they don't cry out for attention. Instead, only by listening long and carefully can we hope to grasp them.

Just as Bach's music is a metaphor for creation, therefore, so is augury a (constructed) metaphor for the rain forest and the principles that govern it.

The continuities among Bach, Diamond and the Kantu' are based on a common, underlying element – the challenge of trying to understand our environment. Interpretations of the challenge vary: Diamond (and Bach) thinks that it comes from the wealth of detail in the world and, by inference, the human penchant for missing the patterns for the detail; while Kantu' augury (according to my analysis here) suggests that the challenge comes from a human penchant for just the opposite – seeing false patterns where there is only detail. Bateson, on the other hand, thinks that the challenge comes from an inability (a 'curious twist' of human consciousness) to see true, systemic patterns. The differences among these interpretations are not without interest, but of more importance here is the fact that they all focus on the same issue: human sentience and culture appear both to facilitate and frustrate the 'knowing' of nature. Bateson asked the question, How can consciousness be used to comprehend phenomena that transcend it? The Kantu' system of augury, I suggest, evolved in response to the question, How do we make sense of nature in cultural terms? Both questions, therefore, address the same fundamental philosophical quandary: in a world dichotomized between ego and other, how can the former know the latter?

This question is much the same as the one that originally inspired this study (and the others in this volume), namely, how do we progress beyond a simply dichotomization of the world between nature and culture? The questions that have just been raised suggest that it is, in fact, no easy thing to get beyond this dichotomy. But more importantly, they also remind us that concern over this question is not the province of anthropologists alone. The belief that we ask different questions of life than the people that we study is remarkably unexamined and quite likely highly misleading. The realization that we are asking the *same* questions can have only salutary effects on our science, and ourselves.[25]

Acknowledgements

This analysis was originally prepared for the MOA International Symposium on 'Beyond Nature and Culture: Cognition, Ecology and Domestication', held in Kyoto and Atami, Japan, in February 1992. The author is grateful to the symposium organizers, Katsuyoshi Fukui and Roy F. Ellen, and to MOA International for the opportunity to participate in this stimulating symposium. The field research upon which the analysis is based was successively supported by the National Science Foundation (Grant #GS-42605), the Rockefeller Foundation, the Ford Foundation and the East-West Center (EAPI). A draft version of this paper was read at the Second Biannual Conference of the Borneo Research Council, Kota Kinabalu, Malaysia. The author is grateful to Carol Carpenter, Neil Jamieson and James Nickum for constructive comments. None of the afore-mentioned organizations or persons necessarily agrees with the analysis and opinions presented in the paper, for which the author alone is responsible.

Notes

1. See William of Ockham [1991], e.g. *Entia non sunt multiplicanda praeter necessitatem* 'No more things should be presumed to exist than are absolutely necessary'.
2. For a detailed analysis of swidden cultivation among the Kantu', see Dove 1985.
3. I agree with Ortiz [1979: 77–8] that 'much can be gained by examining agricultural ritual in the context of agricultural uncertainty.'
4. Perhaps the primary source of the inter-annual variation in Borneo's climate is the El Nino-Southern Oscillation (ENSO) phenomenon, which refers to periodic changes in ocean temperatures in the eastern equatorial Pacific accompanied by global climate anomalies (viz., drought at one extreme of the cycle versus heavy rainfall and flooding at the other extreme). ENSO events and their attendant drought or flooding occur, and affect Borneo, every few years on average, but the exact interval is variable and still defies scientific modeling and prediction (Nicholls 1993; Salafsky 1994). The dry ENSO

event, the year-long El Nino, is often followed by the wet, year-long El Nina, in what amounts to a biannual ENSO pattern, but this too varies unpredictably (ibid.). In any case, in order for the family mentioned in the text to take advantage of this biannual pattern, they would have to have based their prediction of flooding for the coming year on the occurrence of drought during the past year – not on the absence of flooding during the preceding several years. Since this family's prediction of environmental conditions did not match the cyclic patterns that do exist, they would have been best off if they had proceeded on the assumption of no cyclic variation at all.

5. This interpretation can be contrasted with that developed by Malinowski [1925] in his classic study of the correlation between ritual activity and environmental risk. Malinowksi argued that people need ritual in uncertain and high-risk environments in order to cope with anxiety. I am arguing that people in such environments need ritual to reduce the likelihood of exacerbating the risk *because* of their anxiety. Anxiety heightens the desire for deterministic solutions, for achieving success, whereas precisely the opposite – indeterministic strategies, avoiding failure – is often called for in high-risk environments. I am arguing, therefore, that it is not just direct interaction with such an environment that is problematic, but also the way that people respond to this problematic. I interpret ritual as not just as a statement about the degree of uncertainty involved, but as a response to this uncertainty. I interpret ritual as a means of reducing this uncertainty, not just in perception but in fact.
6. The resulting strategy is not *no* strategy, therefore; augury does not result in not choosing, but in randomizing.
7. Ortiz [ibid.] analyses farmer's estimates of future price variation and shows that they tend to be based on the most recent price movements, not on an average of all past movements.
8. The importance of not just stating the cultural norm, but seeing where it is followed and where not, has been emphasised in recent 'anti-essentialist' analyses like that of Vayda [1990].
9. These beliefs are instilled in a classic pattern of instrumental learning. This can be illustrated with reference to the Javanese technique for teaching infants *not* to use their left (and thus 'polluted') hands (I am grateful to Carol Carpenter for this

analogy). The infant is customarily held with its right side – and arm – against the mother and with its left arm hanging free. When food or toys are offered to the infant, it naturally tries to take them with its free hand, the left, which then is summarily slapped down. The key to this learning process, therefore, is that the 'wrong' hand is made available. The augural system does something similar by allowing the farmer to exercise his or her own judgement in making the initial selection of a swidden site and then (on occasion) arbitrarily reversing it. (Note that the initial selection of swidden sites is not random; it is the final selection, after augury has been carried out, that is randomized.)

10. There is a similar, more explicit emphasis on the processual character of travel in Buddhism, in the sense that the meaning of pilgrimage lies in the process as opposed to the destination or end [LaFleur 1989: 197–9].
11. Indeed, the meaning of the augural rules is that they *have* no meaning in this direct, instrumental, empirical sense. Such meaningful meaninglessness is characteristic of systems of divination. Thus, Maclean [1980: 53] writes of the traditional system of bird augury on the island of St Kilda, in Scotland's Outer Hebrides: 'On the unmeaning actions or idleness of such silly birds,' wrote Macaulay, 'on their silence, singing, chirping, chattering and croaking, and their feeding or abstinence, on their flying to the right hand or left, was founded an art.'
12. Errors have a direct relationship with learning. As Bateson [1972: 286] says: 'Wrong choices are appropriately called "error" when they are of such a kind that they would provide information to the organism which might contribute to his future skill.'
13. The Kantu' system of augury is not static, nor are other such systems in the region; see Lawless 1975.
14. Cf. Freeman [1960: 89, 92], Gomes [1911: 157], Sandin [1980: 133–4].
15. The continued, monistic character of contemporary science is reflected in the periodic 'discovery' of one 'magic bullet' or another to solve the world's problems. A recent example was heralded in a publication of the US National Academy with the title, 'Neem: A Tree for Solving World Problems' [Turner-Lowe 1992 :5]. Referring to the lack of familiarity with

systemic thinking in such quarters, Bateson writes [1972: 437], somewhat irreverently but probably correctly: 'In general, governmental decisions are made by persons who are as ignorant of these matters as pigeons.'

16. Individual solutions to environmental challenges do not represent systemic understanding of the environment, and as Bateson [1972: 434] says, a lack of systemic wisdom is always punished.
17. Holling [1978: 107] maintains that deterministic thinking is simply unsuited to environments characterized by uncertainty: 'The engineering-for-safety, fail-safe approach, while appropriate for totally known systems, is not so for uncertain, heterogenous, and partially known ones.'
18. See Gleick [1987] for a popular account.
19. Cf. such previously unlikely concepts as that of 'usable ignorance' [Ravetz 1986].
20. Some scientists are now espousing this 'risk-averse' emphasis as well. Goodland writes [1990: 176]: 'The basis for discussion should be: "How bad would it be if you were wrong?" What will it take for us to learn to err on the side of prudence in the face of so much uncertainty and to choose a course of action likely to do the least damage if our understanding proves faulty?'
21. There is increasing appreciation for the way that traditional societies treat uncertainty. Cf. Low [1990: 254]: 'Many societies treat unpredictable variables not as uncertainties about which little is known, but as risks (i.e., assigning heuristically some probability of failure).'
22. See the accounts by Bateson's daughter, Mary Catherine Bateson [1972: 13–7; 1984: 227–32].
23. An elegant demonstration of the actual character of an apparently closed system is Lévi-Strauss' [1969] analysis of the Indian system of marriage, which shows that it functions only because women from outside the system (from hill tribes) are brought into it at the bottom as wives to the lowest castes, and because women exit it at the top via **suttee** and female infanticide in the highest castes.
24. Cf. a recent collection on the environment entitled 'Learning to Listen to the Land' [Willers 1991].
25. I am especially indebted to one fellow participant in the MOA International Symposium, Tim Ingold, for drawing my attention to this fact.

Bibliography

Bateson, G., *Steps to an ecology of mind*, New York: Ballantine Books, 1972

Bateson, M.C., *Our own metaphor: a personal account of a conference on the effects of conscious purpose on human adaptation*, New York: Alfred A. Knopf, 1972

——, *With a daughter's eye: a memoir of Margaret Mead and Gregory Bateson*, New York: William Morrow, 1984

Braudel, F, *On History*, trans. Sarah Matthews, Chicago: University of Chicago Press, 1980

Diamond, J., 'Bach, God and the jungle', *Natural History* 12(90), 1990, pp.22–7

Dove, M.R., *Swidden agriculture in Indonesia: the subsistence strategies of the Kalimantan Kantu'*, Berlin: Mouton, 1985

——, 'The dialectical history of "jungle" in Pakistan', *Journal of Anthropological Research* 48(3), 1992, pp.231–53

——, 'Uncertainty, and humility and adaptation to the topical forest: the agricultural augury of the Kantu', *Ethnology* 40(2), 1993a, pp.145–67

——, 'Para rubber and swidden agriculture in Borneo: an exemplary adaptation to the ecology and economy of the tropical forest', *Economic Botany* 47(2), 1993b, pp.136–47

Evans-Pritchard, E.E., *Witchcraft, oracles and magic among the Azande*, Oxford: Clarendon Press, 1937

Freeman, J.D., 'Iban Augury', in B.E. Smythies (ed.), *The birds of Borneo*, Edinburgh: Oliver and Boyd, 1960

——, *Report on the Iban*, London School of Economics Monographs on Social Anthropology 41, New York: The Athlone Press, 1970

Geddes, W.R., *The Land Dyaks of Sarawak: a report on a social economic survey of the Land Dyaks of Sarawak presented to the Colonial Social Science Research Council*, Colonial Research Study 14, London: Her Majesty's Stationery Office, 1956

Gleick, J., *Chaos: making a new science*, New York: Penguin Books, 1987

Gomes, E.H., *Seventeen years among the Sea Dyaks of Borneo: a record of intimate association with the natives of the Bornean jungles*, London: Seeley and Co., 1911

Goodland, R., 'Environmental sustainability in economic development – with emphasis on Amazonia', in R. Goodland (ed.),

Race to save the tropics: ecology and economics for a sustainable future, Washington, D.C.: Island Press, 1990

Harrisson, T., 'Men and birds in Borneo', in B.E. Smythies (ed.), *The birds of Borneo*, Edinburgh: Oliver and Boyd, 1960

Holling, C.S., 'Myths of ecological stability: resilience and the problem of failure', in C.F. Smart and W.T. Stanbury (eds), *Studies on crisis management*, Toronto: Institute for Research on Public Policy, 1978

Holling, C.S., Crawford S., Taylor, P. and Thompson M., 'From Newton's sleep to Blake's fourfold vision: why the climax community and the rational bureaucracy are not the ends of the ecological and social-cultural roads' *Annals of Earth* IX(3), 1991, pp.19–21

Hose, C., and W. McDougall, 'The relations between men and animals in Sarawak', *Journal of the Anthropological Institute* 31, 1902, pp.173–213

——, *The Pagan tribes of Borneo*, London: Macmillan and Co, 1912

Jensen, E., *The Iban and their religion*, Oxford Monographs on Social Anthropology, Englewood Cliffs (N.J.): Oxford University Press, 1974

King, V., 'Unity, formalism and structure: comments on Iban augury and related problems', *Bijdragen tot de Taal-, Land-, en Volkenkunde* 133(1), 1977, pp.63–87

LaFleur, W.R., 'Saigyo and the Buddhist value of nature', in B. Callicott and R.T. Ames (eds), *Nature in Asian traditions of thought*, Albany: State University of New York Press, 1989

Lawless, R., 'Effects of population growth and environment changes on divination practices in Northern Luzon', *Journal of Anthropological Research* 31, 1975, pp.18–33

Lévi-Strauss, C., *The elementary structures of kinship*, Boston: Beacon Press, 1969

Low, B.S., 'Human responses to environment extremeness and uncertainty: a cross-cultural perspective', in E. Cashdan (ed.), *Risk and uncertainty in tribal and peasant economies*, Boulder (CO.): Westview Press, 1990

Low, H., *Sarawak: its inhabitants and productions*, London: Richard Bentley, 1848

Ludwig, D., R. Hilborn, and C. Walters, 'Uncertainty, resource exploitation, and conservation: lessons from history', *Science* 260, 1993, pp.17–36

Maclean, C., *Island on the edge of the world: the Story of St Kilda*,

New York: Taplinger Publishing Co., 1980

Malinowski, B., 'Magic, science and religion', in J. Needham (ed.), *Science, religion and reality*, London: Sheldon Press; New York and Toronto: Macmillan, 1925

Malkiel, B.G., *A random walk down Wall Street*, New York: W.W. Norton and Co., 1990 [1973]

Marcus, G.E. and M.M.J. Fischer, *Anthropology as cultural critique: an experimental moment in the human sciences*, Chicago: The University of Chicago Press, 1986

Maruyama, M., 'The second cybernetics: deviation-amplifying mutual causal processes', *American Scientist* 51(2), 1963, pp.164–79

Metcalf, P., 'Birds and deities in Borneo', *Bijdragen tot de Taal-, Land-, en Volkenkunde* 132(1), 1976, pp.96–123

Moore, O.K., 'Divination: a new perspective', *American Anthropologist* 59, 1957, pp.69–74

Nicholls, N., 'ENSO, drought and flooding rain in South-East Asia', in H. Brookfield and Y. Byron (eds), *South-East Asia's environmental future: the search for sustainability*, Tokyo/Kuala Lumper: United Nations University Press/Oxford University Press, 1993, pp.154–75

Norgaard, R.B., 'Coevolutionary development potential', *Land Economics* 60(2), 1984, pp.160–73

Ortiz, S., 'Expectations and forecasts in the face of uncertainty', *Man* 1, 1979, pp.64–80

Rappaport, R.F., *Ecology, meaning, and religion*, Richmond (CA): North Atlantic Books, 1979

Ravetz, J.R., 'Usable knowledge, usable ignorance: incomplete science with policy implications', in W.C. Clark and R.E. Munn (eds), *Sustainable development of the biosphere*, 1986

Richards, A., 'Iban augury', *Sarawak Museum Journal* 20, 1972, pp.63–81

Salafsky, N., 'Drought in the rain forest: effects of the 1991 El Nino-Southern oscillation event on a rural economy in West Kalimantan, Indonesia', *Climatic Change* 27, 1994, pp.373–96

Sandin, B., *Iban adat and augury*, Penang: Universiti Sains Malaysia, 1980

Timmerman, P., 'Mythology and surprise in the sustainable development of the biosphere', in W.C. Clark and R.E. Munn (eds), *Sustainable development of the biosphere*, 1986

Tsing, A.L., *Politics and culture in the Meratus Mountains*, Ph.D. dissertation, Stanford University, Ann Arbor (MI.): University

Microfilms, 1984

Turner-Lowe, S., 'Can the Neem solve world problems? A tree to benefit everyone/wonder tree', *NRC News Report* XLII(2), 1992, pp.5–7

Vayda, A.P., 'Actions, variations, and change: the emerging anti-essentialist view in anthropology', *Canberra Anthropology* 13(2), 1990, pp.29–45

Willers, B. (ed.), *Learning to listen to the land*, Washington, D.C.: Island Press, 1991

William of Ockham, *Quodlibeta Septem, English 'Quodlibetal Questions'*, translated by A.J. Fredoso and F.E. Kelley, New Haven: Yale University Press, 1991

Appendix I – Principles of Augural Interpretation

The basic rules for interpreting the omens borne by just one of the seven birds, the Rufous piculet, are as follows:

(i) *Species of Bird* In general, the 'older' the bird within the ranking of the seven major omen birds, the more potent, 'hot' and negative its omen message. As the second-youngest bird, the omens conveyed by the Rufous piculet are relatively positive.

(ii) *Type of Call* The Rufous piculet has three distinct calls: a normal short call, termed **ketupung**, a loud cry, called **pangkas**, and a trill, termed **bacar**. The first two are generally positive in meaning, while the third is very negative. The different calls do not carry equal weight: one **ketupung** call is said to equal three **pangkas** calls.

(iii) *Number of Calls* One **ketupung** call is almost invariably negative in meaning. (This single cry – along with the **bacar** call, which is said to be equal in potency – ranks among the worst possible omens in the Kantu' universe.) On the other hand, two or more calls have a very positive meaning. Some Kantu' say that hearing one **ketupung** call and then seeing the bird is equivalent to hearing two calls, but others dispute this.

(iv) *Direction of Call/Flight* Hearing the **bacar** call of the Rufous piculet on the right is an extremely bad omen. The Kantu' say

bira' aja, bekalih 'even [going to] defecate, [we would] turn back'. (If heard when going to a cockfight, however, this omen would be positive.) On the other hand, hearing the **bacar** on the left is a good omen, at least when travelling to distant places. Some people say that hearing multiple **ketupung** calls on the left is good, while on the right they are bad; whereas others maintain that just the opposite is true; and still others maintain that the left-right distinction is irrelevant in the case of the **ketupung**. In general, omens on the left have 'soft' and 'wet' connotations, while those on the right have 'hard' and 'dry' connotations.

(v) *Character of Flight* If a Rufous piculet flies across one's path to the right and then back across one's path to the left, which is called **rau' antu padi** 'crossing the rice spirit', this is deemed a very good omen. It means that the recipient will obtain good harvests for the next three years (so long as no one else is told about the omen). Another promise of a coming good harvest is when a Rufous piculet alights on a tree stump in one's swidden.

(vi) *Number of Birds* Whereas the **bacar** call is bad, two **bacar** calls (called **kikih**) are considered to be good, especially if one is heard to the left and one to the right (called **kikih berapit** 'scraping [the sound of the call] becoming close'). The hearing of one **ketupung** call on the left and one on the right (called **burong berandau** 'conversing birds') is also deemed auspicious.

(vii) *Order of Omens* The first omen obtained at an intended swidden site, the 'stem omen', is thought to select the second, the 'stem that gives off roots'. The Kantu' say that an omen bird can only select a bird senior to itself for this purpose. For example, the first omen can come from a White-rumped shama and the second from a Rufous piculet, but not the other way around (see the above point i). Widely favoured combinations for the first and second omens are, respectively, White-rumped shama and Rufous piculet, or Rufous piculet and Scarlet-rumped trogon.

(viii) *Time of Day* If one hears the **bacar** call of the Rufous piculet in the morning, it means that one's swidden will burn well (**nunu ka' alla'**); but if one hears a **bacar** in the afternoon, it means that one's rice will not ripen (**padi ka' matak**).

(ix) *Time Spent Seeking* If one hears the auspicious call of a Rufous piculet on the first day one seeks omens – called **Ketupung turun s'ari** 'Rufous piculet gotten in a day' – one's swidden will be invaded by weeds and one's rice will not ripen (the implication being that the good omen was too easily obtained). If one seeks for a Rufous piculet for seven days without success, on the other hand, this is considered to be the equivalent of having obtained the **ketupung** call as one's omen and a good harvest can be expected. If one seeks the Rufous piculet for ten days and finally hears it, then one can expect an especially good harvest.

(x) *Type of Work* The distinction between swidden and non-swidden work is especially important: for example most of the calls of the Rufous piculet are invested with meaning when felling the forest for a swidden but not when felling trees to make roof shingles. Nor is all swidden work alike in this regard: the Kantu' pay less attention to the Rufous piculet during **mantun** 'weeding' than during **ngetau** 'harvesting', because the welfare of the weeds gathered in weeding is of less concern to them than the welfare of the rice gathered during the harvest. A distinction is also made on pragmatic grounds between swidden stages in which large reciprocal labour groups are used and stages in which they are not: only the most serious omens are heeded in the former case, because of the magnitude of the labour loss entailed by halting work to honour omens.

(xi) *Type of Swidden* The Kantu' accord 'hot' or 'cold' properties to most omen birds, and they interpret the auspiciousness of their omens according to whether these properties mitigate or exacerbate the 'heat' or 'coolness' of particular swidden sites. The Rufous piculet is an exception, however: it is neutral with respect to these qualities, however, and it (or rather, its **ketupung** call) is deemed auspicious in either zone.

(xii) *Type of Observer* While the Kantu' generally regard the **bacar** call of the Rufous piculet as inappropriate for the 'stem omen' for a new swidden, they say that it can be taken by – meaning that it foretells a good harvest for – someone who is poor, someone who has never reaped a good harvest before, or someone who has no elders still alive in his or her household.

Chapter 21

Individual Strategy and Cultural Regulation in Nuaulu Hunting

Roy Ellen

Introduction

Anthropologists writing on ecological subjects during the sixties and early seventies placed new emphasis on the measurement of various costs and benefits in the performance of subsistence activities, particularly in relation to the flow of energy through definable systems. This emergent literature was concerned with the efficiency of extractive processes according to calorific, time and motion, and various other criteria, and with new standards of ethnographically relevant environmental description. Moreover, it sought to demonstrate the extent to which cultural institutions might serve as mechanisms for regulating intake of nutrients or maintaining resources at a sustainable level. The technical and conceptual shortcomings of such approaches have been easy to criticize in retrospect: inadequate samples and indices of what constituted success and adaptation, and a theoretical naivete wedded to a prevailing sociological cybernetics which underestimated the difficulties of defining system boundaries, and thus of demonstrating feedback loops. In emphasising holism it neglected historical contingency, in emphasising mutual causality it avoided explanation [Ellen 1982, 1984; Moran ed. 1990].

One recent response to these perceived deficiencies has been inspired by economic formalism and evolutionary ecology, both characterized by methodological individualism, optimization, deductive modelling and, in the latter case, the intellectual baggage of neo-Darwinism. This is evident in problem-centred hazards research [McCay 1978; Vayda and McCay 1975, 1977], optimal foraging theory [Winterhalder and Smith eds 1981], and in a sociobiology [Standen and Foley eds 1989], where adaptation and change are seen as providing an indefinite number of alternative strategies for enhancing reproductive fitness and survival depending on particular environmental circumstances. Such approaches have, in turn, been criticized for their simple-mindedness at the level of ethnographic analysis and their failure to understand the social significance of subsistence or to consider that the organization of consciousness might be in any sense relevant.

A second response has been to disaggregate post-Stewardian *general* ecology (well-evoked in the term 'socio-ecological'), to speak instead of humans being simultaneously involved in non-conflateable ecological and social relations, and to play-down possible connections between the two. In this model – most clearly stated in the work of Ingold [e.g. 1986] – natural selection operates at the level of ecological rather than social relations, within an effective environment of conscious internal representations, blueprints and scheduling. As consciousness lies beyond (rather than within) the field of ecological relations, social institutions and relationships can never be incorporated within adaptationist explanations. There is much in this approach which I find convincing. What I find problematic is a metaphysic (by definition, untestable) which seems to prevent the effective engagement of the social and ecological and which side steps the issue of adaptation. There is plenty of empirical evidence to suggest that the social and ecological have measurable effects one upon the other. By emphasising the continuity of the social landscape, such an approach does not acknowledge the need in any one analysis to draw boundaries around data-sets and thus create 'systems' of interacting parts which provide insights into existential reality. Such a view places renewed emphasis on the essentially social determination of modes of subsistence acting, at most, as a filter for the expression of an underlying ecological dynamic.

This paper seeks a slightly different angle. It attempts to accommodate within a single framework an analysis which addresses both the cultural specifics of hunting in the reproduction of one society, and competing general theories of resource regulation and evolutionary ecology. It does not deny the relevance of a systems approach, but rather tries to graft into it a consideration of both individual strategizing and social relationships. It does so by providing a critical review of different approaches to the analysis of hunting efficiency and a detailed analysis of the ethnography of Nuaulu hunting on the eastern Indonesian island of Seram. The first part of the paper presents Nuaulu hunting as a set of social relations which not only make some contribution to daily protein requirements, but which – through patterns of meat distribution, exchange and ritual – constitute a necessary part of that process through which 'houses', clans and traditional authority are reproduced. The second part of the paper examines the technical and ecological efficiency of Nuaulu hunting. The data comprise ninety-three trips amounting to some 1199 hours over a three-month period, food intake and work surveys for an overlapping four-month period, and an inventory of trophies deposited in sacred houses.

Theoretical Background

If we look at recent studies of hunting, we can see that the important issues have been: defining it as a distinctive strategy, describing its ecological characteristics [Morren 1977] and measuring its nutritional significance. On the assumption that data were sufficiently adequate to the purpose, there has been much attention paid to its description in energetic terms, and having noted its sometimes spectacular calorific inefficiency, to suggesting that it may make a small but crucial contribution to protein intake [e.g. Dwyer 1974: 290; 1983]. During the eighties, considerable interest was shown in optimal foraging theory [Winterhalder and Smith eds 1981], which predicts that hunters will attempt to optimize their rates of return (often via maximization), and that this has some evolutionary pay-off. In most cases the 'currency' of optimization has been energy, but some workers have also used time, protein or risk (or its converse, security) [Foley 1985: 223], though most would really prefer some measure of reproductive success. But optimality theory presents prodigious problems in its application to social human subjects and has tended to concentrate on the costs and benefits of 'the hunt' narrowly defined. It derives from theory which assumes individual foragers to behave in ways which maximize their own fitness, which may be through group or solitary activities. However, it assumes that the social functions of hunting (e.g. ritual cooking, ceremonial exchange, and the more routine distribution of meat following the hunt) are secondary, always driven by what is ecologically necessary. While optimal foraging theory does not claim to consider the motives which precede a hunt, except in so far as motives are patterned by the pursuit of outcomes, nor deny that what motivates people to go hunting in the first place may not be the same as the motivation which operates during the hunt itself [Bulmer 1976; Durham 1981], for those who approach it from outside the discourse it rekindles debates about formalist assumptions of perfect decision-makers and raises fundamental and unresolved difficulties in terms of defining adaptiveness. There is now evidence that those of a neo-Darwinian persuasion are beginning to explore these issues via kin selection and the positing of a connection between generalized reciprocity of social relationships and reciprocal altruism [Hawkes 1991], partly because of the failure of earlier models to predict patterns of empirical data revealed in the ethnography.

In rejecting these eco-evolutionary explanations, some [e.g. Ingold 1986] have sought to deconstruct 'hunting' as a specific subsistence strategy, pointing to the difficulting of defining it, and instead emphasise its social constitution as a mode of production. Others, less fashionably, have sought the motivation of hunting as some kind of evolutionary survival with little contemporary purpose except the conferment of pleasure [Bulmer 1968: 302; Dwyer 1974: 292]. In Ingold's framework a distinction is to be drawn between the social and ecological relations in which people engage, between the social relation of hunting and the ecological relation of predation. This is fine, and is a distinction I wish to maintain. However, by always approaching hunting through its particular social presentation, comparative analysis of subsistence behaviour becomes near-impossible, while the problem of how the social and ecological engage and how one might constrain the other shifts from centre stage.

It is indeed the case that hunting as a specific subsistence type can be embarrassingly difficult to define. Different writers, with different purposes, have defined hunting variously as something *humans* do (that is, making it a species characteristic), a mode of intentional social production [Ingold 1986: 103], as active direct *searching* and pursuit of larger vertebrates (contrasting it, say, with trapping or scavenging) [Winterhalder 1981: 16], the acquisition by humans of all non-domesticated animal foods [Dwyer and Minnegal, 1991], or as the acquisition of *animal protein*, whether by humans or animals [Potts 1984]. Each definition could be defended in a particular analytic context, though in general terms I would favour a modified version of the third as a basic description for a human cultural technique. It is not intrinsically any better, only more useful in most situations where it is likely to be needed. As it stands, it poses problems in terms of target species and the range of strategies involved. Which animals constitute the proper quarry of hunters, and where does the precise boundary lie between different techniques? Bulmer [1968: 303], for example, perhaps understandably, excludes fish, but less understandably extends this exclusion to crocodiles and other aquatic game. By contrast, Dwyer [1974: 287], also working in New Guinea, seems to include fish, reptiles and frogs. Hunting, like any other analytically isolated mode of subsistence, is composed of a number of more specific techniques or elements, though it is hardly to be expected that these will always be discrete entities. Because the kinds of activities referred to by the conventional labels merge into each other (e.g. hunting > trapping > fishing > collecting), and because the particular combinations of techniques are so varied, simple typologies are dangerous; necessary, but dangerous. For the purpose of this paper I use 'hunting' to refer to the appropriation of all non-domesticated terrestrial vertebrates, and to include such marginal strategies as ambushing, besetting and trapping where the target is non-domesticated terrestrial vertebrates. Moreover, this approximates to most folk and non-technical uses, and in comparative analysis this is no insignificant consideration. I agree with Ingold in distinguishing hunting from *predation*, which must be understood as an ecological relation; but I find it difficult to accept his view that social sharing is so intrinsic to the definition that we cannot consider it as a 'mere' technical practice. As we shall see, Nuaulu employ a variety of social arrangements, only some of which involve distribution outside the immediate domestic unit, while some meat is consumed directly by hunters without further distribution. Moreover, while collective hunting is socially-valued, most hunting (and most effective hunting) is solitary. Hunting cannot simply refer to one social process, or define a particular kind of society or mode of economic production; and it is quite possible to accept the social encapsulation and origins of hunting without restricting it in this way [Ellen 1988; 1990b: 154].

Another problem in coming to grips with 'hunting' as a generic category of human activity, whether technically or socially defined, is that most of the discussion has, inevitably, been in relation to so-called 'hunter-gatherer' societies.[1] Ingold's discussion is about 'hunting society', though his conceptualization has implications when looking at other peoples where pastoral or agricultural production may be ecologically more important. In such societies hunting may or may not be ideologically peripheral. In the case of the Nuaulu – who are also sago-extractors, collectors and swidden cultivators – I would argue that it is central. Furthermore, there sometimes appears to be an assumption that theory developed in relation to hunter-gatherers will apply, or is intended to apply, to hunting in all societies. But in agricultural systems, say, it cannot be right to assume that hunting is necessarily an independent sub-system or variable; and in cost-benefit, optimal foraging or evolutionary pay-off models separating what we call 'hunting' from all other food-getting activities may be quite artificial. What may be more important is the aggregate optimality of a range of subsistence practices with the flexibility and longer term food security which this affords. That less has been written about hunting in non-hunter-gatherer societies may, of course, reflect a presumption that it is relatively unimportant; but it need not necessarily be so, and recent literature has begun to redress this balance [Ellen 1988; Harris and Hillman 1989; Kent ed. 1989].

Having attempted to define hunting as a general subsistence activity, it is necessary to return to the central theoretical concern of this paper: how to construct a theory of hunting which makes sense in terms of the structure of Nuaulu society and which at the same time addresses itself to the articulation of hunting as a socially relevant activity and as an ecologically productive one. A common means of connecting these two levels is to hypothesize that meat is biologically necessary and therefore culturally regulated. However, any persuasive account requires a theory which can show that regulation is not simply a side effect of some other process [Hames 1987: 93], while it must also come to terms with the common-sense assumption of optimal foraging theory that in most circumstances individual hunters do attempt to maximize their output within the constraints set by particular social and ecological parameters [Smith 1988]. I shall argue that in the Nuaulu situation hunting is inefficient by most standard criteria, sufficient for basic needs, with no obvious cosmically sanctioned regulations, and with little evidence of dangerous over-hunting.

Nuaulu Hunting and Social Reproduction

The Nuaulu word **yariana** can be translated as 'hunting'. Focally, it refers to that activity in which humans actively pursue animals in order to kill them for food; in other words, it accords with the common-sense 'core' cross-cultural definition of hunting mentioned in the previous section. Less centrally, it also includes what we would call trapping, but for which the Nuaulu do not have a generic term. **Yariana** refers only to the hunting of 'big' game, which in a Seramese context includes reticulate pythons, monitors, civets, the larger bats, birds, marsupial cuscus, but especially cassowary, pig and deer. These latter three species are so important in Nuaulu thought that they are placed in a single category, **peni**. It would be misleading [Ellen 1993: 115–16] to characterize **peni** simply as a 'meat' category', rather it is simultaneously a category for certain kinds of animals which Nuaulu perceive to have morphological, behavioural and functional affinities, a category which carries a heavy symbolic load, and a lexeme which conforms to certain linguistic regularities typical of Nuaulu animal terms in general.

Nuaulu distinguish three kinds of **yariana: yakahohu, kasari** and **matueu**. The reference of each of these terms overlaps slightly, but summarily **yakahohu** is individual hunting, and by extension hunting carried out by two or three related persons. This is the most common form, accounts for over ninety per cent of individual hunts recorded in the field (Table 21.3), and is largely directed towards basic subsistence needs. **Kasari** is collective hunting involving upwards of thirty individuals and is usually – though not exclusively – directed at fulfilling requirements for a particular social event. **Matueu** is a hunt specifically mounted to collect meat required to fulfil a ritual obligation and is generally focused on particular species.

Those meats having a special social presence are restricted to **peni** and cuscus. Animal protein from any other source is unacceptable in ceremony and ritual. Cuscus is required for certain quite specific occasions, such as male initiation, childbirth rituals and marriage payments [Ellen 1972]. **Peni** is usually required for all other social purposes, though occasionally cuscus will substitute.

Meat enters the social system (that is becomes socially significant) in several ways. First, all meat acquired serves to enhance the status of particular individual males, and indirectly contributes to the making of a 'good' marriage and to the establishing of particular individuals as men of influence. Hunting (including trapping) are activities which cannot and should not be kept secret; one man's hunting – through the consequential distribution of meat – may affect the entire village. There is a constant interest shown in hunting performance, and this finds material reflection in the wide display of all kinds of trophies,[2] in the prominence attached to tales of hunting, and in its metaphorical equation with sexual prowess in the conversation of young unmarried men. On the face of it, this kind of thing should fit well with models drawn from evolutionary ecology.

Secondly, **peni** and cuscus are required for the proper performance of specific rituals, as we have already seen.

Thirdly, meat may enter the social system as a necessary part of a **nasai**. **Nasai** are those feasts which accompany all major ritual occasions, but which are most frequently held at life-crises, and are a means of payment for ritual labour. An example of the latter is work on a sacred house, and on such an occasion the **nasai** is hosted by the guardian of the house (who supplies the meat, but consumes nothing) and consumed by those who have done the work, usually male affines and males from the opposite 'house' (**numa**), the paired half of a clan. In fact, the amount of food consumed at the event is a token, and the residue is distributed to the domestic groups from which the workers are drawn. Thus, **nasai** serve a two-fold purpose: they are part of a necessary process (of which **peni** is an essential ingredient) through which individuals, houses and hence clans are socially reproduced; and they serve to redistribute *cooked* meat from hosts to a hosted group. Since such ritual work is only undertaken when the wherewithal for the required **nasai** is available, it has the effect of distributing meat from those who have to people who are less likely to have it.

The final way in which meat enters the Nuaulu social system, articulating and redefining its constituent groups and relationships, is when **peni** animals are killed for non-ritual purposes, butchered and redistributed as *uncooked* meat. When any **peni** animal is killed, its tongue is severed and placed on top of a betel pouch, on which the person who killed it places his right hand. An invocation is addressed to the ancestors and to the spirit of the dead animal, apologizing for its death. The process of butchery and distribution which follows depends on the social circumstances of the kill and to a certain extent on the species involved.

On solitary hunting expeditions the division of meat at the site of the kill does not arise, but on returning to the village the **monne**, that is sacred parts of each **peni** animal, must be presented to the head of the 'house' to which the hunter belongs. The **monne** parts of the killed animal vary slightly between the three **peni** species; they comprise the lower mandible, tongue, neck and lungs in the case of a pig; the lower mandible, tongue and breast in the case of a deer; and the breastbone and associated meat in the case of a cassowary. The jawbones and cassowary breastbone, collectively termed **penesite** (literally, 'jawbones of **peni**'), are a gift to the 'house' of which the hunter is a member, and the associated meat a gift to the person (**ia onate** or **kapitane**) who guards the sacred house.[3] Such guardians may then redistribute the meat among immediate kin and affines. The tail of a pig is also regarded as **monne**, though this is retained by the hunter to be used as an amulet. The meat remaining once the **monne** parts have been detached is available for consumption by the immediate household of the hunter himself and for redistribution among close kin and affines.

Where two individuals are of unequal status, say married and unmarried or initiated and uninitiated, and where it is the inferior partner who has made the kill, the superior partner must receive the **monne** parts. The inferior partner retains what is left. If the superior partner has made the kill he gives the inferior partner what remains when the **monne** cuts have been taken. In each case the lower mandible and tongue are for the 'house'. If the two individuals are of equal status and from the same clan, the 'house' receives the jaw, the **ia onate** or **kapitane** the breast, the rest being divided between the two individuals concerned. If two clans are represented in a hunt and identification of the killer is certain, the kill goes to that individual's clan. The killer then presents the **monne** parts in the usual way but is also expected to make a gift of meat to the other clan. If rights to the kill cannot be ascertained, the mandible is discarded and the meat split evenly. If three or more clans are represented, the clan of the killer receives the **monne** parts and the remainder is divided equally. Rules concerning the ultimate destination of **monne** parts are adhered to rigorously, though there is some flexibility in apportioning the substantial residues. Thus, on one occasion in 1970 three clans were represented at the kill of a cassowary. Naupate (Matoke), who had shot the animal, took the breast and legs, which came to 3.5 kg butchered weight. The remainder was supposed to be divided equally between Masoli (Somori) and Hitinisi (Sonawe-ainakahata); in practice each received a wing and side, but in the first case this weighed 2.4 kg and in the second 1.9 kg. The meat around the skull and neck was shared by the hunters at the site of the kill. One reason why collective hunts may be so unpopular is because the meat has to be shared so many ways.

Penesite need not be presented to the 'house' straightaway. Sometimes, they may first be left to accumulate in garden huts and ordinary dwellings. Here they are hung in the roof-space or stuck in the walling, where they serve as trophies. **Penesite** often have individual histories which may survive several generations, and there is some evidence that important trophies are likely to be kept longer than others. Once presented to the 'house' they are strung in large numbers across the roofspace, and in some sacred houses over 50 per cent of the roof interior may be covered with strings of **penesite** [cf. Forth 1991: Plate 2; Lewis 1988: 219–30]. Older **penesite** may be stored in baskets, but none are ever discarded unless broken. If a **penesite** is lost it must be replaced by a piece of red cloth; if it is broken it may be thrown away, but must again be replaced by a piece of red cloth. If one is stolen, a large plate must be put in its place or else sickness will follow. **Penesite** constitute part of the sacra of a clan section 'house' and are guarded by the head of the clan section. They remain in the house as long as it stands and become the responsibility of the successor of the guardian who first placed them there. If a guardian moves then the penesite must remain, but if a house comes to the end of its natural life then the **penesite** along with other sacra will be removed to the new sacred house. Some idea of the accumulations of **penesite** in particular sacred houses, ordinary dwellings and garden houses can be gained from Table 21.1. Such accumulations may provide a record of hunting success within the clan section over a period of many years, but inferences to this effect must be drawn cautiously, and numbers of trophies cannot be directly held as an indicator of the contemporaneous levels of game populations [Bulmer 1968: 169]. Bone, of course, deteriorates with age, and cassowary breast-bones in particular are almost certainly under-represented in my lists. In general terms, the largest numbers of trophies reflect the demographic strength of a clan section over time.

Nuaulu distinguish two kinds of valuable: those which may be exchanged between clans (e.g. cloth, porcelain) and those which are simply the accumulated sacra of particular 'houses' and which may never be removed (e.g. sacred shields, the barkcloths from male initiation ceremonies) [Ellen 1990a]. Meat (flesh) is constantly being transferred between clans and 'houses' but **penesite** (bone) must be counted among the second. We may conclude, therefore, that hunting contributes in a significant way to the ritual reproduction of the 'house' through the steady addition to its stock of **penesite.**[4] Certainly, Nuaulu believe that failure to replenish its stock may anger the ancestors and seriously place in jeopardy the welfare of its members. But by the same token, hunting also reproduces traditional patterns of authority and gender relations, through the requirement to distinguish superior and inferior statuses in the hunt, through the allocation of specific cuts for the ritual heads of houses, and by excluding women at all stages. The need to supply meat for other rituals and feasts similarly underwrites, through exchange, the interdependent character of consumption.[5] So, although the meaning of these various ritual transactions makes no sense if we lose sight of the fact that meat is – in the last instance – physically consumed, and is predicated upon an adequate material supply of flesh and bone, there is an important sense in which hunting performs an essential role in social reproduction, even if its ecological impact is wholly unremarkable.

The social reality of hunting is certainly not separate from the

Table 21.1. Accumulations of **peni** trophies in Nuaulu sacred houses, Rohua 1970

	Pig	Deer	Cassoway breastbones	Guardian	Location
1	12	19	0	Hatarai Sonawe	numa kapitane
2	9	8	2	Sorita Matoke	-
	36	18	0	Sorita Matoke	-
3	90	285	0	Komisi Somori	numa onate
4	2	0	0	Paikole Sonawe	-
	32	2	0	Paikole Sonawe	garden hut (Awao)
	14	8	0	Paikole Sonawe	garden hut
5	2	0	0	Retaune Somori	-
6	32	2	0	Tuale Nepane	-
7	66	39	0	Sauute Nepane	-
8	12	1	1	Naupate Matoke	-
9	5	0	0	Maineo Nepane	-
	2	0	0	Maineo Nepane	garden hut
10	29	12	0	Patioka (Kotahatu)	-
	1	1	0	Patioka Sonawe	-
11	7	0	0	Hotena Nepane	-
	134	22	0	Hotena Nepane	-
12	53	13	0	Tuisa Nepane	-
13	145	9	0	Numapena Sonawe	-
14	14	9	0	Tapone Sonawe	numa onate
	125	72	0	Tapone Sonawe	numa onate
15	7	2	0	Sahunete Penisa	-
16	18	18	0	Wairisa Sonawe	-
17	19	11	0	Lohia Penisa	numa onate
18	22	0	0	Naunepe Nepane	-
19	57	7	0	Inane Matoke	-
20	50	50	0	Iako Matoke	numa onate
21	4	0	0	Sahukone Nepane	-
22	18	7	0	Lohia Sonawe	-
23	357	147	0	Konane Nepane	numa onate
Totals	1374	762	3		

Note. In column 5, Sonawe = Sonawe-ainakahata, Matoke = Matoke-pina and Nepane = Nepane-tomoien. In column 6, **numa onate** = 'house' (clan section) of the **ia onate** and **numa kapitane** = 'house' of the kapitane.

ecological reality of predation. The social demand for meat in order to fulfil ritual and social obligations affects the level and periodicity of extraction, particularly of pig, deer, cassowary and cuscus. Indeed, rituals requiring the flesh of these animals are frequent, and this must go some way in explaining the level and species ratio in meat consumption. But that the social prohibitions on hunting constrain – even regulate – ecological transfers is less obvious. In terms of the amount of animal protein consumed, ritual prohibitions probably count for little. Totemic restrictions on big game are negligible [Ellen 1993: Table 6.3]: pig is a secondary totem for one (possibly two) clans, and cassowary and bats are both secondary totems for just one clan. Of reptiles, snakes are a primary totem for three clans and a secondary totem for a further three, monitor lizards are a secondary totem for two clans, testudines (turtles and terrapins) a primary totem for one clan and a secondary totem for another, while crocodiles are a primary totem for two clans and secondary totem for a further two, though they are virtually extinct in the Nuaulu area. Given that 'secondary' totems are respected though not necessarily avoided as food, these restrictions are not so limiting as they might first appear, especially when interpreted in relation to the frequency with which particular species are consumed irrespective of clan. The only prohibition of wide significance is that placed on the male spotted cuscus (*Spilocuscus (Phalanger) maculatus*), and as I first argued twenty years ago [Ellen 1972], what is interesting about this is that the inconvenience is restricted by confining the prohibition to one sex of the two species found on Seram. Having said as much, the impact on the success of particular hunts may sometimes be dramatic. Consider, for example, the following case:

> Bisara Nepane-tomoien and Latulesi Matoke were hunting cuscus in the Lahati area on 9 May 1970. At that time both were aged between 15 and 20, and of the two Bisara was generally reckoned to be the better hunter. Indeed, even at that time Latulesi had a reputation as a rather feckless clown, which continued throughout the period I was acquainted with him. On seeing a female *Phalanger orientalis* Bisara began to pursue his quarry up the tree in which he thought it located, but on getting closer found it to be a male *Spilocuscus maculatus*, taboo for Nepane. He therefore asked Latulesi to climb the tree and shoot it with his bow. Unfortunately Latulesi was both a

bad climber and a bad shot, and could not get close enough to hit it. The cuscus escaped.

Thus, in the consciousness of particular Nuaulu individuals and groups, that totemic prohibitions reduce hunting success is widely accepted.

Data on Nuaulu Hunting Trips and Meat Consumption

All the systematic data sets used in this analysis are based on fieldwork conducted in cooperation with the inhabitants of Rohua between early 1970 and mid-1971, when it had a population of 180. I think it reasonable to claim that similar data for other Nuaulu settlements in the vicinity of Sepa would yield a comparable pattern. The situation elsewhere is undocumented, though my own informal enquiries strongly suggest that Nuaulu hunting yields are higher than those in most Christian and all Muslim settlements, and also than in selected mountain settlements, such as Piliana on the edge of the central mountain ridge.

Between early May and August 1970 I attempted to monitor all hunting trips out of Rohua and all meat coming in. The data are inevitably incomplete: I must have missed entire trips, especially those mounted from garden huts or undertaken while engaging in other activities. Moreover, of the trips reported, I have not always been able to identify the returns (if any) or hunting locations. In addition to data collected specifically on hunting trips, I collected systematic information on (a) the work patterns of 15 mature and effective males over the equivalent of 1,538 person-days between April and September 1970, of which 215 were devoted exclusively or substantially to hunting and trapping; and (b) the food intake of two households over the equivalent of a four-month period between March and December 1970. For a period of sixteen months I kept spot checks on all other households to monitor diet. The specific data on hunting. therefore coincide with the early part of longer data-sets for time-allocation and diet.

Calculations of hunting efficiency require correlations between some measure of input and some measure of output: return for effort. Inputs can be stated in terms of total aggregate time, time adjusted for effort and translated into physiological estimates of calories expended, or number of trips. Outputs can be stated in terms of total aggregate weight of game, that weight translated into calories or weight of protein (using standard equivalents for different species), and in terms of number of carcases. Here I rely largely on time and trip inputs and weight and carcase outputs. This is not because I regard other measures as insignificant (and I will be introducing them into the discussion later), but because they require various assumptions which reduce their accuracy.

In some cases I was able to weigh specimens myself or reconstruct carcase weights from butchered game. This has usually been the case for smaller animals, such as all bats and monitors, and some cuscus. In most cases, however, carcases were too large to make this practicable or, in the case of butchered meat, weighing proved too disruptive of the process of distribution. Where I have not been able to use actual weights, the present calculations are based on the following estimates of average size: *Spilocuscus (=Phalanger) maculatus* and *Phalanger orientalis*, 3.5 kg;[6] pig, 50 kg;[7] deer, 63 kg;[8] and civet, 3 kg.[9] I weighed the total dressed meat of one cassowary on 24 June 1970, which amounted to 7.8 kg, and would estimate the undressed weight at about twice this, say 15 kg. Rand and Gilliard [1967] give 45.35 kg for the largest cassowary reported to them.

I consider efficiency here with respect to the composition and competence of hunting teams, the timing and seasonality of trips, their duration and species hunted. Type of hunting, as defined in terms of the combinations of personnel, timing, equipment and movement strategies, is not treated here as a single variable. I have insufficient data and, anyway, it would be too complex a task. The distinction between group and individual hunting is covered under the first heading (composition) and the distinction between night-time and day-time hunting is treated briefly under 'duration'.

The Composition and Competence of Hunting Teams

My trip data refer to the hunting activities of 36 post-pubertal males, 79 per cent of this category living in Rohua at the time of my survey. Those counted include two elderly men, the ia onate of the clan Matoke, an adolescent at school away from the village, and one virtual invalid. Judging from fieldnotes for the period outside the survey, some of those remaining I know to have had at least average hunting capabilities and experience. Certainly, though, some individuals were more active than others. In order of decreasing aggregate hours, the top four of these were Anarima Sonawe-ainakahata (323 hours), Naupate Matoke (304 hours), Nepinama Sonawe-ainakahata (170) and Napuai Somori (152). The lowest input over the period for those who hunted was five hours, that is no more than one trip over four months. But if we examine the top four hunters in terms of number of trips, the ranking changes to Naupate (17), Masoli (12), Anarima (10), Napuai (8), Nepinama and others (6). However, there is a good correlation between number of trips and total aggregate time.

It is a reasonable prediction that the hunting efficiency of individuals varies, and does so according to age, and also according to amount of time spent hunting (experience) and technical competence. These things are, of course, not independent variables. Table 21.2 shows Nuaulu hunting inputs and reward of adult male age cohorts. Those in the cohort 21–30 are clearly the most active and productive in absolute terms, though experience gained through age would seem to compensate when it comes to measures of efficiency (columns 7 and 8). Data for cohorts 51–60 and 61–70 are clearly not statistically significant and are presented simply in the interests of completeness.

Over 90 per cent of the data upon which I have relied is for individual hunting or hunting in pairs. This is overwhelmingly more significant than hunting conducted using groups of a larger size (Table 21.3), a picture which is more or less reflected in the efficiency data.[10] Given the small number of groups of three or more, computations of efficiency based on such data might hardly be regarded as meaningful. However, there is a strong suspicion that the small number of instances of groups of more than two persons reflect the perceived inefficiency of large groups by Nuaulu. I believe that a more critical factor limiting the number of larger teams is the sheer difficulty in organizing them where individuals have competing uses for their time. It is perhaps not surprising that larger teams are usually those recruited in connection with ceremonial requirements, where the force of social necessity outweighs the objections which might be

Table 21.2. Nuaulu hunting inputs and rewards in terms of adult male age cohorts

1 Age cohort	2 Number	3 Mean number of trips	4 Mean aggregate hunting hours	5 Mean number of carcases	6 Mean kg weight	7 6/4	8 5/3
15–20	10	3.1	72.2	2.7	49.82	0.69	0.9
21–30	10	5.6	89.2	4.27	106.09	1.19	0.7
31–40	9	2.77	26.9	1.45	45.13	1.7	0.5
41–50	5	2	26	1.4	63.1	2.4	0.7
51–60	1	2	22	1	50	2.27	0.5
61–70	1	1	108	1	50	0.5	1.0

Table 21.3. Size of hunting group in relation to return

Size of group	Number of instances	Number of carcases	Aggregate kg weight	Number of carcases per man trip	Mean kg weight per man trip
1	64	40	1159.45	0.6	18.12
2	22	36	567.8	0.8	12.9
3	2	2	18.5	0.5	3.08
4	2	8	74.5	1	9.31
5	1	5	250	1	50
10<	2	0	0	0	0

raised when organizing a collective expedition for more mundane purposes. In short, I would say that hunting success has as much to do with the opportunities afforded, in terms of cooperation (that is, who your companions are) and situations presented, as with regularities in group size and patterns of recruitment. This may in part reflect the availability of suitable companions at particular times, but the number of solitary hunts is so overwhelming that it must to some extent reflect effectiveness.

Timing and Seasonality

Rainfall in south Seram peaks in July with in excess of 450 mm, and the months of May through to August are usually considered to be the rainy season, with monthly averages seldom below 377 mm. The driest months are November through to February, with an average of about 113 mm [Ellen 1978a: 213].

If we look first at the dietary records we can see that the proportion of meals including meat obtained through hunting is markedly lower for the period April to June than in either March or July, though the proportion which has resulted from gifts rises to a high in July (Table 21.4). If we look at work-schedule data, the number of hunting days as a proportion of all days which registered some activity is 13.3 (14 unit days) for April, 22.8 (56 unit days) for May, 10 (38) for June, 32.1 (107) for July, 12.5 (43) for August and 8.6 (19) for September; and if we look at the number of trips, we have 35 for May, 38 for June and 19 for July.

Table 21.4. Meat consumption in Rohua, March to July 1970

	Meal type	March	April	May	June	July
1	Meals without any animal protein	3	15	65	64	14
2	Meals containing hunted animal protein	9	11.5	70.5	100.1	71
3	Meals containing other animal protein	0	1	11.5	57	37
4	Number of meals analysed	12	27.5	147	221.1	122
5	Meals with hunted meat as proportion of total	75	41.8	47.9	45.3	58
6	Gifts as percentage	0	0	8.5	26.9	32.3

It is difficult to know what to make of this. First of all, the pattern is not clear, and data from October through to February absent. However, it does seem that on balance hunting days as a proportion of all days when activity was reported actually *increases* during the wet season. The number of hunting trips peaks in June, but falls in July. On the other hand, meals including meat from the chase as a proportion of the total decline from a high (80) in March, which is checked throughout May through to July, when it increases slightly. Most of this is counter-intuitive: we might reasonably expect there to be less activity during the rainy season and a decline in hunting success. I think we can interpret this pattern by positing that Nuaulu *expect* a decline and therefore increase their activity in order to compensate for it, though the data suggest that this activity is largely confined to dry spells in the wet season. Such increased activity does yield rewards in output, though not in overall efficiency.

Duration of Trips

In Nuaulu time-budgeting, hunting and trapping occupy a significant place, 22.75 per cent of all time devoted to subsistence activities by mature effective males [Ellen 1978a: 227]. In calculating the duration of hunting trips, I have relied mainly upon absences from the village which respondents indicated were devoted to 'hunting'. Of course, we must accept that for some of the time respondents will have been engaged in other activities as well, and sometimes in no productive activity at all. In order to control for this, all measurements err on the conservative side when there has been any doubt. Sometimes preparations for a hunt are made on the previous day, for example trees may be felled to clear a presumed arboreal mammal run or feeding station and thereby increase the chance of a kill. Equally, hunts may be no more than tracking down wounded animals hit the previous night, or collecting dead animals from traps. Successful hunts are usually shorter. That is, successful hunters are inclined to return home while the unsuccessful continue. On the other hand, as we have seen, those who hunt longer tend to hunt more frequently, and success is, in part, a reward for effort. In addition to success or failure, hunts may be terminated due to rain, cold, other social commitments, mystical danger or a strong intuition of failure.

The most time-efficient hunting is that conducted on trips of five hours or less (Table 21.5), and in my experience this is target hunting of animals already located while engaged in other activities. Hunting for periods of between six and ten hours is rare, mainly because if a hunter stays out this long he will usually go out for an entire day or an entire night, that is in the region of twelve hours. This decreases the efficiency of the hunt in relation to time spent hunting but increases the likelihood of success. As can be seen from Table 21.5, returns begin to pick up again after eleven hours. After 36 hours, efficiency over time declines overall, but the likelihood of some kind of success increases. I am convinced that this is an extremely important consideration. Especially when meat is required to fulfil ritual obligations, it is common for hunters simply to continue until the required game has been found. Hunters themselves are not concerned with calculating returns on an hourly basis but rather on the basis of the entire trip.

Individual hunting or hunting undertaken in pairs, where the participants are usually closely related, may extend over several days. On extended hunts wives and children may accompany hunters, though subsistence during such times comes wholly from forest products: the game that has been caught, roots, fruits and so on. Such expeditions may reach considerable distances

Table 21.5. Duration of hunting trip in relation to return

1 Duration in hours	2 Number of trips[1]	3 Number of carcases	4 Aggregate kg weight	5 Range of crude output per hour of trip, carcases (kg weight)
<5	35	29	600.4	0.164–0.82 (3.44–17.2)
6–10	2	0	0	
11–15	37	31	616.15	0.055–0.076 (1.1–1.5)
16–20	0	0	0	
21–25	2	3	176	0.06–0.07 (3.52–4.19)
26–30	0	0	0	
31–35	0	0	0	
36–40	6	13	92	0.05–0.06 (0.38–0.42)
41–120	3	10	552	0.03–0.08 (1.53–4.49)

Note 1: An additional 8 trips recorded no data on duration

from the settlement, in excess of 25 kilometres, though normally they are confined to easily accessible and commonly traversed areas within a radius of six kilometres. This form of hunting is frequently undertaken at night. By contrast, cooperative hunting involving a large number of persons is often undertaken on a clan basis, though recruitment is not invariably along genealogical lines. Up to twenty male individuals may be involved covering a wide range and excluding only the very young, senile, infirm and women. Because of the numbers involved, group hunting lacks mobility and flexibility, is therefore restricted to the area within a radius of the most distant garden, and is essentially diurnal. An expedition rarely exceeds twelve hours, for which time an *ad hoc* leader is usually chosen on the basis of seniority and experience.

Most hunting involving the coordination of three or more persons is conducted during daylight hours, largely because the techniques employed, such as driving, are more effectively undertaken when vision is good. But as we have seen, most hunting is conducted individually or in pairs for periods of between one and fifteen hours, and of these instances I would conservatively estimate at least 50 per cent were deliberately embarked upon as nocturnal hunts. Although time-budgeting is clearly a constraint in all hunting [Smith 1979: 58], night-time is not so limiting in the tropics.[11] A proportion of Nuaulu nocturnal hunting involves preliminary daylight selection of a recently used arboreal feeding station or runway, followed by a night-time ambush.

Overall, Nuaulu hunting yields one animal for every 23 man-hours spent hunting, and 1 kg butchered weight every 57 minutes.

Species Hunted

Based on the data series for specific hunting trips between 2 March and 1 August 1970, the kill rates were as follows:[12] 47 cuscus (167.8 kg), 28 pig (1349.25 kg), 9 deer (567 kg), 2 pythons (138 kg), 2 cassowaries (30 kg), 3 monitors (2.3 kg), 1 civet (3 kg), and 1 fruit bat (0.2 kg). If we look at data on actual meat consumption for the two households surveyed between March and December 1970, the number of occasions particular species were eaten and the mean weight per adult head (in parentheses) are: pig 100 (4975 g), cuscus 65 (3225 g), deer 45 (2225 g), cassowary 9 (450 g), python 3 (150 g), civet 2 (100 g) and bat 2 (75 g) [Ellen 1978a: 70 (Table 13), plus unpublished data]. The rank order by weight for the most important species from both data sets compare quite well, except for cuscus. Cuscus is as high as it is due to the fact that it is technically easier to hunt, comes near the village and can be hunted when doing other things, even by a single person. The reversal of cuscus and pig can easily be accounted for in terms of the bulk of pig weight and its redistribution, which generally translates into an increased number of consumption occasions. In this respect, the actual frequency of consumption matches well the value attached to these animals as sources of food in Nuaulu thought. If we also look at the accumulated ritual trophy data in Table 1, we find that deer and pig jawbones are in a ratio of 1:2 (762 deer, 1374 pig), compared with 1:3 for kill rates in the hunting trip data, 1:2:4 for carcase weight, and 1:2:2 for the consumed meat in the dietary survey. The overall importance of the pig is therefore confirmed, the ratio of pig to deer being remarkably similar irrespective of data set.[13]

The pattern revealed in the above examination of hunting statistics is not easy to interpret. Some hunters are more active than others and some are both more active and more successful; most are in the 15–30 age band, though experience seems to compensate for age, as we might expect. Overwhelmingly, most hunting is conducted by individuals or in pairs, although single hunters do not appear to be any more successful than larger groups, though they probably are so marginally. Seasonal factors are important in that they affect the distribution of game and hinder hunting. The wet season presents difficult hunting conditions and steps are taken to compensate for this; hunting activity actually increases and an expected shortfall in meat supply is contained. The most efficient hunting is that conducted on short trips, though trips will be extended if there has not been any early success, which in turn increases the likelihood of success, but decreases efficiency over time. About 50 per cent of individual hunting is conducted at night, and most is conducted in areas between five and seventeen km from the village, which seems to accord with optimal areas for finding most game.

In the remainder of this paper I try to use this rather equivocal data, and also complementary material on dietary intake, to examine the consequences of Nuaulu hunting for their nutritional condition. I conclude that Nuaulu hunting activity and success is constrained by technical factors and by competing organisational requirements, although the need to engage in ritual redistribution and to supply meat for particular ceremonial events exerts pressure to devote increasing amounts of time to hunting with the effect of raising the amount of animal protein finally consumed.

Efficiency, Nutrition and 'Under-hunting'

It has long been known that hunting often supplies negative caloric returns; that is more energy is spent undertaking it than is gained by consuming the meat. This is especially so for tropical groups, and the Nuaulu case provides yet another instance. Indeed, despite having access to pig, deer and cassowary – 'big' game by Melanesian standards – return for effort in energy terms is low, even compared with other tropical forest hunters [e.g. Dwyer 1983]. Nuaulu hunting has an energy ratio of 0.26, compared with 1.00 for fishing, 1.7 for gathering vegetable products and 2.4 for cultivation. What is more, the energetic profile of Nuaulu hunting effort in terms of its sequential components contrasts with most other subsistence activities, in that 95 per cent of the effort goes into location and transport of food, compared with a mere 5 per cent on actual gathering, nothing on maintenance, and a negligible amount on storage [Ellen 1982: 148, 152]. But the material success of Nuaulu hunting is best measured not by the calories which it produces, but by the supply of protein and nutrients to the diet [Dornstreich 1977: 226; Vayda and Rappaport 1968], and in terms of a currency which lubricates social reproduction. In such circumstances the measurement of calories is only useful because it gives us some idea of the energy subsidy involved.

On the basis of dietary surveys which I have already reported, Nuaulu consume on average 25.43 grams of protein per day (Ellen 1978b: Tables 13, 14 and 33; Appendix D]. This works out at about 34 grams per adult per day (assuming children below 15 years to be half-consumers) or 37.6 grams for males and 28.2

grams for females (counted as three-quarter consumers). Although the actual figures may be raised by a few grams to take into account snacks eaten during the course of the day, which were not systematically recorded, protein intake appears to be on the low side compared with data elicited from highland New Guinea populations. Various workers report 43–55 grams per day for males only, and between 24 and 47 grams for adults overall [Clarke 1971: 178; Norgan, Ferro-Luzzi and Durnin 1974; Rappaport 1967: 20; Waddell 1972]. Judged by these standards and international recommendations [see e.g. Food and Agriculture Organization 1973], Nuaulu intake is inadequate, though there is little direct evidence of protein deficiency amongst adults.[14] Thus, the Nuaulu *seem* to be assimilating adequate protein to maintain their bodies in good order even though measured consumption falls short even of acceptable minimum requirements.[15]

Nuaulu consume approximately 12.23 grams of animal protein per head per day, compared with 13.20 grams of vegetable protein [see Ellen 1978a: 70, 73, 167]; in other words about 48 per cent, although in terms of energy, meat and fish contribute only 182 Cals per head per day, or 7 per cent of the total. Moreover, 67 per cent of Nuaulu animal protein comes from hunting, compared with 6.5 from collecting invertebrates and 25 per cent from fishing, though fish subsidies (partly achieved through purchasing power) are becoming increasingly important. Thus the overall intake of *animal* protein is high by southeast Asian and highland New Guinea standards. Clarke [1971: 181] reports that 90 per cent of Bomagai-Angoiang protein is derived from vegetable sources, and Waddell [1972: 125] reports a figure of around 80 per cent for the Raiapu Enga, though animal protein consumption is higher in 'fringe highland' and near coastal populations [see e.g. Dwyer and Minnegal 1991]. That Nuaulu intake of vegetable protein is so low is almost certainly accounted for by the overwhelming dietary significance of sago, which contains about 0.2 grams of protein per 100 grams. But despite the high significance of hunting in the provision of protein, it fails to compensate for the low vegetable protein intake. This is partly due to time-budgeting constraints, which I shall return to below, but is also partly due to the technical difficulties of rain-forest hunting. In any case, it has little to do with the unavailability of game. Let me explain this a little further.

The greatest part of Seram is covered with mature rain-forest, with man-induced secondary forest and a limited amount of dryfield and swidden cultivation located along the coast and around inland highland settlements. Hunting in tropical forests is now generally acknowledged to be inherently more problematic than hunting in savanna environments [Foley 1982: 398]. Prey species are low in density and patchy in distribution, and most are arboreal and nocturnal [Gross 1975: 527–9]. In addition, an annual precipitation of above 15 cm reduces the resource base for large herbivores, decreasing large mammal biomass. Most mammals are of small body-size, the exceptions on Seram being pig and deer. If we add cassowary (a large non-mammalian biped) we have the entire contents of the Nuaulu category **peni**, all members of which are probably not endemic and introduced by humans. A further peculiarity of Seram arises from its being a relatively small island located in a zone of zoological transition (Wallacea), which gives it a characteristically depauperate fauna rather than the high species-diversity which we might expect of rain-forests. It may be this, along with low levels of human predation, which has allowed for population increase in this non-endemic mega-fauna to a point where densities probably exceed those of neighbouring regions of Melanesia.

By comparison, the human population density of Seram in 1970 was low, 0.07 persons per hectare compared with 2.42 on the nearby island of Ambon [Ellen 1978a: 7]. But population distribution is also highly skewed, with most people concentrated in large settlements (both Muslim and Christian) along the coast.[16] Muslims rarely hunt, and among Christians it is restricted mainly to smaller and highland groups, with some specialists in the larger settlements. Hunting on a routine basis is therefore largely confined to animist groups such as the Nuaulu and some Christian villages. Occasionally Nuaulu hunters are commissioned to hunt deer for Muslims, and the incidental products of the chase for which there is a market are traded: python skins and livers, young cassowaries, cassowary eggs. Only birds are deliberately hunted for live sale. Living in area in which Muslims predominate demographically means that the Nuaulu have little competition for prey, and in so far as they compete at all do so only among themselves. There is a little commercial hunting by non-Nuaulu, particularly of birds, and a little recreational hunting among Chinese traders.

All of this has considerable ecological consequences, maintaining high levels of game populations and high biodiversity. Indeed, in some areas wild pig are considered a major pest and have led to the virtual abandonment of gardening. Only the massive destruction of rain forest which accompanies logging activities is likely to have much affect on this picture, and at present there is no evidence for this. Though wider availability among active subsistence hunters might well pose a threat, the present use of firearms has a negligible affect on game in general. The periods when Nuaulu have had access to firearms have been brief, and I suspect that any increase in the efficiency of hunting during these periods has had very little mid- or long-term impact. We cannot assume, however, that such a situation has always existed, and might reasonably hypothesize that game populations have risen as hunting has decreased in importance over the last hundred years or less. In particular, growth in the human population, movement to the coast and conversion to Islam have all led to a decrease in hunting and in pig consumption in particular. As the Nuaulu population has increased, with more time devoted to cash-cropping, there has been less time for hunting. Any increase in levels of predation resulting from a rise in the population is more than offset by new constraints on the allocation of time, though the decrease in animal protein has to some extent been checked by the purchase of fish from neighbouring Butungese and indigenous Sepanese. Thus, there is no evidence, historical or ethnographic, that the Nuaulu have ever systematically over-hunted or experienced shortages due to a decrease in the size of game populations in recent years.

Hunting output could, of course, be increased, through, for example, less selective hunting, concentrating perhaps on smaller game at the expense of **peni**, and relying more on firearms. Trapping is in some respects a more time-efficient practice than other forms of hunting, though its success rate is lower and even traps have to be checked regularly. But the main constraining factor is that Nuaulu appear to have reached the maximal time allocation for hunting. Particular individuals have time by virtue of being single, but most hunters are constrained in their inputs by the necessity to do other things. The competition is not only between subsistence activities but also with prioritized ritual, and long periods of hunting inactivity due to ritual performance sometimes erode valuable stored reserves. Considerations of time allocation are, as I have argued, reflected in the low level of cooperative hunting, despite its ideological prominence. In particular, hunting declines when sago is short, and vice versa. As already indicated, one of the main factors triggering increased hunting activity is the demand for meat to fulfil ritual requirements. But this requires that time be withdrawn from other activities, the most important being for males sago extraction. Sago extraction and processing is an exclusively male activity, as is hunting. This seriously jeopardizes opportunities to hunt during times when sago is being processed. If we look overall at the provision of carbohydrate and protein in relation to division of labour, a 'steady trickle' of animal protein is supplied by women (freshwater crustacea, shellfish and smaller fish) and marine fish are generally available for purchase from Sepa. By contrast, although hunting for **peni** is economically prominent and its yields dominate patterns of protein intake, returns (and this may apply to hunting as a whole) are erratic (both overall seasonally, and between individual hunters). In other words, it is high risk, may impose a boom-bust pattern of protein availability, and may therefore not provide a good basis for nutritional planning. **Peni** are more likely to be targeted when there are social obligations to fulfil. Moreover, as sago possesses a high cultural value – equal if not surpassing that attached to meat – very little effort can be withdrawn from its extraction while maintaining production at a culturally acceptable level. So, demand for sago in turn feeds back to limit any further growth in activities connected with the supply of meat. The entire process is accentuated at times of important festivals, when demand for both products rises. This hypothesis finds support in the general inelasticity of Nuaulu time-allocation and in the fact that hunting and sago-extraction combined represent about 57 per cent of all time allocated to male food-getting activities. This allows little scope for withdrawing male labour from gardening activity, which is anyway highly seasonal [Ellen 1982: 180–81].

In view of all this, we really have to look no further for those factors limiting Nuaulu hunting and meat consumption. 'Under-hunting' rather than 'over-hunting' is the structural problem, and it would be wholly superfluous to reinforce this through a cosmologically embedded ethic of conservation. The logic of Nuaulu cosmological beliefs provide no evidence for a conservation ethic, neither does it suggest that regulated hunting is mythically prescribed. Of course, any cosmology motivated by animism must generate respect for other species, a respect often reinforced by prohibitions of various kinds. But this need not be inconsistent with hunting, and I suspect that there may long have been a contradiction between the doctrine of infinite renewal and the recognition that hunters could exterminate animals locally [Brightman 1987: 137]. Consider, for example, the logic of two ritual practices: the planting of **asumate** and the implementation of a **sasi**. When an animal is butchered it is generally skewered on a sharpened wooden stake (an **asumate**). After the meat has been cooked the stake is planted in the ground and the chip which was removed in making the pointed end tied to it. This is supposed to represent the re-uniting of the soul and body of the killed animal, and the whole serves as an outward indicator of prestige, as an offering to the ancestors, to return the spirit to the cosmos and thus ensure that finite stocks of the animal are not depleted. Every time a hunter fails to plant an **asumate** the wild stock of that species is thought to be depleted by a factor of one. Such a practice can hardly be interpreted as encouragement for regulated hunting [cf. Brightman 1987: 130]. On the other hand, the **ia onate** Matoke, or some other clan chief, may from time-to-time establish a **sasi**, that is a ritually sanctioned closed season during which a particular species may not be taken. These are generally established with respect to fruit trees, but are occasionally met in connection with hunting. But the establishment of a **sasi** in no way reflects a general cosmological ethos, rather it is a contingent and dynamic answer to a particular perceived short-term problem. And quite apart from ritual, it is impossible to exclude purposeful variation in Nuaulu hunting practice in response to perceived shortages, to ignore the palpable evidence that Nuaulu do regulate their predatory impact on game populations through conservation. They do this quite consciously by reducing the overall intensity of hunting and decreasing its intensity in different habitats or patchs [cf. Hames 1987: 93]. There is evidence – which lack of space prevents me from presenting here – that Nuaulu allocate less time to depleted patches with low rates of return and more to rich patches. This may allow depleted patches to recover; that is, as long as the extra travel time to richer patches does not lower the overall rate of return. Thus, Nuaulu pursue a strategy of short- to mid-term maximization and are not as yet constrained by possible long-term consequences; indeed in the past they have had no reason to be. Equally, Nuaulu may respond to decreases in consumption and/or increases in acquisition costs by intensifying hunting and/or extending hunting ranges [ibid.: 104].

Conclusion

To summarize, the patterns revealed in the data sets on Nuaulu hunting examined in this chapter are to a considerable extent irregular, at times quite unexpected and often baffling. Part of the reason for this is undoubtedly the force of historical contingency. By this I mean that hunting is affected by so many factors that short-term fluctuations cannot be expected to reveal consistent patterns. At an ecological level hunting is inefficient by most usual criteria, though it is more efficient than in some other cultivating populations which also engage in hunting. Adopting the usual indirect measures Nuaulu are receiving adequate levels of protein, though from the direct analysis of food intake this is low. As game is not in short supply overall, there is a gap between what it is technically and socially possible to extract and the ecological limit. In other words, whatever their efficiency, the level of Nuaulu hunting does seem to be nutritionally critical, especially given the relatively low contribution of vegetable protein, which is largely due to the importance of protein-deficient sago. Given the crucial character of animal protein, the existence of rules of social distribution might be thought to ensure a wider spread of risks and benefits, though the positive discrimination in favour of clan-section 'houses' with respect to **peni** makes one wonder about the differential effects this has within the population, in particular the probability that meat is less likely to benefit immediate relatives in the way kin-altruistic versions of neo-Darwinism might predict. At the very least, the assumptions sometimes made by optimal foraging theorists, and even in some agnostic accounts, that redistribution is irrelevant, occurs 'fairly freely' [e.g. Dwyer 1974: 291], or is positively functional all have to be examined much more carefully.

There is, then, a strong suspicion that without the imperative to hunt in order to fulfil social obligations, with its cultural underpinnings, Nuaulu intake of animal protein would be much lower than it is, with deleterious consequences for health. Cycles in hunting activity do seem to be related to the requirements of particular ritual events, and such events usually require precisely those species which provide most protein: pig, deer, cassowary and cuscus. So, returning to the very general issues with which we started, it becomes clear that the frequency of individual hunting as opposed to collective hunting is consistent with a methodological individualism identified with early models of optimal foraging theory and provides a convenient means of generating various indices of efficiency; while the contribution which hunting makes to status indicators means that there need be no disconnection with the interests of evolutionary ecologists. However, the Nuaulu case also demonstrates that hunting may radically disengage ecological from social relations, illustrating the importance of a social dynamic both in constraining what is ecologically possible in terms of hunting activity and success, and in determining the overall level of meat consumption and its circulation. While we must be careful not to repeat the excesses of social cybernetics, the significance of hunting can only be assessed through its *simultaneous* appearance in both ecological and social systems, motivated both by the necessity to provide food and to meet social expectations. It is because of this that it permits us to focus on that point at which these different analytical levels meet through the mutual constraints environmental and social factors impose. This is in no sense to subscribe to a simple adaptationist model, but to describe an empirical situation. In the Nuaulu case 'under-hunting', and therefore effective conservation, has little to do with purposeful cultural regulators, though these undeniably operate intermittently with respect to particular resources. Neither has it anything to do with the serendipitous effect of some ethic built into their general conceptualization of nature. In the last instance, what determines Nuaulu meat consumption are constraints imposed by a particular environment on the technical conduct of hunting and the necessity to engage in other kinds of activity in order to reproduce effectively, both biologically and socially. The compromise so achieved may well have Darwinian consequences, but it is difficult to see how the pattern of hunting itself is part of an optimizing Darwinian solution.

Acknowledgements

The greater part of the data upon which this paper is based were a product of fieldwork conducted for 18 months in 1970 and 1971, under the auspices of The Indonesian Academy of Sciences (LIPI) and with financial support from the British Social Science Research Council, and London-Cornell Scheme for East and Southeast Asia, the Galton Foundation and the Central Research Fund of the University of London. Field data assembled on subsequent shorter trips in 1973, 1975, 1980 and 1990 have also been drawn upon as appropriate. An earlier version of this paper was delivered to an audience at the Institute of Biological Anthropology at the University of Oxford, and I am grateful to Professor G. Ainsworth Harrison for that invitation, and for some useful advice. Peter Dwyer has commented extensively, and with considerable attention to detail, on those parts of the paper relating to optimal foraging theory and the measurement and interpretation of meat yields. He must be singled-out for special thanks, but can take no responsibility for the views as they have been finally expressed. It should be emphasised that the general picture presented here is for the early 1970s. There is evidence that in recent years the availability of game has sharply decreased as a result of deforestation.

Notes

1. Optimal foraging theory has been seen as being particularly appropriate for such peoples, as they 'exhibit a strong degree of interaction with their environment', with subsistence patterns 'fairly analogous to those of other species' [Durham 1981: ix].
2. In addition to **penesite** (the mandibles of deer and pig and cassowary breastbone, discussed further below), other kinds of trophy are widely displayed inside and outside houses and in garden huts. These include dugong skulls (*Dugong dugon*, sawfish saws (*Pristiopsis* sp.), stingray whips (*Dasyatis kuhli*, garfish jaws (*Hemiramphus marginatus*), cuscus mandibles (*Phalanger* and *Spilocuscus* spp.), cassowary claws (*Casuarius casuarius* and deer antlers (*Cervus timorensis*).

3. Those recognized as village headmen, who are also invariably **ia onate**, also receive additional meat to compensate for time lost in their official duties. These payments usually arise from occasional surpluses of individual households, and the only sanction enforcing them is the general belief that everyone should make them whenever possible for fear of appearing stingy.
4. To a lesser extent, hunting provides the means of replenishing other subjects required for the effective performance of ritual: cassowary quills for armbands, cassowary, cockatoo and lory plumes for head-dresses, deer skin for drum heads, pig grease for rubbing on the body, and so on.
5. The appearance of the market mechanism may alter this. Meat has for a long time been traded on a small scale with outsiders and sometimes Nuaulu are actually employed to hunt by Muslims from Sepa. I first recorded sales of meat amongst Nuaulu in 1975. Although it was at that time unusual, it had probably been going on intermittently for a number of years before that. In 1990 I noticed that meat was even being sold occasionally within the clan. This clearly undermines the existing character of clan solidarity but is justified by those who engage in it as a means of increasing the amount of meat one might anyway expect through normal distributive mechanisms. Thus, for the time being at any rate, it is seen not as replacing traditional exchange but as a way of manipulating it to one's advantage.
6. Alastair Macdonald [personal communication] reports an upper weight of 4.3 kg and an average of 2–3 kg for *P. orientalis* from central Seram, and a range of 1–2 kg for *S. maculatus*. I have records of several cuscus weights which average out at 4.6 kg, while Flannery [1990: 130] indicates that the average body-weight of adult *Spilocuscus* in New Guinea is between 3–4 kg, with the largest reported at 6 kg. Bearing in mind that Nuaulu will usually discard immature animals or keep them live as pets, my estimate of 3.5 kg is, therefore, a best guess based on partial and apparently conflicting data.
7. Gathorne-Hardy (Lord Medway) [1978: 104] gives 127 kg maximum for *Sus scrofa*. Macdonald [personal communication] reports sow weights from highland Seram in the region of 40–50 kg. Boars are, on average, an additional kilogram heavier, and lowland animals (which are what the Nuaulu

mostly hunt) heavier than highland animals. I have two weights for butchered meat: 29.25 and 35 kg. It is now clear, as a result of the comments of Peter Dwyer, that my earlier figure overestimated the likely average weight of killed pig. Using a graph of the relationship between pig-weight and mandible-length devised by Dwyer for New Guinea data, I have been able to provide six Nuaulu weights based mandibles deposited in the British Museum (Museum of Mankind) and in my own collection which average out at 46.25 kg. The figure of 50 kg used here is a compromise based on these data. A much better estimate of average weight could be obtained by using the Dwyer regression equation on measurements of trophy mandibles listed in Table 21.1. Regrettably, though these measurements could easily have been obtained, I do not have them.

8. Van Bemmel [1949] gives a range of 162 kg (female) to 260 kg (male) for *Cervus timorensis*. Recent weights for introduced members of the species in northern Australia are in the range 58–115 kg (males) and 50–75 kg (females), for a mixture of *C. t. russa* and *C. t. moluccensis*. *C. t. moluccensis* is a small deer and the Van Bemmel weights must almost certainly be adjusted downwards. My figure is an average of the mean male and female weights provided by Strahan [Strahan (ed.) 1984].
9. Gathorne-Hardy [1978: 91] gives a range of 2.4 to 4 kg for *Paradoxurus hermaphroditus*; Pocock and Gathorne-Hardy [ibid.: 90] provide figures within the range 4.7 to 9 kg for *Viverra tangalunga*.
10. Where hunting trips have involved more than one person, I have divided meat by the number of persons present, on the assumption that they all made some kind of contribution to the outcome, even a negative one. Sometimes such data are provided in terms of the person who actually shot the animal. Sometimes, however, this is not always clear from the data available. As we have seen, Nuaulu accord greatest status to the killer, but also recognize the contribution of others present in the division of the meat.
11. On the equator daytime is a little over 12 hours and dawn and dusk are brief. Nuaulu going on a non-opportunistic hunt will spend a whole day, leaving before 6 and 7 in the morning and returning between 5 and 6 in the evening.

Torrence [1983] has argued that time-stress affects technological responses to hunting, such that in higher latitudes there is a limited time-window for hunting due to, among other things, limited daylight in winter.

12. Aggregate kilogram weights are provided in parentheses, usually for intact meat.
13. It should be noted that there is extreme variation in the ratio pig:deer for trophies in different sacred houses. This may reflect both local patterns of availability and selected target species of different hunters, but also the fact that trophies may be moved between houses, particularly from semi-permanent (garden) houses to village houses.
14. I have used comparative data from New Guinea. Abdoellah [1985], reporting on the pattern of consumption amongst Sundanese peasants (Indonesia), provides average daily adult protein intakes of between 31.9 and 58.3.
15. Food consumption is increasingly considered a poor way of assessing dietary adequacy under field conditions, largely due to problems of measurement. Reliable inferences require data on the nutritional needs of *individual* subjects, and their age, sex and physical status; and on the *quality* of food consumed. In the circumstances, and short of more invasive forms of assessment, anthropometry is probably the best means available [see e.g. Bailey and Peacock 1988: 106–7].
16. The geographic disposition of settlements may affect game availability. A settlement located entirely within forest will have a game potential through 360 degrees; a settlement on the coast bounded on one side by a large area of sago swamp may effectively limit its big game area to less than 180 degrees. Rohua is about 180 degrees. Moreover, coastal villages tend to have high populations.

Bibliography

Abdoellah, O.S., 'Food consumption of Sundanese in Kampung Salamungkal, West Java', in S. Suzuki, O. Soemarwoto and T. Igarashi (eds), *Human ecological survey in rural West Java in 1978 to 1982*, Tokyo: Nissan Science Foundation, 1985

Bailey, R.C. and N.R. Peacock, 'Efe pygmies of northeast Zaire: subsistence strategies in the Ituri forest', in I. Garine and G.A.

Harrison (eds), *Coping with uncertainty in food supply*, Oxford: Clarendon Press, 1988

Brightman, R.A., 'Conservation and resource depletion: the case of the boreal forest Algonquians', in B.M. McCay and J.M. Acheson (eds), *The question of the commons: the culture and ecology of communal resources*, Tucson: The University of Arizona Press, 1987

Bulmer, R., 'The strategies of hunting in New Guinea', *Oceania* 38, 1968, pp.302–18

——, 'Selectivity in hunting and in the disposal of animal bone by the Kalam of the New Guinea Highlands', in F. Sieveking, I.H. Longworth and K.E. Wilson (eds), *Problems in economic and social archaeology*, London: Duckworth, 1976

Clarke, W.C., *Place and people*, Los Angeles: University of California Press, 1971

Dornstreich, M., 'The ecological description and analysis of tropical subsistence patterns: an example for New Guinea', in T. Bayliss-Smith and R.F. Feachem (eds), *Subsistence and survival: rural ecology in the Pacific*, London: Academic Press, 1977

Durham, W.H., 'Overview: optimal foraging analysis in human ecology', in B. Winterhalder and E.A. Smith (eds), *Hunter-gatherer foraging strategies: ethnographic and archaeological analyses*, Chicago: University of Chicago Press, 1981

Dwyer, P.D., 'The price of protein: five hundred hours of hunting in the New Guinea highlands', *Oceania* 44(4), 1974, pp.278–93

——, 'Etolo hunting performance and energetics', *Human Ecology* 11, 1983, pp.143–72

——, 'A hunt in New Guinea: some difficulties for optimal foraging theory', *Man* 20, 1985, pp.243–53

—— and M. Minnegal, 'Hunting in lowland tropical rain forest: towards a model of non-agricultural subsistence', *Human Ecology* 19, 1991, pp.187–212

Ellen, R.F., 'The marsupial in Nuaulu ritual behaviour', *Man* 7, 1972, pp.223–38

——, *Nuaulu settlement and ecology: an approach to the environmental relations of an eastern Indonesian community* (Verhandelingen van het Koninklijk Institut voor Taal-, Land- en Volkenkunde 83), The Hague: Martinus Nijhoff, 1978a

——, 'Problems and progress in the ethnographic analysis of small-scale human ecosystems', *Man* 13, 1978b, pp.290–303

——, *Environment, subsistence and system: the ecology of small-scale social formations*, Cambridge: Cambridge University Press, 1982

——, 'Trade, environment and the reproduction of local systems in the Moluccas', in E.F. Moran (ed.), *The ecosystem concept in anthropology* (AAAS Selected Symposium 92), Boulder CO: American Association for the Advancement of Science, 1984

——, 'Foraging, starch extraction and the sedentary lifestyle in the lowland rainforest of central Seram', in T. Ingold et al. (eds), *Hunters and gatherers 1: history, evolution and social change*, London: Berg, 1988

——, 'Nuaulu sacred shields: the reproduction of things or the reproduction of images?', *Etnofoor* 3(1), 1990a, pp.5–25

——, 'Conceptualising the unique in human ecology and evolution', *Reviews in Anthropology* 19, 1990b, pp.145–58

——, *The cultural relations of classification: an analysis of Nuaulu animal categories from central Seram*, Cambridge: Cambridge University Press, 1993

Flannery, T., *Mammals of New Guinea*, The Australian Museum, Sydney: Robert Brown and Associates, 1990

Foley, R., 'A reconsideration of the role of predation on large mammals in tropical hunter-gatherer adaptations', *Man* 17(3), 1982, pp.393–402

——, 'Optimality theory in anthropology', *Man* 20, 1985, pp.222–42

Food and Agriculture Organization, *Energy and protein requirements* (Series 52), WHO Nutrition Meetings Report: FAO and WHO, 1973

Forth, G., *Space and place in Eastern Indonesia* (Occasional Paper No. 16), University of Kent at Canterbury: Centre of South-East Asian Studies, 1991

Gathorne-Hardy (Lord Medway), G., *The wild mammals of Malaya (peninsula Malaysia and Singapore*, Kuala Lumpur: Oxford University Press, 1978

Gross, D.R., 'Protein capture and cultural development in the Amazon basin', *American Anthropologist* 77(3), 1975, pp.526–49

Hames, R., 'Game conservation or efficient hunting?', in B.M. McCay and J.M. Acheson (eds), *The question of the commons: the culture and ecology of communal resources*, Tucson: University of Arizona Press, 1987

Harris, D. and G. Hillman, *Foraging and farming*, London: Allen

and Unwin, 1989
Hawkes, K., 'Showing off: tests of an hypothesis about men's foraging goals', *Ethology and sociobiology* 12, 1991, pp.29–54
Ingold, T., *The appropriation of nature: essays on human ecology and social relations*, Manchester: Manchester University Press, 1986
Kent, S. (ed.), *Farmers as hunters*, Cambridge: Cambridge University Press, 1989
Lewis, E.D. *People of the source: the social and ceremonial order of Tana Wai Brama on Flores*, (Verhandelingen van het Koninklijk Instituut voor Taal-, Land- en Volkenkunde 135), Dordrecht-Holland/Providence-USA: Foris Publications, 1988
McCay, B., 'Systems ecology, people ecology and the anthropology of fishing communities', *Human Ecology* 6, 1978, pp.397–422
Moran, E.F. 'Ecosystem ecology in biology and anthropology: a critical assessment', in *The ecosystem approach in anthropology: from concepts to practice*, E.F. Moran (ed.) Ann Arbor: University of Michigan Press, 1990
Morren, G.E.B., 'From hunting to herding: pigs and the control of energy in montane New Guinea', in T.P. Bayliss-Smith and R.G. Feachem (eds), *Subsistence and survival: rural ecology in the Pacific*, London: Academic press, 1977
Norgan, N.G., A. Ferro-Luzzi and J.V.G.A. Durnin, 'The energy and nutrient intake and the energy expenditure of 204 New Guinean adults', *Philosophical Transactions of the Royal Society, London*, B 268, 1974, pp.309–48
Potts, R., 'Home bases and early hominids', *American Scientist* 72, 1984, pp.338–47
Rand, A.L. and E.T. Gilliard, *Handbook of the New Guinea birds*, London: Weidenfeld and Nicolson, 1967
Rappaport, R., *Pigs for the ancestors: ritual in the ecology of a New Guinea people*, New Haven: Yale University Press, 1967
Smith E.A., 'Human adaptation and energetic efficiency', *Human Ecology* 7, 1979, pp.53–74
——, 'Risk and uncertainty in the "original affluent society": evolutionary ecology of resource sharing and land tenure', in T. Ingold, D. Riches and J. Woodburn, (eds), *Hunters and gatherers: history, evolution and social change*, Oxford: Berg, 1988
Standen, V.S. and R.A. Foley (eds), *Comparative socioecology: the behavioural ecology of humans and other mammals*, Oxford: Blackwell, 1989

Strahan, R. (ed.), *The complete book of Australian mammals*, Australian Museum Trust, Sydney: Angus and Robertson, 1984

Torrence, R., 'Time-budgeting and hunter-gatherer technology', in G. Bailey (ed.), *Hunter-gatherer economy in prehistory: a European perspective*, Cambridge: Cambridge University Press, 1983

Van Bemmel, A.C., 'Revision of the rusine deer in the Indo-Australian archipelago', *Treubia* 20(2), 1949, pp.191–262

Vayda, A.P. and R.A. Rappaport, 'Ecology: cultural and non-cultural', in J. Clifton (ed.), *Introduction to cultural anthropology*, Boston: Houghton Mifflin, 1968

—— and B. McCay, 'New directions in ecology and ecological anthropology', *Annual Review of Anthropology* 4, 1975, pp.293–306

——, 'Problems in the identification of environmental problems', in T. Bayliss-Smith and R.G. Feachem (eds), *Subsistence and survival: rural ecology in the Pacific*, London: Academic Press, 1977

Waddell, E., *The mound builders: agricultural practices, environment and society in the central highlands of New Guinea*, Seattle: University of Washington Press, 1972

Winterhalder, B., 'Optical foraging strategies and hunter-gatherer research in anthropology: theory and models', in B. Winterhalder and E.A. Smith (eds), *Hunter-gatherer foraging strategies: ethnographic and archaeological analyses*, Chicago: University of Chicago Press, 1981

—— and E.A. Smith (eds), *Hunter-gatherer foraging strategies: ethnographic and archaeological analyses*, Chicago: University of Chicago Press, 1981

Notes on Contributors

Tomoya Akimichi is Professor at the National Museum of Ethnology in Osaka. He has undertaken research on marine resources management, folk biology, traditional navigation and food use in the Pacific, Indonesia and southern Japan. He is the author of several books, including *Man and whale in cultural contexts* (1994, in Japanese), *Territoriality in Japanese Culture and History* (1995, in Japanese) and *Maritime Anthropology* (1995, in Japanese), and has edited (with K. Ruddle) *Maritime Institutions in the Western Pacific* (1984).

James Shilts Boster is Associate Professor in the Department of Anthropology at the University of California, Irvine. His main interests are in cognitive anthropology, intracultural variation, ethnobiology, human ecology and ethnopsychology. He has published a series of influential papers on indigenous knowledge and classification, based largely on fieldwork conducted amongst the Aguaruna and Jivaro.

Harold C. Conklin is presently Professor of Anthropology and Franklin Muzzy Crosby Professor of the Human Environment at Yale University. In a long and distinguished career he has published numerous seminal papers arising from his fieldwork among the Hanunóo and Ifugao of the Philippines. He is also the author of *Bamboo literacy* (1949), *Hanunóo agriculture* (1957) and *Ethnographic Atlas of Ifugao* (1980).

Michael R. Dove is Senior Fellow and Coordinator of the Biodiversity Program at the East-West Center in Honolulu. His main research interests are the swidden societies of Kalimantan, highland agricultural communities in Java, lowland peasants in Pakistan, and development and conservation policy and ideology. He is the author of *Swidden agriculture in Indonesia* (1985) as well as a number of important papers. He has edited

The real and the imagined role of culture in development (1988) and co-edited *The sociology of natural resources in Pakistan* (1992).

Peter David Dwyer is Reader in Zoology at the University of Queensland in Brisbane. His main anthropological interests are in the ethnozoology and socio-ecology of small-scale subsistence populations in Papua New Guinea. Among his varied publications are *The pigs that ate the garden: a human ecology from Papua New Guinea* (1990).

Roy Ellen is Professor of Anthropology and Human Ecology at the University of Kent at Canterbury. His main research interests are currently ethnobiology, classification, subsistence ecology and the regional organisation of trade in eastern Indonesia. He is the author of *Environment, subsistence and system* (1982) and *The cultural relations of classification* (1993), and has written and edited various other monographs, collections and papers.

Steven Feld is Professor of Anthropology at the University of California at Santa Cruz. His main research interests are in ethnomusicology, aesthetics and linguistic anthropology, and he has conducted fieldwork amongst the Kaluli of Papua New Guinea. He is the author of *Sound and sentiment* (1990), and has edited the CD/cassette *Voices of the rainforest* (1991) and (with Charles Keil) *Music grooves: essays and dialogues* (1994).

Charles O. Frake is Samuel P. Capen Professor of Anthropology at the State University of New York at Buffalo, and Emeritus Professor of Anthropology at Stanford University. His main research interests are in cultural ecology and cognitive and linguistic anthropology in the southwestern Philippines, where he has conducted fieldwork. He has recently extended his interests to northwestern Europe and the cultural construction of the past. He is the author of *Language and cultural description* (1980), a collection of his influential essays, and (among others) of 'Cognitive maps of time and tide among Medieval seafarers' (*Man*, 1985).

Katsuyoshi Fukui is Professor of Anthropology in the Graduate School of Human and Environmental Studies of Kyoto University. His research has focused on pastoral, agro-pastoral and swidden societies in Japan, Ethiopia and Sudan, especially in relation to ecology, folk knowledge, ethnicity and conflict. He is the author of *Cognition and culture: ethnography of colour and pattern among the Bodi* (1991, in Japanese) among other monographs, and has edited a number of collections, including *Warfare among East African herders* (1979, with D. Turton).

David Harris is Professor of Human Environment and Director of the Institute of Archaeology, University College London. His main interests are in the ecology and evolution of agricultural and other subsistence systems, plant and animal domestication and the origins of agriculture. He has conducted fieldwork principally in the Caribbean, northern Australia and Papua New Guinea and, most recently, Turkmenistan. He is the author of a series of seminal papers and has edited *Foraging and farming: the evolution of plant exploitation* (1989) and *The origins and spread of agriculture and pastoralism in Eurasia* (1996).

Mitsuo Ichikawa is Professor at the Centre for African Area Studies at Kyoto University. His main research interests are in the ecological anthropology of Mbuti and other hunter-gatherer societies of the African tropical rainforest. He is the author of *Hunters in the forest* (1982, in Japanese) and papers on Mbuti ethnography.

Tim Ingold is Max Gluckman Professor of Social Anthropology at the University of Manchester. His main interests are in economic and ecological anthropology (particularly hunting and gathering and pastoralism), evolutionary theory, and the relations between biological, psychological and anthropological approaches to culture and social life. He has conducted fieldwork amongst the Skolt Saami and in a Finnish farming community. He is the author of *Hunters, pastoralists and ranchers* (1980), *Evolution and social life* (1986) and other books, and has edited several collections of essays.

Junzo Kawada is Professor at the Institute for the Study of Languages and Cultures of Asia and Africa at Tokyo University of Foreign Studies. His main interests are in oral tradition, bodily techniques and the history of non-literate peoples. He has conducted fieldwork in Burkina Faso, Mali and Nigeria, and is the author of *The voice* (1988, in Japanese) and *Studies of the oral tradition* (1992, in Japanese), and various other publications in Japanese, French and English.

Emilio F. Moran is Professor of Anthropology and Professor of Environmental Studies at Indiana University, Bloomington. He is Director of the Anthropological Center for Training and Research on Global Environmental Change. His main research interests have been in Amazonian ecology and resource management, resettlement and research methods. He is the author of *Human adaptability* (1979), *Developing the Amazon* (1981), *Through Amazonian eyes* (1993) and other books and papers, and has edited *The Ecosystem approach in anthropology* (1990).

Ryutaro Ohtsaka is Professor in the Department of Human Ecology of the University of Tokyo. His main interests focus on the connections between subsistence, behaviour, nutrition and demography, and he has conducted fieldwork in Papua New Guinea and in Japanese fishing communities. His publications include *Oriomo Papuans: ecology of sago-eaters in Lowland Papua* (1983), *Human ecology of health and survival in Asia and the South Pacific* (1987, edited with T. Suzuki), and *Population ecology of human survival: bioecological studies of the Gidra in Papua New Guinea* (1990, edited with T. Suzuki).

Paul Richards holds a joint appointment as Professor of Technology and Agrarian Development at Wageningen Agricultural University, the Netherlands, and as Professor of Anthropology, University College London. His main research has been on social and technological aspects of agriculture, agro-ecology and environment in West Africa, focused on Nigeria and Sierra Leone. He is author of *Indigenous agricultural revolution* (1985) and *Coping with hunger* (1986), and edited *African environment* (1975) for the International African Institute.

Sadao Sakamoto was formerly Director and Professor at the Plant Germ-plasm Institute, Faculty of Agriculture, Kyoto University. He is presently Professor of Food Culture and Ethnobotany at the Institute of Intercultural Communication, Ryukoku University, Kyoto. His main research interests are in crop origins and evolution, ethnobotany and the exploration of plant genetic resources, mainly in relation to cereals. His publications include *Waxy-endosperm and -perisperm starch food culture* (1989, in Japanese), and *Domesticated plants and animals of the Southwest Eurasian agro-pastoral culture complex. I. Cereals* (1987, edited with Y. Tani).

Masayoshi Shigeta is Associate Professor at the Centre for African Area Studies at Kyoto University. His main research interests are in agricultural domestication as a human-plant relationship and in the ethnobotany of indigenous African plants. He has worked in southern Sudan, Kenya and southwestern Ethiopia. He is the author of various publications, including 'Folk *in-situ* conservation of ensete (*Ensete ventricosum*): towards the interpretation of indigenous agricultural science of the Ari, Southwestern Ethiopia, *African Study Monographs* (1990).

François Sigaut is Directeur d'Etudes at the Ecole des Hautes Etudes en Sciences Sociales in Paris. His main interests are in the history of technology in agriculture and neighbouring and related activities, especially in eighteenth- and nineteenth-century Western Europe. His main publications include *L'agriculture et le feu* (1975), and he has edited (with M. Gast) *Les techniques de conservation de grains à long terme* (1979–85, 4 vols.).

Yutaka Tani is Professor at the Institute for Research in Humanities at Kyoto University. His main research interests are in Middle Eastern and Mediterranean pastoralism and in discourse and talk. He has conducted fieldwork on the managerial techniques of shepherds in Italy, Greece, Romania, Afghanistan and India. His publications include *One day of a shepherd Francesco, an ethnography of a mountain village in central Italy* (1976, in Japanese) and *Sex, victim and passion: symbolic and discourse analysis of the Biblical narratives* (1984, in Japanese).

Hiroyasu Tomoeda is Professor Emeritus of Ethnology at the National Museum of Ethnology, Osaka. His main research interests are in Andean ethnography and ethnohistory (he has conducted fieldwork in Peru and Bolivia) and in Amazonian mythology, religion and ritual. Among his various publications are *The bull and the condor: rituals and lives in Andean society* (1986, in Japanese). He is editor of *El homfre y su ambiente en los Andes Centrales* (1982).

Name Index

Subject Index

For Product Safety Concerns and Information please contact our EU representative GPSR@taylorandfrancis.com
Taylor & Francis Verlag GmbH, Kaufingerstraße 24, 80331 München, Germany

www.ingramcontent.com/pod-product-compliance
Ingram Content Group UK Ltd.
Pitfield, Milton Keynes, MK11 3LW, UK
UKHW021500240926
475702UK00035B/30
* 9 7 8 0 3 6 7 7 1 9 1 8 0 *